CHICAGO PUBLIC LIBRARY
HAROLD WASHINGTON LIBRARY
BUSINESS /SCIENCE/TECHNOLOGY
400 S. STATE ST. 4TH FLOOR
CHICAGO IL 60605

Advances in the Biology and Management of Modern Bed Bugs

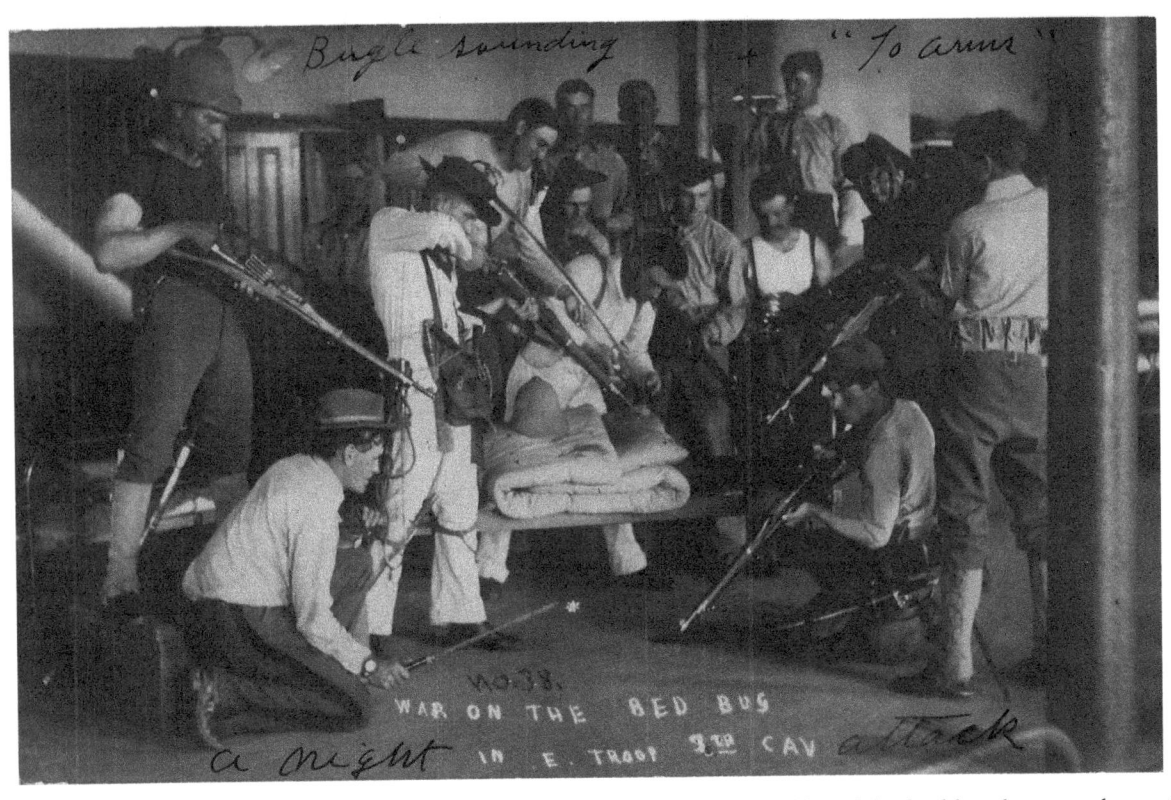

Caption: "War on the bed bug". Postcard c. 1916. Clearly humanity's dislike of the bed bug has not changed through the years!

Advances in the Biology and Management of Modern Bed Bugs

Edited by

Stephen L. Doggett
NSW Health Pathology
Westmead Hospital
Westmead, Australia

Dini M. Miller
Department of Entomology
Virginia Tech,
Blacksburg, Virginia, USA

Chow-Yang Lee
School of Biological Sciences
Universiti Sains Malaysia
Penang, Malaysia

WILEY Blackwell

This edition first published 2018
© 2018 John Wiley & Sons Ltd.

All rights reserved. No part of this publication may be reproduced, stored in a retrieval system, or transmitted, in any form or by any means, electronic, mechanical, photocopying, recording or otherwise, except as permitted by law. Advice on how to obtain permission to reuse material from this title is available at http://www.wiley.com/go/permissions.

The right of Stephen L. Doggett, Dini M. Miller, Chow-Yang Lee to be identified as the author(s) of the editorial material in this work has been asserted in accordance with law.

Registered Office(s)
John Wiley & Sons, Inc., 111 River Street, Hoboken, NJ 07030, USA
John Wiley & Sons Ltd, The Atrium, Southern Gate, Chichester, West Sussex, PO19 8SQ, UK

Editorial Office
9600 Garsington Road, Oxford, OX4 2DQ, UK

For details of our global editorial offices, customer services, and more information about Wiley products visit us at www.wiley.com.

Wiley also publishes its books in a variety of electronic formats and by print-on-demand. Some content that appears in standard print versions of this book may not be available in other formats.

Limit of Liability/Disclaimer of Warranty
While the publisher and authors have used their best efforts in preparing this work, they make no representations or warranties with respect to the accuracy or completeness of the contents of this work and specifically disclaim all warranties, including without limitation any implied warranties of merchantability or fitness for a particular purpose. No warranty may be created or extended by sales representatives, written sales materials or promotional statements for this work. The fact that an organization, website, or product is referred to in this work as a citation and/or potential source of further information does not mean that the publisher and authors endorse the information or services the organization, website, or product may provide or recommendations it may make. This work is sold with the understanding that the publisher is not engaged in rendering professional services. The advice and strategies contained herein may not be suitable for your situation. You should consult with a specialist where appropriate. Further, readers should be aware that websites listed in this work may have changed or disappeared between when this work was written and when it is read. Neither the publisher nor authors shall be liable for any loss of profit or any other commercial damages, including but not limited to special, incidental, consequential, or other damages.

Library of Congress Cataloging-in-Publication data applied for

9781119171522 [Hardback]

Cover Design: Wiley
Cover Image: © John-Reynolds/Gettyimages

Set in 10/12pt Warnock by SPi Global, Pondicherry, India

Printed in the United States of America

10 9 8 7 6 5 4 3 2

Dedicated to the works of luminaries such as J.R. Busvine, A.A. Girault, A. Hase, C.G. Johnson, H. Kemper, K. Mellanby, K. Newberry, N. Omori, and R.L. Usinger. We are merely dwarfs standing of the shoulders of giants.

Contents

List of Contributors *xix*
Foreword *xxiii*
Acknowledgments *xxv*

Introduction *1*
Stephen L. Doggett, Dini M. Miller and Chow-Yang Lee

Part I Bed Bugs in Society *7*

1 Bed Bugs Through History *9*
Michael F. Potter
1.1 Introduction *9*
1.2 Origins and Spread *9*
1.3 Early Extermination Methods *11*
1.4 Propagation Within Cities (1880s–1950s) *13*
1.5 Determination – and a Silver Bullet *15*
1.5.1 Bed Bug Insecticides *16*
1.5.2 Bug Proof Design and Construction *21*
1.5.3 Bed Bug Traps *21*
1.5.4 Lethal Temperatures *21*
1.6 Past is Present *23*
References *23*

2 Bed Bugs in Popular Culture *27*
Stephen L. Doggett and David Cain
2.1 Introduction *27*
2.2 Bed Bugs in Poetry *27*
2.3 Bed Bugs in the Figurative Arts *28*
2.4 Bed Bugs in Theatre *30*
2.5 Bed Bugs in Literature *32*
2.6 Bed Bugs in Music *34*
2.7 Bed Bugs in Television *35*
2.8 Bed Bugs in Linguistics *36*
2.9 Bed Bugs in Erotica *37*
2.10 The Use of Bed Bugs in Popular Culture Through Time *37*
References *38*

Part II The Global Bed Bug Resurgence 43

3 The Bed Bug Resurgence in North America 45
Dini M. Miller
3.1 Introduction 45
3.2 Bed Bug Resurgence in the United States 45
3.3 Bed Bug Resurgence in Canada 46
3.4 Summary 48
References 48

4 The Bed Bug Resurgence in Latin America 51
Roberto M. Pereira, Ana Eugenia de Carvalho Campos, Joao Justi (Jr.) and Márcio R. Lage
4.1 Introduction 51
4.2 Bed Bugs in Brazil 51
4.3 The Resurgence of Bed Bugs in Brazil 53
4.4 Elsewhere in Latin America 55
References 56

5 The Bed Bug Resurgence in Europe and Russia 59
Richard Naylor, Ondřej Balvín, Pascal Delaunay, and Mohammad Akhoundi
5.1 Introduction 59
5.2 History of Bed Bugs in the Region 59
5.3 Historical Laws to Control Bed Bugs Pre-resurgence 60
5.4 Documented Evidence for the Bed Bug Decline 61
5.5 Early Evidence for the Resurgence 62
5.6 The Degree of the Resurgence 63
5.7 Region-specific Factors in the Resurgence 64
5.8 Strategies to Combat the Resurgence 65
References 66

6 The Bed Bug Resurgence in Asia 69
Chow-Yang Lee, Motokazu Hirao, Changlu Wang, and Yijuan Xu
6.1 Introduction 69
6.2 History of Bed Bugs in Asia 69
6.3 Laws, Regulations and Policies for Bed Bug Control Prior to the Resurgence 70
6.4 Modern Resurgence of Bed Bugs in Asia 71
6.5 Specific Factors Related to the Bed Bug Resurgence in Asia 73
6.6 Bed Bug Management Strategies in Asia 74
References 76

7 The Bed Bug Resurgence in Australia 81
Stephen L. Doggett and Toni Cains
7.1 Introduction 81
7.2 History of Bed Bugs in Australia 81
7.3 Documented Evidence for the Bed Bug Decline 82
7.4 The Modern Resurgence 83
7.5 Strategies to Combat the Bed Bug Resurgence 84
References 85

8 The Bed Bug Resurgence in Africa *87*
Josephus Fourie and Dionne Crafford

8.1 Introduction *87*
8.2 History of Bed Bugs in Africa *87*
8.3 Laws, Regulations, and Policies for Bed Bug Control Pre-resurgence *88*
8.4 Documented Evidence for the Bed Bug Decline *88*
8.5 Early Evidence for the Resurgence *89*
8.6 The Extent of the Resurgence *89*
8.7 Region- and Country-specific Reasons for the Resurgence *91*
8.8 Strategies to Combat the Resurgence and Relative Success *92*
References *93*

9 The Bed Bug Resurgence in the Indian Subcontinent *95*
Anil S. Rao and Joshua A. Rao

9.1 Introduction *95*
9.2 History of Bed Bugs in the Indian Subcontinent *95*
9.3 Laws to Control Bed Bugs Pre-resurgence *96*
9.4 Documented Evidence for the Bed Bug Decline *96*
9.5 The Modern Resurgence *96*
9.6 Strategies to Combat the Bed Bug Resurgence *97*
References *98*

10 The Bed Bug Resurgence in the Middle East *101*
Odelon Del Mundo Reyes

10.1 Introduction *101*
10.2 History of Bed Bugs in The Middle East *101*
10.3 Regional Reports of the Bed Bug Resurgence *102*
10.4 Impact of the Bed Bug Resurgence *103*
10.5 Pest Management Professionals Close Encounters with Cimicids *103*
References *104*

Part III Bed Bug Impacts *107*

11 Dermatology and Immunology *109*
Shelley Ji Eun Hwang, Stephen L. Doggett and Pablo Fernandez-Penas

11.1 Introduction *109*
11.2 Bed Bug Saliva *109*
11.3 Cutaneous Reactions *110*
11.4 Dermatological Complications from Bed Bug Bites *111*
11.5 Systemic Reactions *112*
11.6 Immunological and Pathogenic Mechanisms *112*
11.7 Dermatopathology *112*
11.8 Differential Diagnoses of Bed Bug Bites *113*
11.9 Clinical Management *114*
11.10 Conclusion *114*
References *114*

12 Bed Bugs and Infectious Diseases *117*
Stephen L. Doggett
- 12.1 Introduction *117*
- 12.2 Vectors and Transmission Pathways *117*
- 12.3 Bed Bugs and Infectious Diseases: an Overview *118*
- 12.4 Why do Bed Bugs not Transmit Infectious Diseases? *120*
- 12.5 The Future Hunt for Pathogens: A Cautionary Note *121*
- 12.6 Conclusion *122*
- References *122*

13 Mental Health Impacts *127*
Stéphane Perron, Geneviève Hamelin and David Kaiser
- 13.1 Introduction *127*
- 13.2 Methods *127*
- 13.3 Main Findings *129*
- 13.4 What Can Be Inferred from the Current State of the Literature? *130*
- 13.5 Limitations and Future Research *130*
- 13.6 Conclusion *131*
- References *131*

14 Miscellaneous Health Impacts *133*
Stephen L. Doggett
- 14.1 Introduction *133*
- 14.2 Respiratory Issues *133*
- 14.3 Blood Loss *134*
- 14.4 Sleep Loss *134*
- 14.5 Chemical Exposure *135*
- 14.6 Miscellaneous Health Impacts *135*
- References *136*

15 Fiscal Impacts *139*
Stephen L. Doggett, Dini M. Miller, Karen Vail and Molly S. Wilson
- 15.1 Introduction *139*
- 15.2 Types of Cost *139*
- 15.3 Costs to the Multi-Unit Housing Industry *140*
- 15.4 Cost to the Hospitality and Travel Industry *142*
- 15.5 Cost to the Retail Industry *143*
- 15.6 Brand Damage in the Housing, Hospitality, and Retail Industries *143*
- 15.7 Legal Expenses *144*
- 15.8 Cost to Pest Management Companies *144*
- 15.9 Bed Bug Management Revenues *145*
- 15.10 Conclusions *145*
- References *146*

Part IV Bed Bug Biology *149*

16 Bed Bug Biology *151*
Sophie E.F. Evison, William T. Hentley, Rebecca Wilson, and Michael T. Siva-Jothy
- 16.1 Introduction *151*
- 16.2 Hematophagy *151*

16.3	Anachoresis	*152*
16.4	Flightlessness	*153*
16.5	Reproduction	*153*
16.6	Egg Laying	*155*
16.7	Host-seeking Behavior	*155*
16.8	Harborage Seeking Behavior and Aggregation	*156*
16.9	Dispersal	*156*
	References	*158*

17   Chemical Ecology   *163*
Gerhard Gries

17.1	Introduction	*163*
17.2	Olfaction and Contact Chemoreception	*163*
17.2.1	General Introduction	*163*
17.2.2	Olfactory Sensilla on Bed Bug Antennae and their Responses to Odorants	*164*
17.2.3	Molecular Mechanisms Underlying Olfaction	*164*
17.3	Pheromones	*165*
17.3.1	Alarm Pheromone	*165*
17.3.2	Aggregation Pheromone	*165*
17.3.3	Sex-attractant Pheromone	*168*
17.3.4	Anti-mating and Anti-aphrodisiac Pheromones	*168*
17.4	Host Seeking	*169*
	References	*170*

18   Population Genetics   *173*
Warren Booth, Coby Schal and Edward L. Vargo

18.1	Introduction	*173*
18.2	The Evolution of Modern Bed Bugs	*173*
18.3	Genetic Variation Within Populations	*174*
18.4	Genetic Variation Among Populations	*175*
18.5	Mitochondrial Heteroplasmy	*177*
18.5.1	Variation in Heteroplasmy Across Host Lineages and Among Populations	*178*
18.5.2	Implications of Heteroplasmy	*179*
18.5.3	Insecticide Resistance, *kdr*, and Geographic Variation	*179*
18.6	Future Directions in Bed Bug Population Genetics	*180*
	References	*180*

19   Physiology   *183*
Joshua B. Benoit

19.1	Introduction	*183*
19.2	Stress Tolerance and Starvation Resistance	*183*
19.3	Blood Feeding	*187*
19.4	Reproduction and Development	*187*
19.5	Summary and Future Directions	*187*
	References	*188*

20   Symbionts   *193*
Mark Goodman

20.1	Introduction	*193*
20.2	Identity of Endosymbionts	*193*

20.3 Impact of Symbionts on Bed Bug Biology *194*
20.4 Transmission of Symbionts *195*
20.5 Symbionts and Bed Bug Management *195*
References *195*

21 Bed Bug Laboratory Maintenance *199*
Mark F. Feldlaufer, Linda-Lou O'Connor and Kevin R. Ulrich
21.1 Introduction *199*
21.2 General Colony Maintenance *199*
21.2.1 Containers and Harborages *199*
21.3 Feeding Techniques *200*
21.3.1 *In-vivo* and *In-vitro* Blood Sources *200*
21.3.2 *In-vitro* feeding units *202*
21.4 Need for Plasma *203*
21.5 Development of an Artificial Blood Source *204*
References *204*

Part V Bed Bug Management *209*

22 Bed Bug Industry Standards: Australia *211*
Stephen L. Doggett
22.1 Introduction *211*
22.2 Why was the Code Required? *211*
22.3 The History and Aims of the Code *211*
22.4 The Key Elements of the Code *212*
22.5 The Benefits of the Code *213*
References *214*

23 Bed Bug Industry Standards: Europe *217*
Richard Naylor
23.1 Introduction *217*
23.2 Why was the Code Required? *217*
23.3 The History and Aims of the Code *218*
23.4 The Benefits of the Code *219*
References *220*

24 Bed Bug Industry Standards: USA *221*
Jim Fredericks
24.1 Introduction *221*
24.2 History and Development of the NPMA Best Management Practices for Bed Bugs *221*
24.3 Target Audience *221*
24.4 Key Elements of the NPMA Best Management Practices *222*
24.5 Marketing and Adoption of the NPMA Best Management Practices *223*
24.6 Acceptance of the BMP *223*
References *224*

25 A Pest Control Company Perspective 225
Joelle F. Olson, Mark W. Williams and David G. Lilly
- 25.1 Introduction 225
- 25.2 The Resurgence 225
- 25.2.1 Rapid and Sustainable Growth 225
- 25.2.2 Lack of Industry Preparedness 225
- 25.2.3 Lack of Public Awareness 226
- 25.3 Responsibility and Liability 226
- 25.3.1 Educating the Client 226
- 25.3.2 Liability for Services 227
- 25.3.3 Financial Burden 227
- 25.4 Inspection and Control Methods 228
- 25.4.1 Inspection and Monitoring Tools 228
- 25.4.2 Chemical Applications and Resistance 228
- 25.4.3 Non-chemical Applications 229
- 25.4.4 Training and Maintaining Service Consistency 229
- 25.5 Conclusion 230
- References 230

26 Prevention 233
Molly S. Wilson
- 26.1 Introduction 233
- 26.2 Education 234
- 26.3 Monitors 234
- 26.4 Mattress Encasements 235
- 26.5 Desiccant Dusts 236
- 26.6 Heat 236
- 26.6.1 Clothes Dryers 237
- 26.6.2 Heat Chambers 237
- 26.7 Bed Bug Management Policy 237
- References 238

27 Detection and Monitoring 241
Richard Cooper and Changlu Wang
- 27.1 Importance of Detection and Monitoring 241
- 27.2 Detection of Bed Bugs 241
- 27.2.1 Resident Interviews or Surveys 242
- 27.2.2 Visual Inspection 242
- 27.2.3 Canine Scent Detection 243
- 27.2.4 Bed Bug Monitors 244
- Passive Monitors 247
- Active Monitors 248
- 27.3 Field Comparison of Detection Methods 249
- 27.4 Bed Bug Inspections 250
- 27.4.1 Detection 250
- Proactive inspections 251
- Reactive inspections 252

27.4.2	Inspections to Guide and Evaluate Treatment	*252*
27.4.3	Inspections to Determine Elimination	*252*
27.4.4	Inspections in Non-traditional Settings	*253*
	Conflict of Interest Statement	*253*
	References	*254*

28 Non-chemical Control *257*
Stephen A. Kells

28.1	Introduction	*257*
28.2	Excluding Bed Bugs	*258*
28.3	Physically Removing Bed Bugs	*259*
28.4	Creating Adverse Environmental Conditions against Bed Bugs	*261*
28.4.1	Basic Conditions Causing Lethality	*261*
28.4.2	Basic Conditions for High-temperature Control	*261*
	Equipment for Delivery of High Temperatures through Steam Applications	*262*
	Equipment for Delivery of High Temperatures through Dry Heat Applications	*263*
	Basic Conditions for Delivery of Low-temperature Treatments	*266*
	Considerations for Delivery of Low Temperatures	*267*
28.5	Biological Agents Tested Against Bed Bugs	*268*
28.6	Other Non-Chemical Control Methods	*268*
28.7	Conclusion	*269*
	References	*269*

29 Insecticide Resistance *273*
Alvaro Romero

29.1	Introduction	*273*
29.2	Insecticides and Insecticide Resistance in Bed Bugs	*275*
29.3	Metabolic Resistance	*276*
29.4	Reduced Penetration Resistance	*277*
29.5	Target-site Resistance	*277*
29.6	Evolution of Resistance and Fitness Costs	*279*
29.7	Conclusions	*279*
	References	*280*

30 Chemical Control *285*
Chow-Yang Lee, Dini M. Miller and Stephen L. Doggett

30.1	Introduction	*285*
30.2	Insecticide Classes used Against Bed Bugs	*285*
30.2.1	Pyrethroids	*285*
30.2.2	Organophosphates and Carbamates	*298*
30.2.3	Neonicotinoids	*298*
30.2.4	Halogenated Pyrroles	*299*
30.2.5	Insect Growth Regulators	*299*
30.2.6	Inorganic and Mineral Compounds	*299*
30.2.7	Botanical Insecticides	*300*
30.2.8	Poisonous Gases	*300*
30.3	Insecticide Formulations for Bed Bug Management	*300*
30.3.1	Liquid Sprays	*300*
30.3.2	Pressurized Aerosols	*301*
30.3.3	Dusts	*301*

	30.3.4	Fumigants *301*
	30.3.5	Permethrin-impregnated Fabrics *302*
	30.3.6	Insect Repellents *302*
	30.3.7	Total Release Foggers *302*
30.4		Factors Affecting Insecticide Efficacy *302*
	30.4.1	Test Method *303*
	30.4.2	Test Substrate *303*
	30.4.3	Test Arena *303*
	30.4.4	Experimental Details *303*
	30.4.5	Strain and Bed Bug Species *304*
		Susceptible, Resistant, or Both Strains *304*
		References *305*

31   Limitations of Bed Bug Management Technologies *311*
Stephen L. Doggett and Mark F. Feldlaufer

31.1	Introduction *311*
31.2	Bed Bug Detection *311*
31.2.1	Traps *311*
31.2.2	Canines *312*
31.2.3	Novel Detection Methods *313*
31.3	Bed Bug Control *313*
31.3.1	Housing Types *313*
31.3.2	Chemical Control Methods *313*
31.3.3	Other Chemical Control Products *315*
31.3.4	Non-chemical Bed Bug Control Methods *316*
31.3.5	Heat *316*
31.3.6	Other Non-chemical Controls *316*
	References *317*

32   Bed Bug Education *323*
Jody Gangloff-Kaufmann, Allison Taisey Allen and Dini M. Miller

32.1	Introduction *323*
32.2	Strategies and Successes in Bed Bug Education *323*
32.3	Educational Programs Focusing on Bed Bugs *324*
32.4	The Media *326*
32.5	The Effect of Social Media on Bed Bug Education *327*
32.6	Identifying the Target Audience *327*
32.7	Effective Adult Education *327*
32.8	Measuring the Impacts of Bed Bug Education *328*
32.9	Conclusion *328*
	References *329*

Part VI   Bed Bug Control in Specific Situations *331*

33   Low-income Housing *333*
Richard Cooper and Changlu Wang

33.1	Introduction *333*
33.2	Management of Bed Bugs in Low-income Housing *334*
33.3	Components of a Successful Building- or Complex-wide IPM Program *334*

33.3.1 Education *334*
33.3.2 Identification of Apartments with Existing Bed Bug Activity *335*
33.3.3 Preparation of Apartments for Treatment *336*
33.3.4 Treatment of Apartments with Bed Bug Activity *336*
33.3.5 Follow-up Service Visits *336*
33.3.6 Contracts and the Role of Property Management *337*
33.4 The Future *337*
References *338*

34 Multi-Unit Housing *341*
Dini M. Miller
34.1 Introduction *341*
34.2 Challenges Unique to the Human Living Environment *341*
34.3 Obstructions to Control Success *342*
34.3.1 Challenges Unique to Multi-unit Housing *342*
34.3.2 Human Host Behavior *342*
34.3.3 Financial Limitations for Multi-unit Housing Managers and Owners *342*
34.3.4 Pest Management Limitations *343*
34.4 Future Prospects for Success *344*
References *345*

35 Shelters *347*
Molly S. Wilson
35.1 Introduction *347*
35.2 Challenges in Shelters *347*
35.3 Obstacles to Successful Control *348*
35.4 Methods of Control in Shelters *348*
35.5 Key Elements to Successful Control *349*
References *349*

36 Hotels *351*
David Cain
36.1 Introduction *351*
36.2 Bed Bugs in Hotels: The Challenges *351*
36.3 Successful Bed Bug Management in Hotels *352*
References *354*

37 Healthcare Facilities *357*
Stephen L. Doggett
37.1 Introduction *357*
37.2 The Challenges Bed Bugs Pose to Healthcare Facilities *357*
37.3 The History of Bed Bugs in Healthcare Facilities *358*
37.4 Bed Bugs in Healthcare Facilities with the Modern Resurgence *358*
37.5 Bed Bug Management in Healthcare Facilities *359*
References *360*

38 Aircraft *363*
Adam Juson and Catherine Juson
38.1 Introduction *363*
38.2 Aviation Entomology – a Brief History *363*

38.3	Bed Bug Management on Aircraft: The Challenges *364*
38.4	Bed Bug Management on Aircraft *365*
38.5	Improving the Pest Management Protocol *365*
	References *366*

39 Cruise Ships and Trains *369*
David G. Lilly and Garry Jones
39.1	Introduction *369*
39.2	Cruise Ships and Ferries *369*
39.3	Trains *371*
39.4	Conclusion *372*
	References *373*

40 Poultry Industry *375*
Allen Szalanski
40.1	History – Cimicids and Poultry *375*
40.2	Mexican Chicken Bug, *Haematosiphon inodorus* *375*
40.3	Brazilian Chicken Bug, *Ornithocoris toledoi* *376*
40.4	Tropical Bed Bug, *Cimex hemipterus* *376*
40.5	The Common Bed Bug, *Cimex lectularius* *377*
40.6	Biology and Impact of *Cimex lectularius* on Poultry *377*
40.7	Dispersal *378*
40.8	Bed Bug Control in Poultry Facilities *378*
40.9	Insecticide Assays *379*
	References *380*

Part VII Legal Issues *383*

41 Bed Bugs and the Law in the USA *385*
Jeffrey Lipman and Dini M. Miller
41.1	Introduction *385*
41.2	Registration of Pesticides *386*
41.3	Legal Requirements Regarding Who Can Apply Pesticides in the USA *386*
41.4	Legal Requirements for PMPs Regarding the Standard of Care *387*
41.5	Public Health Acts Regarding Bed Bugs *387*
41.6	Bed Bug-related Statutes, Laws, and Ordinances *387*
41.7	Laws Addressing Bed Bug Remediation *389*
41.8	Tenants and Public Housing *391*
41.9	Legal Standing Clients Encountering Bed Bugs in Temporary Occupancies *392*
41.10	Bed Bug Lawsuit Landscape *392*
41.11	Conclusion *393*
	References *394*

42 Bed Bugs and the Law in the United Kingdom *397*
Clive Boase
42.1	Introduction *397*
42.2	Training of Pest Management Professionals *397*
42.3	Bed Bug Pesticide Approval *398*

42.4	Tenants, Guests and Bed Bugs	*399*
42.5	Local Authority Duties and Powers Regarding Bed Bugs	*399*
	References	*400*

43 Bed Bugs and the Law in Australia *403*
Toni Cains, David G. Lilly and Stephen L. Doggett

43.1	Introduction	*403*
43.2	Registration and Use of Bed Bug Management Products	*403*
43.3	Legal Requirements of Pest Management Professionals	*404*
43.4	Public Health Laws Regarding Bed Bugs	*404*
43.5	Tenancy and Public Housing	*405*
43.6	Bed Bug Legal Cases	*406*
	References	*406*

44 Bed Bugs and the Law in Asia *409*
Andrew Ho-Ohara and Chow-Yang Lee

44.1	Introduction	*409*
44.2	Registration and Use of Bed Bug Management Products	*409*
44.3	Legal Requirements for Pest Management Professionals	*410*
44.4	Bed Bug Legal Cases	*411*
44.6	Future	*412*
	References	*412*

45 On Being an Expert Witness *413*
Paul J. Bello and Dini M. Miller

45.1	Introduction	*413*
45.2	What is an Expert Witness?	*413*
45.3	The Expert's Role	*414*
45.4	Providing Expert Testimony	*414*
45.5	Bed Bugs in a Court Case	*415*
45.6	Summary	*416*
	References	*417*

Part VIII Bed Bugs: the Future *419*

46 Bed Bugs: the Future *421*
Chow-Yang Lee, Dini M. Miller and Stephen L. Doggett
Summary *426*
References *426*

Index *429*

List of Contributors

Mohammad Akhoundi
Service de Parasitologie-Mycologie, Centre Hospitalier Universitaire de Nice – Hôpital de l'Archet, Nice, France

Ondřej Balvín
Czech University of Life Sciences, Prague, Czech Republic

Paul J. Bello
PJB Pest Management Consulting, Alpharetta, Georgia, USA

Joshua B. Benoit
Department of Biological Sciences, University of Cincinnati, Cincinnati, Ohio, USA

Clive Boase
The Pest Management Consultancy, Haverhill, United Kingdom

Warren Booth
Department of Biological Science, The University of Tulsa, Tulsa, Oklahoma, USA

David Cain
Bed Bugs Limited, London, United Kingdom

Toni Cains
Sydney Local Health District Public Health Unit, Camperdown, Australia

Ana Eugênia de Carvalho Campos
Unidade Laboratorial de Referênciaem Pragas Urbanas, Instituto Biológico, São Paulo, Brasil

Richard Cooper
Cooper Pest Solutions and Bedbug Central, Lawrenceville, New Jersey, USA

Dionne Crafford
Clinvet International (Pty) Ltd, Bloemfontein, South Africa

Pascal Delaunay
Service de Parasitologie-Mycologie, Centre Hospitalier Universitaire de Nice – Hôpital de l'Archet, Nice, France

Stephen L. Doggett
Department of Medical Entomology, NSW Health Pathology, Westmead Hospital, Westmead, Australia

Odelon Del Mundo Reyes
Ecovar, Dubai, United Arab Emirates

Sophie E.F. Evison
Department of Animal and Plant Sciences, University of Sheffield, Sheffield, United Kingdom

Mark F. Feldlaufer
USDA-ARS, Invasive Insect Biocontrol and Behavior Laboratory, Beltsville, Maryland, USA

Pablo Fernandez-Peñas
Dermatology, Sydney Medical School, The University of Sydney, Westmead, Australia

Josephus Fourie
Clinvet International (Pty) Ltd, Bloemfontein, South Africa

Jim Fredericks
National Pest Management Association (USA), Fairfax, Virginia, USA

Jody Gangloff-Kaufmann
New York State Integrated Pest Management Program, Cornell University, Geneva, New York, USA

Mark Goodman
Varment Guard Environmental Services Inc., Columbus, Ohio, USA

Gerhard Gries
Department of Biological Sciences, Simon Fraser University, Burnaby, Canada

Geneviève Hamelin
Public Health Department of Montreal, Montreal, Canada

Harold J. Harlan
US Army Medical Entomologist (retired), USA

William T. Hentley
Department of Animal and Plant Sciences, University of Sheffield, Sheffield, United Kingdom

Motokazu Hirao
Hirao Biological Institute, Shizuoka, Japan

Andrew Ho-Ohara
Eden Law Corporation, Singapore

Shelley Ji Eun Hwang
Department of Dermatology, Westmead Hospital, Westmead, Australia

Garry Jones
Consultant Pest Management Professional, Buff Point, Australia

Adam Juson
Merlin Environmental Solutions Ltd, Carshalton, United Kingdom

Catherine Juson
Merlin Environmental Solutions Ltd, Carshalton, United Kingdom

João Justi (Jr.)
Unidade Laboratorial de Referênciaem Pragas Urbanas, Instituto Biológico, São Paulo, Brasil

David Kaiser
Public Health Department of Montreal, Montreal, Canada

Stephen A. Kells
Department of Entomology, University of Minnesota, St. Paul, Minnesota, USA

Márcio R. Lage
Faculdade de SaúdePública, Universidade de São Paulo, São Paulo, Brasil

Chow-Yang Lee
School of Biological Sciences, Universiti Sains Malaysia, Penang, Malaysia

David G. Lilly
Ecolab Pest Elimination, Macquarie Park, Australia

Jeffrey Lipman
Lipman Law Firm, West Des Moines, Iowa, USA

Dini M. Miller
Department of Entomology, Virginia Tech, Blacksburg, Virginia, USA

Richard Naylor
The Bed Bug Foundation, Chepstow, United Kingdom

Linda-Lou O'Connor
AirRx Antimicrobial Sciences Inc., Baltimore, Maryland, USA

Joelle F. Olson
Pest Elimination Division, Ecolab Research Center, Eagan, Minnesota, USA

Roberto M. Pereira
Entomology and Nematology Department, University of Florida, Gainesville, Florida, USA

Stéphane Perron
Public Health Department of Montreal, Montreal, Canada

Michael F. Potter
Department of Entomology, University of Kentucky, Lexington, Kentucky, USA

Anil S. Rao
Pest Control (India) Private Ltd, Mumbai, India

Joshua A. Rao
Pest Control (India) Private Ltd, Mumbai, India

Alvaro Romero
Department of Entomology, Plant Pathology and Weed Science, New Mexico State University, Las Cruces, New Mexico, USA

Coby Schal
Department of Entomology, North Carolina State University, Raleigh, North Carolina, USA

Mike T. Siva-Jothy
Department of Animal and Plant Sciences, University of Sheffield, Sheffield, United Kingdom

Allen Szalanski
Department of Entomology, University of Arkansas, Fayetteville, Arkansas, USA

Allison Taisey Allen
National Pest Management Association, Fairfax, Virginia, USA

Kevin R. Ulrich
USDA-ARS, Invasive Insect Biocontrol and Behavior Laboratory, Beltsville, Maryland, USA

Karen Vail
Department of Entomology and Plant Pathology, The University of Tennessee, Knoxville, Tennessee, USA

Edward L. Vargo
Department of Entomology, Texas A&M University, College Station, Texas, USA

Changlu Wang
Department of Entomology, Rutgers University, New Brunswick, New Jersey, USA

Mark W. Williams
Ecolab Pest Elimination, Northwich, United Kingdom

Molly S. Wilson
Department of Biochemistry, Virginia Tech, Blacksburg, Virginia, USA

Rebecca Wilson
Department of Animal and Plant Sciences, University of Sheffield, Sheffield, United Kingdom

Yijuan Xu
Department of Entomology, South China Agricultural University, Guangzhou, China.

Foreword

Harold J. Harlan

Bed bugs were a serious pest of human communities long before recorded history. However, for millenia, information about bed bugs, and advice on how to deal with them, has been in evidence across many lineages of cultural lore. This fact is most obvious when you consider that bed bugs are known to have at least 71 common names originating from 36 different languages across the world (Usinger, 1966). *The Monograph of Cimicidae* (Usinger 1966) was the first comprehensive publication on bed bugs that combined extensive worldwide information from historic, cultural, scientific, pest management, and general public resources.

Advances in the Biology and Management of Modern Bed Bugs (ABMMBB) updates and expands much of Usinger's (1966) information, with emphasis on the worldwide resurgence of both the Common bed bug, *Cimex lectularius* L., and the Tropical bed bug, *Cimex hemipterus* (F.). ABMMBB incorporates extensive new information from a wide range of basic and applied research, as well as the recently observed medical, legal and regulatory impacts of bed bugs.

Today there are many new, extremely precise technologies, and laboratory tools that could not have been imagined in 1966. Recent innovations, especially in molecular biology and genetics, offer a fascinating range of potential applications. In addition, we have new systems for information gathering, processing, and sharing with international colleagues. These technologies have opened up whole new fields of scientific investigation over the past 20 years.

Stephen L. Doggett, Dini M. Miller, and Chow-Yang Lee have done a terrific job of assembling and coordinating more than 60 contributing authors who are highly experienced and widely recognized as experts in their topic areas. The contributing authors offer new information on basic science and advice on using applied management strategies and bed bug bioassay techniques. The authors also present cutting-edge information on the major impacts that bed bugs have had on the medical, legal, housing, and hotel industries across the world, as well as their impacts on public health.

ABMMBB is the most comprehensive compilation yet produced about these bugs that includes historic, technical, and practical information. It will certainly be the most thorough single reference on bed bugs for many decades to come. I believe that ABMMBB will be an essential reference for anyone who is engaged in managing bed bugs, be it in an academic, basic or applied scientific setting, or in a public outreach or pest management role, worldwide. I am very honored, and humbled, to have been asked to provide this foreword. I can hardly wait to buy my own personal copy.

Respectfully,
Harold J. Harlan
Harold J. Harlan, PhD, BCE, LTC (Ret.) was a US Army Medical Entomologist

Reference

Usinger, R.L. (1966) *Monograph of Cimicidae (Hemiptera – Heteroptera)*, Entomological Society of America, College Park.

Acknowledgments

The editors, Stephen L. Doggett, Dini M. Miller and Chow-Yang Lee, would like to express gratitude to all the authors and reviewers that contributed towards the development of *Advances in the Biology and Management of Modern Bed Bugs* (ABMMBB). We would especially like to acknowledge Harold Harlan for writing the foreword, and for contributing so much to so many. The editors also wish to acknowledge Wiley-Blackwell and staff (notably Sonali Melwani, Ramya Raghaven, Bella Talbot, David McDade, Emma Strickland, Gunalan Lakshmipathy, and Ward Cooper) for their assistance, advice, editorial, and production efforts. A special thank you to Junichiro Katayama from Semco Co. Ltd. in Japan, who kindly hosted the editors of ABMMBB in mid-October 2016 so that we could finalize much of the text (thank you also for all the wonderful meals!).

The following are acknowledged for particular chapters:

Chapter 3: The Global Resurgence in North America The author wishes to acknowledge the assistance of Dennis Monk (Director, Bedbug Solutions) and Sean Rollo (Regional Manager, Orkin Canada) for providing information on the current bed bug situation in Canada.

Chapter 5: The Global Resurgence in Europe and Russia The authors are especially grateful to the contribution of Václav Rupeš, former head (retired) of the Department of Desinfection, Desinsection and Deratization, State Health Institute, Prague.

Chapter 8: The Global Resurgence in Africa The authors would like to thank Henda Pretorius and Reinier Zwiegers [Clintest (Pty) Ltd], and Ashleigh Caddick (PES Africa) for their assistance in running the online survey cited in the chapter. The authors would also like to thank Carmen Neethling [Clinvet (Pty) Ltd] for assistance with literature searches.

Chapter 9: The Global Resurgence in India and the Subcontinent The editors greatly appreciate the assistance of K.P. Jayanth (PCI India) in the development of this chapter.

Chapter 17: Chemical Ecology of Bed Bugs The author wishes to thank John Borden for comments, Sharon Oliver for word processing and comments, Michael Hrabar for photographs, and Stephen DeMuth for graphical illustration. The Chapter is dedicated to Regine Gries who was the lead investigator in the research resulting in the identification of the bed bug aggregation pheromone and who in the process endured >180 000 bed bug bites for the maintenance of a thriving bed bug laboratory colony.

Chapter 18: Population Genetics The published and unpublished population genetics studies mentioned in the chapter would not have been possible without the generous cooperation of many colleagues, too many to mention individually, who provided bed bug samples. Ron Harrison is especially acknowledged who facilitated the collections of samples from Orkin and Orkin International, and Rick Santangelo who managed the collections.

Chapter 19: Bed Bug Physiology The author wishes to acknowledge that funding while developing this chapter was provided by the University of Cincinnati.

Chapter 21: Bed Bug Laboratory Maintenance The authors wish to thank; A. Aak, J.F. Anderson, O. Balvin, B. Campbell, S.L. Doggett, C.Y. Lee, D.G. Lilly, J. Olson, R. Naylor, A. Romero, K. Reinhardt, A. Vander Pan, and C. Wang, for completing the rearing survey. Those individuals are acknowledged that contribute to the rearing efforts in all of the laboratories currently maintaining bed bugs. Members of the Armed Forces Pest

Management Board (Silver Spring, Maryland USA) are also thanked, along with the personnel at the Walter Reed National Military Medical Center (Bethesda, Maryland USA) for their assistance in obtaining blood products for the senior author.

Chapter 41: Bed Bugs and the Law in the USA The authors wish to express gratitude to Kolby Warren, Drake University Law School, Des Moines, Iowa, USA, who assisted in the production of the chapter.

Chapter 43: Bed Bugs and the Law in Australia The authors wish to acknowledge the following individuals for providing important information pertaining to environmental health laws pertaining to bed bugs, from the different states; Keith Rogers (Australian Capital Territory, Health), Rebecca Feldman (Department of Health & Human Services, Victoria), Michaela Hobby (Department of Health, South Australia), Nicola Slavin (Department of Health, Northern Territory), Rebecca Richardson (Department of Health, Queensland), and Mike Lindsay and Donald Howell (Department of Health, Western Australia). Keith Farrow (Rapids Solutions) generously provided information pertaining to the licensing of professional pest managers. Nicholas Cowdery AM QC (University of Sydney, Australia) kindly reviewed the chapter for legal accuracy.

Chapter 44: Bed Bugs and the Law in Asia The authors wish to thank the following individuals for providing important information pertaining pesticide registration and the licensing of pest management professionals in their respective countries: Erh-Lieh Hsu (National Taiwan University), Pascal Cai (Chinese Pest Control Association, China), Suchart Leelayouthyotin (King Service Center, Bangkok, Thailand), Sulaeman Yusuf (Indonesia Institute of Science), and Motokazu Hirao (Japan Pest Control Association).

The following individuals have generously supplied images for use in ABMMBB or critically reviewed manuscripts:

Zachary Adelman (Texas A&M University, College Station, Texas, USA)
Joshua Benoit (University of Cincinnati, Cincinnati, Ohio, USA)
Clive Boase (the Pest Management Consultancy, United Kingdom)
Kaci Buhl (National Pesticide Information Center, Oregon State University, Oregon, USA)
David Cain (Bed Bugs Limited, United Kingdom)
Richard Cooper (Cooper Pest Solutions and Bedbug Central, Lawrenceville, New Jersey, USA)
Richard deShazo, (University of Mississippi Medical Center, Mississippi, USA)
Keith Farrow (Rapid Training, Australia)
Mark Feldlaufer (USDA-ARS, Invasive Insect Biocontrol and Behavior Laboratory, Maryland, USA)
Toby Fountain (Uppsala University, Sweden)
Adam Juson (Merlin Environmental Solutions Ltd, United Kingdom)
David Lilly (Ecolab Pest Elimination, Macquarie Park, Australia)
Oliver Madge (Bed Bug Foundation, United Kingdom)
Frank Meek (Orkin LLC, USA)
Mike Merchant (Texas A&M, AgriLife Extension Service, Texas, USA)
Kosta Y. Mumcuoglu (Hebrew University, Israel)
Richard Naylor (Bed Bug Foundation, United Kingdom)
Faith Oi (Entomology and Nematology Department, University of Florida, Florida, USA)
Christopher Orton (PiMACs, Australia)
Lawrence Pinto (Pinto & Associates, Mechanicsville, Maryland, USA)
Alvaro Romero (New Mexico State University, New Mexico, USA)
Bob Rosenberg (National Pest Management Association, Virginia, USA)
Veera Singham (Centre for Chemical Biology, Universiti Sains Malaysia, Malaysia)
Coby Schal (North Carolina State University, North Carolina, USA)
Allison Taisey Allen (National Pest Management Association, Virginia, USA)
Changlu Wang (Department of Entomology, Rutgers School of Environmental and Biological Sciences, New Jersey, USA)
Jeff White (Bed Bug Central, New Jersey, USA)

Introduction

Stephen L. Doggett, Dini M. Miller and Chow-Yang Lee

"...*misery acquaints a man with strange bedfellows...*"

William Shakespeare, *The Tempest*, 1610

"...*intellects vast and cool and unsympathetic, regarded this earth with envious eyes, and slowly and surely drew their plans against us...*"

H.G. Wells, *The War of the Worlds*, 1898

The quotation above from one of William Shakespeare's greatest works could easily read as an allegory for the personal suffering one experiences with a bed bug infestation. However, Shakespeare's play was written in 1610, only a short time after the UK produced their first reliable bed bug record in 1583 (Mouffet, 1634). Thus it was quite possible that Shakespeare himself never acquainted himself with the misery of bed bugs. Yet it was not long before the insect became so common in the country that companies appeared which specialized in bed bug extermination, such as the famous Tiffin & Son, founded in 1690 (Potter, 2011).

In fact, bed bugs have a long history of inflicting their misery upon humanity. The remains of bed bugs have been found in Egyptian settlements dating back some 3565 years (Paragiotakopulu and Buckland, 1999). With the discovery in 1939 that dichloro-diphenyl trichloroethane (better known as DDT) had a powerful insecticidal action, suddenly the world had the magical solution that could rid humanity of bed bugs forever. Subsequently DDT (and other organochlorines), and the organophosphates where widely employed to control bed bugs, and infestations became rare in the developed world for many years after World War II (Usinger, 1966). Yet forever was not to be. The late 1990s saw a worldwide re-emergence of both the Tropical bed bug, *Cimex hemipterus* (F.), and the Common bed bug, *Cimex lectularius* L. Not unlike the Martians in H.G. Wells' classic novel, suddenly nowhere on earth was exempt from bed bugs.

In the early days of the modern resurgence, infestations mainly occurred in the hospitality sector and bed bugs were more limited to premises with high guest turnover. Then people started to take the insect with them wherever they went and, in the process, spread bed bugs into the wider community. Infestations began to appear in such diverse locations as in private homes, on public transport, within the retail sector, in cinemas, at the office, in schools and universities, and even in healthcare facilities. Thus wherever a person went, they could be potentially exposed to bed bugs and take them elsewhere. The greatest concern however, has been the proliferation of bed bugs amongst the socially disadvantaged, a group that often does not have the economic resources to pay for control. As a result, infestations can go uncontrolled and spread throughout a building complex. With a lack of public and government support to ensure that infestations are successfully eradicated, it is unfortunate that such groups have become bed bug reservoirs for the wider society. Thus support should be provided for bed bug eradication programs in low income housing…even if it is only to selfishly protect ourselves from future infestations.

Advances in the Biology and Management of Modern Bed Bugs, First Edition.
Edited by Stephen L. Doggett, Dini M. Miller, and Chow-Yang Lee.
© 2018 John Wiley & Sons Ltd. Published 2018 by John Wiley & Sons Ltd.

There have been a number of reasons postulated for the modern bed bug resurgence (Pinto *et al.*, 2007; Doggett *et al.*, 2012). This includes globalization and the ease with which people move around the world, thereby enabling the spread of modern bed bug strains. Changes in pest management have meant that hotel rooms are no longer routinely treated for pests. The insecticides available today do not have the same residual life as the organochlorines and are simply not as efficacious. Poor knowledge of bed bugs, in particular how to control modern strains, has clearly been a major factor in the degree of the resurgence, as treatment failures often result in the infestation spreading to other apartments in multiple occupancy dwellings. This particular phenomenon has been aided by the ever increasing concentration of people into high density living due to the uncontrolled growth in the world's population. These days it is much easier for bed bugs to spread from an infestation to invade other premises. A variety of other reasons have been given for the resurgence, but it appears that the key contributing factor to the modern bed bug resurgence is the development of insecticide resistance within the insect.

In many ways, the bed bugs that our grandparents experienced are very different to those that we are exposed to now. The contemporary bugs have developed multiple defences against the insecticides we use against them. They have thicker skins, which slows down the penetration of insecticides into the body. They produce a range of complex enzymes that can break down insecticides, and they possess mutations that prevent the insecticides from acting at the target sites. Thus the modern bed bug is truly the superbug of the 21st century. The challenge for modern scientists is to find ways to circumvent this range of extraordinary adaptations.

One aspect that has been the subject of debate is the geographical origin of the modern (and resistant) bed bug. Some US based researchers have suggested that resistant strains of *C. lectularius* arrived from Eastern Europe, although somewhat amusingly, locals in Eastern Europe have the opposing view of blaming American tourists (Borel, 2015). In spite of the contrary opinions, neither belief can explain the origin and simultaneous rise of *C. hemipterus*. This species has a more tropical distribution, has never been reported from Eastern Europe, and only recently reappeared in the USA (Campbell *et al.*, 2016). Arguably a more simple answer to the origin of resistance in both species is that resistant strains originated from a region where the two species are sympatric and where insecticide selection pressures were equally applied. This would help to explain the synchronous revival of the two species. The most logical locations would be from areas of Africa where infestations of *C. hemipterus* and *C. lectularius* coexisted even when infestations elsewhere in the world had become uncommon. In KwaZulu, South Africa, during the late 1980s it was observed that bed bug infestations (of both species) were more common in human dwellings that were sprayed annually with DDT to combat malaria than unsprayed dwellings (Newberry *et al.*, 1987, 1990; Newberry, 1991). The increase in nuisance biting from bed bugs meant that householders often refused chemical treatments despite the potential risk of death from malaria as they believed that spraying may have contributed to greater bed bug numbers! Even some 20 years earlier, a report from the early 1970s noted an increase in bed bug infestations occurred in spite of regular treatments with the organochlorines (Rafatjah, 1971). In both cases, the development of insecticide resistance was the suspected cause of these increases. In fact, by the early 1970s, insecticide resistance had been long known in South African bed bug populations.

The first report of resistance to DDT in *C. lectularius* was from Hawaii in 1947. This was only three years after the pesticide was first employed (Johnson and Hill, 1948). Within a relatively short time thereafter, reports of resistance to both bed bug species had become widespread (Busvine, 1957). Thus it appears that resistance evolved rapidly, but this may not have been the case. Natural pyrethrins were used for bed bug control after the mid-1800s, and resistance to this class of insecticides confers cross resistance to the organochlorines. Perhaps some degree of resistance in bed bugs had developed long before the late 1940s but was simply not identified. The organochlorines being more efficacious than pyrethrins may have helped to rapidly eliminate the non-resistant and less resistant insects. Furthermore, presumably the bed bugs that disappeared in developed nations post World War II were all susceptible strains (or had low-level resistance). Yet in pockets of the world, high levels of resistance had evolved (and presumably continued to evolve). These "superbugs" just required the means of escape in order to spread elsewhere. The means were provided by humans, with our modern tendency to move about the world. Widespread global travel is a relatively recent behavior in human history, yet rapidly growing. The World Bank estimated that in 1970 there were some 310 million people movements globally, and by 2015 this number had risen to 3.4 billion (World Bank, 2016). Thus insecticide-resistant bed bugs had ample means to spread from their original source to a new location. The constant rise in human population combined with the increased movement of people across the globe, the recent changes in pesticides and pest management practices mentioned above, and the presence of resistant bed bugs in certain parts of the world, makes it seem that, perhaps in hindsight, the global bed bug resurgence was inevitable and should have been anticipated.

As a consequence of the modern resurgence, there has been a renewed interest in bed bug research. In recent years, a plethora of publications relating to bed bugs have appeared. For example, a search on Pubmed (www.ncbi.nlm.nih.gov/pubmed/) using the term "Cimex", reveals 259 peer reviewed publications between 1912 and 1999. That number rose to 492 between 2000 to the end of 2016. If industry publications and newspaper reports were included, the increase in bed bug publications would be even more dramatic. Interestingly, the number of scientific publications from 2010 onwards have plateaued at around 50 per year, which probably reflects how minimally funded bed bug research is today. While Usinger's (1966) seminal manuscript, the *Monograph of Cimicidae (Hemiptera – Heteroptera)*, continues to be the key reference for taxonomy of the Cimicidae, other areas of research are now much further advanced. Thus there is a need for the distillation of all the contemporary information into a modern academic text and hence the birth of *Advances in the Biology and Management of Modern Bed Bugs* (ABMMBB).

ABMMBB is a synthesis of bed bug information from the past to the present. It aims to serve as a reference book for academic researchers and students alike. It is a valuable text for those in the hospitality sector and accommodation managers, who are tasked with the job of minimizing the risk of bed bugs in their facility, or who organize the eradication of active infestations. With the growth in bed bug litigation, both the litigant and defendant legal teams will find ABMMBB an essential source of contemporary information. Finally, ABMMBB provides up-to-date information for the pest management professional on bed bug biology and management. In recent years, most bed bug research has focused on *C. lectularius*, as this species has impacted the more economically advantaged nations of Europe and North America. However, ABMMBB is aiming to be global in context, and where possible, both bed bug species are discussed and key differences highlighted. With the aim of having an international appeal, ABMMBB has over 60 contributing authors, spanning some 14 nations.

There are seven main parts within ABMMBB. These cover Bed Bugs in Society, the Global Resurgence, Bed Bug Impacts, Biology, Management, Control in Specific Situations, and Legal Issues. Finally, ABMMBB ends with a discussion on the future of bed bugs in society and research needs.

The first part, Bed Bugs in Society, contains two chapters. Chapter 1, "Bed Bugs Through History", reviews the early methods of extermination. By examining past control methodologies when synthetic insecticides had yet to be discovered, perhaps insights can be gained in how to successfully eradicate insecticide-resistant strains now; thus the importance of including this work in ABMMBB. The chapter also covers the origins and spread of bed bugs throughout the world. While the bed bug spread was documented historically to a reasonable extent, few bothered to determine which species was involved. It can be surmised (based on past records and current distributions) that it was mainly *C. lectularius* that was introduced and established in Europe, North America, and initially into Australia. However, few records exist regarding the global spread of *C. hemipterus*. Thus it is necessary to review the early taxonomic descriptions for this species, as these texts contain information on the site of collection (reviewed in Usinger, 1966). Fabricius first described *C. hemipterus*, which was captured from South America houses in 1803 (Fabricius, 1803), while other early records include "Ile Bourbon" (1852) [now known as Réunion], "Ost-Indien" (1861) [East India], Colombia (1854), and Sokotra (1899) [also spelled Socrota, which is part of Yemen]. These references indicate that *C. hemipterus* had spread around the world by the mid-19th century as had *C. lectularius*.

Chapter 2, "Bed Bugs in Popular Culture", highlights the intimate relationship that humans had with bed bugs throughout history via the depiction of the insect in various forms of media. Bed bugs have appeared over the years in poetry, art, the theatre, literature, music, and more recently, in television.

Part 2, on the global bed bug resurgence, contains contributions from all the major regions across the world, including chapters from the Americas, Europe and Russia, Asia, Australia, Africa, the Indian subcontinent, and the Middle East. Each chapter discusses the history of bed bugs in their respective region, the resurgence, and strategies employed to combat the return of the bed bug.

Bed Bug Impacts (Part 3) reviews the dermatological, mental health, and miscellaneous heath impacts associated with bed bugs. The potential for bed bugs to transmit infectious diseases is also considered. It is difficult to argue the fact that the health-related impacts of bed bugs are relatively minor compared with insects, such as mosquitoes, that are known to transmit vector-borne diseases. However, the fiscal impacts of bed bugs are significant and the resurgence has probably cost the world economy billions of dollars. The monetary effects of bed bugs are many and varied, and detailed in the final chapter of this section.

The basis of integrated pest management (IPM) is a comprehensive understanding of the biology of the pest as it relates to effective control strategies, hence the need to include such issues within ABMMBB. Topics covered within Part 4 include host-seeking and blood-feeding behaviors, harbourage selection and aggregation, dispersal, chemical ecology, population genetics, physiology, symbionts, and laboratory maintenance. The bizarre reproductive behaviour of bed bugs known as "traumatic insemination" is also discussed.

Part 5 focuses on bed bug management. The first three chapters of this section review the industry standards on bed bug control that have been developed in recent years in Australia, Europe, and the USA. These standards are followed by a chapter on how a large multinational pest management firm approaches bed bug management to ensure consistency of treatments and a positive outcome for the clients. The following chapters focus on prevention (in terms of minimizing the risk of bed bug establishment and spread); detection and monitoring; non-chemical management; insecticide resistance; and chemical control. In recent years with the modern bed bug resurgence, many technologies have appeared on the market, but only few are truly efficacious. Thus Chapter 31 reviews the inherent limitations in bed bug management technology. The bed bug management coverage finishes with a focus on education. This involves sending the correct message to the community on how to minimize the risk of bed bugs, and how an infestation should be properly managed.

Part 6, on bed bug control in specific situations, reviews the experiences of industry leaders who are responsible for bed bug management in particular environments. These include low-income housing, multi-unit housing, shelters, hotels, healthcare facilities, aircraft, trains and cruise ships, and within the poultry industry. In many cases, little has been published in these areas, so the authors often had to recount their own personal experiences and the challenges involved in achieving a successful outcome.

In recent years, bed bugs have been the cause of legal action, especially in the USA, where some cases involve settlements of several million dollars. Thus Part 7 deals with bed bugs and the law, covering legal aspects from the USA, the UK, Asia, and Australia. There is also a chapter on the challenges of being an expert witness involved in bed bug litigation.

The final chapter of ABMMBB undertakes some crystal-ball gazing to imagine what the future will look like in terms of bed bug and human interaction, and reviews the strategies and research required to reverse the resurgence. Perhaps one of the greatest challenges ahead is to make control affordable for all people. However, only considerable technological and methodological advancements will make this happen. Ultimately the real question is, can we defeat bed bugs again, or are bed bugs set to plague human society forever? Only time will tell.

References

Borel, B. (2015) *Infested*, University of Chicago Press, Chicago.

Busvine, J.R. (1957) Insecticide-resistant strains of insects of public health importance. *Transactions of the Royal Society of Tropical Medicine and Hygiene*, **51** (1), 11–31.

Campbell, B.E., Koehler, P.G., Buss, L.J. and Baldwin, R.W. (2016) Recent documentation of the tropical bed bug (Hemiptera: Cimicidae) in Florida since the common bed bug resurgence. *Florida Entomologist*, **99** (3), 1–3.

Doggett, S.L., Dwyer, D.E., Peñas, P.F. and Russell, R.C. (2012) Bed bugs: clinical relevance and control options. *Clinical Microbiology Reviews*, **25** (1), 164–192.

Fabricius, J.C. (1803) *Systema Rhyngotorum Secundum Ordines, Genera, Species, Adiectis Synonymis, Locis, Observationibus, Descriptionibus*, Apud Carolum Reichard, Brunsvigae.

Johnson, M.S. and Hill, A.J. (1948) Partial resistance of a strain of bed bugs to DDT residual. *Medical News Letter*, **12** (1), 26–28.

Mouffet, T. (1634) Insectorumsive minimorum animalium theatrum, Thomas Cotes, London, UK. (Transl. by Topsel, E. (1658) in: *The History of Four-footed Beasts, Serpents and Insects*, pp. 10, 889–1130. E. Cotes., London, UK.)

Newberry, K. (1991) Field trials of bendiocarb, deltamethrin and fenitrothion to control DDT-resistant bedbugs in KwaZulu, South Africa. *International Pest Control*, **33** (3), 64–68.

Newberry, K., Jansen, E.J. and Thibaud, G.R. (1987) The occurrence of the bedbugs *Cimex hemipterus* and *Cimex lectularius* in northern Natal and Kwazulu, South Africa. *Transactions of the Royal Society of Tropical Medicine and Hygiene*, **81** (3), 431–433.

Newberry, K., Mchunu, Z.M. and Cebekhulu, S.Q. (1990) The effect of bedbug control on malaria control operations. *South African Journal of Science*, **86** (4), 211–212.

Panagiotakopulu, E. and Buckland, P.C. (1999) *Cimex lectularius* L., the common bed bug from Pharaonic Egypt. *Antiquity*, **73** (282), 908–911.

Potter, M.F. (2011) The history of bed bug management-with lessons from the past. *American Entomologist*, **57** (1), 14–25.

Pinto, L.J., Cooper, R. and Kraft, S.K. (2007) *Bed Bug Handbook – the Complete Guide to Bed Bugs and Their Control*, Pinto & Associates, Inc., Mechanicsville, MD.

Rafatjah, H. (1971) The problem of resurgent bed-bug infestation in malaria eradication programmes. *Journal of Tropical Medicine and Hygiene*, **74** (2), 53–56.

Usinger, R.L. (1966) *Monograph of Cimicidae (Hemiptera – Heteroptera)*, Entomological Society of America, College Park.

World Bank (2016) *Air Transport, Passengers Carried*, http://data.worldbank.org/indicator/IS.AIR.PSGR?end=2015&start=1970&view=chart (accessed 10 January 2017).

Part I

Bed Bugs in Society

1

Bed Bugs Through History

Michael F. Potter

Among all the night enemies which often perturb our sweet quiet sleep, there is none more cruel than bedbugs.

Andrea Matthioli, 1557

1.1 Introduction

Bed bugs and humans have had long and interesting relations. Few pests throughout our history have been more detested, or inspired such innovation in pursuit of a solution. Much of humanity had a respite from bed bugs during the second half of the 20th century. Now that the reprieve is over, the past can provide insight as to what lies ahead.

1.2 Origins and Spread

Bed bugs have been biting us pretty much from the beginning. Evidence suggests the parasites first fed on bats, turning their attention to humans after we began inhabiting the same caves (Usinger and Povolny, 1966; Booth *et al.*, 2015). Relations between bed bugs and humans were probably intermittent back then, because hunters and herdsmen were wanderers. Life for the parasites became simpler with the formation of villages and cities, making it easier for infestations to become established. Bed bug remains have been unearthed from sites dating back to the Pharaohs (more than 3500 years ago), when they were considered both pest and potion (Panagiotakopulu and Buckland, 1999). The Roman scholar Pliny the Elder wrote that the Egyptians drank a bed bug "cocktail" as a cure for snakebite, while the Greeks and Romans burned the bugs to make leeches loosen their hold. Ingesting bed bugs was thought to cure maladies ranging from lethargy to urinary infections (Busvine, 1976). Bed bugs continued to be used for medicinal purposes well into the 20th century in Europe and North America. Included in the fifth (1896) edition of the *American Homeopathic Pharmacopoeia* are directions for making a "tincture of bed bug" to be used as a remedy for malaria (Riley and Johanssen, 1938). The ancients also devised creative (if not entirely effective) measures to defeat them. The Greek philosopher Democritus (400 BCE), for example, recommended hanging the feet of a hare or stag at the foot of the bed, while others suggested suspending a bear skin (Cowan, 1865).

As civilization and trade expanded, bed bugs spread north and east through Europe and Asia, reaching Italy by 77 CE, China by 600 CE, and Germany and France, respectively, in the 11th and 13th centuries (Usinger, 1966). Warmth produced by sleeping and cooking fires enabled the bugs to thrive in castles of the

Advances in the Biology and Management of Modern Bed Bugs, First Edition.
Edited by Stephen L. Doggett, Dini M. Miller, and Chow-Yang Lee.
© 2018 John Wiley & Sons Ltd. Published 2018 by John Wiley & Sons Ltd.

wealthy and huts of the working class (Figure 1.1). However much like today, the poor and disadvantaged suffered the most: "They infest both the chambers of rich and poor, but are more troublesome to the poor... For they do not breed in beds of which the linen and straw is frequently changed, as in the houses of the rich" (Aldrovandi, 1603).

Bed bugs were first reported in England in 1583 and became common by the 17th and 18th centuries. They hitchhiked their way to the Americas aboard the ships of the first European explorers and settlers. Aided by commerce, infestations initially arose in seaport towns, appearing farther inland later on (Marlatt, 1916). The current resurgence of bed bugs in North America has followed a similar pattern, with initial reports of infestation in the late-1990s appearing in such "gateway" cities as New York, Chicago, Toronto, and San Francisco.

The global spread of bed bugs can also be traced to their naming. In ancient Rome, bed bugs were called *Cimex* (meaning "bug"). The species designation (*lectularius*), assigned centuries later by Linnaeus, referred to a bed or couch. The early Greek term for bed bug was *coris,* meaning "to bite," from which the word coriander comes. One of civilization's oldest spices, coriander (cilantro) was probably so named because when the leaves are crushed the pungent smell resembled that of bed bugs. Ancient Chinese who ground the bugs up to treat wounds called them *chòu chong,* or "stinky bug," and the obnoxious odor prompted a similar christening (*punaise,* "to stink") in medieval France. The Japanese once called bed bugs, *Nankin mushi,* or "insects from Nanjing (China)", an expression anecdotally said to have been coined during the Sino-Japanese war (1937–1945). However, this term is considered derogatory and now *Tokojirami* (bed louse) is more often

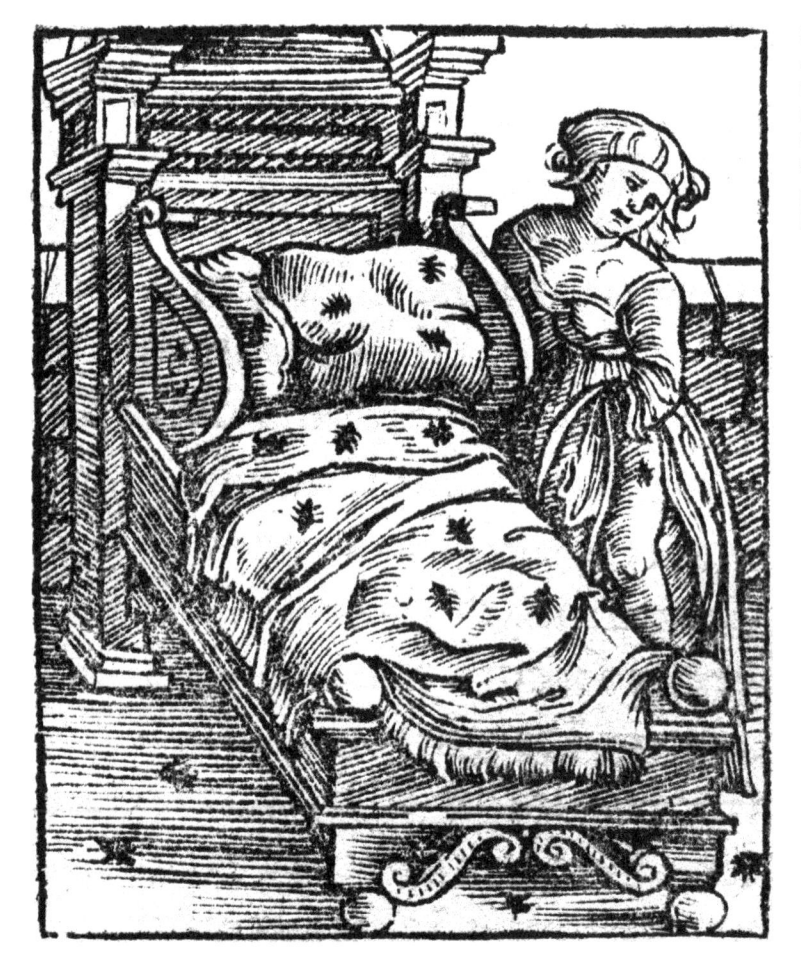

Figure 1.1 Bed bugs depicted in *Hortus Sanitatis* (Anonymous, 1536). This was the first ever encyclopedic compilation of natural history, originally published in Germany in 1485. Credit: Wellcome Images CC (https://www.diomedia.com/stock-photo-hortus-sanitatis-image19956256.html).

used. Another disparaging moniker, *venerschen* (little venereal), was used in Germany, presumably because of the pest's infectious disposition (Borel, 2015). In England, bed bugs were simply called "bugs." The early Spanish word for bed bug was *chinche,* and Spanish-speaking people today often refer to them as *chinches* or *chinche de cama*; literally, "bug of the bed." Other descriptive names originating from Europe and North America included "bed louse," "wall louse," "wallpaper flounder," "night rider," "red coat," "mahogany flat", and "crimson rambler." Bed bugs presumably did not occur in North America before the arrival of European settlers, thus there is no definitive native word for them in the language of indigenous Americans (Usinger, 1966). Although the Hopi people of the southwestern USA do have a native word for bed bugs (*pesets'ola*), it is unclear if the word referred to the Common bed bug, *Cimex lectularius* L., or another Cimicid species perhaps associated with cliff-dwelling birds (Reinhardt, 2012).

1.3 Early Extermination Methods

People crushed, swatted, and hand-picked bed bugs long before they relied on pesticides (Figure 1.2). The close "hand-to-hand combat" could help explain the bugs' oft-reported loathsome smell: "This insect if it be crushed or bruised emits a most horrid and loathsome stench, so that those that are bitten by them are often in doubt whether it be better to endure the trouble of their bitings, or kill them and suffer their most odious and abominable stink" (Ray, 1673).

Figure 1.2 "Summer Amusement, Bugg Hunting" (Cruikshank, 1782). People hand-picked bed bugs long before the use of insecticides. Credit: Library of Congress, Washington, DC (https://www.loc.gov/item/00652100/).

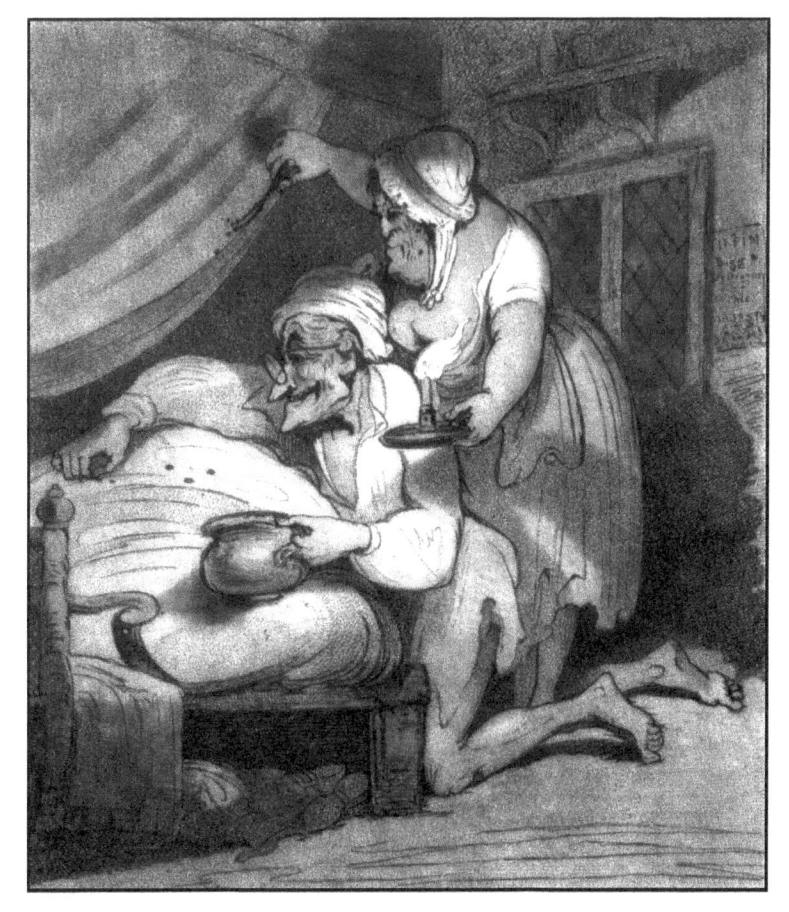

Many of the modern methods for managing bed bugs today can be traced to early European exterminators. Among the most famous were Tiffin & Son of London, who formed a business in 1690 to exterminate bed bugs for the nobility. In the 17th and 18th centuries, the affluent became very concerned that the working classes would transport bed bugs into their homes. Recognizing the need for vigilance, Tiffin noted: "We do the work by contract, examining the house every year. It's a precaution to keep the place comfortable. You see, servants are apt to bring bugs in their boxes" (Mayhew, 1861). Tiffin mentioned finding the most bed bugs in beds, but cautioned "if left unmolested they get numerous, climb to the tops of rooms, and about the corners of the ceilings, and colonize anywhere they can." Centuries later, pest management professionals are again advocating prevention for bed bugs, although the public is not always willing to pay for such services (Potter et al., 2015).

Another of England's early exterminators, John Southall, published a treatise on the bug, which he referred to as "that nauseous venomous insect." Published in 1730, the 44-page manual contained observations on bed bug behavior and advice for eliminating infestations (Figure 1.3). Like Tiffin & Son, Southall advocated vigilance and cautioned against bringing in infested belongings: "In taking of Houses, new or old, and in buying Bedsteds, Furniture, &c. examine carefully if you can find Bugg-marks. If you find such, though you see not the Vermin, you may assure yourself they are nevertheless infected. If you put out your

Figure 1.3 Title and facing page from John Southhall's "A Treatise of Buggs" (London, 1730).

Linnen to wash, let no Washer-woman's Basket be brought into your Houses; for they often prove as dangerous to those that have no Buggs…" (Southall, 1730).

To simplify treatment, Southall recommended that beds be plain, easy to disassemble, and as free from woodwork as possible. The evolution of the bed in modern society has been shaped by the bed bug. Additional influences on design will be discussed later in this chapter.

Exterminator Southall also gained notoriety for his "Nonpareil Liquor," a secret, supposedly sensational bed bug killer. The formula for the liquid has been lost, but may have been derived from quassia wood, a tropical tree with insecticidal properties (Busvine, 1976). A bottle of the stuff could be had for two shillings (about the cost of a nice dinner at the time). Many other "secret" bed bug formulas have been marketed throughout history, a trend continuing to this day. Tiffin had a pragmatic view of such remedies, noting that "secret bug poisons ain't worth much, for all depends upon the application of them" (Cowan, 1865). Some of the early advertisements for killing bed bugs were extreme. One recipe described in an English edition of the French *Dictionaire Oeconomique*, suggested mixing the drippings from a roasted cat with egg yolks and oil to form an ointment, which could then be rubbed onto infested furniture (Chomel, 1727). Other counsel, appearing in *The Compleat Vermin-Killer* (Anonymous, 1777), instructed the reader to boil a handful of wormwood and white hellebore, a poisonous flower, in "a proper quantity of urine" and wash the beds with it; or fill the cracks of the bed with gunpowder and set it on fire (the latter tome had similarly fervent advice for treating headaches — bleeding the person with leeches attached to one's temples.)

As noted earlier, bed bugs became plentiful in North America with the coming of European settlers. As a deterrent, beds were often made from sassafras wood (presumed to be repellent), and the crevices doused with boiling water, arsenic, and sulfur. According to Kalm (1748) this gave only temporary relief. Ships afforded ideal accommodation for bed bugs, and there are accounts of voyagers being fed upon during passage to the Americas, including on the *Mayflower*. Completion of the transcontinental railway system in the latter half of the 1800s afforded rapid transit to inland cities where the bugs had not been seen before. Hotels and boarding houses were especially buggy and travelers unwittingly carried them from place to place in their trunks and satchels. Vigilant travelers learned to pull beds away from walls and immerse the legs in pans of oil. Others relied on pyrethrum powder: "Dusted between the sheets of a bed, it will protect the sleeper from the most voracious hotel bug" (Osborne, 1896).

By the mid-1800s, bed bugs had become a particular problem in poor, overcrowded areas with low standards of hygiene. As in Europe, wealthy households with an abundance of domestic help discovered that bed bugs could be kept in check with vigorous housecleaning, especially in respect to beds. Washing bedding and dousing the slats, springs, and crevices with boiling water or grease from salt pork or bacon proved helpful. Another benefit from such efforts was detection of infestations in their more manageable initial stages: "The greatest remedy is cleanliness, and a constant care and vigilance every few days to examine all the crevices and joints, to make sure that none of the pests are hidden away" (USDA, 1875). Watchfulness and vigilance were oft-repeated recommendations throughout the annals of bed bug management, a refrain being emphasized again today.

1.4 Propagation Within Cities (1880s–1950s)

Bed bugs received a big reproductive boost during the late 1800s and early 1900s. Cities in the USA and Europe grew at a frenetic pace due to expansion of industry and the pursuit of jobs. Mass influxes of people seeking a better life afforded the parasites easy access to sustenance. Builders and architects packed as many housing units as possible into available space, facilitating building-to-building movement of infestations. Propagation of bed bugs in the early 1900s was also aided by central heating of buildings. By the turn of the century, cast iron radiators were delivering warmth to every room in the house, a process made easier in the 1930s by electricity, fans, and forced air heating. Whereas bed bug populations had previously followed a more seasonal trend, increasing as the weather warmed, this enabled the bugs to thrive year-round (Johnson, 1942; Trustees of the British Museum, 1973).

In Europe in the 1930s and 1940s, an estimated one-third of dwellings in major cities had bed bugs. Half the population of Greater London encountered them at some point during the year, and in some areas, nearly all households were affected to some extent (UK Ministry of Health, 1934; Hartnack, 1939). George Orwell's *Down and Out in London and Paris* (1933) depicted bed bugs as enemy combatants: "near the ceiling long lines of the bugs marched all day like columns of soldiers, and at night came down ravenously hungry..." During this time, bed bugs became a community-wide problem like rats. In some cases, infestations were so severe that the bugs were seen crawling from house to house, escaping through exterior windows and doors and traveling along walls, pipes, and gutters (Matheson, 1932).

Infestations were similarly horrendous in cities within the USA. In 1939, the National Association of Housing Officials (NAHO) held an emergency meeting of housing managers from across the country. Opining on the bed bug menace, they lamented that: "If the furnishings of but one family moving into the new building are verminous – the percentage is usually much higher – there is every possibility that the woodwork, cupboards, and fixtures of the apartments will have become infested before the management can bring the condition under control (NAHO, 1939).

Although no social stratum was spared, the scourge was worse in poor, overcrowded neighborhoods. In England, the bugs became synonymous with slum living conditions, leading to the belief (even among some health officials) that bed bugs were one of the factors helping to create slums by attracting those who tolerated them and had acquired a degree of immunity (UK Ministry of Health, 1934). Consequently, slum clearance and the supervised transfer of tenants to new housing became an important means of combating the bed bug problem throughout much of Europe. According to Millard (1932), "Part of a complete campaign against the bed-bug must be to organize propaganda with a view to arousing an 'anti-bug conscience.'" Articles were written to focus attention on especially dilapidated, bed-bug-ridden communities: "Here nearly every house is a haunted house. After dark there is no place more eerie, no torture more prolonged and blood-curdling than that enacted here year after year, no atrocity more revolting than the nightly human sacrifice. For there are vampires. I have seen them. I have smelt them" (England, 1931). Slum clearance campaigns were sometimes accompanied by fabrication and subsequent burning of large bed bug "effigies" as a means of consciousness-raising in the community (Campkin, 2013). Such public displays are eerily similar to efforts to elevate awareness about bed bugs today.

Rigid disinfestation protocols were instituted in Europe to minimize the chance of people transporting bed bugs from old to new housing. In England, families were taken to bed bug "cleansing stations," where their clothing and bedding were passed through a steam disinfector. Concurrently, furniture and other belongings were loaded into vans and fumigated with hydrogen cyanide (UK Ministry of Health, 1934). In Sweden, citizens were housed in tents while their premises and belongings were being fumigated and several cities contemplated building hotels for this purpose. In Germany, some landlords required a written testimonial from an exterminator, stating that the apartment being vacated showed no signs of an infestation (Hartnack, 1939). Today, in similar fashion, some property managers are asking about bed bugs during pre-screening of prospective renters, although tenants' rights are greater today than they were back then. In New York City, for example, legislation was passed requiring lessors to provide bed bug infestation history for the prior year to any renter before the lease of such property (Buckley, 2010).

A more comprehensive approach to preventing dissemination of bed bugs was taken by the Department of Health in Scotland. This approach, known as the Glasgow System, placed emphasis on educating newly relocated tenants on the importance of household cleanliness and the habits of vermin. Within a few days of occupancy, specialists within the Public Health Department trained in the detection of bed bugs inspected the dwelling and provided instruction on prevention and treatment. All tenants were visited at least monthly during the first three months to ensure that no bed bugs were introduced and preventative measures were proceeding satisfactorily (UK Ministry of Health, 1934; Hartnack, 1939).

Disinfestation protocols were also deployed by public housing authorities in the USA. Challenged by the bed bug's mobility (and restrictions on using cyanide in multi-occupancy buildings), communal initiatives were undertaken patterned after those being used concurrently in Europe. In such cities as Chicago and New

York, cooperating managers and tenants received federal and local funds to de-infest their communities. Paradoxically, the communal pest control programs disappeared soon after DDT (dichloro-diphenyl trichloroethane) became available for householder use in 1945. Exuberance over the sensational new pesticide caused housing managers to abandon the need for a community-wide approach, since each tenant could now slay their own bed bugs affordably and efficiently (Biehler, 2009). Unfortunately, the quick and easy "technological fix" proved unsustainable, and communities are facing the same bed bug challenges today.

During wartime, bed bugs were transported on bedding into many public air-raid shelters. They also fed on sleeping soldiers in barracks and battlefront trenches, and were spread on belts, backpacks, canteens, and helmets. Matheson (1950) reported one such account from World War I: "In the East African campaign the bugs invaded the cork lining of the sun helmets of the soldiers. As the helmets were piled together at night, all soon became infested and the soldiers complained of bugs attacking their heads."

During World War II, bed bugs were so abundant they became a morale issue for the US Army. Families of soldiers who were being feasted upon by bed bugs in their bunks pressed their representatives in Congress for a solution. Hearings were held, and as a stop-gap measure, hundreds of barracks were fumigated with hydrogen cyanide (Whitford, 2006). Soon thereafter, DDT was discovered to be a safer, more economical method of controlling infestations in military sleeping quarters. Bed bugs were also common on warships and even in the nooks and crannies of submarines.

In the first half of the 20th century, bed bugs also infiltrated all aspects of civilian life. Besides households and hotels, infestations were common in dressing rooms, restaurant seating areas, furniture upholstery shops, and laundry services (Herrick, 1914; Mallis, 1945). Theaters had big problems with bed bugs and sometimes had to tear out entire rows of seats and install new ones. Coat rooms and lockers in schools and businesses were also commonly infested. All modes of transport including trains, buses, taxicabs, and airplanes were carriers of bed bugs, and passengers unwittingly picked them up and transported them home or to work. In the 1930s, a survey of 3000 moving vans in Stockholm, Sweden found bed bugs on 47% of the vans inspected. A subsequent survey in Finland showed that bed bugs were often found inside televisions and radios being serviced by appliance repair shops (Markkula and Tiittanen, 1970). Not surprisingly, infestations were also a persistent problem in hospitals. Professional pest managers today are battling bed bugs in virtually all the same places (Potter *et al.*, 2010, 2015).

Seeking monetary compensation because of bed bugs is not just a modern-day phenomenon. Bed bug bites have in fact triggered lawsuits for more than a century. In 1895, a Chicago jury ruled that "no man shall be required to pay rent for a house infested with bedbugs." Editorializing on the verdict, the news media noted that if the ruling held, "the great majority of Chicagoans would be relieved of their rent bills." In another early case involving a hotel (Bly vs Sears), the court ruled that the presence of bed bugs did not furnish grounds for the recovery of damages because the plaintiff must have known that the hotel (like so many others in the day) was previously "buggy" (Anonymous, 1902). Railroads were also defendants in bed bug lawsuits. In 1913, a Milwaukee man sued the St. Paul Railroad for $10 000 (a lot of money in those days), claiming the bites made him so ill that it interfered with his business trip. When the man returned home he stepped off the train carrying one arm in a sling (Potter, 2011). Suits involving malevolent (intentional) introduction of bed bugs are not new either. In 1733, a porter was accused of purposefully seeding a London bathhouse with bed bugs. The same year another instance of a "Person whose Head had a very Mischievous Turn" was reported in Dublin (Sarasohn, 2013).

1.5 Determination – and a Silver Bullet

Humans have long sought to make their habitations less favorable to bed bugs. Heavy, wooden beds laden with cracks and crevices were replaced with metal frames that were less congenial to the pests and easier to inspect. Bed bug-proof building construction was also stressed (see Section 1.5.2). Most importantly, people took measures to prevent bed bugs from entering and establishing themselves within the home. This involved

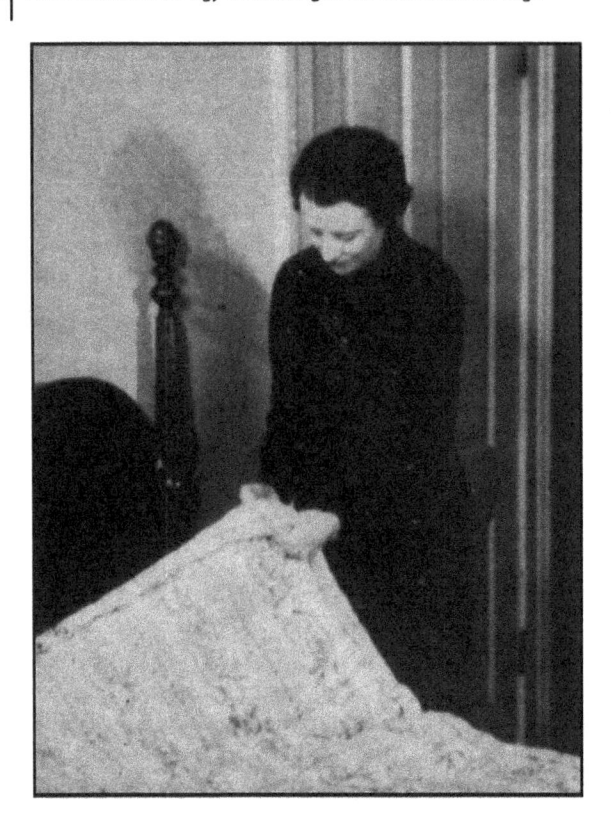

Figure 1.4 Bed bug inspections used to be important in maintaining a clean and healthful home. Credit: Clemson Agricultural Bulletin 101 (1941).

checking such things as clothes sent to the laundress, blankets returning from summer camps, and suitcases after traveling. Frequent and careful examination of beds was advised to aid in finding the first bed bug (Figure 1.4).

Because bed bugs were so difficult to keep out of the home, the housewife often battled them during spring cleaning. An advantage of such timing back then was that in unheated homes, bed bug populations tended to be lower at the end of winter due to the effects of cold temperatures. De-bugging the home was laborious; measures often included boiling anything that was washable, re-stuffing beds with new filling, scalding walls and floors with hot water, setting the bedposts in cans of oil, and setting off sulfur candles. Oftentimes such measures needed to be repeated since the effects were short-lived.

1.5.1 Bed Bug Insecticides

Insecticides used for bed bug control have a long and interesting history. All manners of concoctions were employed – gaseous, liquid, and dust – and some were as toxic to people as to pests. Typical bed bug remedies during the 1800s and early 1900s included arsenic and mercury compounds prepared by the local druggist. The poisons were often mixed with water, alcohol, or spirits of turpentine and applied with a brush, feather, syringe, eyedropper, or oil can, wherever the bugs were found. Mercury chloride, popularly known as "Bed Bug Poison" (Figure 1.5), was a common remedy used by both exterminators and the general public (the toxic compound was also used widely to treat syphilis). One way to apply it was with the whites of an egg, beaten together and then laid with a feather (Kinsley, 1893). Unfortunately, a number of these products were also toxic to people, killing some accidentally, or perhaps by intent. Many early bed bug sprays, such as kerosene and gasoline, were also highly flammable. Consequently, buildings sometimes caught on fire if a match was struck too soon after treatment.

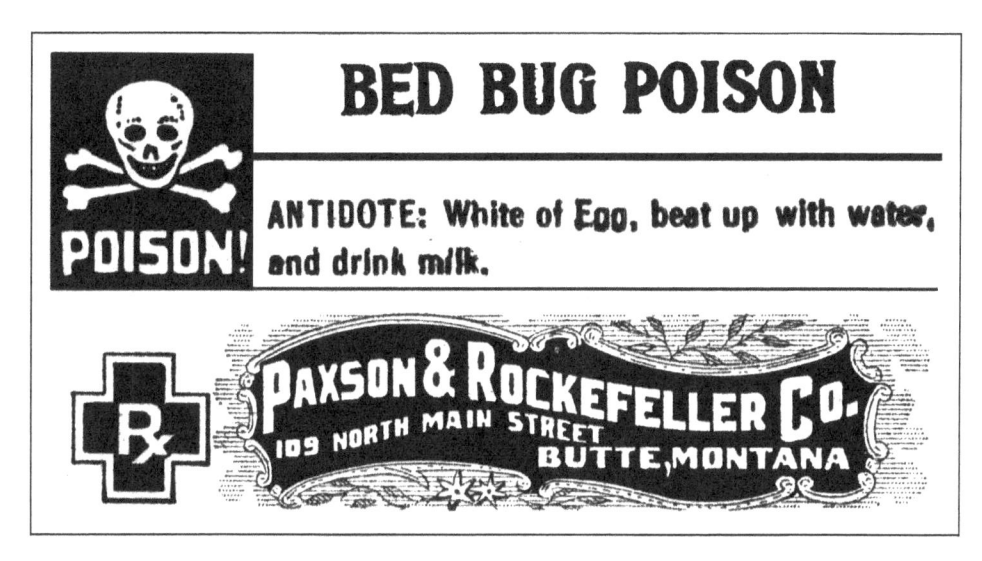

Figure 1.5 Mercury chloride, popularly known as "bed bug poison," was a common remedy for bed bugs. Many people died from accidentally or intentionally ingesting the poison.

Pyrethrum, prepared from dried chrysanthemum flowers, was a much safer material that was used from the mid-1800s to treat bed bug infestations. Pyrethrum was included in many early bed bug preparations formulated as sprays and powders. During wartime, when pyrethrum was in short supply, many other bed bug-killing compounds were used, including rotenone, cresol, and naphthalene. Kerosene, turpentine, benzene, and gasoline were also widely used, as was alcohol, which is still being sprayed onto bed bugs today. The effect of all these materials, however, was short-lived, seldom lasting beyond a day. Since the sprays lacked residual action and did not kill bed bug eggs, treatment had to be thorough enough to contact the insects directly. Lacking effectiveness as a dry deposit, follow-up spraying one or two weeks later was necessary to kill emerging eggs and any adults or nymphs that were missed. A recurring theme of bed bug treatment has been the need for thoroughness (Hockenyos, 1940a). Mallis (1945) succinctly cautioned "It should be remembered that amateur efforts usually produce amateur results," which the pest management industry is finding to be just as true today.

Advertisements for early bed bug insecticides were often entertaining (Figure 1.6). Despite having persuasive-sounding names like "Bed Bug Poison", "Bed Bug Killer", and "Bed Bug Murder", experts cautioned against putting too much confidence in their claims: "It is foolish to place too much reliance on the very numerous preparations on the market which claim to get rid of bed bugs. The efficacy of some of these is doubtful since the chemicals they contain must come in actual contact with the bug in order to destroy it. This is extremely difficult to achieve on account of the bug's power of concealment" (Hunter, 1938). It would be prudent to heed such advice again since many products being marketed for bed bugs today have similar limitations.

Lacking in residual action, early bed bug sprays were most effective against smaller infestations. For heavy infestations (before availability of DDT) fumigation was recommended. Early bed bug fumigation often involved burning sulfur, sometimes called the "fire and brimstone" method (brimstone was the ancient word for sulfur). A kettle or dish of powdered sulfur was placed in the center of the room, surrounded by a larger pan to keep the molten mass from spattering and setting fire to the floor (Hockenyos, 1940b). Alcohol was often added to enhance ignition and burning. Ready-made sulfur candles could also be used but were more expensive. Metal fixtures prone to tarnishing and corrosion were removed or coated with lard or Vaseline. The sulfur fumes also bleached and damaged wallpaper and fabrics. In order to confine the fumes, cracks around windows and doors were sealed with strips of old newspapers coated with thin flour paste or soaked in water. Fireplaces and chimneys were sealed off with sacks or blankets, while the keyholes were stuffed with rags (Herrick, 1914; Matheson, 1950). Apart from the damage to household items and the stench from the

Figure 1.6 Promotions for bed bug products were common and often entertaining. The cartoonist for this 1928 advertisement was Theodore Geisel (Dr Suess). Credit: Standard Oil Company of New Jersey.

burning sulfur, the procedure was comparatively simple and affordable, making it a viable control option for both householders and professionals. The sulfur fumes were lethal to all bed bug life stages, including eggs, but had poorer penetration than some other gases, and the process sometimes had to be repeated.

The gold standard for bed bug fumigation during the first half of the 20th century was hydrocyanic acid (HCN, cyanide) gas. Fumigating with cyanide was highly effective, but costlier and more dangerous than other methods. As with modern-day fumigations, the entire building had to be vacated, which was not essential when burning sulfur. Due to the danger, cyanide fumigations were best performed by professionals, but this was not always the case. Many people without the proper training and safety equipment were killed or seriously injured, and even professionals had mishaps using the effective but lethal material.

Various commercial preparations of hydrogen cyanide were available, including *Zyklon B* pellets and powder used in the gas chambers during the Holocaust. The most popular and convenient formulation used by pest control firms were "discoids," consisting of fibrous absorbent discs saturated with liquid cyanide, packed in gastight metal containers. When exposed to air, the liquid cyanide quickly volatilized into toxic gas, necessitating the use of a gas mask. Applicators worked in teams with one person opening cans while the other scattered the discs onto layers of cardboard or newspaper. Special care was needed, post-fumigation, to adequately ventilate the building and its contents (Mallis, 1945).

Despite the dangers and other drawbacks, cyanide fumigation was long considered the most efficient means of eliminating serious bed bug infestations. Fumigation chambers and vans were widely used for disinfesting furniture and other belongings. But all that changed after the start of World War II when a new and more potent chemical spray became available: DDT.

The discovery and development of DDT for battling bed bugs and other pests is legendary. DDT was originally synthesized in 1874 by a young German chemistry student working on his thesis, but the compound stayed in obscurity until 1939, when Paul Muller, a Swiss scientist with the Geigy Company, discovered its remarkable insecticidal properties (Muller was awarded the Nobel Prize for the discovery in 1948). Initial quantities were under sole allocation of the War Production Board, to protect US armed forces during World War II from disease-carrying lice, flies, and mosquitoes. Beginning in 1942, DDT was also evaluated against

Figure 1.7 In 1945, suppliers began advertising the availability of DDT for civilian (non-military) uses, including control of bed bugs. Credit: M.F. Potter.

bed bugs in hopes of finding a more effective and economical method of control in military barracks. Test results by the USDA Bureau of Entomology and Plant Quarantine in Orlando, Florida, were deemed phenomenal and DDT was proclaimed "the perfect answer to the bed bug problem" (US Bureau of Entomology, 1945). By the end of 1945, chemical companies were also heralding the availability of DDT for civilian use, giving the public a potent new weapon in the war on bed bugs (Figure 1.7).

What made DDT special was its long-lasting effectiveness as a dry deposit. No longer did bed bug sprays have to contact the insects directly, as was required with other materials. For the first time, bed bugs residing in hidden locations and nymphs hatching from eggs succumbed by resting or crawling on previously treated surfaces. While some studies reported a residual effect lasting at least six months (Madden *et al.*, 1944, 1945), Mallis (1954) noted that samples of wallpaper sprayed with DDT continued killing bed bugs three years later, eliminating the need for reapplication in the event that some bugs were missed or reintroduced. Experiments further showed that DDT had no repellency and did not disperse bed bugs throughout a room or building, like pyrethrum and some other materials.

DDT applied as a 5% oil-base spray (typically blended with deodorized kerosene) or 10% powder was so effective that all the bed bugs in a room could eventually be killed by thoroughly treating the bed and nowhere else, since the bugs eventually had to come there to feed (Stenburg, 1947). In practical use, most other locations in the room were also treated to hasten eradication. Thorough treatment of the entire mattress (Figure 1.8), pillows, bed springs, and frame was recommended (US Bureau of Entomology, 1945; USDA, 1953). One application usually did the job, in contrast to the recurring treatments previously needed (and being experienced today).

Interestingly there was little mention of having to prepare for extermination by de-cluttering and washing bedding and clothing. This is quite different from current methods, which place great importance on such preparatory measures. Years ago, many households had fewer furnishings, clothing, knickknacks, and clutter. Contaminating people's belongings with pesticide was also less of a concern at the time.

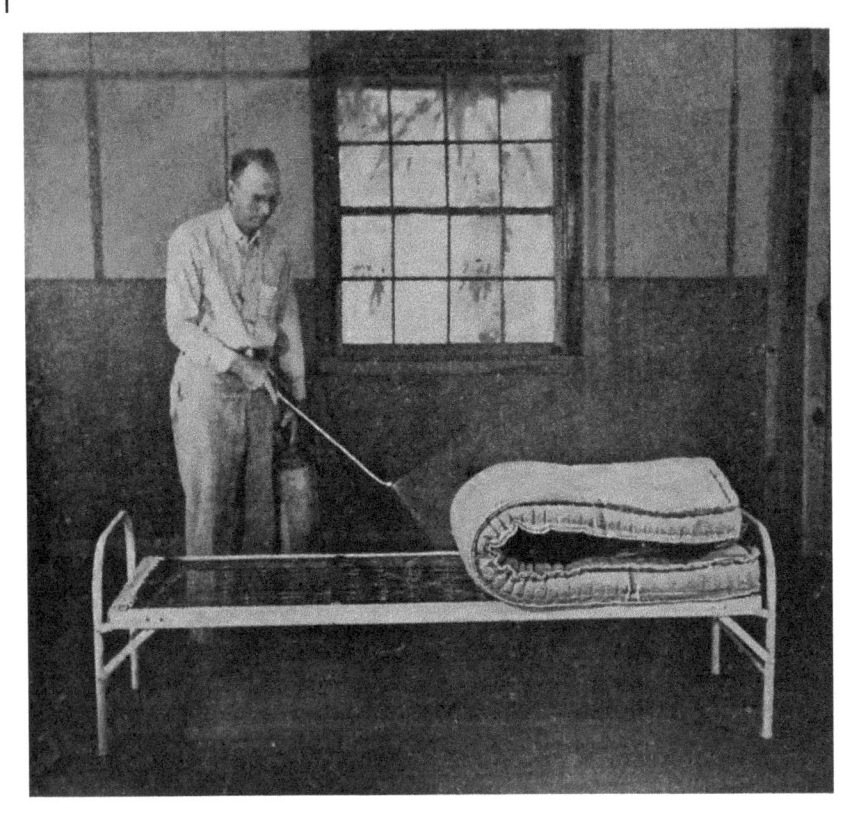

Figure 1.8 When controlling bed bugs with DDT, treatment of the entire bed was recommended, including the entire mattress. Credit: US Department of Agriculture (1953).

Another factor that helped hasten the bed bug's demise was that DDT was relatively inexpensive and could be bought and used by anyone. DDT in various preparations could be purchased at most drug, hardware, and department stores, and at some food markets (USDA, 1953). Unlike most fumigants, the material could be applied by householders and professionals alike with successful results. A few ounces of spray or an ounce of the powder was enough to treat a full-size bed and prevent re-infestation for at least a year. For added convenience, total-release DDT "bombs" (the same ones used in wartime by the military) were sold. The insecticide was also incorporated into paints and wallpaper. The all-out civilian assault with DDT was so effective and widespread that within five to seven years, it became difficult to find populations of bed bugs on which to do further research (J.V. Osmun, Purdue University, West Lafayette, unpublished results).

As bed bugs were disappearing, reports began surfacing that some populations had become DDT-resistant. Failures were first noted in barracks of the Naval Receiving Station at Pearl Harbor in 1947, only a few years after the product was first used (Johnson and Hill, 1948). During the next ten years, other reports of bed bug resistance to DDT were confirmed, especially in tropical areas of the world (Busvine, 1958). Spraying inside houses during malaria-eradication efforts probably contributed to the onset of resistance in bed bugs (Rafatjah, 1971). With growing reports of DDT resistance, insecticides such as malathion, diazinon, and lindane were used as alternatives. As with DDT, a single application usually did the job, provided spraying was thorough. Pyrethroids were subsequently used as replacements, but resistance to these insecticides has also been documented in bed bugs throughout the world (Romero *et al.*, 2007; Zhu *et al.*, 2010; Davies *et al.*, 2012).

1.5.2 Bug Proof Design and Construction

Throughout history, modifications were made to make beds and buildings less habitable to bed bugs. In the 16th and 17th centuries, mattresses were typically stuffed with straw and placed atop a latticework of ropes that needed regular tightening by twisting a wooden dowel. When the bed bugs became intolerable, the straw ticking was burned and replenished. Beds were long considered a status symbol for the wealthy. During the 14th through 18th centuries, they often were fashioned of ornately carved wooden timbers, which afforded countless places for the bugs to hide. Such beds also tended to be draped in fabric to keep out dust and drafts. Because of the bed bugs, exterminators began discouraging such constructions.

By the mid-18th century, heavy crack-laden wooden beds were being replaced with cast iron, which was less attractive to bed bugs and easier to dismantle and inspect. Another advantage of metal over wood was that alcohol or kerosene could be poured over the joints and ignited with a lighted match. The mid-18th century introduction of cotton mattresses also made it easier to de-infest bedding since the bugs "could be boiled to death without spoiling the fabric" (Wright, 1962). Mattresses were also redesigned with fewer buttons, folds, and creases.

Bed bug deterrent construction was also encouraged in design of buildings. In the 1930s and 1940s, hospitals and hotels in Europe were being constructed with metal windows and doors and little or no woodwork. Floors were of cement or other tight composition with no baseboards. Walls were smoothly painted in lieu of peeling-prone wallpaper, and cracks and crevices were filled with soap, putty, or other sealants (UK Ministry of Health, 1934; Hartnack, 1939). Today, such measures have been abandoned in favor of aesthetics and comfort. The coziness of the modern sleeping room is testament to how long it's been since bed bugs were top of mind. At-risk entities such as hotels, hospitals, and college dormitories may eventually need to re-think the way they design and furnish their rooms to make them less habitable to bed bugs.

1.5.3 Bed Bug Traps

Devices have long been used for trapping and removing bed bugs. Dishes, pans, and the like were placed under bed legs to discourage the vermin from scaling the bed and biting the sleeper. Oftentimes the saucers were filled with a liquid such as oil or kerosene. Similar pitfall traps are being marketed today to deter and monitor for bed bugs. In the 1700s, peasants also fashioned simple bed bug traps from planks of wood punched full of small holes. Placed under the mattress, the trap afforded convenient harborage for wandering bed bugs, which were removed and killed the following morning. Another trap for revealing bed bugs' presence utilized a wooden board and a flap of felt (Busvine, 1976). More intricate "lobsterpot"-sized bed bug traps were concocted of wicker by 19th century basket makers: "The trap was placed behind the bolster and between it and the head of the bed… the little anthropophagi after their nightly meal would retire to digest between the interstices of the wicker trap. The housemaid in the morning would take the trap into the yard or garden and shake out the victims, who would meet a violent death under her feet" (Wright, 1962).

In the Balkan countries of southeastern Europe, common bean leaves (*Phaseolus vulgaris* L.) were used for centuries to entrap bed bugs. The leaves were spread on the floor of infested rooms, and the following morning, the leaves with the bugs on them were removed and burned (Bogdandy, 1927). The bean leaves have no attractant effect on bed bugs, but the bugs become ensnared in the hooked hairs (trichomes) on the leaves while wandering at night (Richardson, 1943). Recent studies have attempted to fabricate synthetic versions of the tiny hooks so that the ancient approach might one day be used for management (Szyndler, *et al.* 2013).

1.5.4 Lethal Temperatures

Heat has been used to kill bed bugs for centuries. Boiling water was used to scald bugs residing in bedding, bed slats, springs, and other locations. Candles were also deployed: "I can still recall the acrid smell of roasting bedbugs in bedsprings with a candle, when I was a youngster in the 1920s. Candling bedsprings was what my

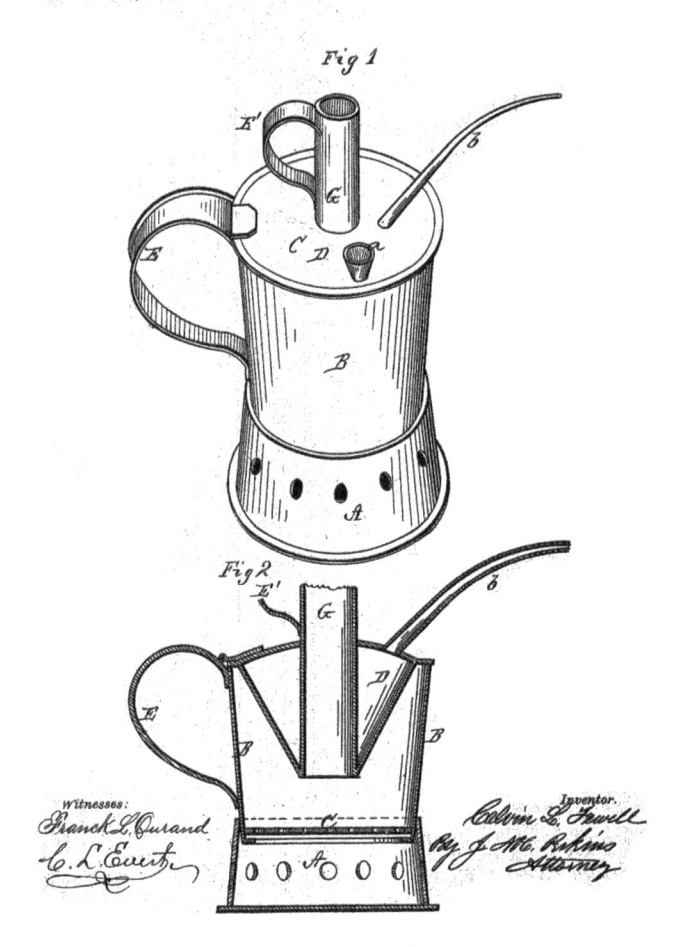

Figure 1.9 Early patent for a bed bug steamer, published in 1873. Credit: US Patent and Trademark Office.

mom learned when she lived in Russia at the turn of the century. We also put bottle caps filled with oil under the bed legs." (H.L. Katz, pers. comm. to R.D. Kozlovich, Safeway Pest Control, Cleveland). Others, including the US Military, used more drastic measures: "Flaming the cracks of steel cots with a blowtorch is quite effective" (US War Department, 1940).

C.L. Fewell received a patent in 1873 for the first portable bed bug steamer, which was fashioned like a tea kettle with an underlying fire and ash box (Figure 1.9). "The manner of using the exterminator is by moving the spout along crevices in furniture or walls, as the case may be, when the jet of steam issuing from the spout penetrates to the lurking places of the vermin and carries with it instant destruction" (Fewell, 1873). More sophisticated bed bug steamers powered by electricity are being used by the pest control industry today.

A more comprehensive way of controlling bed bugs with heat was adapted from methods developed in the early 1900s to de-infest granaries and flour mills. In an article entitled "Eradication of the bedbug by superheating," investigators in Canada showed that it was possible to de-infest a two-story house by stoking up the furnace and other stoves during summer to a temperature of 160 °F (Ross, 1916). Similar success was reported in another study where steam was used to heat a 350-room dormitory on a college campus in Mississippi

(Harned and Allen, 1925). In this case, maximum temperatures in bed bug-infested rooms ranged from about 110 to 125 °F, over a heating period lasting a few days. The authors concluded that very high mortality can be achieved at temperatures as low as 110 °F when maintained for two days, and from a few hours exposure to 120 °F. Mallis (1945) mentioned using superheating to eliminate a severe infestation of bed bugs in an animal-rearing laboratory. He reported that after eight hours of heating, "the mortality was so terrific, that a carpet of bedbugs covered the floor, and a slight draft through the room piled up windrows of the bugs against several objects on the floor."

Interest in using heat to control bed bugs all but vanished after the discovery of DDT. Today's renewed utilization reflects the lack of effective management options and greater concerns over pesticides.

1.6 Past is Present

History reveals both insights and concerns about bed bugs and their management. For much of the developed world, the modern-day resurgence of this pest serves as a reminder that it is not a birthright to live free of parasitic vermin. There will be new challenges this time around, including unprecedented movement of people locally and globally; more clutter and belongings in which the bugs can hide; less potent pesticides for home and professional use, and yet more restrictions on how liberally they can be used. Perhaps most challenging will be instilling again a mindset of societal vigilance. The foundation of bed bug management still consists of hard work, public education, and preventing or detecting infestations in the initial stages. It will be interesting to see if humanity is up to the challenge.

References

Aldrovandi, U. (1603) *De Animalibus Insectis*, Bonon, Bologna.
Anonymous (1777) *The Compleat Vermin-Killer: A Valuable and Useful Companion for Families in Town and Country*, Fielding & Walker, London.
Anonymous (1536) *Hortus Sanitatis (1536 edition)*, Jacob Meydenbach, Mainz, Germany.
Anonymous (1902) Bedbugs no cause for damage. *Daily Iowa State Press* (January 20).
Biehler, D.D. (2009) Permeable homes: a historical political ecology of insects and pesticides in US public housing. *Geoforum*, **40** (6), 1014–1023.
Bogdandy, S. (1927) Ausrottung von bettwanzen mit bohnenblattern. *Naturwissenschaften*, **15** (22), 474.
Booth, W., Balvin, O., Vargo, E.L., Vilimova, J. and Schal C. (2015). Host association drives genetic divergence in the bed bug, *Cimex lectularius. Molecular Ecology*, **24** (5), 980–992.
Borel, B. (2015) *Infested: How the Bed Bug Infiltrated Our Bedrooms and Took Over the World*, The University of Chicago Press, Chicago.
Buckley, C. (2010) *Legislature Passes Bed Bug-Notification Law, The New York Times* (24 June), http://www.nytimes.com/2010/06/25/nyregion/25bedbugs.html (accessed 11 November 2010).
Busvine, J.R. (1958) Insecticide-resistance in bed-bugs. *Bulletin of the World Health Organization*, **19** (6), 1041–1052.
Busvine, J.R. (1976) *Insects, Hygiene and History*, The Athlone Press, University of London, London.
Campkin, B. (2013) *Remaking London: Decline and Regeneration in Urban Culture*, I.B. Tauris & Co., New York.
Chomel, N. (1727) *Dictionaire Oeconomique: or the Family Dictionary. Containing the Most Experience'd Methods of Improving Estates and of Preserving Health*, trans. R. Bradley, Dublin.
Cowan, F. (1865) *Curious Facts in the History of Insects*, J.B. Lippincott & Co., Philadelphia.
Cruikshank, I. (1782) *Summer Amusement – Bugg Hunting*, https://www.loc.gov/item/00652100/ (accessed 15 May 2017).

Davies, T.G.E., Field, M. and Williamson M.S. (2012) The re-emergence of the bed bug as a nuisance pest: implications of resistance to the pyrethroid insecticides. *Medical and Veterinary Entomology*, **26** (3), 251–254.

England, K.M (ed.) (1931) *Housing: A Citizen's Guide to the Problem*, Chatto & Windus, London.

Fewell, C.L. (1873) Improvement in bed-bug exterminators. US patent 139,562, issued 3 June 1873.

Harned, R.W. and Allen, H.W. (1925) Controlling bedbugs in steam-heated rooms. *Journal of Economic Entomology*, **18** (2), 320–330.

Hartnack, H. (1939) *202 Common Household Pests of North America*. Hartnack Publishing Co., Chicago.

Herrick, G.W. (1914) *Insects Injurious to the Household and Annoying to Man*. The MacMillan Company, New York.

Hockenyos, G.L. (1940a) Bedbug spraying. *Pests*, **8** (5), 12–16.

Hockenyos, G.L. (1940b) Sulfur dioxide fumigations. *Pests*, **8** (11), 23–25.

Hunter, L. (1938) *Domestic Pests: What They Are and How to Remove Them*. John Bale, Sons & Curnow, Ltd, London.

Johnson, C.G. (1942) The ecology of the bed-bug, *Cimex lectularius* L. in Britain. *Journal of Hygiene*, **41** (4), 345–361.

Johnson, M.S. and Hill A.J. (1948) Partial resistance of a strain of bedbugs to DDT residual. *Medical News Letter*, **12** (1), 26–28.

Kalm, P. (1748) *Travels into North America*, T. Lowndes, London.

Kinsley, C. (1893) *The Circle of Useful Knowledge: for Farmers, Mechanics, Merchants, Surveyors, Housekeepers, Professional Men, etc*. Kinsley Publishing, Clinton.

Madden, A.H., Lindquist, A.W. and Knipling, E.F. (1944) DDT as a residual spray for the control of bedbugs. *Journal of Economic Entomology*, **37** (1), 127–128.

Madden, A.H., Lindquist, A.W. and Knipling, E.F. (1945) DDT and other insecticides as residual-type treatments to kill bedbugs. *Journal of Economic Entomology*, **38** (2), 265–271.

Mallis, A. (1945) *Handbook of Pest Control*, Mac Nair-Dorland Company, New York.

Mallis, A. (1954) *Handbook of Pest Control*, 2nd edn, Mac Nair-Dorland Company, New York.

Markkula, M. and Tiittanen, K. (1970) Prevalence of bed bugs, cockroaches and human fleas in Finland. *Annales Entomologici Fennici*, **36** (2), 99–107.

Marlatt, C.L. (1916) The bedbug. *UDSA Farmers' Bulletin*, 754, Washington, DC.

Matheson, R.M. (1932) *Medical Entomology*, Charles C. Thomas, Springfield.

Matheson, R.M. (1950) *Medical Entomology*, 2nd edn, Comstock Publishing Associates, Ithaca.

Mayhew, H. (1861) *London Labour and the London Poor (Vol. 3)*, Harper & Brothers, New York.

Millard, C.K. (1932) Presidential address, on an unsavoury but important feature of the slum problem. *The Journal of the Royal Society for the Promotion of Health*, **53**, 365–372.

NAHO (1939) *Disinfestation of Dwellings and Furnishings: Problems and Practices in Low-rent Housing*, National Association of Housing Officials, Chicago.

Orwell, G. (1933) *Down and Out in Paris and London*, Victor Gollancz, Ltd., London.

Osborne, H. (1896) The common bed bug. insects affecting domestic animals, *Bulletin of Entomology*, USDA Government Printing Office, Washington, DC.

Panagiotakopulu, E. and Buckland, P.C. (1999) *Cimex lectularius* L., the common bed bug from Pharaonic Egypt. *Antiquity*, **73** (282), 908–911.

Potter, M.F. (2011) The history of bed bug management-with lessons from the past. *American Entomologist*, **57** (1), 14–25.

Potter, M.F., Rosenberg, B. and Henriksen, M. (2010) Bugs without borders: defining the global bed bug resurgence. *PestWorld*, **Sept/Oct**, 8–20.

Potter, M.F., Haynes, K.F. and Fredrickson, J. (2015) Bed bugs in America: the 2015 national bed bug survey. *PestWorld*, **Nov/Dec**, 4–14.

Rafatjah, H. (1971) The problem of resurgent bed-bug infestation in malaria eradication programmes. *Journal of Tropical Medicine and Hygiene*, **74** (2), 53–56.

Ray, J. (1673) *Observations Topographical, Moral, & Physiological; Made in a Journey through part of the Low-Countries, Germany, Italy, and France*, The Royal Society, London.

Reinhardt, K. (2012) *Pesets'ola:* Which bed bug did the Hopi know? *American Entomologist*, **58** (1), 58–59.

Riley, W.A. and Johanssen, O.A. (1938) *Medical Entomology*, McGraw-Hill, New York.

Richardson, H.H. (1943) The action of bean leaves against the bedbug. *Journal of Economic Entomology*, **36** (4), 543–545.

Romero, A., Potter, M.F., Potter, D.A. and Haynes, K.F. (2007) Insecticide resistance in the bed bug: a factor in the pest's sudden resurgence? *Journal of Medical Entomology*, **44** (2), 175–178.

Ross, W.A. (1916) Eradication of the bedbug by superheating. *Canadian Entomologist*, **48** (3), 74–76.

Sarasohn, L.T. (2013) That nauseous venomous insect: bedbugs in early modern England. *Eighteenth-Century Studies*, **46** (4), 513–530.

Southall, J. (1730) *A Treatise on Buggs*, J. Roberts, London.

Stenburg, R.L. (1947) The techniques of application and the control of roaches and bedbugs with DDT. *Pests*, **15** (8), 16–22.

Szyndler, M.W., Haynes, K.F., Potter, M.F., Corn, R.M. and Loudon, C. (2013) Entrapment of bed bugs by leaf trichomes inspires microfabrication of biomimetic surfaces. *Journal of the Royal Society Interface*, **10** (83), 1–9.

Trustees of the British Museum (Natural History) (1973) *The Bed-Bug*, Economic Series Number 5, The British Museum, London.

UK Ministry of Health (1934) *Report on the Bed-bug, Reports on Public Health and Medical Subjects*, His Majesty's Stationary Office, London.

US Bureau of Entomology (1945) Suggestions regarding the use of DDT by civilians. *Pests*, **12** (10), 24–26.

USDA (1875) *Report of the Commissioner of Agriculture*. US Department of Agriculture, Washington, DC.

USDA (1953) Bed Bugs: How to Control Them. Leaflet no. 337. US Department of Agriculture. US Government Printing Office, Washington, DC.

US War Department (1940) *Basic Field Manual: Military Sanitation and First Aid*. US Government Printing Office, Washington, DC.

Usinger, R.L. (1966) *Monograph of Cimicidae (Hemiptera - Heteroptera)*, Entomological Society of America, College Park.

Usinger, R.L. and Povolny, D. (1966) The discovery of a possibly aboriginal population of the bedbug (*Cimex lectularius* Linnaeus, 1758), *Acta Musei Moroviae*, **51**, 237–242.

Whitford, M. (2006) Bed bug war stories. *Pest Control*, **74** (5), 30–40.

Wright, L. (1962) *Warm and Snug: the History of the Bed*, Routledge and Kegan Paul, London.

Zhu, F., Wigginton, J., Romero, A., *et al.* (2010) Widespread distribution of knockdown resistance mutations in the bed bug, *Cimex lectularius* (Hemiptera: Cimicidae), populations in the United States. *Archives of Insect Biochemistry and Physiology*, **73** (4), 245–257.

2

Bed Bugs in Popular Culture

Stephen L. Doggett and David Cain

2.1 Introduction

For almost as long as a word could be written down, a play performed, a poem recited, or a song sung, bed bugs have appeared in popular culture. From poetry, to figurative art, to theatre, to literature, to music, to television, and even in everyday language, bed bugs have made an appearance. They have even worked their way into the sensual world of erotica. Yet over the years, the use of bed bugs in popular culture has fluctuated and changed quite dramatically. This chapter explores this theme and examines the evolution of bed bugs in popular culture from the time when the insect was very much a part of our life, to the decline post World War II, to the now with the modern bed bug resurgence.

Due to limitations of space, it is not possible to provide a complete list all appearances of bed bugs in popular culture. For more examples, the following outstanding texts should be consulted: *Des Compagnons de Toujours* (Doby, 1997), *Literarische Wanzen Eine Anthologie* (Reinhardt, 2014), and *Infested* (Borel, 2015).

2.2 Bed Bugs in Poetry

The earliest known appearance of bed bugs in prose was by the Roman physician, Quintus Sammonicus Serenus, who wrote cures in verse. In his famous text, *De Medicina Praecepta Saluberrima*, a squashed bed bug could stop nose bleeds, and when combined with garlic, cured fevers (Ruffato, 2004).

Without a question the most well-known (and certainly most cited) poem featuring bed bugs is "Night, Night Sleep Tight and Don't Let the Bed Bugs Bite", a version of which even appears in the lyrics of the song 'Good Night' within the Beatles' *White Album* in 1968 as "Good Night, Sleep Tight" (Martin, 2016). Many believe that the term "sleep tight" related to rope beds that were regularly tightened for reasons of comfort (the rope would sag overtime), but this appears to be a common myth (Anonymous, 2015). In fact the origin of the saying is obscure, with the first appearance of "sleep tight" being recorded in a diary from 1866 (Eppes, 1926). In this context, it is suggested to mean have a good sleep (Martin, 2016). Presumably the bed bug tag was added later as a humorous piece of rhyming slang.

In a French fable from 1813, a silkworm attempts to convince a bed bug to be beneficial to society (Arnault, 1813). The worm makes the accusation that "biting is a habit and torment is a pleasure."

The following is another famous poem featuring bed bugs where the origins are obscure and often attributed simply to an "American poet", (Anonymous, 1893; Shipley, 1914; Millard, 1932). Several variations of this rhyme exist (Burns, 1991; Borel, 2015)

Advances in the Biology and Management of Modern Bed Bugs, First Edition.
Edited by Stephen L. Doggett, Dini M. Miller, and Chow-Yang Lee.
© 2018 John Wiley & Sons Ltd. Published 2018 by John Wiley & Sons Ltd.

> The Junebug has the wings of gauze, the lightning bug the flame,
> The bedbug has no wings at all, but gets there just the same,
> The Junebug leaves the last of June, the lightning bug in May,
> The bedbug takes his bonnet off and says, "I've come to stay."

The children's rhyme "Ballad of a Bedbug" (Barton, 2012) tells the story of a bed bug with marital issues, who eventually departs his mattress leaving behind his wife and children.

2.3 Bed Bugs in the Figurative Arts

Many of the early works of art that depict bed bugs can be found in Doby (1997). One of the best known is the satirical depiction from 1782 by Isaac Cruikshank, titled *Summer Amusement Bugg Hunting*. Here a rather decrepit elderly couple are picking bed bugs off their bed and curtains, and dropping them into a chamber pot (Cruikshank, 1782; see Figure 1.2). Interestingly, in the background is a poster for the famous company Tiffin & Son, the so called "Bugg-destroyers to his Majesty", a company that operated exclusively to control bed bugs for almost 300 years. Later in 1811, the prodigious English satirical caricaturist, Thomas Rowlandson, drew a version of *Bugg Hunting*, making the female character considerably more buxom, as was his style (Rowlandson, 1811). Rowlandson by then had created other bed bug-related images, including *Bug Breeders in the Dog Days* from 1806 showing a rotund couple in bed ready to feed the bugs (Rowlandson, 1806), and the less than subtle *A Tit Bit for the Buggs* depicting a rather robust woman scratching her left breast, presumably after being bitten by bed bugs (Rowlandson, 1793).

Scenes and cartoons on postcards depicting people hunting for bed bugs were quite common from the late 1800s to around the 1940s (Figures 2.1 and 2.2; Doby, 1997, Doggett, 2016) and can be readily found in online auction sites. The question lost to history is why were these so common? Were these just amusing curios or did they represent an early form of social media whereby people sent these cards to friends and family warning that a particular establishment was infested? In the UK, many of these postcards were sent from popular seaside holiday destinations (Stephen Doggett, unpublished results) and so the cards were probably just a bit of silly fun.

One of the postcards in Figure 2.1 (Figure 2.1G) depicts a small island, labelled as "Bed Bug Island, Lake Hopatcong, N.J.". Several versions of the card exist, some being postmarked as early as 1917 (CardCow, 2016a), while others are later and postmarked 1951 (CardCow, 2016b). Apparently, this island was named in the early 1900s and was given the title because bed bugs are known to be very small, as was the island. Therefore, calling it Bed Bug Island was appropriate given its size (R. Cooper, personal communication, via the Lake Hopatcong historian). However, the naming of locations after bed bugs was not unique; there was also a Bedbug Hill in the same state of New Jersey (Beck, 1937). There was the town of Bedbug in California (Williams, 2014), and a Bed-bug Alley in Central Falls, Rhode Island (National Library of Congress, 2016).

In 1960, a series of stamps depicting the "Patriotic Health Campaign" was issued in China. One series was "To Wipe out the Four Pests", and this included rats, mosquitoes, flies", and bed bugs (Zhao, 2008).

Bed bugs have also appeared in comic strips. One appeared as a topic in Calvin & Hobbes in 2008 (an imaginary giant bed bug appeared under the bed; Watterson, 2008) and the Florida based comic, *Save the Bed Bug* (Borel, 2015).

In 1985, the toy manufacturer, Milton Bradley, released a board game called "Bed Bugs" (YouTube, 2013). The aim was to pick up with tweezers all the bed bugs of a particular color off a vibrating bed in the fastest possible time. This game was rereleased in 2013; presumably the manufacturer saw a new marketing opportunity with the modern resurgence. Other bed bug games and children toys have been produced in recent years, with most depicting an unrealistic cute version of the insect (see BedBugBeware, 2012, for a number of examples). Across the years, imitation bed bugs have been produced for use as practical jokes (Figure 2.3).

Figure 2.1 Bed bug postcards through history. (A) Stereogram dated 1897 from Missouri, USA. (B) "A hunting we will go", 1910 from Maine, USA. (C) Also "A hunting we will go", postmarked 1907 (clearly the same illustrator as B). (D) "I've seen all the big bugs of this place", postmarked 1907. (E) "A one night stand", 1908, New York, USA.

Figure 2.1 (Cont'd) (F) "It's the little things that count! Drop me a line", c1910. (G) "Bed bug Islands (where poets dream), Lake Hopatcong, N.J.", c1940. From the private collection of David Cain, except (C) from the private collection of Stephen Doggett.

Bed bugs have even made it into jewellery. One vintage piece (date unknown) is a replica of a bed, where the base of the bed can be tilted back. Inside of the bed a bed bug is revealed (Figure 2.4). Other more recent jewellery includes a necklace with a replica bottle of bed bug poison (Figure 2.4).

2.4 Bed Bugs in Theatre

Bed bugs have performed on the world stage. The first known play that features bed bugs is the classic Greek Comedy *The Clouds* by Aristophanes, written in 423 BCE (The Internet Classics Archive, 2000). In this play, the somewhat slow-witted student, Strepsiades, laments to his tutor Socrates about being forced to sit on a bed bug infested couch, "What cruel fate! What a torture the bugs will this day put me to!" Almost 2500 years later, we can all still relate to this comment.

J.M. Doby describes a French comedy from the late 1800s titled, *La Grande Symphonie des Punaises*, the English translation being *The Great Symphony of Bed Bugs* (Doby, 1997). The story is about a traveller staying at an inn who is almost devoured by an army of bed bugs. However, he fights back and the bed bugs are ultimately forced to flee.

Vladimir Mayakovsky in 1929 published the famous play *Klop* (or *The Bedbug*, in English), which was a satire on communist society (Anonymous, 1963a). The musical score was written by the legendary Russian composer, Dmitri Shostakovich (Wikipedia, 2014). The play focuses on a man who is resurrected in a laboratory along with an object of the past, a bed bug (Gerin, 2000; Sova, 2004). Unfortunately, satires of the communist regime at the time were not welcomed, and the play was quickly banned. Sadly, the author committed suicide soon afterwards. The play did eventually undergo a revival in Russia during the 1950s and was taken around the world in the early 1960s (Anonymous, 1963b). It was even performed in Berlin in 1999 (Steinberg, 1999).

Bedbug Celebration by John Blay was a musical drama with the score composed by the Australian music legend, Red Symonds, of the 1970s glam rock band, Skyhooks. It was first performed in Australia in 1981 and was a tribute to Mayakovsky's *The Bedbug*. In the story, a pop idol is brought back to life 50 years after his death. Somewhat ironically, the play is set in the early 2030s, when the bed bug is a threatened species (The Playwrights Database, 2003). While we may hope, this is unlikely to be the case even by then. More recently, a sci-fi rock opera entitled *BEDBUGS!!!* opened in New York. The story involved a scientist who inadvertently caused bed bugs to mutate into flesh-eating beasts (Borel, 2015).

Figure 2.2 More postcards through history. (A) "It's very lively", undated. (B) A postcard of an early photograph depicting a couple searching for bed bugs, undated. (C) "Bed-bugs outwitted!", c. 1900. (D–I). A series of late-Victorian postcards, c. 1900; see text for details. From the private collection of David Cain.

Figure 2.3 Bed bugs as practical jokes across the ages. (A) Imitation Bed Bugs, 1920s. (B) Fake Bed Bugs, 2016. (C) A close up of the "bed bugs", which are surprisingly similar despite almost a century difference in age, ~5 mm in length. Both originate from the US. From the private collection of Stephen Doggett.

2.5 Bed Bugs in Literature

Bed bugs have featured prominently in literature over the last century, and the references are too numerous to completely list here. The comprehensive review of Reinhardt (2014) should be consulted for more information.

Historically, bed bugs have been used in literature to set a background atmosphere of poverty or misery, which endears the characters to the reader by creating a feeling of empathy. For example, in Anton Chekhov's, *The Cherry Orchard*, (Chekhov, 1904), the author highlights the downtrodden worker, "…anyone can see that the workers eat abominably, sleep without pillows, thirty or forty to a room, and everywhere there are bedbugs, stench, dampness, and immorality". In *The Eye of Zeiton* by Talbot Mundy (1920), the main protagonists enter their hotel room and notes "the dull red marks on the walls…where bed-bugs had been slain with slipper heels…we were not in search of luxury…" In *Elmer Gantry* by Sinclair Lewis (1927), a seminary was thus described: "The walls were of old plaster…marked with the blood of… bed-bugs slain in portentous battles long ago…" Later, Lewis reinforced the relationship between bed bugs

Figure 2.4 Bed bug jewellery. (A&B) A small vintage 14 K charm; folding back the bed base reveals a bed bug (date unknown), approx. 1 cm in length; from the private collection of Stephen Doggett. (C) A modern necklace of a replica bed bug poison bottle, c. 2007; from the private collection of David Cain.

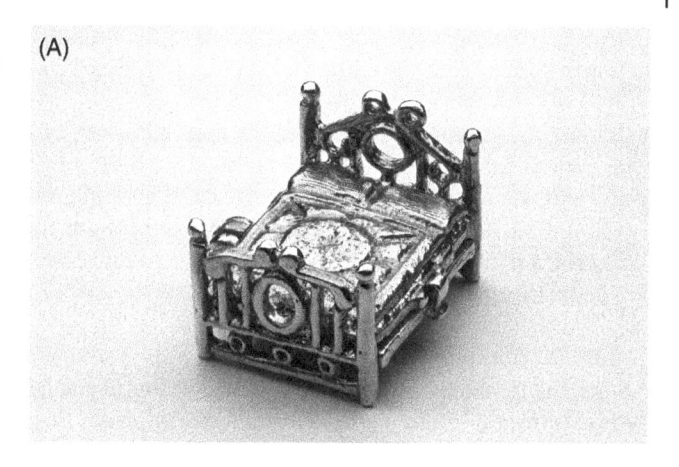

and poverty in *Work of Art* (Lewis, 1934): "…this carpetless, paperless room was frequented by bedbugs, and the beds were hard canvas cots without sheets…" Other authors also emphasize a setting of destitution by using bed bugs, such as in Talbot Mundy's *Jimgrim and Allah's Peace*; "So I knew about the bedbugs and the stench of the citadel moat" (Mundy, 1936).

Arguably the most evocative description in the literature of bed bugs inflicting their misery on the disadvantaged comes from the author Ruth Park in *The Harp in the South* (Park, 1948):

> As the darkness grew deeper, the bugs came out of their cracks in the walls, from under the paper and out of the cavities in the old bedsteads where they hung by day in grape-like clusters. They were thin and flat and starved, but before dawn they would return to their foul hiding places round and glistening and bloated with blood. Captain Phillip brought them in the rotten timbers of his first fleet and ever since they have remained in the old tenement houses of Sydney, ferocious, ineradicable, the haunters of the tormented sleep of the poor.

This description has to send a shiver down the spine of anyone familiar with the insect!

Since the modern resurgence, authors have employed bed bugs as a literary tool in a very different fashion. Bed bugs are no longer used as a simile to emphasis despair as part of a wider storyline; rather, they have become the key element themselves. This is particularly exemplified in the horror genre play *BEDBUGS!!!* mentioned above. Similarly, *Bed Bugs Can You See Them* (Taylor, 2012) was a horror film where an alien race of bed bugs runs amok in a small town murdering the inhabitants. In *Bedbugs* by Rick Hautala (1999), one short story focuses on a woman who believes she has bed bugs in her new apartment. However, there something even more sinister about the property. In *Bedbugs* by Ben H. Winters (2011), a young couple move into an apartment and the woman wakes up every morning with fresh bites, although neither her husband nor her child receive bites. The family soon realizes that they are literally confronting supernatural bed bugs that have come straight from Hell (perhaps this is an allegory of people's actual experience of the insect?). *Sleep Tight* by Jeff Jacobson (2013) makes the current crop of real bed bugs appear utter wimps. Jacobson's super bed bugs can multiply a hundred fold within a week and they transmit a virus, deadlier than Ebola, that turns people into homicidal maniacs. In many ways, these books are focusing on the current fad for horror and the zombie genre, as much as the resurgence of bed bugs.

Before the modern resurgence, horror books featuring bed bugs did not exist. The other genre that is now featuring bed bugs as the key element is in children's literature, and the works are largely fanciful. For example, in David Carter's *Bed Bugs* (Carter, 1998), the book proclaims "Good night, sleep tight, these pop-up bugs will never bite." *The Bedbug Who Wouldn't Bite* (Rhodes, 2009) is in a similar vein, while in the *Night of the Bedbugs* (Fricke, 2010) a young girl fears at night are calmed by a friendly group of bed bugs who sing lullabies to help her sleep; nonsensical notions indeed!

2.6 Bed Bugs in Music

Evidence of the use of bed bugs in music predating physical recordings was reviewed by J.M. Doby (1997). Early examples (1700s) of bed bugs in music included a Christmas carol, a bawdy song from Germany, and a sea shanty from France.

The first actual musical recording featuring bed bugs was from 1927, being *Mean Old Bed Bug Blues* by Walter E. Lewis (Lewis, 1927). This song was later recorded by multiple artists (Peterson, 2014; Borel, 2015) and the lyrics detail the desire a bed bug has to bite its victim:

> Man, a bedbug sure is evil, he don't mean me no good.
> Man, a bedbug sure is evil, he don't mean me no good.
> He thinks I'm a woodpecker, and he takes me for a chunk of wood.
> When I lay down at night, I wonder how can a po' man sleep.
> When I lay down at night, I wonder how can a po' man sleep.
> When some holdin' your hands, while the other ones eat your feet.
> Bedbugs big as a jackass, will bite you and stand and grin.

> Bedbugs big as a jackass, will bite you and stand and grin.
> And drink up all the bedbug poison, then come back and bite you again.
> Something was moanin' in a corner, I tried my best to see.
> Something was moaning in a corner, then I walked over to see.
> It was the mother bedbug prayin' to the good Lord, for some mo' to eat.
> I have to set up all night long, my feet can't touch the floor.
> I have to set up all night long, my feet can't touch the floor.
> 'Cause the mean old bedbug told me, that I can't sleep there no more.

After Johnson (1927), the lyrics were often feminized for female singers; see Pinto *et al.* (2007) for such a version.

A well-known musical recording was from 1953 by Lord Invader & His Calypso Rhythm Boys. Lord Invader was born Rupert Grant in Trinidad, and wrote "Rum & Coca-Cola", made famous by the Andrew Sisters. The song that features bed bugs is "Reincarnation (The Bed Bug)", where the artist sings about being reincarnated as a bed bug with a desire to bite only "big fat women" (SonicHits, 2016).

Brook Borel in her book *Infested,* and on her website *Bed Bug Songs* features a number of bed bug songs, some written as recently as 2013 (Borel, 2015, 2016). Additional music not mentioned in the Borel resources include "Bed Bug" by Roaring Jelly (1981) (YouTube, 2014), "Bad Bed Bug Song" by the Matthew Skoller Band (1996) (BadBedbugs.com, 2016), "Bedbug" by Cherrystones (2002) (45cat, 2002), and music from the TV series "Bedbug Party" (2006) (Sweeney *et al.*, 2006).

If one was to count the number of musical creations from the above listed resources there would be a total of 89 pieces that mention bed bugs. Broken down by decade, the number of bed bug themed songs: is 1920s (6), 1930s (3), 1940s (2), 1950s (3), 1960s (4), 1970s (2), 1980s (5), 1990s (8), 2000s (13), and 2010s (43). Thus, more than half of these songs (56/89) were created after the start of the modern bed bug resurgence, and interestingly, most have a comedic theme.

2.7 Bed Bugs in Television

As TVs in homes did not become common until after the 1950s, bed bugs obviously were not featured in TV programs prior to their decline (post World War II). However, similar to other forms of media, bed bugs became a hot topic for TV after the modern resurgence.

Perhaps the first appearance of bed bugs on TV was *My Bedbugs*, a children's TV series from the USA that aired in 2004 (Wikipedia, 2016a). Shortly thereafter, the New York based sitcom *The King of Queens* featured bed bugs in Season 8, Episode 15, in 2006 (S08E15, 2006). The premise of this episode was that bed bugs were introduced into the home from hotel sheets. Then, subsequent attempts to control the infestation using a "bug bomb" made the bed bugs (according to the exterminator in the show) "horny", causing them to breed prolifically (IMDb, 2016).

In another New York based TV sitcom, *30 Rock* (S04E04, 2009), one of the key characters' home becomes infested with bed bugs and they are ostracized by co-workers (Wikipedia, 2016b). In the US drama *Suits*, episode "Dirty Little Secrets" (S01E04, 2011), a lawyer takes on a client who is being pressured to move out of his apartment by a landlord who will not control a bed bug infestation (Wikia, 2016a). Another American Sitcom, *Parks and Recreation* (S03E11, 2011), based in Indiana, featured bed bugs, in an episode in which one character found an infestation in his hotel room and was forced to share a room with a colleague (Hatton, 2016). The American version of the classic British comedy, *The Office*, based in Pennsylvania, also featured bed bugs (S08E16, 2012). In this episode, one character claimed that there were bed bugs in his bed in an attempt to avoid the attention of an amorous colleague (Wikipedia, 2016c). In *The Simpsons* episode "Pulpit Friction" (S24E18, 2013), the family replace their old couch with an infested sofa purchased from New York (Wikipedia, 2016d). In *The Big Bang Theory*, a situation comedy based in southern California (S07E01, 2013), one character is concerned that their partner may have brought the bed bugs home after staying in a hotel (Wikia, 2016b). Similar to the Simpson's episode, the comedy, *The Michael J. Fox Show* (S01E08, 2013), featured a couch purchased from New York that turned out to be infested with bed bugs (Club, 2013). *Supernatural* is a fantasy horror series, and in one episode (S10E09, 2014), a character gives the following advice when asked what to do about bed bugs, "you should sleep tight and not let them bite" (YouTube, 2015). In the prison comedy-drama series *Orange is the New Black* (S03E02, 2015), inmates panic as bed bugs invade the penitentiary (Club, 2015).

There are several common themes that run through these TV shows. Almost all of the shows are comedies and attempt to make fun of the real-world fears that the American public has regarding bed bugs. All are set in the USA and most depict, or at least reference, New York City (NYC), which has had a major problem with bed bugs. Interestingly, some of the most popular TV series of all time, which where based in NYC, were in production at the beginning of the modern bed bug resurgence. This includes *Friends* (1994–2004), *Seinfeld* (1989–1998), and *Sex and the City* (1998–2004). However, the bed bug problem was not yet widespread enough to warrant any mention in these productions. Regarding the association between bed bugs and NYC, the famous New Yorker and show host, David Letterman summed it up best when he said, "Instead of rude and irritable, New Yorkers are now rude, irritable and itchy" (Newsmax, 2010).

2.8 Bed Bugs in Linguistics

The origin of the term "bed bug" and its use in different languages has been reviewed by Usinger in the *Monograph of Cimicidae* (Usinger, 1966) and by Michael Potter in the preceding chapter of this textbook. However, the term "bed bug" has crept into language throughout the ages, and has commonly been used as a derogatory insult.

Even as far back as Horace in *Satires*, around 35 BCE, bed bugs were used as a personal slur. In a reference to an acquaintance, it was stated "Do you think I am irritated by this bug Pantillus" (Doby, 1997). One early insult was from 1654 with the term "*cimex Grammaticus*" or "bedbug grammarian", which was considered a term of reproach in literature (Ronnick, 2001). In the USA, bed bugs were called "red coats" which was probably a derisive term given as a result of the British Army uniform being red during the War of Independence (Doby, 1995). In Yiddish, an insult for being small (vantzel) has its origin from the term used for bed bug, "vantsn" (Borel, 2015). The term "bedbugs" was used for those authors during the 1950s who, struggling to eke out a living, would sleep in the back of a famous book store Le Mistral (later called "Shakespeare and Company"), located on the left bank of Paris (Mercer, 2005). In the book *Bedbugs* by Clive Sinclair (1982), the title is a metaphor for the sexual activities of those in his story. The term "bedbugs" was used in Australia during the 1950s as a term for the boys who lived in a boarding school in Sydney (Pollock, 2007). In the novel *The Dream* by the legendary science fiction writer H.G. Wells, characters in the book called bed bugs the "Norfolk-Howards". This was after a Mr Bugg decided to change his name to the more pompous sounding "Norfolk-Howard" (Wells, 1924).

The father of the famous actor, Peter Ustinov, was known as "Klop Ustinov" (Day, 2015). "Klop" was a British spy who used guile to gain the confidence of those from whom he was tasked to extract information. As noted above, the term "Klop" is Russian for bed bug, and was given to him by his understanding wife. This was due to Klop Ustinov having to bed many women in order to learn "secrets" whilst in the service of the king.

"Crazy as a bed-bug" was an expression commonly employed in the USA (Doby, 1995). In one newspaper it was used in a reference to a mass murderer in New York (Anonymous, 1937). Around the same time, the term was used by various authors including Zane Grey in *Robber's Roost* (Grey, 1932), Arthur Zagat in his 1935 novel, *Riverfront Horror* (Zagat, 1935), and Robert Howard in *High Horse Rampage* (Howard, 1936).

"Bed bug" has also been used as a derogatory metaphor. For example, the newspaper article "The NSW bed bug bites hard" (Anonymous, 1998), has nothing to do with bed bugs. Instead "bed bug" in this case is used as a simile for a tax that the government had recently imposed on hotel beds.

2.9 Bed Bugs in Erotica

Considering the extraordinary nature of bed bug mating, a process known as "traumatic insemination", perhaps it is somewhat surprising that they do not feature more prominently in adult entertainment. Probably the only example in popular culture that has attempted to emulate this sexual behavior is Isabella Rossellini's performance art piece, *Seduce Me.* This piece, while not intending to be erotic, is clearly very sexual in nature (Rossellini, 2014). Despite that fact that few people have focused on their mating behavior, the use of bed bugs in erotica is not a recent phenomenon. During the conservative late Victorian and Edwardian eras, bed bugs were used in sensual ways. In a series of racy postcards, a heroine's story unfolds (Figure 2.2);

> A lady took apartments, which looked clean and snug,
> And felt herself quite happy – never thinking of a bug,
> Tired out she sought her virtuous couch covered over with a rug,
> And started up in a great affright, and said "I smell a bug",
> She shrieked aloud with horror, curled up her pretty mug,
> An swore to have her sweet revenge upon that filthy bug,
> She hunted every crevice, in every hole she dug,
> She searched most diligently for that filthy stinking bug,
> She hunted all her linen well, her shoulders gave a shrug,
> Within the hems she looked in vain, to catch that nasty bug,
> At last she pounced upon her toe, and put it in a jug,
> And let it have a cooling swim, "That terrifying bug".

Naturally the cards depict a scantily clad pretty young lady, and the innuendoes in the text are screamingly obvious. Similar postcards from the early 1900s can be readily found in online auction sites, depicting skimpily attired women searching themselves for bed bugs, fleas, or flies.

The Bedbugs' Night Dance and Other Hopi Sexual Tales originate from the Hopi people of northeastern Arizona, USA (Malotki, 1995). There are two narrations involving bed bugs. The first, *Bedbug Woman and Louse Woman*, deals with the desire of a certain male bed bug to feed in the vagina of a girl who is popular with other male bed bugs and lice. This desire eventually this leads to his death. Two widows, one of the bed bug and the other a louse, then fall into a steamy sexual relationship. The second tale, *The Bedbugs's Night Dance*, is about a young bed bug that witnesses the dance of a local Hopi tribe and becomes so engrossed in the dance that he has to tell the other bed bugs. The bed bugs then form a troupe to produce their own dance for the villagers and, in the process, seduce the womenfolk.

The corollary to the Hopi stories is that bed bugs are thought not to have been introduced to North America prior to the arrival of Europeans (Usinger, 1966). Yet Malotki (1995) has stated that these stories were thought to have predated the European invasion. Klaus Reinhardt in an eloquent argument using deductive reasoning settles this question by suggesting that the Hopi bed bug was actually the Mexican chicken bug, *Haematosiphon inodorus* (Dugés) (Reinhardt, 2012).

As mentioned above, the song "Mean Old Bed Bug Blues" was released in 1927 and initially had a quite innocent, somewhat comical set of lyrics about a bed bug wanting to bite its victim. However, a slightly later version by Bessie Smith had a more bawdy tone, using metaphors for sexual intercourse (Peterson, 2014). Similarly, in the song "Muriel and the Bug" by Lord Kitchener from 1961, the bed bug seeks out Muriel's "treasure" (Borel, 2015).

2.10 The Use of Bed Bugs in Popular Culture Through Time

As evidenced from the examples depicted above, the number of references to the use of bed bugs in popular culture has fluctuated over the years according to the prevalence of the insect in society. Before their decline in the 1950s, bed bugs were often featured, and then during the quiet years from around 1950 to 2000, they were rarely mentioned. Naturally this is to be expected because popular culture is a reflection of what is current (and topical) in society. With the recent resurgence, references to bed bugs in all forms of popular culture have exploded.

It is interesting to compare the changes in how bed bugs have been used in literature through the years. Since the resurgence, bed bugs have featured prominently, either playing on our fears, or as a mythical beast to allay our fears (at least in children's literature). As discussed above, this is very different to how bed bugs were portrayed in literature when the insect was a common part of everyday life. Perhaps the temporary bed bug reprieve changed our thinking and attitudes towards this pest. Once, the bed bug was considered to be a general nuisance, but an unfortunate part of our life. Hence the sufferers of the time received our empathy and this sympathetic perspective was prevalent in writing prior to WWII. Now, however, bed bugs are completely reviled; after all, a defeated enemy should not be on the offensive! It is probable that our modern hatred of bed bugs is enforced by modern marketers; a home with bacteria or insects is usually depicted as unclean and not safe for the family. Maybe the current use of bed bugs in the horror genre is simply a reflection of the modern perspective on bed bugs, and household pests in general.

Clearly many authors have found inspiration from the modern bed bug resurgence. Some may argue that these authors are simply exploiting current trends to help advertise and sell their own work, however popular culture is about dealing with contemporary issues. Whatever the reason for bed bugs' reappearance in popular culture, one thing has been clear across the ages: no matter how much misery and angst they cause, bed bugs are still the inspiration for a good laugh.

References

45cat (2002) *Cherrystones Bedbug*, http://www.45cat.com/record/tn041 (accessed 27 May 2016).
Anonymous (1893) Another lost opportunity. *Auckland Star* (19 August), p. 2.
Anonymous (1937) My client's crazy. *Truth* (18 July), p. 18.
Anonymous (1963a) "The Bedbug" makes a comeback. *The Canberra Times* (23 July), p. 2.
Anonymous (1963b) Premiere of "The Bedbug". *The Canberra Times* (23 July), p. 10.
Anonymous (1998) The NSW bed bug bites hard. *Business Review Weekly*, **20** (35), 107–109.
Anonymous (2015) *Mythbuster Friday: "Sleep Tight, Don't Let the Bedbugs Bite"*, https://chaddsfordhistorical.wordpress.com/2015/07/04/mythbuster-friday-sleep-tight-dont-let-the-bedbugs-bite/ (accessed 2 May 2017).
Arnault, A.V. (1813) *Fables*, Eymery, Paris.
BadBedbugs.com (2016) *Bed Bug Song*, http://www.badbedbugs.com/bed-bug-song/ (accessed 27 May 2016).
Barton, K. (2012) *Ballad of a Bedbug and Other Tales*, Xlibris Corporation, Great Britain.
Beck, H.C. (1937) *More Forgotten Towns of Southern New Jersey*, Rutgers University Press. New Jersey.
BedBugBeware (2012) *Bed Bug Beware Museum Gallery*, http://www.bedbugbeware.com/MuseumGallery.html (accessed 26 June 2016).
Borel, B. (2015) *Infested*, University of Chicago Press, Chicago.
Borel, B. (2016) *Bed Bug Songs*, http://bedbugsongs.tumblr.com/ (accessed 30 May 2016).
Burns, D.A. (1991) A potpourri of parasites in poetry and proverb. *British Medical Journal (Clinical Research Edition)*, **303** (6817), 1611–1614.
CardCow (2016a) *Bed Bug Island*, https://www.cardcow.com/315177/bed-bug-island-lake-hopatcong-new-jersey/ (accessed 15 June 2016).
CardCow (2016b) *Bed Bug Islands*, https://www.cardcow.com/161965/bed-bug-islands-lake-hopatcong-new-jersey/ (accessed 27 May 2016).
Carter, D. (1998) *Bed Bugs: A Pop-up Bedtime Book*, Little Simon, New York.
Chekhov, A. (1904) *The Cherry Orchard (translated by Stephen Mulrine)*, Nick Hern Books, London.
Club, A. (2013) *The Michael J. Fox Show*; "Bed Bugs", http://www.avclub.com/tvclub/the-michael-j-fox-show-bed-bugs-105327 (accessed 27 May 2016).

Club, A. (2015) *Orange Is The New Black; "Bed Bugs and Beyond"*, http://www.avclub.com/tvclub/orange-new-black-bed-bugs-and-beyond-220799 (accessed 27 May 2016).

Cruikshank, I. (1782) *Bugg Hunting*, http://loc.gov/pictures/resource/cph.3g08188/ (accessed 27 May 2016).

Day, P. (2015) *The Bedbug: Klop Ustinov, Britain's Most Ingenious Spy*, Biteback Publishing, London.

Doby, J.M. (1995) Les ectoparasites de l'homme dans le langage II. Punaise des lits, moustique, gale et son acarien [The ectoparasites of man in language: II. Bed bugs, mosquitoes, scabies and the scabies mite]. *Bulletin de la Societe Francaise de Parasitologie*, **13**, 253–279.

Doby, J.M. (1997) *Des Compagnons de Toujours… III – Punaise des Lits, Moustiques, Gale et Son Acarien*, J.M. Doby, Bayeux.

Doggett, S.L. (2016) Bed bug bites through time. *Pest*, **46**, 30–31.

Eppes, S.B. (1926) *Through Some Eventful Years*, J.W. Burke Company, Macon.

Fricke, P. (2010) *Night of the Bedbugs*, Image Comics, Fullerton.

Gerin, A. (2000) Stories from Mayakovskaya Metro Station: the production/consumption of Stalinist monumental space, 1938. PhD thesis, University of Leeds.

Grey, Z. (1932) *Robbers' Roost*, Harper & Brothers, United States.

Hatton, N. (2016) *TV Fanatic; Parks and Recreation Review: "Jerry's Painting"*, http://www.tvfanatic.com/2011/04/parks-and-recreation-review-jerrys-painting/ (accessed 27 May 2016).

Hautala, R. (1999) *Bedbugs*, Cemetery Dance Publications, Forest Hill.

Howard, R.E. (1936) *High Horse Rampage*, Benediction Classics, United States.

IMDb (2016) *The King of Queens; "Buggie Nights"*, http://www.imdb.com/title/tt0763465/plotsummary?ref_=tt_ov_pl (accessed 30 May 2016).

Jacobson, J. (2013) *Sleep Tight*, NY Pinnacle Books, New York.

Johnson, L. (1927) *"Mean Old Bedbug Blues"*, http://www.letssingit.com/lonnie-johnson-lyrics-mean-old-bedbug-blues-1927-tspbjzt#axzz4WYC7d9yj (accessed 23 January 2017).

Lewis, S. (1927). *Elmer Gantry*, Harcourt Trade Publishers, United States.

Lewis, S. (1934) *Work of Art*, Middlebury College Publications, United States.

Lewis, W.E. (1927) *"Mean Old Bed Bug Blues" (recording)*, Vocalion Label, Chicago.

Malotki, E. (1995) *The Bedbugs' Night Dances and Other Hopi Sexual Tales*, University of Nebraska Press, United States.

Martin, G. (2016) *Sleep Tight (The Phrase Finder)*, http://www.phrases.org.uk/meanings/sleep-tight.html (accessed 30 June 2016).

Mercer, J. (2005) *Books, Baguettes and Bedbugs: the Left Bank World of Shakespeare and Co*, Weidenfeld & Nicolson, London.

Millard, C.K. (1932) Presidential address, on an unsavoury but important feature of the slum problem. *The Journal of the Royal Society for the Promotion of Health*, **53**, 365–372.

Mundy, T. (1920) *The Eye of Zeitoon*, Bobbs-Merrill, Indianapolis.

Mundy, T. (1936) *Jimgrim and Allah's Peace*, Adventure Magazine, United States.

National Library of Congress (2016) *View of Privies, Garbage Dumps, etc., in Back Yards Near Bed-bug Alley and High Street, Central Falls, R.I. Location: Central Falls, Rhode Island*, http://www.loc.gov/pictures/item/ncl2004003779/PP/ (accessed 27 May 2016).

Newsmax (2010) *The Best of Late Nite Jokes*, http://www.newsmax.com/Jokes/310/ (accessed 15 August 2016).

Park, R. (1948) *The Harp in the South*, Angus & Robertson, Sydney.

Peterson, R.K.D. (2014) Mean old bed bug blues. *American Entomologist*, **60**, 241–243.

Pinto, L.J., Cooper, R. and Kraft, S.K. (2007) *Bed Bug Handbook-the Complete Guide to Bed Bugs and Their Control*, Pinto & Associates, Inc., Mechanicsville, MD.

Pollock, A. (2007) *"Bedbugs, Daygos and Slabs" A Century of Memories 1907–2007*, Hurlstone Agricultural High School, Glenfield.

Reinhardt, K. (2012) Pesets'ola: which bed bug did the Hopi know? (A present for Robert Leslie Usinger's 100th birthday). *American Entomologist*, **58** (1), 58–59.

Reinhardt, K. (2014) *Literarische Wanzen Eine Anthologie*, Neofelis Verlag, Berlin.

Rhodes, M. (2009) *The Bedbug Who Wouldn't Bite*, CreateSpace Independent Publishing Platform, Charlestown.

Ronnick, M.V. (2001) *Cimex, Tinea, and Blatta: insect imagery in Milton's Pro Populo Anglicano Defensio Secunda*. Notes and Queries, March, 20–21.

Rossellini, I. (2014) *Seduce Me: Bed Bug, Green Porno*, https://www.youtube.com/watch?v=tVpSoHubwTY (accessed 30 May 2016).

Rowlandson, T. (1793) *A Tit Bit for the Buggs*, http://brbl-dl.library.yale.edu/vufind/Record/3954936 (accessed 30 June 2016).

Rowlandson, T. (1806) *Bug Breeders in the Dog Days*, http://brbl-dl.library.yale.edu/vufind/Record/3955194 (accessed 27 May 2016).

Rowlandson, T. (1811) *Summer Amusement. Bugg Hunting*, http://brbl-dl.library.yale.edu/vufind/Record/3955804 (accessed 27 May 2016).

Ruffato, C. (2004) *Q. Sereno Sammonico Liber Medicinalis*, Poetry Wave, Napoli.

Shipley, A.E. (1914) Insects and war: II.—The bed bug (*Cimex lectularius*). British Medical Journal, **2**, 527–529.

Sinclair, C. (1982) *Bedbugs*, Allison & Busby, London.

SonicHits (2016) *"Reincarnation" by Lord Invader & His Calypso Rhythm Boys*, https://sonichits.com/video/Lord_Invader_%26_His_Calypso_Rhythm_Boys/Reincarnation_%28The_Bed_Bug%29 (accessed 27 May 2016).

Sova, D.B. (2004) *Banned Plays: Censorship Histories of 125 Stage Dramas*, Infobase Publishing, New York.

Steinberg, S. (1999) *Mayakovsky's The Bedbug at the Maxim Gorki Theatre in Berlin: a Missed Opportunity*, https://www.wsws.org/en/articles/1999/10/maya-o21.html (accessed 19 July 2016).

Sweeney, C., Arnold, R., Kressler, C., Horner, J. and Jones, F. (2006) *My Bed Bugs Bedbug Party!*, Greenestuff Incorporated, Clinton Township.

Taylor, L.A. (2012) *Bedbugs. Can You See Them?*, Lulu.com, North Carolina.

The Internet Classics Archive (2000) *The Clouds* by Aristophanes, http://classics.mit.edu/Aristophanes/clouds.html (accessed 27 May 2016).

The Playwrights Database (2003) *John Blay*, http://www.doollee.com/PlaywrightsB/blay-john.html (accessed 27 May 2016).

Usinger, R.L. (1966) *Monograph of Cimicidae (Hemiptera - Heteroptera)*, Entomological Society of America, College Park.

Watterson, B. (2008) *Calvin and Hobbes*, http://www.gocomics.com/calvinandhobbes/2008/12/21 (accessed 27 May 2016).

Wells, H.G. (1924) *The Dream*, Jonathan Cape, London.

Wikia (2016a) *Suits*; "Dirty Little Secrets", http://suits.wikia.com/wiki/Dirty_Little_Secrets (accessed 31 May 2016).

Wikia (2016b) *The Big Bang Theory*; "The Hofstadter Insufficiency", http://bigbangtheory.wikia.com/wiki/The_Hofstadter_Insufficiency (accessed 31 May 2016).

Wikipedia (2014) List of compositions by Dmitri Shostakovich, https://en.wikipedia.org/wiki/List_of_compositions_by_Dmitri_Shostakovich (accessed 27 May 2016).

Wikipedia (2016a) *My Bedbugs*, https://en.wikipedia.org/wiki/My_Bedbugs (accessed 30 June 2016).

Wikipedia (2016b) *30 Rock*, https://en.wikipedia.org/wiki/30_Rock_%28season_4%29 (accessed 30 May 2016).

Wikipedia (2016c) *The Office*; "After Hours", https://en.wikipedia.org/wiki/After_Hours_%28The_Office%29 (accessed 30 May 2016).

Wikipedia (2016d) *The Simpsons*; "Pulpit Friction", https://en.wikipedia.org/wiki/Pulpit_Friction (accessed 27 May 2016).

Williams, N.K. (2014) *Haunted Hotels of the California Gold Country*, Haunted America, Charleston.

Winters, B.H. (2011) *Bedbugs*, Quirk Books, Philadelphia.

YouTube (2013) *Milton Bradley Bed Bugs Board Game*, https://www.youtube.com/watch?v=AG2DR2BIfcM (accessed 26 June 2016).

YouTube (2014) *Roaring Jelly* – Bed Bug, https://www.youtube.com/watch?v=_2a29iYgIx4 (accessed 27 May 2016).

YouTube (2015) *Supernatural*; "Castiel tries to get Claire out of group home", https://www.youtube.com/watch?v=lwvP7rgqp7Y (accessed 31 May 2016).

Zagat, A.L. (1935) *Riverfront Horror*, Terror Tales, United States.

Zhao, J. (2008) Thirty years of landscape design in China (1949–1979): the era of Mao Zedong. PhD thesis. The University of Sheffield.

Part II

The Global Bed Bug Resurgence

3

The Bed Bug Resurgence in North America

Dini M. Miller

3.1 Introduction

Like the rest of the modern world, the USA and Canada have a history of bed bug infestations. Because of the temperate climate in the North American latitudes, infestations have been almost exclusively limited to the Common bed bug (*Cimex lectularius* L.). While there has been a recent discovery of several fossil cimicids in the caves of Southern Oregon, USA (Adams and Junkins, 2107), and records (both recent and historical) of the Tropical bed bug, *Cimex hemipterus* (F.) identified within the state of Florida (Hixson 1943; Usinger, 1966; Ebeling, 1975; Campbell *et al.*, 2016), *C. lectularius* is still the predominant human pest species in North America.

Cimex lectularius is thought to have been originally introduced to the continent by European explorers (Usinger, 1966; Pinto, 2016). The basis for this assumption is that *C. lectularius* is known to be of Old World origin. With the spread of the (US) human population from the east coast toward the west, bed bug populations became widely distributed. By the 1900s, nearly every home in the USA had experienced an infestation (Usinger, 1966; Ebeling, 1975).

A revolution in bed bug remediation occurred with the use of DDT (Usinger, 1966; Ebeling, 1975; Potter, 2011; Pinto, 2016; see also Chapter 1). DDT was originally used for mosquito control by the military but was quickly found to have excellent bed bug efficacy. By 1947, the US Army technical manual recommended replacing all hydrocyanic acid applications with DDT (Potter, 2011). Bed bug populations in North America appear to have developed resistance to DDT (and other chlorinated hydrocarbons) within eight years of its initial use (and most authors would say within five) (Usinger, 1966; Ebeling, 1975). However, bed bug infestations had already been greatly reduced. In 1958, the National Pest Control Association of the USA recommended the organophosphate, malathion, and carbamates, be used to treat DDT-resistant bed bug populations (Busvine, 1958). By this time, bed bugs had been almost completely eradicated from North America (Usinger, 1966).

3.2 Bed Bug Resurgence in the United States

Although the eventual return of the bed bug might have been predicted, Americans were taken by surprise when bed bugs began to reappear in the late 1990s. Major urban centers like New York, Chicago, and Cincinnati were among the first cities to become re-infested with this pest. Bed bug infestations spread quickly and infestations soon began to be reported all over the country (Potter *et al.*, 2011). The pest management industry saw an exceptionally rapid increase in the annual number of bed bug complaints (Potter *et al.*, 2010, 2011, 2013, 2015; Potter, 2011; Potter and Haynes, 2014). Orkin Pest Control, a nationwide pest

Advances in the Biology and Management of Modern Bed Bugs, First Edition.
Edited by Stephen L. Doggett, Dini M. Miller, and Chow-Yang Lee.
© 2018 John Wiley & Sons Ltd. Published 2018 by John Wiley & Sons Ltd.

management company (PMC) in the USA, reported a 534% increase in bed bug calls between 2002 and 2003 (DeNoon, 2003). By 2006, a national survey of PMCs documented bed bug infestations within all 50 states (Gangloff-Kaufmann et al., 2006). By 2016, the total number of bed bug jobs performed in the USA was estimated to be 907 875 (Anonymous, 2017), up from 815 000 in 2015 (Curl et al., 2016).

Because bed bugs had not been a problem for almost 40 years, most PMCs had no experience in controlling infestations, and did not know which products would kill bed bugs. In fact, the only individual in the nation who even had a colony of bed bugs (captured from Fort Dix, New Jersey in 1972) was Dr. Harold Harlan, a military medical entomologist. Dr Harlan generously donated his "Harlan strain" to researchers so that they could quantify the insecticide susceptibility of field strains.

In the USA, the official "first responder" to the bed bug resurgence was the pest management industry. In 2004, the National Pest Management Association (NPMA) organized a series of bed bug workshops for members and other participants across the nation. The NPMA also took on the leadership in developing bed bug treatment protocols and released a best practice guide for bed bug management in 2011 (NPMA, 2011; see also Chapter 24).

The federal agency that has been most involved in the resurgence has been the Environmental Protection Agency (EPA). The EPA's efforts included the formation of the Federal Bed Bug Workgroup, which involved the Centers for Disease Control and Prevention (CDC), the Department of Defense, the Division of Housing and Urban Development, and the Department of Agriculture. As part of this working group, the EPA and CDC joined forces to officially declare bed bugs a public health pest in 2007. This declaration was intended to allow national research funding agencies (like the National Institute of Health) to justify the funding of bed bug research.

Although many states in the USA have now passed laws regarding responsibility for bed bug remediation (physically, financially, or both), there is still much to learn about which treatment techniques are the most effective. Approximately 95% of PMPs apply spray formulation insecticides for bed bug control (Potter et al., 2010, 2011, 2013, 2015; Potter, 2011) because it is the least expensive treatment option. However, as researchers have continued to document more resistance, the pest management industry has started to employ additional non-chemical methods. Whole home heat treatment is being used, although the efficacy of some heating systems is questionable. Steam, applied using professional equipment, is widely used for treating infested furniture. The vacuuming of live bed bugs, their cast skins, and eggs off infested surfaces, has been found to be important, particularly for the removal of small nymphs that hide in the shed skins of their larger siblings, and which are thus protected from insecticide sprays. Desiccant dusts are becoming the standard for bed bug prevention. Although these dusts cannot prevent bed bugs from infesting a home, they can limit bed bug spread from one apartment unit to another.

While all of the above listed bed bug remediation methods are in widespread use, eradication has still proven very difficult. Declaring a home "bed bug free" can only be done with ongoing monitoring (Cooper et al., 2016). However, monitoring has yet to become a widely used "early detection" method. This is because housing owners are reluctant to identify (new or recurring) bed bug infestations.

3.3 Bed Bug Resurgence in Canada

The history of the *C. lectularius* in Canada has been very similar to that of the USA. Bed bugs were thought to have been first introduced by European explorers, and were a known pest of human dwellings for many decades until the advent of DDT and other cyclodiene insecticides (Ebeling, 1978).

In general, Canada has fewer pesticide products registered for indoor use than the USA. In fact, only a few liquid pyrethroid formulations are available for bed bug control, in addition to a single pyrethroid dust formulation, and diatomaceous earth (Sean Rollo, Orkin Canada, Ontario, personal communication).

Interestingly, there have been major differences in how the resurgence was initially addressed between Canada and the USA. In the USA, the pest control industry, specifically the NPMA (representing the private sector), took the leadership role in commissioning researchers and other industry experts to deliver bed bug educational programs. The government was only tangentially involved. In Canada, the Canadian Pest Management Association and the local provincial pest management associations were also working on bed bug education. However, the media, the public, and local politicians were not familiar with the pest management industry, and were looking for direction from their local health departments.

Similar to New York City, Toronto was one of the first Canadian cities to be hit hard by the resurgence. Toronto is the most populated city in Canada, with over 6.1 million residents. Prior to 2003, bed bugs were not considered a common pest (Zavys and Cassin, 2009). However, by 2005, requests for pest control services in low-income housing situations had increased and pressure was put on the Toronto Public Health Department (TPHD) to look into the situation. In 2005, TPHD conducted a survey of local housing authorities to determine the number of bed bug infestations. Over a two-year study, TPHD found that bed bugs had infested many types of residences, including single family homes, student dormitories, hostels, shelters, and multi-unit housing (Zavys and Cassin, 2009).

One difficulty for TPHD (and other Canadian health departments) when attempting to address the situation, was that bed bugs were not considered a public health issue, because they do not transmit disease. However, following pressure from the media, TPHD took the lead in bed bug management (Shum *et al.*, 2012). TPHD established a group of stakeholders to develop the Toronto Bed Bug Project in 2008. However, TPHD had to rely on their own funds, and partner with other organizations, to finance the project's remediation activities (Shum *et al.*, 2012).

By 2010, TPHD had conducted 3000 assessments of apartments, 80 of which had undergone extreme cleaning at THPD expense. THPD had also provided bed bug education to over 2000 individuals and produced 17 bed bug fact sheets in 14 different languages. In 2011, the province of Ontario implemented a C$5 million plan to educate people in that province about bed bugs. This program allowed the TPHD to contract with private agencies to help citizens with treatment preparation and "extreme cleaning". However, this program ended in 2012 and sustainable (permanent) funding still has not been dedicated to bed bug remediation.

The 2010 Canadian Public Health Association Conference examined the re-emergence of bed bugs in Canada, and compared the management approaches of municipal and public health authorities in four large cities. Below is a review of how bed bugs have been addressed in recent years in different provinces (Shum *et al.*, 2012).

Montreal The Montreal Health Service, Montreal's Housing Corporation, and other non-profit organizations, provide bed bug education. However, responding to bed bug complaints is the responsibility of the city inspectors. In 2007, a city inspector in Montreal recorded that 20 buildings of the 700 managed by the municipal housing corporation were infested with bed bugs. By 2011, a survey of Montreal Island residents found that more than 24 000 households had dealt with a bed bug infestation within the previous few years (Shum *et al.*, 2012). In spite of the increasing number of infestations, it is interesting to note that these municipal inspectors received no training in bed bug biology or management methods. Therefore, the inspectors close the case once they have document that the landlord (the party responsible for control) has hired a PMP. There are no quality assurance checks to determine if the infestation has been eliminated or even reduced.

Winnipeg Winnipeg has also suffered from rapidly spreading infestations. One PMC received 2800 bed bug calls in 2010. City by-law enforcement officers (CBEOs) are responsible for contacting landlords that choose not to address infestations in their buildings. If landlords fail to comply, the CBEOs issue a written order forcing landlords to undertake extermination or face possible criminal charges. In 2011, the Province of Manitoba invested C$770 000 in a comprehensive bed bug management program. The funding allowed for the creation of a public education campaign and a bed bug call-line. The call-line not only offers eradication information, but also allows the province to track infestations. In addition, the funding provides grants to community-based organizations for prevention efforts. The success of this program has resulted in complaints decreasing by 60% (to 2008 levels).

Vancouver Prior to 2008, the Vancouver Coastal Health Authority (VCHA) was the lead agency involved with bed bug control. The VCHA began in 2004 to develop educational materials and hold training seminars specifying best management practices. By 2006, the British Columbia Structural Pest Management Association started hosting public information sessions on bed bugs. In 2008, the Property Use Division of Vancouver took over the enforcement of bed bug remediation.

Since 2015, very little has been published in Canada regarding the current bed bug situation. Thus two bed bug experts were contacted to discuss contemporary issues. Dennis Monk, owner of BedBug Solutions (a company that specializes in bed bug inspection and non-chemical remediation), highlighted some of the current limitations in the City of Toronto's approach to bed bug remediation in city managed housing. Bed bug eradication resources are only provided for "the most marginalized residents of the community." The city employs its own PMPs, and the "IPM program" consists of spray insecticide application only, and as a consequence, there is a need for pre-treatment preparation. This includes de-cluttering apartments, but the "marginalized" residents are frequently hoarders, and often hinder the process. This delays control efforts, allowing the infestations to grow and spread.

Sean Rollo, Regional Manager for Orkin Canada, has additional concerns. He notes that certain municipalities (including Toronto, Vancouver, and Montreal) initially allotted some funding for educational materials but that this has now ceased. Much of the bed bug information was taken from the Internet, with little effort to consult experts to ensure the information was correct. The lack of communication was due to the fact that the public health authority was unfamiliar with the local pest management associations. This relationship has since improved.

With regards to the City of Toronto's bed bug control methods, Rollo states that vacuuming is an essential tool for controlling bed bugs. Hence, it is very important that the time must be taken to vacuum, particularly as insecticides have little effect. Rollo mentions that heat treatment is not widely used in Canada, but has proven to be a very effective bed bug management tool.

3.4 Summary

In conclusion, both Canada and the USA, are hard at work fighting bed bug infestations, which continue to increase both in size and frequency. In both nations, the application of spray formulation insecticide is still the most widely used method of control, in spite of known resistance. We can only hope that over the next decade, PMPs, government agencies, landlords, and residents will become more aware of how to manage bed bugs effectively.

References

Adams, M.E. and Jenkins, D.L. (2017) An early Holocene record of *Cimex* (Hemipteran: Cimicidae) from western North America. *Journal of Medical Entomology*, **54** (4), 934–944.

Anonymous (2017) *Speciality Consultants: US Structural Pest Control Market Surpasses $8 Billion*, http://www.pctonline.com/article/specialty-consultants-research-2017-market-report/ (accessed 11 April 2017).

Busvine, J.R. (1958) Insecticide-resistance in bed-bugs. *Bulletin of the World Health Organization*, **19** (6), 1041–1052.

Campbell, B.E., Koehler, P.G. Buss, L.J. and Baldwin, R.W. (2016) Recent documentation of the tropical bed bug (Hemiptera: Cimicidae) in Florida since the common bed bug resurgence. *Florida Entomologist*, **99** (3), 549–551.

Cooper, R.A., Wang, C. and Singh, N. (2016) Effects of various interventions, including mass trapping with passive pitfall traps, on low-level bed bug populations in apartments, *Journal of Economic Entomology*, **109** (2), 762–769.

Curl, G.D. (2016) *U.S. Structural Pest Control Market to Reach $10 Billion in 2020*, http://www.pctonline.com/article/sc-research-pest-control-market-report/ (accessed 11 February 2017).

DeNoon, D.J. (2003) *Bedbugs Back in U.S. Beds*, http://www.webmd.com/skin-problems-and-treatments/news/20031003/bedbugs-back-in-us-beds#1 (accessed 1 May 2017)

Ebeling, W. (1975) *Urban Entomology*, University of California Division of Agricultural Sciences, Los Angeles, CA.

Gangloff-Kaufmann, J., Hollingsworth, C., Hahn, J., *et al.* (2006) Bed bugs in America; a pest management industry survey. *American Entomologist*, **52** (2), 105–106.

Hixson, H. (1943) The tropical bedbug established in Florida. Records of *Cimex hemiptera*, F., in houses in several localities in 1938–1942. *Florida Entomologist*, **26**(3), 47.

NPMA (2011) NPMA library update: best management practices for bed bugs. *PestWorld*, **March/April**, I–XVI.

Pinto, L. (2016) *Bed Bug History*, http://www.techletter.com/Archive/Technical%20Articles/bedbugcomeback.html, (accessed 28 June 2016).

Potter, M.F. (2011) The history of bed bug management-with lessons from the past. *American Entomologist*, **57** (1), 14–25.

Potter, M.F. and Haynes, K.F. (2014) Bed bug nation: Is the United States making progress? In: *Proceedings of the Eighth International Conference on Urban Pests*, 2014, *July 20–23, Zurich, Switzerland*. OOK-Press, Veszprém.

Potter, M.F., Haynes, K.F., Rosenberg, B. and Henriksen, M. (2011) The 2011 bed bugs without borders survey. *PestWorld*, **Nov/Dec**, 4–15.

Potter, M.F., Haynes, K. F., Fredericks, J. and Henriksen, M. (2013) Bed bug nation: are we making progress? *PestWorld*, **Sept/Oct**, 4–11.

Potter, M.F., Haynes, K.F. and Fredericks, J. (2015) *The 2015 Bed Bugs Without Borders Survey*, http://www.pestworld.org/all-things-bed-bugs/bed-bug-facts-statistics/ (accessed 1 May 2017).

Potter, M.F. Rosenberg, B. and Henriksen, M. (2010) Bed bugs without borders: defining the bed bug resurgence. *PestWorld*, **Sept/Oct**, 1–12.

Shum, M., Comack, E., Stuart, D.T., *et al.* (2012) Bed bugs and public health: new approaches for an old scourge. *Canadian Journal of Public Health*, **103**, (6), 399–403.

Usinger, R.L. (1966) *Monograph of Cimicidae (Hemiptera – Heteroptera)*, Entomological Society of America, College Park.

Zavys, R., Cassin, P, Tersigni, C., *et al.* (2009) *Bed Bugs in Toronto: Developing an Effective Response*, http://www.toronto.ca/legdocs/mmis/2009/hl/bgrd/backgroundfile-25110.pdf (accessed 4 June 2016).

4

The Bed Bug Resurgence in Latin America

Roberto M. Pereira, Ana Eugênia de Carvalho Campos, João Justi (Jr.) and Márcio R. Lage

4.1 Introduction

Among Latin American Spanish-speaking countries, bed bugs are known as "chinches" or "chinches de la cama". In most of these nations, there is no official data on the incidence of this urban pest, but bed bugs appear to have been on the increase in recent years. Pest control companies in Latin America (many of them specializing in bed bugs) have treated infestations in most of the large cities, and continue to emphasize bed bug awareness and prevention during their communication with customers (Chemotecnica, 2011, 2012). Unfortunately, public health authorities have been very slow to respond to the proliferation of this urban pest in Latin America.

A history of parasitology and entomology in Latin America in the 20th century (Schenone, 2000) identifies bed bugs as a common pest in the 1950s. In many countries in Latin America, control campaigns for disease vectors, particularly mosquitoes, led to a decline in bed bug populations. However, since the 1990s, bed bug infestations have been reappearing in many of these countries.

The public health importance of bed bugs in the region may be exaggerated due to the common fear that they possibly may have a connection to Chagas disease. Salazar *et al.* (2015) recently reintroduced the idea that bed bugs are potential vectors of *Trypanosoma cruzi* Chagas, even though the competence of these insects as vectors under laboratory conditions was intimated more than 30 years ago (Jörg, 1982). In spite of these results, epidemiological evidence suggests that bed bugs are unlikely to play a role in the domestic transmission of the pathogen (Tonn *et al.*, 1982) and there is, so far, no evidence that they actually do so. Hence, large bed bug infestations seem to pose more of a non-disease related health threat (bites, sleep deprivation, and stress for the host) than the threat from potential pathogen transmission (Criado *et al.*, 2011; Doggett *et al.*, 2012; see also Part 3, Bed Bug Impacts).

4.2 Bed Bugs in Brazil

Bed bugs, and their hematophagous habits, have been reported in Brazil for more than a century. During a journey on the São Francisco River and its tributaries, Lutz and Machado (1915) reported finding the Common bed bug, *Cimex lectularius* L., in the small village of Jacaré in the state of Minas Gerais, a settlement of around 100 houses, and in Bom Jardim in the state of Bahia, a town with around 300 houses. In the village in Minas Gerais, bed bugs were reported to be a common occurrence. At that time of that trip, the researchers were searching for *Triatoma sordida* (Stal) (Heteroptera) in order to understand the epidemiology of Chagas disease. The authors reported that, due to the high populations of bed bugs in the village, they could not be associated with Chagas disease transmission, otherwise the disease would be much more widespread.

Some years later, bat bugs were reported feeding on human blood in Brazil. In 1927, Cesar Pinto described the first bat bug in Brazil, namely *Cimex limai* on bats (which he named *Propicimex limai* (Pinto); see Usinger (1966)). This species had the distinction of being the first representative of the Cimicinae subfamily described as having the organs of Ribaga and Berlese in the center of the abdomen. The species was found harboring in bat roosts that occurred on the roof of houses and was known to feed on human blood at night. The same feeding habit was reported later for another bat bug species, *Cimex pipistrelli* Jenyns, 1839 (Lent and Proença, 1937). At that time of these descriptions, there was a concern that such insects might transmit pathogens from bats to humans, since the role of these flying mammals in the zoonosis transmission chain of certain pathogens was already known.

Concerns about the transmission of bat bug diseases to humans continued, but later the attention turned to the bed bugs that were actually associated with humans, namely *C. lectularius* and the Tropical bed bug, *Cimex hemipterus* (F.). Due to the hematophagous habits of these insects, scientists in the region investigated the competence of bed bugs for transmitting various disease-causing pathogens, including:

- nematodes (filariasis)
- protozoans (Chagas disease, *Leishmania*)
- rickettsias (tick-borne typhus, spotted fever, Q fever)
- bacteria (bubonic plague, leprosy, typhoid fever, brucellosis)
- spirochetes (recurrent fever, Weil's disease)
- viruses (Yellow fever, acute disseminated encephalomyelitis, poliomyelitis).

Despite all the extensive experimental work, no evidence has yet been found that bed bugs transmit any human pathogen.

Costa Lima (1940) stated that *C. lectularius* and *C. hemipterus* were the two prevalent bed bug species in Brazil, but that *C. hemipterus* was the most common at that time, even in the most populous city centers. In the city of São Paulo, however, perhaps due to intense European migration, *C. lectularius* appeared to be the most abundant. The presence of *C. lectularius* in a second class hotel in the center of São Paulo city was reported by Blatchley (1934), who stated that, during his trip to South America, he only saw or felt bed bugs in São Paulo, Brazil, and in Chile. The author remarked that these observations "spoke well for the cleanliness of the beds in the many hotels he occupied."

Interestingly, the use of heat for control of bed bugs was already a recommendation for bed bug control at the time of these observations. Costa Lima (1940) wrote that it was economical to treat infested rooms by holding the temperature at 50°C for 6h. The high temperature would result in the destruction of all the bed bug developmental stages.

In 1990, Oswaldo Paulo Foratinni, a Brazilian researcher at the Public Health University (São Paulo) published a review on the public health importance of Cimicidae, emphasizing that *C. lectularius* was the most common species in the south and southeast regions of Brazil, and in urban centers. *Cimex hemipterus* was restricted to warmer regions and rural areas, and usually not found in urban centers. Foratinni pointed out that, in general, the poor conditions of households in the peripheral areas of large urban centers favored bed bug spread, and that very high levels of infestations could be found among the people living at the lowest economical level. Infestations found in new and more affluent housing were usually acquired from furniture purchased at antique shops. However, in such circumstances, infestations hardly ever reached high levels.

In Brazil, bed bugs have historically been associated with poor housing, poor hygiene, promiscuity, living with domestic animals, and a high human population density. However, this situation started to change during the 1950s with the improvement of social health and living conditions amongst the population. As in other countries around the world, infestations became smaller and rare for the next 40 years, and bed bugs were no longer considered a serious public health nuisance in Brazil.

As a consequence of the campaigns focused on controlling insect vectors – yellow fever, dengue, malaria, and Chagas disease – bed bug infestations were also controlled, albeit indirectly. Nevertheless, there was a drastic reduction in bed bug occurrence associated with these campaigns (Negromonte *et al.*, 1991). The vector control campaigns started in 1903 when Oswaldo Cruz organized battalions of "mosquito-killers" to eradicate outbreaks of insect vectors. The campaigns were triggered by a bubonic plague epidemic in the Port of Santos. However, vector control efforts were also focused on control of yellow fever in Rio de Janeiro. This vector control effort not only resulted in the eradication of yellow fever from Rio de Janeiro in 1907, but also the subsequent elimination of other pests.

4.3 The Resurgence of Bed Bugs in Brazil

An increase in the number of bed bug infestations in Brazil has occurred over recent years. Unlike in the past, today's bed bugs find favorable conditions for their proliferation in newer and more affluent housing. Recent scientific information reveals infestations among more developed socioeconomic regions in Brazil. This is due to increased international travel, which results in the passive transportation of bed bugs on personal belongings and luggage. Also evident is a possible underreporting of infestations, masking the actual levels of bed bug infestations in Brazil (Nascimento, 2010; Justi Jr. and Campos, 2014; Lage, 2014; Bocalini, 2015). One study early in the resurgence examined bed bug infestations in the Magalhães area of Brazil, where there are many tourist cities. The majority of infestations in this area were found in hostels (Faúndez and Carvajal, 2014).

Since late September 2012, the Instituto Biológico in São Paulo maintains bed bug information on its webpage (see link to "Percevejos de Cama" from http://www.biologico.sp.gov.br). A short text explains where to look for bed bugs and how to identify them, and what steps are necessary to eliminate the bed bugs. In order to better understand the bed bug problem in Brazil, two questionnaires are also available on the website, one for the resident and another for pest management professionals (PMPs). Beside demographic information, the resident questionnaire asks about bed bug control measures, and any information on how the infestation may have been acquired, for example, if the resident had undertaken any travel abroad recently. The PMP questionnaire asks about treatment methods, the type of facility treated (hotel, household, hostel, movie theater, and so on), the geographic location, and the potential source of infestation.

Between September 2012 and March 2016, 357 people completed the Instituto Biológico resident questionnaire, but it was completed by only 18 PMPs. Most of the PMPs indicated that the majority of their bed bug control efforts were focused on residences (55.6%), followed by construction worker lodging and detention facilities. The 357 resident questionnaires represented 14 out of 26 states, and the Federal District, or around 52% of the 27 federal units. Interestingly, most of the responses came from the state of São Paulo (Figure 4.1).

From the zip codes obtained, the Human Development Index (HDI, a measurement of household economic level) was calculated for the residences infested with bed bugs. Infested residences included those from low HDI (0.622 – Francisco Badaró, MG) to a very high HDI (0.961 – Vila Nova Conceição neighborhood in the city of São Paulo). This wide range indicates that modern bed bug infestations are very different from those of the past. Current infestations affect the more affluent residences in addition to those of a low socioeconomic level. A similar trend has been seen in other countries around the world, where global commerce, international travel, and misguided insect control efforts are important factors influencing the dispersal of bed bug infestations.

Among the 95% of responses to resident questionnaires, only 14% indicated that they had travelled abroad, suggesting that bed bugs were well established in Brazil by the time of the questionnaire. Respondents were asked to name their (potential) sources of infestation, and many responses were related to travel but not all. Responses included:

- travel as a missionary visiting several states in Brazil
- the purchase of American-style spring box and mattress combinations
- travel on cruise ships

Figure 4.1 Overall distribution of responses to resident questionnaires in Brazil. The numbers represent the number of counties represented in the survey. The percentages represent the percentage of counties responding in relation to the total number of counties in each state within the geographical regions of Brazil. Data compiled from September 2012 to March 2016.

- stays in hotels and hostels with foreign clientele
- former prisoners or foreigners in the household
- chicken houses in the yard
- residents working at an international airport or as airplane pilots
- household items brought home from used furniture/clothing stores
- adopted street dogs and cats
- bed bugs moving through the apartment's electrical conduits

- neighbors working in hostels or as stewards
- pigeon infestations
- visiting friends who travel abroad frequently.

It is interesting to note that no increase in bed bug complaints was observed after the soccer World Cup was held in Brazil during 2014.

With the bed bug resurgence in Brazil, PMPs had to rapidly assimilate new information on the biology and behavior of these insects. A study of PMPs in Brazil (Bocalini, 2015) determined that few pest management companies were prepared to control bed bugs. Infestations treated by the pest management industry have been mainly in hotels and residences. However, a comprehensive assessment of where bed bug populations are typically found within the entire country has not been undertaken.

Currently, Brazil has pesticide products that are registered for professional bed bug control. The active ingredients in these products include

- beta-cyfluthrin
- imidacloprid and beta-cyfluthrin
- alpha-cypermethrin and flufenoxuron
- pyrethrin and piperonyl butoxide
- propoxur
- lambda-cyhalothrin
- bifenthrin.

However, based upon what is known about bed bug resistance to pyrethroids in other nations, it remains to be seen if these products can limit bed bug proliferation within Brazil.

4.4 Elsewhere in Latin America

In Venezuela, few bed bug infestations have been reported in the scientific literature. The first cases to be reported for three decades occurred in 2001, in the city of Baruta and in the municipality of Libertador, where bed bugs were found infesting residential premises. In both locations, there were indications that the insects were brought in on either luggage or furniture (Universidad del Zulia, 2016). In 2002, the first bed bug infestation was reported in Maricaibo City, when a college professor detected them in his house two years after taking a trip to Europe. According to the case publication, port authorities were informed about these infestations. The port authorities promised to enforce a control policy in Venezuelan sea ports and airports, where they would inform passengers about the possibility of passive bed bug dispersal (Reyes-Lugo and Rodríguez-Acosta, 2002).

In Argentina, a medical study reported three cases of bed bug bites that, according to the patients, occurred while they were traveling through Uruguay (Vera *et al.*, 2012). Chilean publications have indicated that *C. lectularius* is the main species infesting that country (Faúndez *et al.*, 2014).

For Mexico, a survey undertaken in 2010 indicated that most respondents from the country felt that bed bug infestations were "staying about the same" and that the resurgence was not as dramatic as in other nations (Potter *et al.*, 2010). In this survey, respondents felt that bed bugs were not overly difficult to control and that eradication could be achieved in under two applications. This was attributed to the local availability of more efficacious products including organophosphates and carbamates, which are no longer used for bed bug management in most parts of the world. More contemporary data on bed bug incidence in Mexico is not available. It is interesting to note that the Mexican pest control association (Asociación Nacional de Controladores de Plagas Urbanas) contains no information on bed bugs on their website (http://www.ancpuac.org/), nor are there any government programs or directives in place to combat the resurgence (Frank Meek, Orkin Pest Control, Atlanta, unpublished results). Perhaps these older insecticide chemistries are still effective in controlling bed bug infestations.

In addition to indicating the range of bed bug infestations in Latin America, the above listed studies suggest that incidental transport of bed bugs is the primary dispersal mechanism (Faúndez and Carvajal, 2014). The studies also documented that *C. lectularius* was the main species collected and identified in all recent infestations, even in areas where, according to the literature, *C. hemipterus* might be expected.

In general, Latin American countries have seen an increase in the presence of bed bugs over the last ten years. References to these insects have been recently cited in many publications authored by universities, research institutes, and health officials. The information provided usually includes descriptions of the signs of a bed bug infestation, an overview of the situation in developed nations, and the potential for the pest to infest the current nation. However, most publications emphasize that bed bugs are an old problem that has not been seen for many decades and is now a reemerging pest problem around the world (Jörg, 1982; Schenone, 2000).

Population, ecological, and genetic studies are needed to trace the origin of the recent infestations in Latin American countries, while epidemiological analyses are required to examine the extent of the bed bug resurgence and to determine which sectors of society are most impacted. These studies will hopefully encourage the establishment of effective public policies for the prevention, monitoring, and the control of bed bugs, and to identify where resources should be most appropriately applied. It is especially important to conduct local research on these insects as most current knowledge has been obtained from studies on *C. lectularius* in the northern hemisphere. While this information is valuable, the epidemiological and institutional characteristics of northern nations may prove different from those of Latin American countries (Lage, 2014). Differences may affect the prevalence of bed bugs and the effects of these pests on the human population.

References

Blatchley, W.S. (1934) *South America as I Saw it: The Observation of a Naturalist on the Living Conditions of its Common People, its Topography and Products, its Animals and Plants*. The Nature Publishing Co., Indianapolis. http://babel.hathitrust.org/cgi/pt?id=uc1.$b721928;view=1up;seq=225 (accessed 14 April 2016).

Bocalini, S.S. (2015) Identificação de métodos e produtos utilizados, bem como o perfil das empresas especializadas no controle de vetores e pragas urbanas, com ênfase em percevejos de cama (Hemiptera: Cimicidae). Master thesis, São Paulo: Faculdade de Saúde Pública.

Chemotecnica (2011) El regreso de las chinches de cama. *Enfoques de Salud Ambiental: Boletín de Chemotecnica*, **11**, 1.

Chemotecnica (2012) Chinches de cama: Consejos al cliente. *Enfoques de Salud Ambiental: Boletín de Chemotecnica*, **25**, 2.

Costa Lima, A. (1940) Insetos do Brasil. 2°. Tomo, Capítulo XXII, Hemípteros. Série didática **3**, 1–351.

Criado, P.R., Belda, Jr., W., Criado, R.F.J., Silva, R.V. and Vasconcellos, C. (2011) Bedbugs (Cimicidae infestation): the worldwide renaissance of an old partner of human kind. *Brazilian Journal of Infectious Diseases*, **15** (1), 74–80.

Doggett, S.L., Dwyer, D.E., Peñas, P.F. and Russell, R.C. (2012) Bed bugs: clinical relevance and control options. *Clinical Microbiology Reviews*, **25** (1), 164–192.

Faúndez, E.I. and Carvajal, M.A. (2014) Bed bugs are back and also arriving is the southernmost record of *Cimex lectularius* (Heteroptera: Cimicidae) in South America. *Journal of Medical Entomology*, **51** (5),1073–1076.

Foratinni, O.P. (1990) Os cimicídeos e sua importância em saúde pública (Hemiptera: Heteroptera; Cimicidae) [The cimicidae and their importance in public health (Hemiptera: Heteroptera; Cimicidae)]. *Revista de Saúde Pública*, **24** (Suppl l), 1–37.

Jörg, E. (1982) *Cimex lectularius* L. (la chinche común de cama) trasmisor de *Trypanosoma cruzi*. Prensa Medica Argentina, **69** (13), 528.

Justi Jr., J. and Campos, A.E.C. (2014) Perguntas e respostas sobre percevejos de cama (Hemiptera: Cimicidae): consultas atendidas pelo instituto biológico. *O Biológico*, **76**, 63–67.

Lage, M.R. (2014) Descrição das infestações e estudo morfométrico de percevejos de cama (Hemiptera: Cimicidae) do Estado de São Paulo. Masters thesis, São Paulo, Faculdade de Saúde Pública.

Lent, H. and Proença, M.C. (1937) *Cimex limai*, 1927, parasito de morcegos no Brasil. *Memórias do Instituto Osvaldo Cruz*, **32** (2), 211–219.

Lutz, A. and Machado. A. (1915) Viajem pelo rio S. Francisco e por alguns dos seus afluentes entre Pirapora e Joazeiro. *Memórias do Instituto Oswaldo Cruz*, **7** (1), 5–50.

Nascimento, L.G.G. (2010) Investigação da ocorrência de infestação por Cimicidae (Heteroptera: Cimicomorpha) na região metropolitana de São Paulo, no período de 2004 a 2009. Masters thesis, São Paulo: Faculdade de Saúde Pública.

Negromonte, M.R.S., Linardi, P.M. and Nagem, R.L. (1991) *Cimex lectularius* L., 1758 (Hemiptera, Cimicidae): Sensitivity to commercial insecticides in labs. *Memórias do Instituto Oswaldo Cruz*, **86** (4), 491–492.

Potter, M. F., Rosenberg, B. and Henriksen, M. (2010) Bugs without borders. Defining the global bed bug resurgence. *Pestworld*, **Sep/Oct**, 1–12.

Reyes-Lugo, M. and Rodríguez-Acosta, A. (2002) Se ha extinguido la infestación por chinche de cama (*Cimex lectularius* Linnaeus, 1758) en Venezuela?, *Revista Científica* FCV-LUZ, **12**, (3), 182–185.

Salazar, R., Castillo-Neyra, R., Tustin, A. W., et al. (2015) Bed bugs (*Cimex lectularius*) as vectors of *Trypanosoma cruzi*. *American Journal of Tropical Medical and Hygiene*, **92** (2), 331–335.

Schenone, H. (2000) La parasitología y la entomología en la narrativa latinoamericana del siglo XX. *Boletin Chileno de Parasitologia*, **55**, 3–4.

Tonn, R.J., Nelson, M., Espinola, H. and Cardozo, J.V. (1982) Notes on *Cimex hemipterus* and *Rhodnius prolixus* from an area of Venezuela endemic for Chagas disease. *Bulletin of the Society of Vector Ecologists*, **7**, 49–50.

Vera, C.I., Orduna, T., Bermejo, A., Leiro, V. and Maronna, E. (2012) Dermatosis por picaduras de cimícidos (chinches de cama). *Dermatologia Argentina*, **18** (4), 295–300.

Universidad del Zulia (2016) *Fagroluz Reporta Primer Caso de Chinche de Cama en Maracaibo*, http://www.agronomia.luz.edu.ve/index.php?option=com_content&task=view&id=415&Itemid=141 (accessed 12 December 2016).

Usinger, R.L. (1966) *Monograph of Cimicidae (Hemiptera – Heteroptera)*, College Park, Maryland, Entomological Society of America.

5

The Bed Bug Resurgence in Europe and Russia

Richard Naylor, Ondřej Balvín, Pascal Delaunay, and Mohammad Akhoundi

5.1 Introduction

As in other parts of the world, Europe and Russia have experienced a resurgence of the Common bed bug, *Cimex lectularius* L. following years of little to no activity. However, there is a long history of *C. lectularius* in the region, with some of the first records of the insect dating back to early Greek literature. Strategies to combat the modern resurgence of insecticide-resistant bed bugs have been largely *ad hoc* and quite variable between nations. However, the development of the European Code of Practice for Bed Bug Management has increased the use of "best practices" for bed bug control. This chapter reviews the history of bed bugs in Europe and Russia, and discusses the modern resurgence, including the strategies implemented in response to the rise in bed bug activity.

5.2 History of Bed Bugs in the Region

Cimex lectularius was first associated with bats (Horváth, 1913), so it is conceivable that the bed bug's arrival in Europe predated that of humans. However, the primary bat host species in Europe appears to be *Myotis myotis* (Borkhausen) and *Myotis emarginatus* (Geoffroy) (Povolný, 1957; Balvín *et al.*, 2014), which roost in large, deep caves. At present, such caves in Europe are too cold for bed bug reproduction (Simov *et al.*, 2006) and all known bat-associated bed bug populations come from synanthropic roosts (Balvín *et al.*, 2014). The climate, at least during Holocene, likely did not allow for the establishment of cave-dwelling bed bug populations (Davis *et al.*, 2003). Such cave-dwelling populations are only known from Afghanistan (Povolný and Usinger, 1966), and therefore Middle Eastern caves are believed to be the habitats in which bed bugs made the switch from bat to human hosts (see, for example, Usinger, 1966). Nevertheless, the bat- and human-associated bed bug lineages have lived in mutual isolation for tens of thousands of years (Balvín *et al.*, 2012). These lineages had most likely split long before humans settled in permanent dwellings, following the Neolithic rise in agriculture. Based on these assumptions, it seems most plausible that bed bugs arrived in Europe in the belongings of early humans, rather than developing within the caves of that region.

There are also several apocryphal stories and legends (Busvine, 1976) about bed bugs infesting early civilizations in Europe. In addition, bed bugs have been identified in the fossil record, from both the Roman (1st century) and medieval periods (10th century), as discussed by Kenward and Allinson (1994).

Interestingly, the climate in northern parts of Europe most likely limited the initial bed bug spread into the British isles. This is because bed bugs (that is, *C. lectularius*) are unable to reproduce below 13 °C (see, for example, Kemper, 1936). Consequently, the bed bug spread into cooler locations was only made possible by

Advances in the Biology and Management of Modern Bed Bugs, First Edition.
Edited by Stephen L. Doggett, Dini M. Miller, and Chow-Yang Lee.
© 2018 John Wiley & Sons Ltd. Published 2018 by John Wiley & Sons Ltd.

improvements in housing construction and in-house heating that occurred during the post-medieval period, which also coincided with a time of increasing urbanization (Busvine, 1980; Krinsky, 2002).

Geographical isolation may also have played a part in the relatively late arrival of the bed bugs in the British Isles. The exact date of bed bug arrival in Britain is uncertain. In fact, the first reliable evidence comes from an account of two noble ladies who were bitten while staying in Mortlake in 1583 (Mouffet, 1634). Bed bugs most likely arrived on cargo ships, in the belongings of sailors who travelled to the UK from other European locations. Evidence for the bed bugs being transported in from other regions is presented in Southall's *A Treatise of Buggs* (Southall, 1730), in which he states "not one sea-port in England is free [of bed bugs]; whereas in inland-towns, buggs are hardly known."

Wherever they came from, once the bed bugs arrived, they reproduced prolifically. In 1690, Tiffin and Son of London was established (a company that continues to this very day). This was one of the first pest management companies (PMCs) to specialize in bed bug control for the nobility in the British isles (Potter, 2011). By the start of the 19th century, the bed bug was "a troublesome inhabitant of most houses in large towns [in the UK]" (Bingley, 1803), and remained "a fact of life" in urban centers everywhere well into the 20th century (Crissey, 1981).

Larrouse (1874) states that in the latter 19th century "it seems [bed bugs] are still unknown in the very northern countries such as Sweden, Denmark, Russia." However, World War I led to extensive migrations of people towards urban areas, facilitating the spread of bed bugs throughout Europe. Even in Stockholm, Sweden, one third of dwellings were infested by the end of World War I in 1918 (Usinger, 1966). In fact, a survey of 3000 removal vans in Stockholm in the 1930s revealed that 47% were carrying bed bugs (Potter, 2011). In an attempt to reduce the problem, people were housed in tents while their homes and belongings were fumigated.

A report of the British Royal Commission on Bed Bugs, established in early 1930s, states that "in many areas [of London] practically all the houses [were] to a greater or lesser degree infested with Bed-bugs". The commission also recognized that bed bugs were not only a feature of slums, but also a cause of them (UK Ministry of Health, 1934; Busvine, 1964). As terraced houses became infested, the more affluent tenants moved away, resulting in socioeconomic decline. The severe bed bug situation between the wars, together with the increase of economic power of the middle classes, provided motivation and opportunity for development of new chemicals for bed bug control, including DDT, which opened a new chapter in bed bug history.

5.3 Historical Laws to Control Bed Bugs Pre-resurgence

Throughout the 1930s, laws and protocols were put in place across Europe, which helped to reduce the spread of the bed bug. Some of the legislation was specifically targeted at bed bug infestations, while the majority focused on improving living standards. In Britain, the first step aimed at raising living standards within poorer communities was the 1930 Housing Act (UK Housing Act, 1930). This legislation made subsidies available for clearing slums, while also obliging local authorities to re-house tenants. Over the course of the following decade, almost a quarter of a million slum dwellings were demolished. One result was that families were taken to bed bug cleaning stations prior to being rehoused. Their clothes were steamed and their belongings fumigated with hydrogen cyanide before they moved to a new location (UK Ministry of Health, 1934).

In 1934, a committee of the British Medical Research Council was formed to promote bed bug eradication research (Busvine, 1964). This research, in conjunction with the findings of the British Royal Commission on Bed Bugs, contributed to the passing of the Public Health Act, making it the "duty [of the local authority]…in the case of verminous premises…[to take] such steps as may be necessary for destroying or removing vermin" (UK Public Health Act, 1936).

In Scotland, the Department of Health (DOH) recognized the value of educating tenants in the importance of cleanliness, as it related to the habits of vermin. Within a few days of occupancy, specialists were sent from the DOH to train the new tenants on how to prevent and treat vermin, including bed bugs (Hartnack, 1939; Potter, 2011).

In Germany, some landlords required departing tenants to provide a statement from an exterminator confirming that their apartment was bed bug free (Hartnack, 1939; Potter, 2011). Similarly, in 1944, authorities in one district of Denmark made a bed bug inspection mandatory where tenants were moving from one apartment to another within the same district (Busvine, 1957; Kilpinen *et al.*, 2008).

For most countries under the communist regime, the majority of pest control services were organized by the government and usually provided free of charge for all citizens. In the USSR, the "Sanitary-Epidemic Stations" had a supervisory role, while the "Disinfection Stations" were responsible for conducting pest control (Svetlana Roslavtseva, Scientific Research Disinfectology Institute, Moscow, unpublished results). In Czechoslovakia, during a democratic period between the two world wars, there was a rise in private pest management companies (PMCs). After communism took over in 1948, these companies continued to conduct the majority of the pest control. However, in 1952, PMCs were reorganized under state structures called "Desinfection, desinsection and deratization groups of communal service" and were then placed under the supervision of the State Institute of Public Health (Přívora, 1975).

5.4 Documented Evidence for the Bed Bug Decline

The campaign against *C. lectularius* made significant headway across Britain between 1934 and 1943, driving down infestation rates in British cities from 10.7% to 1.3% (Busvine, 1964). This decline was further aided by the arrival of the synthetic organochloride insecticide DDT in the mid-1940s. DDT revolutionized bed bug control (Busvine, 1957). The combined result of DDT (along with the other organochlorines) and the efforts to improve living conditions in British cities drove bed bug infestation rates down from 1.2% to just 0.11% between 1947 and 1960 (Busvine, 1964). This pattern was similar in Denmark, where infestation rates declined from 3.6% in 1945 to 0.13% in 1955, largely as a result of the introduction of DDT (Busvine, 1957). In Berlin after World War II, a survey of public buildings and private houses revealed infestation rates of up to 40% in the city center, and 2% in the suburbs. A campaign to reduce infestation rates, planned for 1948, was affected by the east–west division of the city. After the division, the eastern representatives refused to cooperate in the bed bug eradication effort. However, neighborhoods in West Berlin successfully dropped their bed bug infestation rates to 0.2% by 1950 (Busvine, 1957). Throughout the 1950s bed bugs became increasingly rare across most of Europe, and infestations were restricted to areas afflicted with "particularly unsanitary conditions" (Ebeling, 1978; Snetsinger, 1997).

Between 1967 and 1973, infestation rates of the *C. lectularius* across Britain remained relatively low but stable. An analysis of pest surveys carried out by Rentokil (a large international PMC) in 1974 indicated that *C. lectularius* accounted for just 0.5% of all pests treated (Cornwell, 1974).

For most of central Europe (countries like Czechoslovakia) almost no data exist on bed bug occurrence after World War II, suggesting that the pest was almost entirely absent in most regions at that time (Václav Rupeš, former head of the Department of Desinfection, Desinsection and Deratization, State Health Instutute, Prague, unpublished results). A single case is described in a local journal on hygiene, documenting new ways that bed bugs could disperse within buildings (Lýsek, 1966). Stories also exist about buildings where Soviet officers were housed during the occupation of Czechoslovakia after 1968. Data from eastern Slovakia, the poorest region of Czechoslovakia with a high proportion of Roma people, as well as Hungarians and Rusyns, suggested that bed bug prevalence was low during the 1970s and 1980s (Kočišová, 2006). Bed bugs accounted for only 2.3% of the arthropod pests found in living situations.

The bed bug decline in Hungary may not have been as dramatic as it was elsewhere in Central Europe. Herczeg (1977) reported that in 1973, particular types of housing in Budapest had infestation rates were as high as 14%.

In the USSR, the decline in bed bug infestation rates began much later than in the rest of Europe. A steep rise in bed bug occurrence was observed after World War II and was attributed to the devastation of the country (Bogdanova *et al.*, 2005). However, the government only began to take measures to resolve this issue in 1958, by which time infestation rates were very severe. In Moscow, for example, 50–70% of apartments were found to be infested with bed bugs. By 1960, this number had decreased to 35%. Infestation rates continued to decline, reaching 12.9% by 1965, and 0.9% by 1987 (Bogdanova *et al.*, 2005).

5.5 Early Evidence for the Resurgence

The first published anecdotal evidence that *C. lectularius* was making a comeback in England came from Birchard (1998), who noted its apparent rise, as well as the difficulty of controlling infestations with insecticides. The first empirical evidence of the resurgence did not come until 2001 when Boase (2001) published a study on bed bug treatment data collected from four sources: two municipalities, one PMC, and one hotel chain. The data indicated that the number of bed bug infestations in the UK had risen approximately tenfold between 1997 and 2000.

Data from the UK Chartered Institute of Environmental Health show that treatments for *C. lectularius* had initially declined throughout the late 1980s and early 1990s, from about 26 treatments per local authority in 1986 to about 8 in 1992 (Boase, 2008). However, Boase (2013) also found that four of the thirteen regions of England and Wales had experienced substantial increases in numbers of bed bug treatments during this same period of overall decline. The largest increases were in Greater London and the South (Buckinghamshire, Oxfordshire, Berkshire and Hampshire), which experienced increases of 140% and 370% respectively between 1990 and 2000.

In 2009, Richards *et al.* (2009) found that enquiries relating to bed bugs received by London local authorities increased between 2000 and 2006 by an average of 28.5% per annum. Similar patterns can be found elsewhere in Europe, though often delayed by a few years. Kilpinen *et al.* (2008) reported that the number of bed bug cases treated by one large PMC in Sweden increased from 383 in 2002 to 770 in 2006. A second large Swedish PMC also experienced a very dramatic rise in the number of bed bug treatments, going from just three treatments in 2003 to more than 3000 by 2011 (Christine Dahlman Jacobsen, Nattaro Labs, Lund, Sweden, unpublished data). Denmark's bed bug problem seems to have begun as early as 1995, but has rapidly increased since 2002 (Kilpinen *et al.*, 2008). Interestingly, Denmark had also experienced an earlier resurgence between the 1950s and mid-1980s, which was attributed to an increased flow of workers and tourists from southern Europe (Kilpinen *et al.*, 2008).

The Health and Hygiene Department for the city of Zurich, Switzerland, has recorded recent bed bug trends through its pest advisory service. In 1994, just two bed bug-related enquiries were received. Between 1995 and 2005 the annual number of enquires rose by an average of 11.5% per annum, and between 2005 and 2015 the annual number of enquiries increased by an average of 25.5% per annum (Gabi Müller, Health and Hygiene Department of Zurich, Zurich, Switzerland, unpublished results).

Pinpointing the start of the bed bug resurgence in Italy is difficult because data is scarce. A single case was recorded in Pisa in 2003, followed by two cases on Italian trains in 2005 (Masetti and Bruschi, 2007). The Istituto Zooprofilattico Sperimentale (IZS) began conducting entomological surveillance of the areas of Piemonte, Liguria, and Valle d'Aosta in north-west Italy in 2001 (Giorda *et al.*, 2013), and no bed bug cases were recorded until 2008 when the IZS received three bed bug specimens. The IZS continued to receive bed bug specimens every year thereafter. Another study by Montarsi (2011) identified 23 additional *C. lectularius* infestations between 2006 and 2011 in the north-eastern region of Italy.

The beginning of the resurgence in Russia appears to be similar. Data from a large PMC based in Moscow, Ne-Kton-Ziber, reported a rise in the proportion of bed bug treatments relative to other pest species, from 0.8% to 6.3% between 1996 and 2005 (Bogdanova *et al.*, 2005).

In central Europe, the earliest evidence of the bed bug resurgence was published by Kočišová (2006). In 1996, bed bugs represented 20% of pest arthropods found in surveyed apartments in eastern Slovakia. From 1996, the occurrence of bed bugs steadily increased, and by 2006 they accounted for 80% of all arthropods cases occurring in apartments. However, the study was based in a geographical region with a well-documented pre-resurgence history of bed bug infestations. Therefore, this bed bug increase may pre-date the increase in neighboring regions. In a personal communication, Alica Kočišova (Parasitology Institute, University of Veterinary Medicine and Pharmacy, Košice, Slovak Republic) also noted a more abrupt rise in bed bug cases from 2006 onwards.

The first indication of the resurgence in Czech Republic comes from a note published in 2002 from the owner of a large Prague-based PMC (Visnicka, 2002). This note described an unspecified increase of bed bug cases in hotels. In 2005, the association of PMCs (Sdružení pracovníků dezinfekce, dezinsekce, deratizace České republiky), responsible for the education of Pest Management Professionals (PMPs), started providing bed bug seminars in response to the growing problem. Treatments records from Adera, a PMC in Prague, documented a slow and stable rise in bed bug occurrence from 2001, followed by a steeper rise after 2006 (Michal David, Adera, Prague, Czech Republic, unpublished results). A very similar increase in bed bug activity was identified by a large PMC based in Budapest (Papp *et al.*, 2014), where a slow increase was indicated from around the year 2000, followed by a sharp upturn in annual bed bug cases from 2006 onwards.

For certain locations, it is difficult to determine if the bed bug resurgence started later than in other places or if the resurgence data was simply not recorded. Multiple sources of evidence from central European countries, including Switzerland, appear to have experienced a rapid increase in the rate of new bed bug cases beginning in 2005–2006. However, this dramatic upturn is not recorded for Denmark or the UK. It is unclear why the rate of resurgence in central Europe appears to have increased so dramatically after 2006. This may indicate multiple factors at play. For example, the sharp upturn could mark the arrival of insecticide resistance in a population of bed bugs that was already increasing.

5.6 The Degree of the Resurgence

No centralized records exist for bed bug remediation conducted by private PMCs in Europe. Data from government agencies and other organizations are collected in different ways (Boase, 2013). Consequently, what little data there is can only be used to plot trends between consecutive years, painting a general picture of the current bed bug infestation trajectory.

The pest management industry has been the primary resource for quantifying the extent of the bed bug resurgence in the UK. The British Pest Control Association's Executive Summary of 2014 showed that bed bug treatments by local authorities between the years 2010/11 and 2013/14 increased by 21% (BPCA, 2014). However, between the years 2012/13 and 2013/14 the number of treatments conducted by local authorities only increased by 1.6%. *Pest* magazine, a UK-based publication, and BASF (an international chemical company and insecticide manufacturer) conduct an annual survey of PMCs in the UK. The result of the most recent survey (data from 2013/14) showed that bed bug jobs accounted for 7%, 3% and 4% of the total number of pest control jobs for PMCs, self-employed PMPs, and local authorities respectively (Anonymous, 2015). Ecolab is a multinational hygiene company, with a large pest control division that operates across the UK, although exclusively in the commercial/hospitality sector. Between 2007 and 2013, Ecolab experienced a 158% increase in bed bug jobs. However, between 2014 and 2015, the number of bed bug jobs remained the same (Mark Williams, Ecolab, Northwich, UK, unpublished results). Overall, the general trend indicated by data from both the large private PMC (working in the hospitality sector) and the local authorities (working in the private residential sector) is that bed bug infestations in the UK were increasing fairly rapidly up until around 2013, but have since begun to plateau. In spite of this, a recent report has suggested that bed bug services in the UK will increase by 4.7% annually from 2016 to 2026 (Anonymous, 2016). The same report predicts an annual increase in Germany by some 5.2% for a similar period.

The National Institute of Public Health in Norway, known as the Folkehelse Instituttetor (FHI) is an administrative body with a wide range of activities, including the recording of bed bug treatments from PMCs in Norway. In 2007, the FHI reported 392 bed bug treatments (Kilpinen *et al.*, 2008). By 2013, the number of reported treatments had risen to around 1600 cases. By 2015, the number of bed bug treatments had risen again, to 2500 (Christine Dahlman Jacobsen, Nattaro Labs, Lund, Sweden, unpublished results).

In Sweden, the two largest PMCs, Anticimex and Nomor, experienced considerable increases in bed bug occurrences. Between 2011 and 2015, the annual number of bed bug treatments by Anticimex rose from nearly 6000 to 13 500. Similarly, treatments conducted by Nomor increased from 3000 to 16 000 (Christine Dahlman Jacobsen, Nattaro Labs, Lund, Sweden, unpublished results). Interestingly, in a recent survey of 33 Swedish student housing companies, 66% reported current problems with bed bugs (Christine Dahlman Jacobsen, Nattaro Labs, Lund, Sweden, unpublished results).

In France, a survey of the Municipal Health and Safety Services (MHSS) and PMCs indicated that every Département in metropolitan France has encountered bed bugs (Jourdain et al., 2016). Furthermore, 74% of the MHSS, and 80% of PMCs indicated that the number of bed bug call-outs had increased over the past five years.

Little data is available on the current extent of the resurgence in central Europe. However, personal communication with several central European PMCs suggests that infestation rates have been stable for the past few years. Adera Company (Michal David, Adera, Prague, Czech Republic, unpublished results) claims to have undertaken a fairly constant number of treatments since 2013, while Asana PMC (Roman Šimák, Asana, Prague, Czech Republic, unpublished results) even recorded a drop in the number of bed bug treatments in 2012. Bed bug treatment numbers also decreased according to a PMP from Rožnava (Oskar Lorinz, Rožnava, East Slovakia, Czech Republic, unpublished results). But they remained the same for a PMP from Košice (Maroš Bačo, East Slovakia, Czech Republic, unpublished results). Only one PMP from Swinoujscie (Łukasz Brożek, Swinoujscie, Poland, unpublished results) indicated that there was an increase in bed bug abundance in 2014.

In conclusion, the resurgence of the *C. lectularius* across western and central Europe (at this time) appears to be stable or increasing relatively slowly. However, in the Nordic countries, recent data indicates that the resurgence continues to grow rapidly. Furthermore, increasing levels of insecticide resistance has resulted in repeated call backs, forcing companies to charge more for their services and to switch to alternative treatment options. These options tend to be more labor-intensive and consequently more expensive (R. Naylor, unpublished results). As the price of bed bug treatment increases, people from lower socioeconomic backgrounds are forced to deal with the problem themselves. This can often exacerbate the spread, while also hiding the extent of the bed bug problem, as do-it-yourself treatments are rarely reported.

To date, almost all of the bed bug activity in Europe has involved *C. lectularius*. Only recently has the Tropical bed bug, *Cimex hemipterus* (F.) been detected. Four incidents involving the Tropical bed bug were detected in southern Sweden between 2014 and late 2016 (Thomas P. Vinnersten, Anticimex, Stockholm, unpublished results). However, *C. hemipterus* has been detected many times in Russia between 2015–2016, most notably in Moscow and St Petersburg (Gapon, 2016). It is probable that additional introductions of *C. hemipterus* have gone unnoticed as few PMPs bother to identify the bed bug species that are being treated.

5.7 Region-specific Factors in the Resurgence

The recent resurgence of bed bugs within Europe appears to coincide with the increase of international trade and traffic resulting from the creation of the European Economic Community in 1958. This was later followed by the formation of the European Union (EU) in 1993, and the opening of national borders within the Schengen area since 1985. The EU has increased the movement of labor, sometimes considerably: Poles to the UK, or in 2007, Romanians to France. However, large migrations from Russia and the Ukraine, where large bed bug reservoirs were known to have existed, began with the breakup of the USSR in 1991.

In Eastern Europe, where people maintain strong bonds with their families and native regions, the men often travel abroad for work. These men tend to stay in budget accommodation in order to save money. Consequently, each time these workers visit their homes, there is a high risk of bringing bed bugs with them.

European Union legislation requiring the use of biocidal products has undoubtedly favored the bed bug. For example, most organophosphate insecticides have been withdrawn from the market as of 2001 due to regulations from the European Chemicals Agency. More recently, the Biocidal Products Regulation (Regulation (EU) 528/2012) has resulted in further reductions in the range of products available for bed bug control. While careful regulation of biocides is essential for the safety of people, animals, and the environment, the over-use of a very small number of insecticide classes exacerbates insecticide resistance, and limits control success.

The difference in the way PMCs were organized in communist countries does not seem to have dramatically influenced bed bug infestation rates prior to the resurgence. However, sociological factors associated with the communist period have had a pronounced and lasting effect on the ethos surrounding pest management. Most housing in communist countries was managed by state employees, who had little motivation to resolve problems faced by the inhabitants. Many apartments are still managed under city ownership or by large housing associations. In both cases, it is the manager, rather than the inhabitant, who is responsible for pest control. Friendly relationships between managers and owners of PMCs, often persisting from communist times, continue to hinder control efforts when bed bug infestations occur, as neither party have a vested interest in resolving pest issues.

Paradoxically, the legislative effort to break these landlord/PMC relationships may have made the situation worse. City governments (or the housing-associations) are now obliged to have open public review of pest control contracts. When this occurs, at least in the Czech Republic and Slovakia, price becomes the only criterion taken into consideration. Consequently, PMCs frequently offer prices that are so low that it is not financially viable to provide a quality service.

Lastly, communist city planning and building construction has provided bed bugs with particularly suitable habitats in large, centrally heated apartment complexes. The low-quality prefabricated construction allows for easy dispersal among apartments (Lýsek, 1966), while also attracting several species of crevice-dwelling bats, which brought the bat bug (*Cimex pipistrelli* Jenyns) with them (Šmaha, 1976; Balvín and Bartonička, 2014). It is, however, likely that the isolation of communist countries from other nations and international visitors delayed the bed bug resurgence, at least in central Europe.

While EU-wide legislation has resulted in broadly consistent regulation of insecticides, the same cannot be said for the regulation of PMP qualifications. In central Europe, PMCs are organized into professional "guilds", which provide mandatory education and regular exams (for example, every five years in Czech Republic), to meet licensing requirements. By contrast in the UK, no specific qualifications are required. However, non-mandatory training courses are provided by insecticide suppliers and trade associations.

5.8 Strategies to Combat the Resurgence

To date there have been few, if any, nationally adopted strategies for combating the resurgence of *C. lectularius* across Europe. Management tools, techniques, and approaches vary widely between PMCs (Boase, 2007). Where a company feel they have developed an effective bed bug management strategy, there is a tendency towards secrecy, which is born out of a desire to maintain the commercial edge over competing companies (R. Naylor, unpublished results). However, in the Czech Republic, the association of PMCs responsible for certification for the pest management industry, officially recommend chlorfenapyr (Mythic) in combination with pyrethroids. This method is widely used and, if applied responsibly, is currently effective. Similarly in France, a working group was created in 2014 to produce a nationally recognized code of practice for bed bug management (Jourdain *et al.*, 2016; Centre National d'Expertisesur les Vecteurs, 2015). Overall, attempts have been and continue to be made to identify best practice for bed bug management in Europe. These attempts have resulted in the creation of a European Code of Practice for Bed Bug Management (BBF, 2013). This code of practice has been reasonably successful at reducing the bed bug resurgence in Europe and is discussed in detail elsewhere within this book (see Chapter 23).

References

Anonymous (2015) The National UK Pest Management Survey. *Pest Magazine*, **39**, 19–21.

Anonymous (2016) *U.K. and Germany Market Study on Bed Bug Control Services: Chemical Control Service Type Segment Expected to Gain Significant Market Share By 2026*, http://www.persistencemarketresearch.com/market-research/uk-and-germany-bed-bug-control-services-market.asp (accessed 21 April 2017).

Balvín, O., Munclinger, P., Kratochvíl, L. and Vilímova, J. (2012) Mitochondrial DNA and morphology show independent evolutionary histories of bedbug *Cimex lectularius* (Heteroptera: Cimicidae) on bats and humans. *Parasitology Research*, **111** (1), 457–469.

Balvín, O. and Bartonička, T. (2014) Cimicids and bat hosts in the Czech and Slovak Republics: ecology and distribution. *Vespertilio*, **17**, 23–36.

Balvín, O., Bartonička, T., Simov, N., Paunovic, M. and Vilímová, J. (2014) Distribution and host relations of species of the genus *Cimex* on bats in Europe. *Folia Zoologica*, **63** (4), 281–289.

BBF (2013) *European Code of Practice for Bedbug Management, Version 2*. Bed Bug Foundation, http://bedbugfoundation.org/en/ecop/ (accessed 15 April 2016).

Bingley, W. (1803) *Animal Biography; or Anecdotes of the Lives, Manners, and Economy, of the Animal Creation, Vol III*, Wilks and Taylor, London, UK.

Birchard, K. (1998) Bed bugs biting in Britain: only rarely used pesticides are effective. *Medical Post*, **34**, 55.

Boase, C. (2001) Bedbugs: Back from the brink. *Pesticide Outlook*, **12** (4), 159–162.

Boase, C. (2007) Bedbugs – research and resurgence. *Emerging Pests and Vector-Borne Diseases in Europe*, **2007**, 261–280.

Boase, C. (2008) *Bed bugs (Hemiptera: Cimicidae): An evidence-based analysis of the current situation*. In: Proceedings of the Sixth International Conference on Urban Pests, *July 13–16, 2008, Budapest, Hungary*. OOK-Press, Veszprém.

Boase, C. (2013). Bed bugs in the UK – before the upsurge. *Outlooks on Pest Management*, **24** (1), 11–15.

Bogdaneva, E.N., Roslavtseva, S.A. and Slobodin, A.Z. (2005) Bed bugs (Hemiptera: Cimcidae). Contemporary situation in Russian Federation and disinsection measures against them. *Dezinfekcionioje Delo*, **4**, 55–59.

BPCA (2014) *National Survey Executive Summary*. British Pest Control Association, http://www.bpca.org.uk/assets/National%20Survey%20Exec%20Summary%2014.pdf (accessed 26 April 2016).

Busvine, J.R. (1957) Recent progress in the eradication of bed bugs. *The Sanitarian*, **65**, 365–369.

Busvine, J.R. (1964) Medical entomology in Britian. *Annals of Applied Biology*, **53** (2), 190–199.

Busvine, J.R. (1976) *Insects, Hygiene and History*, Athlone Press, London, UK.

Busvine, J.R. (1980) *Insects and Hygiene. The Biology and Control of Insect Pests of Medical and Domestic Importance*, 3rd edn. Chapman & Hall, London, UK.

Centre National d'Expertise sur les Vecteurs (2015) *Punaises de Lit en France: État des Lieux et Recommandations*, http://www.ars.paca.sante.fr/fileadmin/PACA/Site_Ars_Paca/Sante_publique/Sante_environnement/punaises_de_lit/CNEV_2015_09_rapport_punaises_de_lits.pdf (accessed 3 August 2016).

Cornwell, P.B. (1974) The incidence of fleas and bedbugs in Britain. *International Pest Control*, **16** (4), 17–20.

Crissey, J.T. (1981) Bedbugs. An old problem with a new dimension. *International Journal of Dermatology*, **20** (6), 411–414.

Davis, B.A.S., Brewer, S., Stevenson, A.C. and Guiot, J. (2003) The temperature of Europe during the Holocene reconstructed from pollen data. *Quaternary Science Review*, **22**, 1701–1716.

Ebeling, W. (1978) Bed bugs and allies (Cimicidae). In: *Urban Entomology*, (ed. W. Ebeling), University of California's Division of Agricultural Science. Berkeley, California, USA, pp. 463–475.

Gapon, D. A. (2016) First records of the tropical bed bug *Cimex hemipterus* (Heteroptera: Cimicidae) from Russia. *Zoosystematica Rossica*, **25** (2), 239–242.

Giorda, F., Guardone, L., Mancini, M., *et al.* (2013) Cases of bed bug (*Cimex lectularius*) infestations in northwest Italy. *Veterinaria Italiana*, **49**, 335–40.

Hartnack, H. (1939) *202 Common Household Pests of North America*, Hartnack Publishing Co., Chicago, IL.

Herczeg, T. (1977) Insektenbefall von Budapest und seine Sanierung [The incidence and control of insects] (in German) *Wiadomosciparazytologiczne*, **23**, 167–170.

Horváth, G. (1913) La distribution géographique des cimicides et l'origine des punaises des lits. In: *Extrait du IXe Congres International de Zoologie Tenu a Monaco* (ed L. Joubin), March 25–30, 1913, Monaco, pp. 294–299.

Jourdain, F., Delaunay, P., Bérenger, J.M., Perrin, Y. and Robert, V. (2016) The common bed bugs (*Cimex lectularius*) in metropolitan France. Survey on the attitudes and practices of private- and public-sector professionals. *Parasites*, **23** (38), 1–8.

Kemper, H. (1936) *Die Bettwanze und ihre Bekampfung, Deutsche Gesselschaftfur Kleintier- und Pelztierzucht*. Leipzig, Germany.

Kenward, H.K. and Allinson, E.P. (1994) Rural origins of the urban insect fauna. In: *Urban-Rural Connections: Perspectives from Environmental Archaeology*, (eds A.R. Hall and H.K. Kenward), Oxbow, Oxford, UK, pp. 55–58.

Kilpinen, O., Jensen, K.M.V. and Kristensen, M. (2008) *Bed bug problems in Denmark, with a European perspective*. In: Proceedings of the Sixth International Conference on Urban Pests, *July 13–16, 2008, Budapest, Hungary*. OOK-Press, Veszprém.

Kočišová, A. (2006) Výskyt ploštice posteľnej (*Cimex lectularius*) na Východnom Slovensku. *Dezinfekce, Dezinsekce, Deratizace*, **15** (2), 70–72.

Krinsky, W.L. (2002) True bugs (Hemiptera). In: *Medical and Veterinary Entomology*, 2nd edn, (eds G.R. Mullen and L.A. Durden), Academic Press, London, UK, pp. 80–87.

Lýsek, H. (1966) Neobvyklé způsoby šíření *Cimex lectularius* [Unusual ways of dispersal of *Cimex lectularius*] (in Czech). *Československá Hygyiena*, **11**, 617–620.

Masetti, M. and Bruschi, F. (2007) Bedbug infestations recorded in central Italy. *Parasitology International*, **56** (1), 81–83.

Montarsi F. (2011) *Osservazioni del laboratorio di Parassitologia (IZS delle Venezie) sulla distribuzione della cimice dei letti (Cimex lectularius, Linneaus) nell'area del Nord-est*. In: Proceedings of the Congresso Nazionale Italiano di Entomologia, June 13–16, 2011, Nazionale Italiano di Entomologia, Genova.

Mouffet, T. (1634) *Insectorumsive Minimorum Animalium Theatrum*, Thomas Cotes, London, UK . (Transl. by Topsel, E. (1658) in: *The History of Four-footed Beasts, Serpents and Insects*, pp. 10, 889–1130. E. Cotes., London, UK.)

Papp, G., Madaczki, L. and Bajomi, D. (2014) *Occurrence of bed bugs in Budapest, Hungary*. In: Proceedings of the Eighth International Conference on Urban Pests, *2014*, July 20–23, Zurich, Switzerland. OOK-Press, Veszprém.

Potter, M.F. (2011) The history of bed bug management-with lessons from the past. *American Entomologist*, **57** (1), 14–25.

Povolný, D. (1957) Kritickástudie o štěnicovitých (Het. Cimicidae) v Československu. [Review study on cimicids (Het. Cimicidae) in Czechoslovakia] (in Czech). *Folia Zoologica*, **6** (10), 59–80.

Povolný, D. and Usinger, R.L. (1966) The discovery of a possibly aboriginal population of the bed bug (*Cimex lectularius* Linnaeus, 1958). *Acta Musei Moraviae Scientiae Naturales*, **51**, 237–242.

Přívora, M. (1975) Die Schadlingsbekampfung in der ČSSR. *Praktische Schadlingsbekampfer*, **27**, 9–11.

Richards, L., Boase, C.J., Gezan, S. and Cameron, M.M. (2009) Are bed bug infestations on the increase within greater London? *Journal of Environmental Health Research*, **9** (1), 17–24.

Simov, N., Ivanova, T. and Schunger, I. (2006) Bat-parasitic *Cimex* species (Hemiptera: Cimicidae) on the Balkan Peninsula, with zoogeographical remarks on *Cimex lectularius*, Linnaeus. *Zootaxa*, **1190**, 59–68.

Šmaha, J. (1976) Die Fledermauswanze, *Cimex dissimilis* (Horváth) (Heteropt., Cimicidae), als Lästling in Paneeltafelhäusern. *Anzieger Für Schädlingskunde Pflanzenschutz Umweltschutz*, **49** (9), 139–141.

Snetsinger, R. (1997) Bedbugs and other bugs, in: *Mallis' Handbook of Pest Control*, 9th edn, (ed. S. Hedges), GIE Publishing, Cleveland, Ohio, USA, pp. 392–424.

Southall, J. (1730) *A Treatise of Buggs*, J. Roberts, London, UK.

UK Housing Act (1930) *United Kingdom Housing Act*, http://www.legislation.gov.uk/ukpga/1930/39/pdfs/ukpga_19300039_en.pdf (accessed 13 April 2016).

UK Ministry of Health (1934) *Reports on Public Health and Medical Subjects No. 72. Report on the Bed-bug.* Ministry of Health, London, UK.

UK Public Health Act (1936) *United Kingdom Public Health Act*, http://www.legislation.gov.uk/ukpga/1936/49/pdfs/ukpga_19360049_en.pdf (accessed 22 April 2016).

Usinger, R.L. (1966) *Monograph of Cimicidae (Hemiptera – Heteroptera)*, Entomological Society of America, College Park.

Višnička, J. (2002) Štenice v luxusníchhotelích. *Dezinfekce, Dezinsekce, Deratizace*, **11** (2), 65.

6

The Bed Bug Resurgence in Asia

Chow-Yang Lee, Motokazu Hirao, Changlu Wang, and Yijuan Xu

6.1 Introduction

The resurgence of bed bugs in Asia occurred during the late 1990s, but it was not until the period from 2005 to 2010 that the general public began to show concern and pay more attention to these insects. Today, bed bugs have become one of the most important urban insect pests managed by pest management professionals (PMPs) in Asia. In this chapter, the history of bed bugs in Asia is reviewed, along with the historical laws, regulations, and policies pertaining to the control of these insects, the extent of the modern resurgence, the factors behind the resurgence, and the management strategies employed for the control of bed bugs in Asia. Special emphasis is made on the past and present bed bug situations in Japan, China, and southeast Asia.

6.2 History of Bed Bugs in Asia

Bed bugs have been a long and persistent issue in Asia. In Japan, bed bugs were first brought into the nation around the 1860s, at end of the Edo period, probably on Dutch warships. The first location to receive bed bugs was Nagasaki, which was the only port in the southern region of country at that time that was open to international trade. As foreign trade increased, bed bugs spread throughout the rest of Japan. There are no records of the bed bug species, but it was suspected to be the Common bed bug, *Cimex lectularius* L., as this is the main species that affects the nation today. During the Satsuma Rebellion in 1877, there was a record of bed bugs in Kokura military barracks. Additional records exist for Osaka in 1880, and Nagoya and Tokyo in 1882 (Konishi, 1977). By the early 1900s, major bed bug infestations were reported in cities that had international ports, such as Nagasaki, Kobe, Nagoya, Yokohama, and Niigata (Aoki, 1901). During the 1940s, only a few PMPs were operating in Tokyo, and one of their major targeted pests in the urban environment was bed bugs. The control methods used were steam and the application of an insecticidal dust containing a pyrethrin-rotenone mixture (Hirao, 2012).

After World War II, many demobilized Japanese soldiers returned home from southeast Asia infested with a range of pests, including fleas, lice, and bed bugs. In Kobe, 20% of the 0.4 million houses became infested with bed bugs (Uemura, 1974). Fortunately, the newly discovered organochloride insecticides, such as DDT and BHC, effectively controlled the bed bug infestations. By the 1960s, especially prior to the Tokyo Olympics in 1964, bed bug infestations became rare in Japan. Most current large Japanese pest management companies (PMC) were established in the 1970s, and by then, bed bug treatments were largely no longer undertaken. For the next 30 years, it was not necessary for the Japan Pest Control Association (JPCA) to provide bed bug training courses. But that situation changed in 2008.

Advances in the Biology and Management of Modern Bed Bugs, First Edition.
Edited by Stephen L. Doggett, Dini M. Miller, and Chow-Yang Lee.
© 2018 John Wiley & Sons Ltd. Published 2018 by John Wiley & Sons Ltd.

More recently, besides *C. lectularius*, the Tropical bed bug, *Cimex hemipterus* (F.), was found in a hotel and student dormitory in Okinawa prefecture (Komatsu et al., 2016). The authors of the report suggested that the rapid increase in tourist numbers could be one of the contributing factors to the increase in bed bug infestations in Okinawa.

In China, because of the years of insurrections and wars, there were no studies or reports on the incidence of bed bugs prior to the establishment of the People's Republic of China in 1949. In fact, there are few studies prior to 1976. Currently, both *C. lectularius* and *C. hemipterus* are found in China (Deng and Feng, 1953). *Cimex lectularius* is distributed in most parts of China, whereas *C. hemipterus* is mostly found in Guangdong province and Guangxi Zhuang Autonomous Region in the south. Overall, bed bug infestations were common in the 1950s and 1960s, but declined in the 1970s and 1980s, and became rare in the 1990s, according to interviews of residents (Wang et al., 1997). In Sheldon Lou's memoir of his time as a student at Tsinghua University in Beijing during the late 1950s (Lou, 2005), he recalled that his dormitory was heavily infested with bed bugs, as were most other student dormitories in Tsinghua. In Shanghai in 1977, a survey of 3596 residents revealed at least 3.3% were infested, with a rate of 15.4% being reported from one community of 169 homes (Liang et al., 1981). A later survey (1982) in Shanghai, found that 0.21% of 145 695 homes had bed bugs (Shanghai Huangpu District Health and Epidemic Prevention Station, 1983). Recognizing bed bugs as an important public health pest, the Chinese government organized city-wide bed bug control campaigns from the 1960s to the 1980s. These campaigns reduced the infestation rates dramatically. There were only two bed bug infestations reported from 1983 to 1989 (Jiang, 1991; Chi et al., 1994). One of these was in worker dormitories and the other was in a train station. There were eight bed bug surveys conducted over 1990–1999. The sites where the surveys indicated infestations included military stations, homes, school dormitories, hotels, prisons, and hospitals. In Jiaxing city (Zhejiang province) a city-wide survey of 50 584 homes in 1992 only found eight homes with bed bugs (Hong et al., 1993). A similar survey in 1996 of the same city, including 528 homes, 114 hotels, schools, and hospital beds, did not find any bed bugs (Wang et al., 1997).

In southeast Asia, bed bugs were an important insect pest associated with people from the 1940s to the 1970s. The bed bug species was identified as *C. hemipterus*. A limited number of published papers documented the importance of bed bugs during that era, but several papers described their prevalence in hospitals and worker quarters on crop plantations in Malaya, and their insecticide-resistance status (Reid and Lim, 1959; Reid, 1960). The Pest Control Association of Malaysia reported that in the 1960s and 1970s, bed bugs were the major pest targeted in cinemas in Malaysia. Most control efforts in the 1960s consisted of applications of residual insecticide sprays containing DDT and dieldrin. In the 1970s, organophosphates and carbamates such as diazinon and propoxur were the most widely used insecticides for bed bug control. By the late 1970s, the frequency of bed bugs as a pest in the urban environments of Southeast Asia had gradually diminished.

6.3 Laws, Regulations and Policies for Bed Bug Control Prior to the Resurgence

With the exception of China, there were few laws or policies enacted to combat bed bugs in Asia. The Chinese government designated bed bugs as one of the four main urban pests in 1960 (Wang and Wen, 2011). The other three pests were rodents, flies, and mosquitoes. Bed bug elimination campaigns were carried out nationwide by city governments. Trained personnel were sent to neighborhoods to carry out surveys and treatments. Residents were provided with insecticides so that they could undertake bed bug control. These campaigns were highly successful in eliminating bed bugs. In Shanghai, the application of 0.5% or 1.0% fenthion resulted in total bed bug elimination in various neighborhoods (Liang et al., 1981). Applying 0.98% fenthion liquid reduced the bed bug infestation rate from 9.6% in 1979 to 0.01% in 1985 (Ma, 1986). In Hunan, bed bug infestation rates reduced from 1.2% to 0.07% after treatment with 1% fenthion spray between 1982–1985 (Yan et al., 1986).

6.4 Modern Resurgence of Bed Bugs in Asia

The modern bed bug resurgence in Asia occurred in the late 1990s, at approximately the same time as the resurgence of bed bugs in the US, UK, and Australia. In Japan, suspected bed bug bites after decades of absence were first reported by dermatologists in 1996 and then again in 1997 (Ohtaki *et al.*, 1996; Nose, 1997). By 2000, a Japanese hotel industry newsletter reported that bed bugs were being detected occasionally at some low-cost guest houses used mainly by backpackers (Sawa, 2009). For the pest management industry, bed bugs started becoming a more high-profile pest from around 2007. With a high demand from the pest management industry, the JPCA initiated the first bed bug management training course in 2008. The JPCA later published a bed bug treatment methods guide in 2010 (Hirao, 2010). Bed bug infestations became a more serious problem in Tokyo after 2008 (Figure 6.1). Similarly, the JPCA reported a fourfold increase in the number of inquiries between 2010 and 2014 (Japan Pest Control Association, 2015). Two large PMCs in Japan also reported an increase in the number of bed bug treatment services over 2004–2014 (Figure 6.2). A survey of PMCs conducted by the JPCA in 2010 indicated that 60% of PMCs had encountered bed bug infestations. Bed bug treatments were mostly conducted in hotels (471 cases), followed by low-cost guest houses (207 cases), homes (128 cases), apartments (71 cases), hospitals (8 cases), and others (35 cases), such as psychiatric rehabilitation facilities, nursing homes, elderly care facilities, public baths, internet cafés, and ferry boats. The hospitality sector has been consistently reluctant to report bed bug infestations, fearing that it would damage their reputation. Although the actual number of bed bug infestations may be under-reported, bed bug infestations are known to have occurred in 97% of Japanese prefectures (provinces) according to a 2012 survey (Hirao, 2013). This percentage had increased to 100% in a survey conducted in 2015 (M. Hirao, Japan Pest Control Association, unpublished results).

In China, there were 15 publications documenting bed bug infestations between 2001 and 2010, compared with eight articles in the decade prior. Among these, eight articles reported bed bugs in army dormitories (Di Wu *et al.*, 2003; Xue *et al.*, 2003; Zhang and Di Wu, 2003; Li G. *et al.*, 2007; Liu *et al.*, 2007; Zhou *et al.*, 2009). The increased number of reports from the army, at least in part, was probably a result of the overall increase in bed bug infestations. The highest percentage of infested beds from army dormitories was 50.8% (Xue *et al.*,

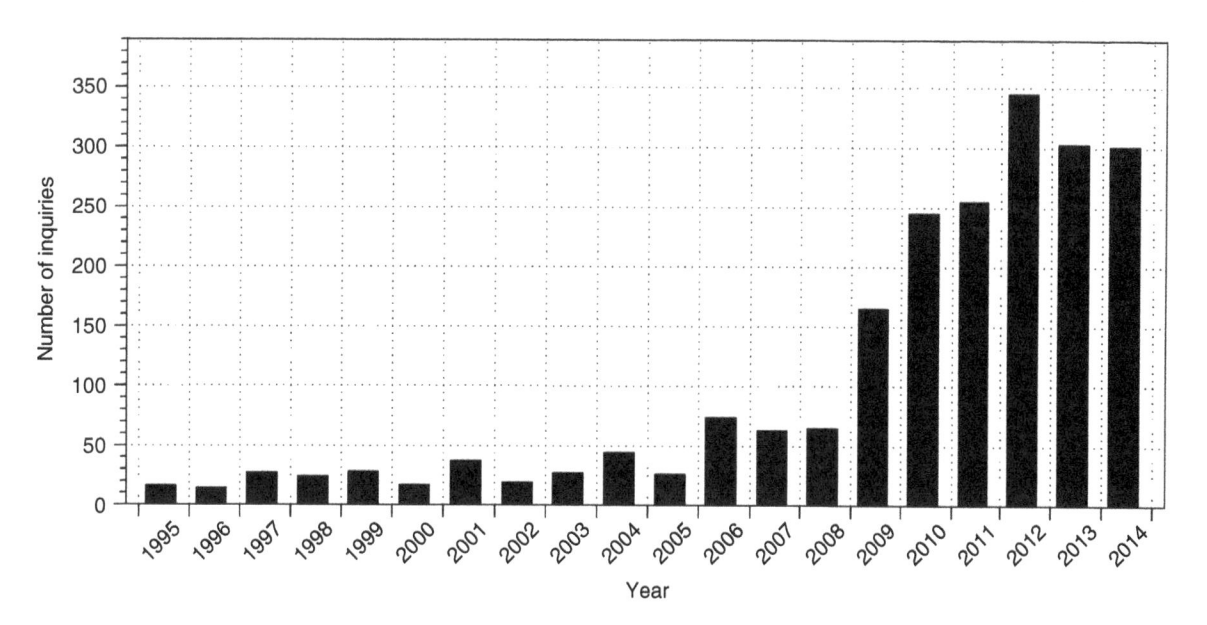

Figure 6.1 The number of bed bug enquiries to the government of Tokyo, 1995–2014 (*Source:* Bureau of Social Welfare and Public Health, Tokyo Metropolitan Government, Japan, unpublished results).

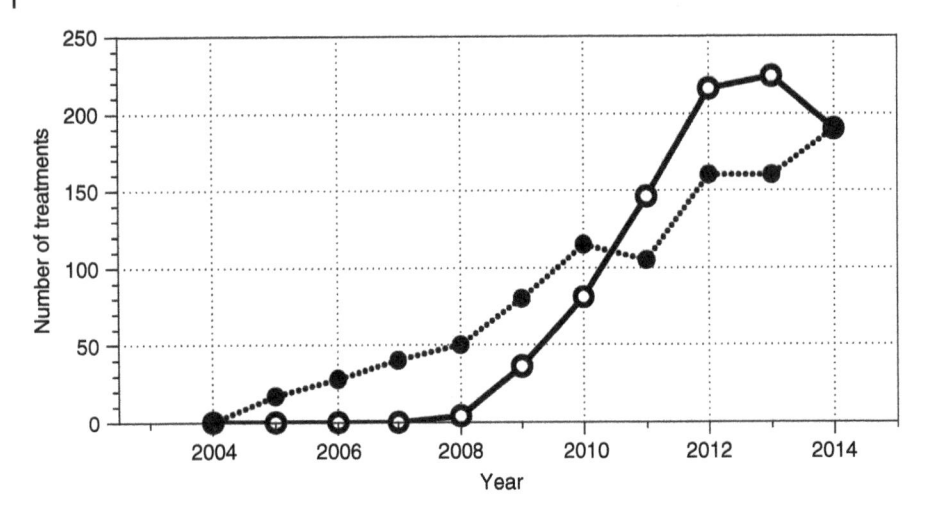

Figure 6.2 The number of bed bug treatments per year reported by two different leading pest control companies in Japan from 2004 to 2014 (Motokazu Hirao, unpublished results).

2003). An additional 12 bed bug surveys were reported between 2011 and 2014 (Li G.M. *et al.*, 2011; Chen *et al.*, 2012; Ye *et al.*, 2012; Bo *et al.*, 2013; Chen and Huo, 2013; Wang L. *et al.*, 2013a; Qi *et al.*, 2014; Wang X.Y. *et al.*, 2014; Zhang et al, 2014; Liu and Fan, 2015; Wang, Z.X. 2015; Yuan *et al.*, 2015). Li X.W. *et al.*, (2013) investigated the extent of the bed bug resurgence between October 2011 and April 2012 in Shenzhen city (near Hong Kong). The authors surveyed worker dormitories, college dormitories, and apartments. Up to 59.8% of the rooms in the surveyed buildings were infested. Worker dormitories in industry parks had the highest infestation rates, followed by older rental homes, and school dormitories. Among those facilities experiencing bed bug infestations, 63.2, 26.1, 6.0, and 4.7% said bed bugs had occurred within the last <1, 1–2, 3–5, and ≥6 years, respectively. These numbers indicate that the majority of the bed bug infestations occurred after 2009. Based on pest control records from two PMCs that service Shenzhen and Dongguan cities, bed bug infestations are currently very common in worker sleeping quarters (Wang L. *et al.*, 2015).

Prior to 2008, all published reports of bed bugs in China were infestations that occurred in homes or dormitories. However, between 2008 and 2015, nine of the 14 published articles reporting bed bugs dealt with infestations in passenger trains. Incidences of bed bug infestations in trains were reported from the following railway bureaus; Kunming, Guangzhou, Chengdu, Shanghai, Shenyang, Zhengzhou, Xian, Lanzhou, and Urumqi (Wang X.C. *et al.*, 2013b; Zheng *et al.*, 2015). In addition, trains from Fujian, Shijiazhuang, Guilin, and Nanning, were also reported to be infested with bed bugs (Zheng *et al.* 2015). These reports reflected a clear resurgence of bed bugs on passenger trains. A survey of sleeper cars in Fujian Railway Bureau during 2009 and again in 2010, found 6.1 and 6.4% of the sleeper cars were infested with *C. lectularius*, respectively (Li G.M. *et al.*, 2011). Inspections of sleeper train cars managed by the Guangdong Railway Bureau in 2011 and 2012 found that 2.45% and 14.97% of the train cars had *C. hemipterus*, respectively (Qi *et al.*, 2014). Sixteen percent of the beds (11 out of 66 beds) in a sleeper train car in Shijiazhuang (Hebei province) were also found to be infested with bed bugs (Chen and Huo, 2013).

The resurgence of bed bugs in southeast Asia likely started around the mid-2000s, when PMPs reported that bed bug control was becoming increasingly common in their daily operations (Table 6.1). Most of these reports were anecdotal in nature and published in local newspapers and other trade magazines (How and Lee, 2010). With the exception of Thailand (which had both *Cimex* species), all samples collected in southeast Asia were *C. hemipterus*. How and Lee (2010) surveyed 54 hotels, public accommodations, and residential premises in Malaysia and Singapore over 2005–2008, and found *C. hemipterus* to be the only bed bug species present. Previously, Lee H.L. *et al.* (2006) also reported that *C. hemipterus* was among the ectoparasites received for

Table 6.1 The number of bed bug treatments in Singapore from 2006 to 2011.

Year	No. jobs
2006	200
2007	265
2008	375
2009	450
2010	468
2011	470

Source: National Environment Agency, Singapore based on survey carried out on selected pest management companies.

identification at the Institute for Medical Research, Malaysia, and that these samples were collected from hotels and industrial plants. Subsequent samples sent by PMPs in Malaysia and Singapore between 2009 and 2013 revealed that a small number of samples that were *C. lectularius* (Lee C.Y., 2013). A later larger survey, conducted from November 2013 to December 2014 in apartments and worker dormitories in Peninsular Malaysia, revealed that only *C. hemipterus* was present throughout the 185 sites sampled (Zulaikha *et al.*, 2016). In Thailand, Tawatsin *et al.* (2011) identified *C. hemipterus* in hotels in Bangkok, Chonburi, Phuket, and Krabi, whereas *C. lectularius* was found only in hotels in Chiang Mai. The reason for this discrepancy is unknown, but it is likely related to the cooler climate found in Chiang Mai.

6.5 Specific Factors Related to the Bed Bug Resurgence in Asia

One possible major contributory factor that led to the rapid spread of bed bugs in Japan was the availability of cheap accommodation for the homeless and foreign backpacking tourists (Motokazu Hirao, Japan Pest Control Association, unpublished results). In Japan, the Homeless Support Act established in 2002 ensured housing and medical care for homeless people, and cheap accommodation and shelters were made available. In these inexpensive lodgings, foreign backpacker tourists also stayed and probably introduced bed bugs. This was similar to Australia, where it had also been documented that that backpacking lodges were commonly infested during the early days of the resurgence (Doggett and Russell, 2008). Bed bugs spread rapidly with the homeless people who often moved from one shelter to another (Yaguchi and Kasai, 2010). Since the early years of the new millennium, the bed bug problem in Japan has been reported most commonly amongst the socially disadvantaged, where the facility owners have limited or no financial means to pay for PMPs.

In China, it appears that two important and commonly infested environments contributed to the bed bug resurgence: migrant worker dormitories and passenger trains. As an example of the former, an outbreak of *C. hemipterus* was reported in worker dormitories in a factory in Fujian province (Yuan *et al.*, 2015). The source of this infestation was the workers who had transferred from a factory in Guangdong province. A compounding factor was that these migrant workers usually slept together in crowded quarters, which enabled the bed bug infestations to spread rapidly. The second infested environment that contributed to the bed bug resurgence was the common and repeated introduction of bed bugs onto passenger trains (Zheng *et al.*, 2015). Passenger trains are the main source of long distance transportation in China. Staying for hours in a sleeper bed provides the opportunity for bed bugs to relocate while feeding on a new host, and allows the long-distance dispersal of bed bugs between cities and provinces.

As for southeast Asia, several factors may explain the recent resurgence of bed bugs. Over the last 20 years, there has been an increase in the influx of legal and illegal migrant workers to Malaysia (Zulaikha *et al.*, 2016). These workers have come from poorer developing countries such as India, Bangladesh, Myanmar, and Indonesia. In addition, native workers have migrated into the major cities from poorer regions of the country. The prevalence of bed bug infestations in maids' rooms and worker dormitories in Malaysia and Singapore suggest that these pests may have been transported with the influx of workers. Secondly, the increasing number of budget airlines has made international travel very affordable, and increased travel has no doubt led to the spread of bed bugs globally, as evidenced by the number of infested hotels and guest houses (Tawatsin *et al.*, 2011; Zulaikha *et al.*, 2016). Finally, the occurrence of insecticide resistance in populations of *C. hemipterus* (Tawatsin *et al.*, 2011; Dang *et al.*, 2014, 2015) and *C. lectularius* (Suwannayod *et al.*, 2010) has, without doubt, also contributed to the resurgence of these pests in southeast Asia as well as elsewhere in the world.

6.6 Bed Bug Management Strategies in Asia

For detection and monitoring of bed bugs, Japanese PMPs employ visual inspections, bed bug interceptors, sticky traps with or without attractants, and trained canines. Most of the JPCA certified PMPs can control bed bug infestations effectively using an integrated pest management (IPM) approach incorporating steam, vacuuming, heating devices, and insecticides. Organophosphate insecticides are still available in Japan for the control of bed bugs and have been found to be highly effective. While organophosphates and pyrethroids are currently registered in Japan for bed bug control, diatomaceous earth is presently not. One of the challenges that Japanese PMPs had to overcome was the "tatami" mat, which covers the floor of a Japanese-style room. These mats are still commonly used in the majority of the hotels, apartments, homes, and housing facilities in Japan. The tatami mat is made by weaving plant stems of the soft rush, *Juncus effusus* L. into a floor covering, and is a perfect harborage for bed bugs. High-temperature tatami-sanitizing and drying machines have been used to treat mite infestations, but are also effective in controlling bed bugs. The bed bug infestation problem in Japan is slowly beginning to migrate from hotels and shelters to apartments and private homes. The Japanese government and PMPs are concerned that bed bug infestations may spread due to the lack of bed bug awareness amongst the general public. Raising bed bug awareness and promoting methods to reduce infestation risk among the general public will be required to minimize the spread of infestations.

The Japanese National Institute of Infectious Diseases, and affiliated institutions, conducted a genetic screening for knockdown resistance amongst bed bugs collected from 60 different field sites across Japan (Tomita *et al.*, 2012). Similar to the pyrethroid-resistant *C. lectularius* populations reported in the USA (Zhu F. *et al.*, 2010) with haplotype mutation of V419L and L925I (known as "haplotype C" after Zhu F. *et al.*, 2010), 50 out of the 60 populations of *C. lectularius* in Japan had a single point haplotype mutation, L925I ("haplotype B") at the α-subunit of the voltage-sensitive sodium channel gene. In addition, four populations had two point haplotype mutations, V419L and L925I (that is, haplotype C) (Tomita *et al.*, 2012). The L925I mutation (haplotype B) was found in 88% of the populations, the majority of which were homozygous (Tomita *et al.*, 2012). Twenty-five bed bugs out of the twenty-seven individuals tested from seven populations identified with the I925 mutation were pyrethroid-resistant. These individuals survived more than 24 h exposure to 0.13 mg/cm^2 deltamethrin residue on filter paper (Tomita *et al.*, 2012). Fortunately, most bed bug populations in Japan were not resistant to organophosphates (such as fenitrothion and propetamphos) or carbamates (propoxur), which are both still used to treat bed bug infestations.

In China, bed bug management strategies usually involve a combination of non-chemical and chemical treatment options (Wang and Wen, 2011). Commonly used non-chemical methods include picking bed bugs using needles, sealing crevices and cracks in walls and floors, and pouring hot water onto infested furniture and bedding. Less commonly used methods are applying hot steam, and placing infested furniture and bedding in the sun during summer. The effectiveness of placing furniture and personal belongings in the hot sun has not be tested in China, but has been found to be ineffective for treating infested mattresses elsewhere (Doggett *et al.*, 2006). Insecticides are always used in bed bug control, either pyrethroids alone or a combination of a pyrethroid and an organophosphate. Commonly used pyrethroids include cyfluthrin, cypermethrin, deltamethrin, tetramethrin, and d-phenothrin. Pyrethroids tend to be used more often than organophosphates. The following organophosphates for bed bug control have been used: chlorpyrifos, fenthion, malathion, and propoxur (Li G.M. *et al.*, 2011; Bo *et al.*, 2013; Zhang G.P. *et al.*, 2014; Wang X.Y. *et al.*, 2014; Zhu J.H. *et al.*, 2014; Liu *et al.*, 2007). Fenthion has been in use since the 1970s and was found to be more effective than lindane (Department of Epidemiology of Nanjing Military Medical Research Institute, 1979). Fenthion is the most commonly used organophosphate to control bed bugs, and found to be more effective than the pyrethroids. Each infestation is usually treated 2–3 times at 3–7 d intervals (excluding control via fumigation). Sulfuryl fluoride fumigation was successfully used for treating a bed bug-infested cargo ship (You *et al.*, 2014).

Although there are no papers yet published reporting insecticide resistance among bed bug populations in China, in recent years insecticide resistance has been assumed based on preliminary laboratory evaluations (Ren Dong-Sheng, Chinese Center for Disease Control and Prevention, Beijing, unpublished results) and global reports. While bed bug control using the currently available insecticides appears to be effective, there is a lack of public awareness and a community-wide effort to manage bed bugs in China. Other than the institution-initiated surveys in passenger trains, no community-wide surveys have been conducted in homes since 1996.

In southeast Asia, the bed bug is not regarded as a major urban pest. In surveys conducted by pest management associations and the Urban Entomology Laboratory, Universiti Sains Malaysia in 2014, bed bugs were ranked between fourth and ninth amongst the nine most targeted pests (Table 6.2) (Lee C.Y., 2013). Nevertheless, PMPs in both Malaysia and Singapore consider the bed bug to be the most difficult pest to control. This likely is due to the limited knowledge and lack of training regarding bed bug management among PMPs, the scarcity of effective products to manage bed bugs, and insecticide resistance (Lee C.Y., 2013). Use of residual insecticide sprays remains the most common method of controlling bed bugs (Table 6.3), followed by heat treatment, and vacuuming. In spite of the of insecticide effectiveness due to resistance, pyrethroid-based formulations (such as deltamethrin, alpha-cypermethrin, lambda-cyhalothrin, permethrin) are still regularly used. Some PMPs also use an organophosphate (such as chlorpyrifos or fenitrothion) in difficult-to-manage situations. However, the use of organophosphates is limited by their strong odor, and restrictions against indoor use in certain situations. The introduction of the pyrethroid-neonicotinoid mixtures (such as beta-cyfluthrin + imidacloprid and lambda-cyhalothrin + thiamethoxam mixtures) has dramatically improved bed bug management efforts in the region.

Table 6.2 Importance and difficulty of control of bed bugs in southeast Asian countries compared to other major urban pests.

Country	Rank Importance	Difficulty of control
Malaysia	4	1
Singapore	4	1
Thailand	6	4
Indonesia	9	5
Philippines	7	2

Source: Based on surveys carried out by pest management associations and the Urban Entomology Laboratory, Universiti Sains Malaysia (Lee C.Y., 2013).

Table 6.3 The ranking of bed bug related services offered by pest management professionals in southeast Asia.

Treatment method	Rank of services (based on frequency)				
	Malaysia	Singapore	Thailand	Indonesia	Philippines
Residual spray	1	3	1	1	1
Heat treatment	2	1	3	4	3
Vacuuming	3	1	2	5	2
Trapping	4	6	4	2	4
Inspection	6	5	5	6	6
Fumigation	5	4	6	3	5

Source: Lee C.Y., 2013.

Bed bug-infested commercial aircraft are often flown into Malaysia. Fumigation of these aircraft using methyl bromide is carried out by PMPs. There are also a number of PMPs in southeast Asia that use cold-fogging (misting with pyrethroids) to treat bed bugs, but this approach is unlikely to be effective at controlling resistant bed bugs.

To effectively curb the current bed bug resurgence in Asia, it is necessary to implement proactive bed bug monitoring programs in areas that are currently experiencing a bed bug resurgence. Without such efforts, bed bug infestations will continue to expand in the years ahead.

References

Aoki, T. (1901) Possibility of the plague transmission by bed bugs (in Japanese). *The Insect World (Konchu Sekai)*, **5** (48), 328–334.

Bo, J.X., Zhu, S.Y., Yang, C.G., et al. (2013) Investigation and disposal of bed bugs carried by an entry ship. *Chinese Journal of Vector Biology and Control*, **24** (5), 462–463, 466.

Chen, B.S. and Huo, W. (2013) An investigation of bed bug infestation in a passenger train. *Journal of Medical Pest Control*, **29** (10), 1127, 1129.

Chen, S.M., Tang, S.X., Zheng, J.M., et al. (2012) Study on the habit of bed bug and the effectiveness of control measures in passenger trains. *Chinese Journal of Vector Biology and Control*, **23** (1), 86–87.

Chi, Y.L., Dai Y.S., Chen, Z., et al. (1994) Bed bug survey results at Sanming city. *Chinese Journal of Vector Biology and Control*, **5** (3), 39.

Dang, K., Toi, C.S., Lilly, D.G., et al. (2014) Detection of knockdown resistance (*kdr*) in *Cimex lectularius* and *Cimex hemipterus* (Hemiptera: Cimicidae). *Proceedings of the Eighth International Conference on Urban Pests, 2014, July 20–23, Zurich, Switzerland.* OOK-Press, Veszprém.

Dang, K., Toi, C.S., Lilly, D.G., et al. (2015) Identification of putative *kdr* mutations in the tropical bed bug, *Cimex hemipterus* (Hemiptera: Cimicidae). *Pest Management Science*, **71** (7), 1015–1020.

Deng, G.F. and Feng, L.Z. (1953) Geographic distribution of *Cimex lectularius* L. and *Cimex hemiptera* F. in China. *Acta Entomologia Sinica*, **2**, 253–264.

Department of Epidemiology of Nanjing Military Medical Research Institute (1979) Control bed bugs using fenthion in 356 military units. *People's Military Surgeon*, **12**, 33.

Di Wu, J.X, Zhou, L., Yang, J., et al. (2003) Survey of bed bugs in Shaanxi military dormitories. *Chinese Journal of Vector Biology and Control*, **14** (2), 134.

Doggett, S.L. and Russell, R.C. (2008) The resurgence of bed bugs, *Cimex* spp. (Hemiptera: Cimicidae) in Australia. In: *Proceedings of the Sixth International Conference on Urban Pests, July 13–16, 2008, Budapest, Hungary*. OOK-Press, Hungary, Veszprém.

Doggett, S.L., Geary, M.J. and Russell, R.C. (2006) Encasing mattresses in black plastic will not provide thermal control of bed bugs, *Cimex* spp. (Hemiptera: Cimicidae). *Journal of Economic Entomology*, **99** (6), 2132–2135.

Hirao, M. (2010) *Technical Report on Bed Bugs* (in Japanese), Japan Pest Control Association, Tokyo.

Hirao, M. (2012) Bed bug treatments before World War II (in Japanese). *Pest Control*, **32** (4), 50–53.

Hirao, M. (2013) Bed bug survey result (in Japanese). *Pest Control*, **43** (2), 47.

Hong, Q.L., Hong, Y.Q. and Wu, S.B. (1993) Survey and control of bed bugs in Jiaxing city. *Zhejiang Journal of Preventive Medicine*, **3**, 4.

How, Y. F. and Lee, C.Y. (2010) Survey of bed bugs in infested premises in Malaysia and Singapore. *Journal of Vector Ecology*, **35** (1), 89–94.

Japan Pest Control Association (2015) Report of the number of inquiries on urban pests (in Japanese). *Pest Control*, **45** (4), 43–47.

Jiang, X.S. (1991) Observations on the efficacy of bed bug control at Huaihua railway station. *Hunan Journal of Preventive Medicine*, **3**, 42–43.

Komatsu, N., Nakamura, H. and Fujii, K. (2016) Distribution of tropical bedbug *Cimex hemipterus* in Okinawa Prefecture, Japan. *Medical Entomology and Zoology*, **67** (4), 227–331.

Konishi, M. (1977) *Bed Bugs, in Mushi no bunkashi*, The Asahi Shimbun Company (Asahi Shinbunsha), Tokyo, Japan.

Lee, C.Y. (2013) Bed bugs in Asia – Perspective from Southeast Asia. Speech presented at the Global Bed Bug Summit, in Denver, Colorado, December 5, 2013.

Lee, H. L., Krishnasamy, M., Jeffery, J., *et al.* (2006) Notes on some ectoparasites received by the Medical Entomology Unit, Institute for Medical Research. *Tropical Biomedicine*, **23** (1), 131–132.

Li, G., Cao, X. and Huang, C. (2007) Bed bug outbreak in a military unit in Shandong province. *Practical Journal of Medicine and Pharmacy*, **24** (2), 226.

Li, G.M., Wang, X.L. and Zhang, S.M. (2011) Investigation and study of the bugs in the sleeping cars of passenger trains. *Journal of Pest Control*, **27** (9), 813–814.

Li, X.W., Ma, T., Wang, P., *et al.* (2013) Bug hazard investigation in Shenzhen. *Chinese Journal of Hygiene Insecticides and Equipment*, **19** (3), 236–238, 241.

Liang, T.L., Wang, W.Z., Wang, S.Z., *et al.* (1981) Bed bug control experiment. in *Collection of Cockroach and Bed Bug Research Papers* (vol. 1), (ed. X.Y. Hu). National Cooperative Group for Controlling Four Main Urban Pests, China, pp.124–126.

Liu, H., Wang, Y., Xin, Z., *et al.* (2007) Survey of bed bug infestation and control efficacy. *Chinese Journal of Hygiene Insecticides and Equipment*, **13** (3), 218–219.

Liu, P.T. and Fan, C.Y. (2015) Investigation and control of bedbug infestation on a passenger train. *Chinese Journal of Vector Biology and Control*, **26** (5), 536.

Lou, S. (2005) *Sparrows, Bedbugs, and Body Shadows: A Memoir*, University of Hawaii Press, Honolulu, Hawaii.

Ma, S.Z. (1986) Post-treatment survey of bed bugs in Nanshi district of Shanghai. *Medical Animal Control*, **2**, 92–95.

Nose, T. (1997) Biting of bed bugs (in Japanese). *Practical Dermatology (Hifubyoh Shinryoh)*, **19**, 445–448.

Ohtaki, M., Yamamoto, I. and Shinonaga, S. (1996) Two clinical cases of bed bug bites (in Japanese). *Journal of the Japan Organization of Clinical Dermatologists*, **48**, 141–144.

Qi, J.X., Zhu, J.H., Fan, H.L., *et al.* (2014) Investigation on bed bugs in sleeper compartments of passenger trains. *Chinese Journal of Hygiene Insecticides and Equipment*, **20** (5), 490–492.

Reid, J.A. (1960) Resistance to dieldrin and DDT and sensitivity to malathion in the bed-bug *Cimex hemipterus* in Malaya. *Bulletin of the World Health Organization*, **22** (5), 586–587.

Reid, J.A. and Lim, C.S. (1959) A note on the residual life of the insecticide malathion. *Medical Journal of Malaya*, **13**, 239–242.

Sawa, K. (2009) Unexpected bed bug attack (in Japanese). *Japan Tourist Hotel Association Magazine*, New Year's issue, Tokyo, Japan.

Shanghai Huangpu District Health and Epidemic Prevention Station (1983) Report on bed bug survey and control in Huangpu district, Shanghai. In: *Collection of Cockroach and Bed Bug Research Papers (vol. 2)*, (eds X.Y Hu, T.L. Liang, X.F. Jiang, *et al.*), National Cooperative Group for Controlling Four Main Urban Pests, China, pp. 168–172.

Suwannayod, S., Chanbang, Y. and Buranapanichpan, S. (2010) The life cycle and effectiveness of insecticides against the bed bugs of Thailand. *Southeast Asian Journal of Tropical Medicine and Public Health*, **41** (3), 548–554.

Tawatsin, A., Thavara, U., Chompoosri, J., *et al.* (2011) Insecticide resistance in bedbugs in Thailand and laboratory evaluation of insecticides for the control of *Cimex hemipterus* and *Cimex lectularius* (Hemiptera: Cimicidae). *Journal of Medical Entomology*, **48** (5), 1023–1030.

Tomita, T., Komagata, O., Kasai, S., *et al.* (2012) Nationwide study of the bed bug pyrethroid susceptibility (in Japanese). *The Japan Society of Medical Entomology and Zoology*, **63** (Supplement issue), 85.

Uemura, T. (1974) Bed bug story (in Japanese). *Pest Control*, **3** (3), 16–21.

Wang, C.L. and Wen, X.J. (2011) Bed bug infestations and control practices in China: Implications for fighting the global bed bug resurgence. *Insects*, **2** (2), 83–95.

Wang, H.J., Yu, L.S. and Mao, H.S. (1997) Report on bed bug survey in Jiangshan city. *Medical Animal Control*, **13**, 225–226.

Wang, L., Cai, X.Q. and Xu, Y.J. (2015) Status of urban bed bug infestations in southern China: an analysis of pest control service records in Shenzhen in 2012 and Dongguan in 2013. *Journal of Medical Entomology*, **52** (1), 76–80.

Wang, L., Xu, Y.J. and Zeng, L. (2013a) Resurgence of bed bugs (Hemiptera: Cimicidae) in mainland China. *Florida Entomologist*, **96** (1), 131–136.

Wang, X.C., Yang, P., Zhang, Y. (2013b) Bed bug control in passenger trains. *Journal of Disease Monitor and Control*, **7** (12), 769.

Wang, X.Y., He, H.G., Zhang, S.Q., *et al.* (2014) An investigation of a bug infestation. *Chinese Journal of Vector Biology and Control*, **25** (4), 372.

Wang, Z.X. (2015) An investigation of two bed bug bite incidents. *Chinese Journal of Hygiene Insecticides and Equipment*, **21** (4), 436.

Xue. J., Li, J., Zhu, T., *et al.* (2003) Infestation condition and control of bedbug in an army. *Chinese Journal of Hygiene Insecticides and Equipment*, **9**, 36–37.

Yaguchi, N. and Kasai, S. (2010) Current situation and problem of bedbug infestation in Tokyo, Japan: from public health officer's angle (in Japanese). *Medical Entomology and Zoology*, **61** (3), 231–237.

Yan, G.W., Yan, H.D. and Hu, Y.F. (1986) A report on bed bug control in Lengshuitan city in Hubei province. *Medical Animal Control*, **2**, 96–98.

Ye, Y.K., Zhao, H.J. and Yu, Z.C. (2012) Investigation on infestation of bedbug and its control. *Chinese Journal of Hygiene Insect and Equipments*, **18** (3), 268.

You, M.C., Dong, S.D., Qiu, J.L., *et al.* (2014). Fumigation treatment with sulfuryl fluoride in an old vessel seriously infested with bedbugs. *Chinese Journal of Hygiene Insect and Equipments*, **20** (1), 93–94.

Yuan, B., Wang, L. and Chen, Z.J. (2015). Field control of an imported bedbug infestation. *Chinese Journal of Vector Biology and Control*, **26** (4), 428–429.

Zhang, L.Y. and Di Wu, J.X. (2003) Discussions on bed bug control strategies in a military base in Shaanxi. *Medical Animal Control*, **19** (2), 120.

Zhang, G.P., Chi, D., Liang, X.D., *et al.* (2014) Management and advices about an encroachment of bedbugs on passenger train. *Chinese Journal of Hygiene Insecticides and Equipment*, **20** (6), 612.

Zheng, J.M., Chen, S.M., Jiang, X., *et al.* (2015) Passenger trains bedbug prevention and control. *Journal of Pest Control*, **31** (1), 61–65.

Zhou, G.Z., Chen, X.H., Fu, Y., *et al.* (2009) Investigation of bed bug infestation in an archery ground and effectiveness of the control effort. *Chinese Journal of Hygiene Insecticides and Equipment*, **15** (4): 281–282.

Zhu, F., Wigginton, J., Romero, A., *et al.* (2010) Widespread distribution of knockdown resistance mutations in the bed bug, *Cimex lectularius* (Hemiptera: Cimicidae), populations in the United States. *Archives of Insect Biochemistry and Physiology*, **73** (4), 245–257.

Zhu, J.H., Qi, J.X., Fan, H.L., *et al.* (2014) The effect of propoxur and fenthion applied in passenger trains for bedbug control. *Chinese Journal of Hygiene Insecticides and Equipment*, **20** (4), 363–364.

Zulaikha, Z., Hafiz, A.M.A., Hafis, A.R.A., *et al.* (2016) A survey on the infestation levels of tropical bed bugs in Peninsular Malaysia: Current updates and status on resurgence of *Cimex hemipterus* (Hemiptera: Cimicidae). *Asian Pacific Journal of Tropical Disease*, **6** (1), 40–45.

7

The Bed Bug Resurgence in Australia

Stephen L. Doggett and Toni Cains

7.1 Introduction

The history of bed bugs in Australia is largely a reflection of what has happened elsewhere in the world. Infestations rapidly spread through society following their introduction into the nation via the early colonialists, then bed bugs became uncommon following World War II until around the year 2000 with the start of the modern resurgence. Australia was somewhat unique in that the resurgence has involved the Common bed bug, *Cimex lectularius* L., and the Tropical species, *Cimex hemipterus* (F.). Battling bed bugs has involved a three-pronged strategy: the development of the world's first industry standard that promotes best practice in bed bug management, education about the standard, and research to enhance it.

7.2 History of Bed Bugs in Australia

No member of the Cimicidae naturally occurs in Australia and it was likely that *C. lectularius* was introduced with the first British colonialists during the late 18th century (Woodward *et al.*, 1970). The presence of bed bugs on early sailing vessels is typified by the descriptions of Lt. Matthew Flinders, the first explorer to circumnavigate Australia. In his journals from 1803, he bemoans (Becke and Jeffery, 1899):

> Of all the filthy little things I ever saw, this schooner, for bugs…rises superior to them all…I believe that I, as well as my clothes, must undergo a good boiling in the large kettle.

It would appear that by the mid-1800s bed bugs were well established. The American, Charles Wilkes, while touring in Australia around 1840, writes (Wilkes, 1849):

> …at Liverpool, and other hotels in Campbelltown [in Sydney]: a larger supply of…bed-bugs is seldom seen…

While it is difficult to gauge the extent of bed bugs, both in their distribution and frequency, it was clear that they were common throughout Australia during the late 19th and early 20th centuries. This is evident in newspaper reports on bed bugs (Anonymous, 1889) or the many and varied ways described to control them (Anonymous, 1890, 1899, 1901; Hackett, 1916; Froggatt, 1919; Purdy, 1920). One of the most widely advertised products in the 1880s was 'Rough on Rats'. At the time, this arsenic-based poison sadly became the chemical of choice for suicide victims (Johnson, 2010).

Advances in the Biology and Management of Modern Bed Bugs, First Edition.
Edited by Stephen L. Doggett, Dini M. Miller, and Chow-Yang Lee.
© 2018 John Wiley & Sons Ltd. Published 2018 by John Wiley & Sons Ltd.

It is difficult to determine when bed bugs became less of a problem in Australia. Even during the early part of the 1900s, bed bugs were not uncommonly reported in newspapers (Anonymous, 1911, 1914, 1915, 1916) and products to control bed bugs were still being advertised in 1924 (Anonymous, 1924).

In the intra-war period, few reports of bed bugs exist. However, history has the tendency to report the unusual rather than the usual. This is typified in the seminal entomological text of the day, *The Insects of Australia and New Zealand* (Tillyard, 1926):

> [the Cimicidae is] represented in [Australia] by the detestable and all too common…bed bug…which is too well known to need description here.

The first reference to which bed bug species occurred in Australia was from 1924, where it was noted that "…the bed bug (*Cimex lectularius*)…lives in the walls of dirty houses, in the mattresses of beds, and in furniture" (Musgrave, 1924). Bed bugs were still very common in Australia during the 1920s as evident from a report that appeared in the premier medical journal, *The Medical Journal of Australia*: "They [bed bugs] are certainly far too abundant in many of the towns in New South Wales and Queensland…the species being *C. lectularius*" (Ferguson, 1926). Not long after this, the correspondent known as "W.B.S." wrote in a regional Australian newspaper that, "Most people are familiar with that malodorous insect the bed-bug…" (W.B.S., 1928). In the early 1930s it was stated that "The ubiquitous bed-bug… [is] prevalent in dirty houses…" (McKeown, 1930).

Bed bugs were well known and despised by hospital workers in Australia in the late 1930s and early 1940s. Sydney entomologist, Dr Chris Orton, recalls his mother, then a nurse in Perth, Western Australia, describing to him some of the measures taken to control the pest in hospital wards at that time. These included painting ward furniture (especially metal framed beds) with kerosene in the hospital yard and painting gaps in the skirting boards and architraves in the wards with turpentine. Calico mattress encasements were meticulously sewn onto mattresses under the watchful eye of the dreaded Matron and infested bedding was bagged and boiled (Chris Orton, Univerity of NSW, Sydney, unpublished data)

Bed bugs continued to be a problem during the years of World War II, according to a report on bed bugs in military camps in Western Australia (Jenkins, 1942). In a book on Australian insects from 1942, it was stated that the bed bug is a "well-known cosmopolitan domestic pest, which frequents houses (Mckeown, 1942). An agricultural journal from South Australia in 1945 provided extensive advice on how to rid the home of bed bugs (Womersley, 1945). Internment camps were also known to be heavily infested (Splivalo, 1982).

By 1947 it was suggested in one regional paper from the mid-north coast of New South Wales that bed bugs in the wider community were becoming uncommon, "except in homes where slatternly women paid no attention to hygiene" (Anonymous, 1947). Clearly if bed bugs had abated by then, sexism in journalism had not.

After World War II, camps were established for displaced persons from Europe and new immigrants. Bed bugs were an issue at the Holden Immigration Camp at Northam (100 km west of the state capital, Perth, Western Australia) and a photo exists from 1952 with the caption "Dutch women airing the beds to rid them of bed bugs" (Peters, 2001). These camps were perhaps the last major problem site for bed bugs prior to the insects' decline after the war.

7.3 Documented Evidence for the Bed Bug Decline

From the end of World War II to the start of the modern resurgence, reports of bed bugs are virtually non-existent. Ruth Park, a New Zealand born novelist, wrote fiction about the poor in Sydney and how bed bugs fed off their poverty (Park, 1948; see also Chapter 2). It was quite possible that she was reliving her own experiences, having come to Australia with little money and being forced to live in what was then considered a slum area.

Beyond this however, there appear to be no reports of bed bugs in Australia in the scientific literature, popular press, or in advertising, over the years from 1950 to 2000. This indicates that infestations must have become rare during this period.

Over the years 1949 to 1983, the Department of Medical Entomology at Sydney University kept all correspondence from the public. In this there were three references to bed bugs. The first was in March 1963, from Vaucluse in Sydney, and it was believed the bed bugs (species not identified) came via luggage from the residents travelling to Burma and the Middle East. The second was *C. lectularius* in an affluent area of Sydney, and the third was associated with staff of a major airline who presumably acquired the bed bugs (again species not identified) from overseas (Doggett *et al.*, 2011). Clearly bed bug infestations were exceptionally rare during those years. This was reaffirmed in 1988, when, in contrast to only a generation past, it was stated of bed bugs that, "...many Australians have never seen one..." (Southcott, 1988).

7.4 The Modern Resurgence

The bed bug resurgence in Australia probably began around the late 1990s and was reviewed in detail to the period up until 2010 (Doggett and Russell, 2008; Doggett *et al.*, 2011). The first suggestion of a resurgence in the nation was not from Australia, but the UK, when in 2002 it was stated that "From New York to Sydney… bed bugs are making a comeback" (Coghlan, 2002). Interestingly, no Australian attribution for this claim was ever established, even if factually correct (Doggett and Russell, 2008).

In 2003, *C. hemipterus* was identified in Australia for the first time (Doggett *et al.*, 2003). In a very short period the species became well established in latitudes north of 29°S (Doggett and Russell, 2008; Doggett *et al.*, 2011). It is not known when or from where *C. hemipterus* was first introduced into Australia and this will only be resolved in the future with extensive population genetic analyses. The species was known in the Pacific rim, as evidenced by a report from Samoa in 1925 (Lambert, 1925), while in the reference collection of the Department of Medical Entomology at Westmead Hospital, there is a sample of *C. hemipterus* from Papua New Guinea (PNG) collected in 1926. At the time, PNG was administered by Australia and human movements between the countries were not infrequent. A publication from 1942 states that *C. hemipterus* was common in Fiji at the time and that it was also common in 1911 (Lever, 1942).

An article from New Guinea in 1972 indicated that bed bugs (presumably *C. hemipterus*) had increased in number because DDT had killed its predators, such as lizards and spiders (Ewers, 1972). As neither are particularly effective at reducing bed bug numbers, perhaps the real reason for the increase was due to insecticide resistance. This was the conclusion in a publication in the following year discussing the same situation (Bourke, 1973). To date, the oldest samples of *C. hemipterus* in Australia are from museum specimens and include Katherine in 1991 and Darwin in 1994, both locations in the Northern Territory (Doggett and Russell, 2008).

Definitive evidence for the bed bug resurgence in Australia was provided in two publications in 2004. The first documented that samples submitted to the Department of Medical Entomology's pathology service at Westmead Hospital had increased by around 250% from 2001 to 2004 compared with the previous four years (Doggett *et al.*, 2004). This report noted how other groups had also experienced an increase in recent years. This included the Australian Quarantine and Inspection Service, who recorded a notable rise in bed bug interceptions since 1999, and one Sydney-based pest control company in the early years of 2000, which had seen an increase of 700% in the number of treatments undertaken (Doggett *et al.*, 2004).

The second paper was of a survey of bed bugs in short-stay lodges in Sydney (Ryan *et al.*, 2004). These facilities are mainly frequented by young backpackers. Of the 47 lodges surveyed, 79% had bed bugs in the prior year and most lodges believed that the problem was increasing. The survey highlighted the lack of knowledge in controlling bed bugs and the unorthodox methods employed.

In 2006, a survey of the pest management industry was undertaken (Doggett and Russell, 2008). This was to gain a better understanding of the degree of the resurgence, the types of properties affected, the management procedures employed, and to advertise the new industry standard on bed bug management that had just been launched (Doggett, 2005). The survey showed that all states in Australia had experienced an increase in bed bug numbers and an overall rise of 4500% had been experienced since 1999, compared with the previous years. The survey revealed that many unregistered insecticides were being used and methods with no scientific basis were commonly being employed. This highlighted that the industry desperately needed education on best management practices for bed bug control.

In the years to 2016, no further bed bug survey was undertaken, yet from the senior authors' own experience, certain trends were evident. During the early years of the resurgence, infestations were more common in backpacking lodges and hotels, and then spread into the wider community. Bed bugs were subsequently reported in virtually all forms of accommodation, as well as on public transport, in healthcare facilities, cinema complexes, and even brothels (Doggett and Russell, 2008). However, arguably the most significant of these was the growth of infestations amongst the socially disadvantaged. This trend, however, was not unique to Australia, with the USA experiencing similar issues (US Department of Housing and Urban Development, 2012).

A new survey of the pest management industry was conducted in mid-2016, with the aim of establishing the current state of the bed bug resurgence, and to review the management options being undertaken to control infestations (Doggett, 2016). The 2006 survey indicated that bed bug infestations were growing at almost an exponential rate across the country. However, as of 2016, it appears that infestations have stabilized and are no longer increasing in number. Respondents were asked their opinion about the state of bed bug infestations over the last five years (2012–2016) and 51.6% felt that infestations had increased, 16.8% that they had decreased, and 31.6% felt that they were at around the same level. In contrast, when asked about the previous year (2015), only 40.0% felt that bed bugs were increasing, whereas 18.9% considered infestations were decreasing, and 41.1% thought they were around the same level. The 2016 survey did reveal that a small number of pest management professionals were still employing insecticides that were largely ineffective against resistant bed bugs and products that were not registered for bed bug control. Also, a small number failed to undertake a mandatory follow-up inspection following a bed bug treatment. Thus despite the obvious gains achieved, further work in educating the pest management industry is required.

7.5 Strategies to Combat the Bed Bug Resurgence

In Australia, there never have been any specific laws dealing with bed bugs. Environmental health laws tend to be more descriptive rather than prescriptive (see Chapter 43). The impetus therefore to combat the bed bug resurgence has largely come from the pest management industry.

The strategy to combat bed bugs has used a three-pronged approach: the development of an industry standard that promotes best practice in bed bug management, education about best practice, and research to enhance best practice.

Best practice is defined in *A Code of Practice for the Control of Bed Bug Infestations in Australia* (Doggett, 2013a) and this is discussed in greater detail in Chapter 22. Education has been delivered through training courses, industry publications (more than 50 to date), lectures (around 150 to date), and the production of a simple visual guide *Do You Have Bed Bugs? A Help Guide for the Identification of Bed Bug Infestations* (Doggett 2013b), which has now been printed in seven different languages. Research has involved the testing of new insecticides and products for bed bug management, as well as investigations into insecticide resistance. Research has been used to evolve the code of practice.

Like elsewhere in the world, the bed bug resurgence caught Australia quite unprepared, and it is evident that certain groups have been impacted more than others, notably the backpacking industry and the socially disadvantaged. However, the Australian response to the resurgence was rapid compared to other countries. The timely development of a rigorous industry standard assisted the education of those who combat bed bugs, and the emergence on the markets of new insecticides, notably the silicates in the last five years, have made control less challenging. At present in Australia, bed bugs have not been defeated, but their decline is underway.

References

Anonymous (1889) Ancient History. *The Mildura Cultivator* (14 November), p. 6.
Anonymous (1890) You can get rid of them. *The Maitland Mercury and Hunter River General Advertiser* (14 August), p. 3.
Anonymous (1899) Fighting bed bugs. *Leader* (21 October), p. 39.
Anonymous (1901) To destroy roaches and bed-bugs. *Albury Banner and Wodonga Express* (23 August), p. 10.
Anonymous (1911) Man's insect foes. fleas and bed-bugs. *Evening News* (18 August), p. 8.
Anonymous (1914) Deadly diseases spread by one small bug. *The Newsletter* (24 January), p. 6.
Anonymous (1915) Canned courtesy. *Maryborough Chronicle, Wide Bay and Burnett Advertiser* (10 July), p. 6.
Anonymous (1916) Bug inspector required. *Canowindra Star and Eugowra News* (7 April), p. 3.
Anonymous (1924) Lotol advert. *The Telegraph* (13 November), p. 12.
Anonymous (1947) Scratching won't cure scabies. *The Manning River Times and Advocate for the Northern Coast Districts of New South Wales* (30 August), p. 9.
Becke, L. and Jeffery, W. (1899) *The Naval Pioneers of Australia*, John Murray, London.
Bourke, T.V. (1973) Some aspects of insecticide application in malaria control programmes other than the effect on the insect vectors. *Papua New Guinea Agricultural Journal*, **24** (1), 33–40.
Coghlan, A. (2002) Bedbugs bite back. *New Scientist*, **176** (2363), 10.
Doggett, S.L. (2005) *A Code of Practice for the Control of Bed Bug Infestations in Australia (draft)*, Department of Medical Entomology and The Australian Environmental Pest Managers Association, Sydney.
Doggett, S.L. (2013a) *A Code of Practice for the Control of Bed Bug Infestations in Australia, 4th edn*. Department of Medical Entomology and The Australian Environmental Pest Managers Association, Sydney.
Doggett, S. L. (2013b). *Do You Have Bed Bugs? A Help Guide for the Identification of Bed Bug Infestations*, Stephen L. Doggett, Sydney Australia.
Doggett, S.L. (2016) Bed bugs in Australia 2016: are we biting back? Results of the 2016 national bed bug survey. *Professional Pest Manager*, **Aug/Sep**, 28–30.
Doggett, S.L. and Russell, R.C. (2008) The resurgence of bed bugs, *Cimex* spp. (Hemiptera: Cimicidae) in Australia. In: *Proceedings of the Sixth International Conference on Urban Pests, July 13–16, 2008, Budapest, Hungary*. OOK-Press, Veszprém.
Doggett, S.L., Geary, M.J., Crowe, W.J., Wilson, P. and Russell, R.C. (2003) Has the tropical bed bug, *Cimex hemipterus* (Hemiptera: Cimicidae), invaded Australia? *Environmental Health Journal*, **3** (4), 80–82.
Doggett, S.L., Geary, M.J. and Russell, R.C. (2004) The resurgence of bed bugs in Australia: with notes on their ecology and control. *Environmental Health Journal*, **4** (2), 30–38.
Doggett, S.L., Orton, C.J., Lilly, D.G. and Russell, R.C. (2011) Bed bugs: the Australian response. *Insects*, **2** (2), 96–111.
Ewers, W.H. (1972) Parasites of man in Papua – New Guinea. *Southeast Asian Journal of Tropical Medicine and Public Health*, **3** (1), 79–86.
Ferguson, E.W. (1926) The distribution of insects capable of carrying disease in eastern Australia. *Medical Journal of Australia*, **2** (13), 339.
Froggatt, W.W. (1919) Destruction of bugs by fumigation. *Agriculture Gazette of N.S.W.*, **30**, 828.
Hackett, L. (1916) *The Australian Household Guide*, E.S. Wigg & Son Limited, Perth.
Jenkins, C.F.H. (1942) Some insect pests of military camps. *Journal of the Department of Agriculture for Western Australia*, **19** (1), 13–37.

Johnson, A. (2010) Rough on rats: an unsolved mystery in Examiner St, 1889. *Nelson Historical Society Journal*, **7** (2), 36–40.

Lambert, S.M. (1925) *Health Survey of Western Samoa, with Special Reference to Hookworm Infection*. Mandated Territory of Western Samoa, Annual Report of the New Zealand Department of Health for the Year ended 31st March 1925, 25–37.

Lever, R.J.A.W. (1942) The bed bug in Melanesia. *Fiji Agricultural Journal*, **13** (1), 26.

McKeown, K.C. (1930) Insects and disease. *The Australian Museum*, **4** (2), 65–68.

McKeown, K.C. (1942) *Australian Insects an Introductory Handbook*, Royal Zoological Society of New South Wales, Sydney.

Musgrave, A.J. (1924) Some Australian insects injurious to man. *The Australian Museum Magazine*, **2** (2), 59–63.

Park, R. (1948) *The Harp in the South*, Angus & Robertson, Sydney.

Peters, N. (2001) *Milk and Honey – But no Gold*, UWA Publishing, Western Australia.

Purdy, J.S. (1920) The control of insect vectors of disease in war and peace. In: *Transactions of the Eleventh Session of the Australasian Medical Congress, August 21–28, 1920, Brisbane, Queensland*. Anthony James Cumming Government Printer, Brisbane.

Ryan, N., Peters, B. and Miller, P. (2004) A survey of bedbugs in short-stay lodges. *NSW Public Health Bulletin*, **15** (11–12), 215–214.

Southcott, R.V. (1988) Some harmful Australian insects. *Medical Journal of Australia*, **149** (11/12), 656–662.

Splivalo, A. (1982) *The Home Fires*, Fremantle Arts Centre Press, Fremantle.

Tillyard, R.J. (1926) *The Insects of Australia and New Zealand*, Angus & Robertson, Sydney.

US Department of Housing and Urban Development (2012) *Guidelines on Bedbug Control and Prevention in Public Housing*, http://portal.hud.gov/hudportal/documents/huddoc?id=pih2012-17.pdf (accessed 20 July 2016).

W.B.S. (1928) Nature Notes. *The Northern Herald* (4 January), p. 32.

Wilkes, C. (1849) *Narrative of the United States Exploring Expedition*, G. P. Putnam & Co., Philadelphia.

Womersley, H. (1945) Insect and allied pests of the home. *Journal of the Department of Agriculture of South Australia*, **48** (11), 439–440.

Woodward, T.E., Evans, J.W. and Eastop, V.F. (1970) *The Insects of Australia*, Melbourne University Press, Melbourne.

8

The Bed Bug Resurgence in Africa

Josephus Fourie and Dionne Crafford

8.1 Introduction

Africa has the earliest recorded evidence of bed bugs from anywhere in the world, with *Cimex* sp. specimens being found from the time of the pharoahs. While bed bug infestations declined in some countries after World War II, this was not uniform across Africa, with a number of tribal villages experiencing intense infestations of both the Tropical bed bug, *Cimex hemipterus* (F.) and the Common bed bug, *Cimex lectularius* L., during the 1970s and 1980s. Some of the earliest reports of insecticide resistance in bed bugs have been from Africa and appeared to develop in association with insecticide applications during malaria-control programs. To date, no coordinated strategies have been implemented in response to the bed bug resurgence. This chapter reviews the history of bed bugs in Africa and discusses the factors contributing to the modern resurgence in the region.

8.2 History of Bed Bugs in Africa

Bed bugs have a long history in Africa. Bed bug (*Cimex* sp.) evidence has been documented in occupation sites dating back to the time of the pharoahs (c. 1350–1323 BCE) in Tell-e-Amarna, Egypt (Panagiotakopulu and Buckland, 1999). These bed bug remains are the earliest records of this insect living in association with humans, and show that bed bugs have been in Africa for at least 3365 years. Panagiotakopulu and Buckland (1999) state that the species was *C. lectularius*, although from the published images, the possibility of the specimens being *C. hemipterus* can not be excluded (pronotum is not included in the photographs).

In a more recent review of bed bug history, Potter (2011) mentioned how in the early 1900s the cork linings of soldiers' helmets during the East African campaign of World War I were infested with bed bugs. Bed bugs continued to be a problem in Africa even after the widespread use of insecticides began. In the 1950s, Gratz (1959) reported dieldrin resistance in *C. hemipterus* populations in Zanzibar. Other publications focus on bed bug resistance to organochlorine (OC) pesticides. Localities in Africa where resistance to OC pesticides was recorded in populations of *C. lectularius* included Kenya (Mombasa), Somalia (Mogadishu), Gambia (Bathurst) (Busvine, 1958), Zimbabwe (Kariba) (Reid, 1960), South Africa (Johannesburg area) (Whitehead, 1962), Congo, east Africa and west Africa (World Health Organization, 1963), Egypt and Somalia (World Health Organization, 1970). Cross-resistance between OC insecticides and pyrethrins was confirmed from field-collected bed bug samples by the World Health Organization (1963); a 10-fold increase in pyrethrin tolerance was recorded in a population of *C. hemipterus* in Mombasa, Kenya that was known to be resistant to DDT and dieldrin.

Advances in the Biology and Management of Modern Bed Bugs, First Edition.
Edited by Stephen L. Doggett, Dini M. Miller, and Chow-Yang Lee.
© 2018 John Wiley & Sons Ltd. Published 2018 by John Wiley & Sons Ltd.

The continued incidence of bed bugs in African society resulted in the mention of both *C. lectularius* in South Africa (Marshall and Heyl, 1963) and *C. hemipterus* in Zimbabwe, in a publication reviewing the impacts of human parasitic diseases (Goldsmid, 1978). Between the late 1970s and early 1980s, the potential of *C. lectularius* as potential vector of the hepatitis B virus also sparked research interest in South Africa (see, for example, Jupp et al., 1978; Jupp and McElligott, 1979; Jupp et al., 1983) and Zimbabwe (Taylor and Morrison, 1980), suggesting that prevalence of this species was high enough to warrant concern.

While newspaper articles investigated the presence of bed bugs in urban South African prisons in the 1970s (Ancer, 2004), Newberry and Jansen (1986) investigated the size and life-stage composition of *C. lectularius* in traditional rural Zulu mud huts in KwaZulu-Natal (KZN), South Africa. Newberry et al. (1987) indicated that *C. hemipterus* also occured in KZN, and was in fact even more prevalent than *C. lectularius* in the far northern region. Newberry et al. (1987) further intimated the possibility that *C. hemipterus* may be displacing *C. lectularius*, and invading South Africa from the more tropical Mozambique region to the north of KZN. Newberry and Mchunu (1989) suggested that *C. hemipterus* prevalence appeared to be increasing, when compared to *C. lectularius*, in the KZN region adjacent to the Mozambique border. However, Newberry (1990) found that the average degree of infestation of *C. hemipterus* was the same as that of *C. lectularius* in traditional Zulu huts in KZN. The Newberry (1990) study indicated that large populations of *C. hemipterus* could survive out of the tropics, but KZN was considered the southernmost extent of its range. Newberry et al. (1991) later investigated reinfestation rates in these traditional Zulu huts, and found that reinfestation (species not specified) from local sources was estimated at 35% of the huts per year, of which about two-thirds were from local sources. It must be emphasised that the research by Newberry and colleagues was focused on traditional, rural Zulu settlements in South Africa, with comparatively fewer published accounts from other parts of Africa.

8.3 Laws, Regulations, and Policies for Bed Bug Control Pre-resurgence

There appear to be no published accounts of laws, regulations or policies pertaining to control of bed bugs in African countries. In Nigeria, Omudu and Kuse (2010) stated that there have never been organized, comprehensive bed bug control efforts. Consequently, individuals living in infested buildings took responsibility for bed bug control themselves, although control efforts were often inefficiently implemented (Nalwanga and Ssempebwa, 2011). An online survey was conducted by Clintest (Bloemfontein, South Africa) in collaboration with Pathogen and Environmental Solutions (PES) Africa (Brackenfell, South Africa) during June 2016. The survey, conducted amongst 32 pest control companies in South Africa and Namibia, inquired if the companies were aware of any historical laws or policies pertaining specifically to the control of bed bugs. Of the companies surveyed, 50% indicated that there were no such policies. Of the remaining 50% of respondents, 44% indicated that they were not aware of any such policies, and 6% cited only the general standard operating procedures on pest control, or laws governing the registration and selling of pesticides. Interestingly, Newberry et al. (1991) noted that an extensive bed bug control program was indeed implemented in South Africa (KZN) between 1986 and 1987. This program involved the Department of Health spraying huts in rural settlements with fenitrothion. Based on this data from South Africa, it is possible other more localized control programs could have existed within other locations in Africa, but such programs were not always documented or recorded in the scientific literature.

8.4 Documented Evidence for the Bed Bug Decline

The African bed bug decline resulted indirectly from mosquito and malaria control strategies, rather from control strategies aimed specifically at bed bugs (see, for example, Lindsay et al., 1989). An example of this inadvertent bed bug control was recorded in the village of Mng'aza (Tanzania), which was heavily infested with *C. hemipterus* prior to introduction of pyrethroid-treated bed nets in 1988. The introduction of bed nets resulted in the eradication of bed bugs in one village for 14 years. However in many villages, bed bugs began to reappear within six years, and these populations were found to be highly resistant to the pyrethroids (Myamba *et al.*, 2002). While malaria control programs, in many instances, did result in a decline of bed bug numbers, this decline was not uniform throughout Africa.

8.5 Early Evidence for the Resurgence

The global resurgence of bed bugs commenced in the late 1990s (Kolb *et al.*, 2009; Davies *et al.*, 2012; Doggett *et al.*, 2012). Resistance to insecticides has been considered the major contributing factor to the bed bug resurgence (Romero *et al.*, 2007; Kolb *et al.*, 2009; Davies *et al.*, 2012), and certainly insecticide resistance has been observed in Africa (Kweka *et al.*, 2009; see also references above). Bed bug resistance to dieldrin and DDT emerged in the late 1950s, from localities as diverse as West Africa (Holstein, 1959), Zanzibar (Gratz, 1959), and Libya (Shalaby, 1970). Resistance to pyrethroids was identified in Tanzanian villages in the early 2000s, after pyrethroid-treated bed nets for mosquito control were introduced between 1995 and 1996. However, as mentioned above, cross-resistance between the OCs (such as DDT) and pyrethroid insecticides may have already have been present in bed bug populations before modern reports of resistance were documented in Africa. In fact in 1984, Newberry *et al.* (1984) noticed a correlation between Zulu huts in KZN being treated with DDT against malaria and bed bug infestation rates. Those huts that were sprayed regularly had a more severe bed bug problem and perhaps this was due to resistance.

The majority of research papers published after 2001 mentioned that the African bed bug resurgence (Table 8.1) corresponded with both the global bed bug resurgence, and the documentation of pyrethroid insecticide resistance (Davies *et al.*, 2012; also see Chapter 29). Evidence of the bed bug resurgence in Africa has also been documented by the local media (Haw, 2005; Skade, 2009), including reports of bed bug infestations in 4000 homes in Nakuru, Kenya (Damary, 2015).

8.6 The Extent of the Resurgence

Newberry and Jansen (1986) (South Africa) and Temu *et al.* (1999) (Tanzania) published results of general bed bug prevalence surveys. However, a number of surveys (Nigeria, Sierra Leone and Ethiopia) were specifically initiated in response to observed increases in bed bug infestations in the late 1990s (Table 8.1). Prevalence, expressed as the percentage of infested premises, ranged from 9% to 100%. Levels of high bed bug prevalence generally correlated with high human population densities, such as communal housing and refugee camps, with the lowest prevalence reported from apartments and households. Increased prevalence also correlated positively with increasing poverty. Data was also collected from other diverse localities, such as lecture halls and prisons, but the data were not expressed in terms of prevalence (Table 8.2). Considering all the studies represented in Tables 8.1 and 8.2, the total numbers of bed bugs collected per sampling locality type (huts, home, hostel, lecturing hall, prisons, and so on) ranged between 100 and 4906, indicating moderate to severe infestations. However, Omudu and Kuse (2010) demonstrated that the degree of resurgence may be underestimated, as bed bugs can easily escape visual detection (Davies *et al.*, 2012; Vaidyanathan and Feldlaufer, 2013).

The results listed in Table 8.3 are the results of a survey of 32 pest control companies in South Africa and Namibia conducted in 2016, as mentioned above. Pest management companies (PMCs) were asked if they were ever specifically requested to treat bed bug infestations. Thirty-one companies (97%) indicated "yes". Of these companies, 25% were located in Gauteng Province (Johannesburg region of South Africa), 25% in Cape Town (Western Cape Province, South Africa), and 18.8% in Durban (KZN Province, South Africa). All three localities are densely populated and popular tourist destinations. The percentage of companies that reported

Table 8.1 Summary of bed bug surveys conducted in Africa where bed bug prevalence was specifically expressed as percentage of infested sampling localities.

Reference	Country and survey locality	Assessment period	Species (Bed bugs)	Percentage prevalence (additional notes)
Newberry and Jansen (1986)	South Africa, KwaZulu-Natal Province	Regular collections between 1979 and 1982	*C. lectularius*	100% (six out of six huts, total bed bug counts per hut ranged between 404 and 4906)
Temu *et al.* (1999)	Tanzania, Bagomoyo villages	1993, prior to bed net installations	*C. hemipterus*	56.5% (61 out of 108 houses)
Gbakima *et al.* (2002)	Sierra Leone, Freetown, camps for internally displaced persons	Summer of 1999	*C. lectularius* (56.1%) and *C. hemipterus* (43.9%)	98% (233 out of 238 rooms, total of 584 specimens)
Okwa and Omoniyi (2010)	Nigeria, Lagos state, five sub-urban areas	July 2008 to December 2008	*C. hemipterus*	9% (18 out of 200 households)
Omudu and Kuse (2010)	Nigeria, Benue state, Gbajimba town	January 2010 to March 2010	*C. lectularius*	21.8% (26 out of 119 apartments) Signs of bed bug infestation (egg cases, fecal marks) sighted in 62.2% of apartments (74 out of 119)
Emmanuel *et al.* (2014)	Nigeria, Benue state, Gboko town, 9 hostels and 600 homes.	January 2011 to April 2011	*C. lectularius*	14.3% (86 out of 600 homes, 705 specimens) 66.7% (6 out of 9 hostels, 1937 specimens) 86% of hostel inhabitants examined presented with bed bug bites.
Karunamoorthi *et al.* (2015)	Ethiopia, Amuru town, 260 households.	January 2014 to May 2014	Not specified	72.7% (189 out of 260 houses)

Table 8.2 Summary of studies conducted on bed bugs in Africa where occurrence is reported on, but not necessarily in terms of prevalence.

Reference	Country and survey locality	Assessment period	Species (Bed bugs)	Notes on occurence
Aigbodion and Megbuwe (2008)	Nigeria, Benin City, lecture theater University of Benin	August 2005 to September 2005	*C. hemipterus*	Total of 157 bed bugs collected
Angelakis *et al.* (2013)	Rwanda, Miyove and Muhanga prisons	January 2011 to May 2011	*C. hemipterus*	Overall 100 bed bugs were tested for *Bartonella quintana* (Schmincke). It is not clear if more than 100 specimens were collected.

Table 8.3 Percentage of enquiries received per category for years indicated, as obtained from a survey completed by 32 pest management companies or individuals in South Africa and Namibia during June 2016.

Category in terms of number of bed bug related inquiries	Year				
	1980–1995[1]	2000	2005	2010	2015
0	0	3.1	6.3	3.1	0
1 to 5	0	0	3.1	18.8	25.0
6 to 10	0	0	6.3	6.3	21.9
11 to 20	0	0	3.1	6.3	12.5
21 to 30	0	0	0	3.1	12.5
No response	100	96.9	81.2	62.4	28.1

[1] Options provided in five-year intervals from 1980. No responses for period from 1980 to 1995.

treating bed bugs in other localities (three provinces in South Africa and two regions in Namibia) ranged between 3.1% and 12.5%. Whilst the survey is not necessarily representative of the geographic distribution of either bed bug infestations or PMCs in South Africa and Namibia, it goes to show that the degree of the resurgence may be quite extensive geographically.

When survey respondents were asked the percentage of their inquiries that addressed bed bugs, 72% stated that less than 25% of inquiries were related to bed bugs, 6% stated that between 25% and 50% of inquiries related to bed bugs, and 22% failed to respond. The former responses were from Cape Town (South Africa) and Windhoek (Namibia). While the geographic extent of the resurgence may be widespread, it would appear that the severity, at least in South Africa, may still be considered moderate. PMCs were also asked to indicate, how may bed-bug-related enquiries they received in five-year intervals, starting from 1980 (Table 8.3). No responses were indicated for the period from 1980 to 1995.

The above survey documented a temporal increase in bed bug inquires from 2000 through 2015 (excluding the "no response" category). While the increase in the number of bed bug inquiries does not accurately reflect the extent of the bed bug resurgence, it does demonstrate that 71.9% of the companies participating in the survey treated at least one case of bed bugs during 2015.

8.7 Region- and Country-specific Reasons for the Resurgence

While insecticide resistance has been implicated as a contributory factor to the bed bug resurgence in Africa, it cannot be considered primary. Both *C. hemipterus* and *C. lectularius* occur in Africa (Table 8.1), and in spite of the frequent use of pyrethroids in bed nets and as spray formulations insecticides, pyrethroids are still considered effective for control of bed bugs in areas of Tanzania and other locations (Kweka *et al.*, 2009). Gratz (1959) stated that bed bug population variations in different locations cannot be accounted for by resistance alone. In the recent survey, only one participant, from Cape Town (South Africa), specifically commented that they encountered pyrethroid-resistant bed bug populations. While the development of resistance is a reality (Zhu *et al.*, 2013), the socio-economic conditions of human victims and their "do-it yourself" control practices have emerged as important concerns. Human behavior has resulted in greater bed bug dispersal, in that increased human population densities in concentrated locations and the high human turnover in hotels, as well as multi-tenant housing facilities, have contributed to the bed bug population increase. Other factors that have favored bed bug proliferation include poor housekeeping practices, a lack of proper health education, the trading of second-hand furniture, and inadequate knowledge of effective control practices (Kweka, et al. 2009; Okwa and Omoniyi, 2010; Omudu and Kuse, 2010; Angelakis et al., 2013; Emmanuel et al., 2014). Note that these factors are not unique to Africa, but have also been implicated in the UK, the USA, and elsewhere (Boase, 2008).

8.8 Strategies to Combat the Resurgence and Relative Success

There appear to be no published accounts of organised strategies to combat the resurgence in African countries (see, for example, Omudu and Kuse, 2010). At present, many individuals living in infested buildings are taking on the responsibility for bed bug control themselves (Nalwanga and Ssempebwa, 2011). The 2016 survey participants were asked if there were any current laws or policies in their regions pertaining specifically to the control of bed bugs. Fifty-six percent of respondents indicated that there were no such policies, while 38% indicated that they were not aware of any such policies. Six percent of respondents cited general standard operating procedures for pest control or laws governing the registration and selling of pesticides in general.

These responses were very similar to those received when asked the same question regarding historical laws or policies. But, as could be expected, more of the participants were confident in giving a definitive answer ("no" as opposed to "unknown") with reference to the current status of laws and policies. To date, no government has undertaken any proactive measures to limit bed bug spread, nor has any major industry organization developed any guidelines (or standards) for bed bug management.

Interestingly, publications addressing the bed bug surveys conducted after the 1990s' resurgence, provide some recommendations as to what strategies need to be put in place. Multiple authors agree that an integrated management approach is required (Table 8.4). Specifically, current control measures need to be employed within a framework of community empowerment and training initiatives. The integrated pest management approach specifies that education and training (Davies et al., 2012) must be combined, including monitoring and refinement of effective pest methods that are based on research and cooperative control efforts (Koganemaru and Miller, 2013). Whilst there are published accounts of such recommendations, there appear to be no documented accounts that record if such measures were implemented and were successful.

Table 8.4 Specific bed bug recommendations proposed for African countries.

Additional measures recommended	Country	Reference
Additional training to make residents aware of the advantages of effectively using impregnated nets, also to control other pests such as cockroaches and head lice.	Tanzania	Temu et al. (1999)
Health education, good personal hygiene, avoiding overcrowding, and good housekeeping.	Nigeria	Okwa and Omoniyi (2010)
Integrated approach recommended for management involving community participation in inspection, detection, and education. Processes may include physical removal and exclusion as well as pesticide application.	Nigeria	Omudu and Kuse (2010)
Proactive public education and awareness campaigns. Government and non-governmental organizations should be involved in training offered and tailored to relevant parties.	Nigeria	Emmanuel et al. (2014)
Community-based awareness campaigns, implementation of sustainable preventive/containment strategies, educational interventions to ensure translation of knowledge into practices, and the implementation of appropriate poverty alleviation programs.	Ethiopia	Karunamoorthi et al. (2015)

References

Aigbodion, F.I. and Megbuwe, P. (2008) Resting site preferences of *Cimex hemipterus* (Heteroptera ; Cimicidae) in human dwelling in Benin City, Nigeria. *International Journal of Biomedical and Health Sciences*, **4** (1), 19–23.

Ancer, J. (2004) Terrible stuff has now been replaced by great things. *Star* (17 March), p. 3.

Angelakis, E., Socolovschi, C. and Raoult, D. (2013) *Bartonella quintana* in *Cimex hemipterus*, Rwanda. *American Journal of Tropical Medicine and Hygiene*, **89** (5), 986–987.

Boase, C. (2008) Bed bugs (Hemiptera: Cimicidae): An evidence-based analysis of the current situation. In: *Proceedings of the Sixth International Conference on Urban Pests, July 13–16, 2008, Budapest, Hungary.* OOK-Press, Hungary, Veszprém.

Busvine, J.R. (1958) Insecticide-resistance in bed-bugs. *Bulletin of the World Health Organization*, **19** (6), 1041–52.

Damary, R. (2015) *Kenya: Nakuru Loses Sleep as the Bedbugs Bite*, http://allafrica.com/stories/201509091800.html (accessed 25 July 2016).

Davies, T.G.E., Field, L.M. and Williamson, M.S. (2012) The re-emergence of the bed bug as a nuisance pest: implications of resistance to the pyrethroid insecticides. *Medical and Veterinary Entomology*, **26** (3), 241–254.

Doggett, S.L., Dwyer, D.E., Peñas, P.F. and Russell, C. (2012) Bed bugs: clinical relevance and control options. *Clinical Microbiology Reviews*, **25** (1), 164–192.

Emmanuel, O.I., Cyprian, A. and Agbo, O.E. (2014) A survey of bedbug (*Cimex lectularius*) infestation in some homes and hostels in Gboko, Benue State, Nigeria. *Psyche: A Journal of Entomology*, **2014**, 1–5.

Gbakima, A.A., Terry, B.C., Kanja, F., Kortequee, S., Dukuley, I. and Sahr, F. (2002) High prevalence of bedbugs *Cimex hemipterus* and *Cimex lectularis* [sic] in camps for internally displaced persons in Freetown, Sierra Leone: a pilot humanitarian investigation. *West African Journal of Medicine*, **21** (4), 268–271.

Goldsmid, J. (1978) A review of the importance of human parasitic diseases in Rhodesia. *The Central African Journal of Medicine*, **24** (9), 181–187.

Gratz, N.G. (1959) Insecticide-resistance in bed-bugs and flies in Zanzibar. *Bulletin of the World Health Organization*, **20**, 668–670.

Haw, P. (2005) Tiny bugs pose big threat. *Business Day* (31 May), p. 26.

Holstein, M.H. (1959) Resistance to dieldrin of *Cimex hemipterus* Fab, in Dahomey, Occidental Africa. *Bulletin de la Societe de Pathologie Exotique et de Ses Filiales*, **52**, 664–668.

Jupp, P.G. and McElligott, S.E. (1979) Transmission experiments with hepatitis B surface antigen and the common bedbug (*Cimex lectularius* L). *South African Medical Journal*, **56** (2), 54–57.

Jupp, P.G., Prozesky, O.W., McElligott, S.E. and Van Wyk, L.A. (1978) Infection of the common bedbug (*Cimex lectularius* L.) with hepatitis B virus in South Africa. *South African Medical Journal*, **53** (15), 598–600.

Jupp, P.G., McElligott, S.E. and Lecatsas, G. (1983) The mechanical transmission of hepatitis B virus by the common bedbug (*Cimex lectularius* L.) in South Africa. *South African Medical Journal*, **63** (3), 77–81.

Karunamoorthi, K., Beyene, B. and Ambelu, A. (2015) Prevalence, knowledge and self-reported containment practices about bedbugs in the resource-limited setting of Ethiopia: a descriptive cross-sectional survey. *Health*, **7** (9), 1142–1157.

Koganemaru, R. and Miller, D.M. (2013). The bed bug problem: past, present, and future control methods. *Pesticide Biochemistry and Physiology*, **106** (3), 177–189.

Kolb, A., Needham, G.R., Neyman, K.M. and High, W.A. (2009) Bedbugs. *Dermatologic Therapy*, **22** (4), 347–352.

Kweka, E.J., Mwang'onde, B.J., Kimaro, E.E., et al. (2009) Insecticides susceptibility status of the bedbugs (*Cimex lectularius*) in a rural area of Magugu, Northern Tanzania. *Journal of Global Infectious Diseases*, **1** (2), 102–106.

Lindsay, S.W., Snow, R.W., Armstrong, J.R.M. and Greenwood, B.M. (1989) Permethrin-impregnated bednets reduce nuisance arthropods in Gambian houses. *Medical and Veterinary Entomology*, **3** (4), 377–383.

Marshall, J. and Heyl, T. (1963). Skin disease in the Western Cape province. *South African Medical Journal*, **37** (3), 1308–1310.

Myamba, J., Maxwell, C.A., Asidi, A. and Curtis, C.F. (2002) Pyrethroid resistance in tropical bedbugs, *Cimex hemipterus*, associated with use of treated bednets. *Medical and Veterinary Entomology*, **16** (4), 448–51.

Nalwanga, E. and Ssempebwa, J.C. (2011) Knowledge and practices of in-home pesticide use: a community survey in Uganda. *Journal of Environmental and Public Health*, **2011**, 230894.

Newberry, K. (1990) The tropical bedbug *Cimex hemipterus* near the southernmost extent of its range. *Transactions of the Royal Society of Tropical Medicine and Hygiene*, **84** (5), 745–747.

Newberry, K. and Jansen, E.J. (1986) The common bedbug *Cimex lectularius* in African huts. *Transactions of the Royal Society of Tropical Medicine and Hygiene*, **80** (4), 653–658.

Newberry, K. and Mchunu, Z.M. (1989) Changes in the relative frequency of occurrence of infestations of two sympatric species of bedbug in northern Natal and KwaZulu, South Africa. *Transactions of the Royal Society of Tropical Medicine and Hygiene*, **83** (2), 262–264.

Newberry, K., Jansen, E.J. and Quann, A.G. (1984) Bedbug infestation and intradomiciliary spraying of residual insecticide in Kwazulu, South Africa. *South African Journal of Science*, **80** (8), 377.

Newberry, K., Jansen, E.J. and Thibaud, G.R. (1987) The occurrence of the bedbugs *Cimex hemipterus* and *Cimex lectularius* in northern Natal and KwaZulu, South Africa. *Transactions of the Royal Society of Tropical Medicine and Hygiene*, **81** (3), 431–433.

Newberry, K., Mchunu, Z.M. and Cebekhulu, S.Q. (1991) Bedbug reinfestation rates in rural Africa. *Medical and Veterinary Entomology*, **5** (4), 503–505.

Okwa, O.O. and Omoniyi, O.A.O. (2010) The prevalence of head lice (*Pediculus humanus capitus*) and bed bugs (*Cimex hemipterus*) in selected human settlement areas in Southwest, Lagos state, Nigeria. *Journal of Parasitology and Vector Biology*, **2** (2), 8–13.

Omudu, E.A. and Kuse, C.N. (2010) Bedbug infestation and its control practices in Gbajimba: a rural settlement in Benue state, Nigeria. *Journal of Vector Borne Diseases*, **47** (4), 222–227.

Panagiotakopulu, E. and Buckland, E. (1999) *Cimex lectularius* L. the common bed bug from Pharaonic Egypt. *Atiquity*, **73** (282), 908–911.

Potter, M.F. (2011) The history of bed bug management – with lessons from the past. *American Entomologist*, **57** (1), 14–25.

Reid, E. (1960) Insecticide resistance. *The Central African Journal of Medicine*, **6** (12), 528–534.

Romero, A., Potter, M.F., Potter, D.A. and Haynes, K.F. (2007) Insecticide resistance in the bed bug: a factor in the pest's sudden resurgence? *Journal of Medical Entomology*, **44** (2), 175–178.

Shalaby, A.M. (1970) Insecticide susceptibility of the bedbug *Cimex lectularius* (Hemiptera; Cimicidae), in Libya. *Journal of the Egyptian Public Health Association*, **45** (6), 485–499.

Skade, T. (2009) Bed bugs are Creeping back into SA homes – World Cup could bring more than just tourists. *Star* (3 April), p. 10.

Taylor, P. and Morrison, J. (1980) *Cimex lectularius* as a vector of hepatitis B. *The Central African Journal of Medicine*, **26** (9), 198–200.

Temu, E.A., Minjas, J.N., Shiff, C.J. and Majala, A. (1999) Bedbug control by permethrin-impregnated bednets in Tanzania. *Medical and Veterinary Entomology*, **13** (4), 457–459.

Vaidyanathan, R. and Feldlaufer, M.F. (2013) Bed bug detection: current technologies and future directions. *American Journal of Tropical Medicine and Hygiene*, **88** (4), 619–625.

Whitehead, G. (1962) A study of insecticide resistance in a population of bed bugs, *Cimex lectularius* L., and a method of assessing effectiveness of control measures in houses. *Journal of the Entomological Society of Southern Africa*, **25** (1), 121.

World Health Organization (1970) *Insecticide Resistance and Vector Control – Seventeenth Report of the WHO Expert Committee on Insecticides*, 443rd edn., Geneva, World Health Organization.

World Health Organization (1963) *Insecticide Resistance and Vector Control – Thirteenth Report of the WHO Expert Committee on Insecticides*. World Health Organization. Geneva.

Zhu, F., Gujar, H., Gordon, J.R., *et al.* (2013). Bed bugs evolved unique adaptive strategy to resist pyrethroid insecticides. *Scientific Reports*, **3**, 1456.

9

The Bed Bug Resurgence in the Indian Subcontinent

Anil S. Rao and Joshua A. Rao

9.1 Introduction

India and the subcontinent is one of the few regions in the world where both the Tropical bed bug *Cimex hemipterus* (F.) and the Common bed bug, *Cimex lectularius* L. are endemic. During the time of the British colonial rule in the early 20th century, considerable research was undertaken on the potential of bed bugs to transmit infectious diseases. Malaria control programs initiated in the 1950s utilizing DDT appeared to have brought about a concomitant reduction in bed bug populations. However, this also led to the development of insecticide resistance, with reports of resistance from the mid-1950s. Anecdotal evidence has noted an increase in bed bug infestations in urban areas from around 2006. To date, no coordinated national initiative has been undertaken in response to the resurgence. This chapter reviews the history of bed bugs in India and the subcontinent, and reviews the various factors contributing to the modern resurgence.

9.2 History of Bed Bugs in the Indian Subcontinent

Cimex hemipterus, also referred to locally as the Indian or sub-tropical bed bug, is distributed widely across India, particularly in the plains (Wattal and Kalra, 1961). Identified erroneously as *C. lectularius* at the beginning of the 20th century (Girault, 1907), *C. hemipterus* is the most widespread bed bug species in India (Kalra and Krishnamurthy, 1965). It has been recorded from Assam in the north-east (Patton, 1907) to Madras Presidency (Tamil Nadu) in the south (Girault, 1907). Bhat (1974), who reviewed the Indian Cimicidae, reported its occurrence in Uttar Pradesh, West Bengal, Sikkim, Himachal Pradesh, Jammu and Kashmir, and Karnataka. *Cimex hemipterus* has also been reported from bats (Patton, 1908; Kunhikannan, 1912) and the common swift (Distant, 1910) in India.

Cimex lectularius, the cosmopolitan bed bug of temperate regions (Usinger, 1966), is the most prevalent bed bug species in hilly regions of India (Wattal and Kalra, 1961; Bhat, 1974). It has also been reported from Hyderabad (Venkatachalam and Belavady, 1962) in the northern part of the Deccan plateau, while Burton (1962) reported its presence and co-existence with *C. hemipterus* in the coastal district of Alappuzha in Kerala, south India. Although Siddiqui and Raja (2015) referred to the bed bug they were working on in the Vidharbha region as *C. lectularius*, it is evident from the image in the paper that it was *C. hemipterus* (Stephen L. Doggett, Westmead Hospital, NSW, Australia, pers. comm.). This raises the question of the possibility of other Indian researchers misidentifying *C. hemipterus*, highlighting the need to accurately identify bed bugs across the country to categorize the current resurgence.

In addition to *C. hemipterus* and *C. lectularius*, eight more species of Cimicidae have been recorded in India, all from bats. These include *Cimex himalayanus* Bhat, *Cimex insuetus* Ueshima, *Cimex usingeri* Bhat

---

Advances in the Biology and Management of Modern Bed Bugs, First Edition.
Edited by Stephen L. Doggett, Dini M. Miller, and Chow-Yang Lee.
© 2018 John Wiley & Sons Ltd. Published 2018 by John Wiley & Sons Ltd.

et al., *Cimex pipistrelli* Jenyns, *Cacodmus indicus* Jordan & Rothschild, *Aphrania vishnou* (Mathur), *Stricticimex pattoni* (Horvath), and *Leptocimex inordinatus* Ueshima (Bhat, 1974).

The possibility of transmission by bed bugs of various pathogens that cause human disease prevalent in the Indian subcontinent was studied by several authors during the first half of the 20th century. The diseases investigated include plague (Cornwall and Menon, 1917), leishmaniasis or kala-azar (Girault, 1907), and filariasis (Burton, 1962), but all the evidence was either negative or inconclusive (Burton, 1963). However, the role of bed bugs in causing iron deficiency anemia in infants and young children (Venkatachalam and Belavady, 1962), loss of vitality due to deprivation of sleep (Mishra, 1980), and skin conditions (Mendiki *et al.*, 2014) have clearly been established. Bed bug infestations were considered a source of great nuisance to the Indian armed forces, resulting in general deterioration of health and lowered efficiency (Menon *et al.*, 1958).

Reports citing utilization of bed bugs for treating common ailments such as plague, malaria, epilepsy, piles, alopecia, urinary disorders, guinea worms, and even snakebites by traditional healers indicate that they were long present in rural communities in the Indian subcontinent (Sharma, 1990; Oudhia, 1995; Singh *et al.*, 1998; Sajem Betlu, 2013; Verma *et al.*, 2014). It is also interesting to note that such practices were not restricted to a particular community or region but were widely adopted by tribal populations from Manipur (Meetel community) and Assam (Biate tribe) in the north-east to Chhattisgarh (traditional healers in six districts) in central India and Rajasthan (Bhil tribe) in the north-west.

9.3 Laws to Control Bed Bugs Pre-resurgence

There is no documented evidence to indicate that bed bugs were considered a serious pest in India warranting the government to pass legislation, form policies or issue directives for their management.

9.4 Documented Evidence for the Bed Bug Decline

Indoor residual treatment carried out under the five year National Malaria Control Programme (NMCP) initiated in 1953, and the National Malaria Eradication Programme (NMEP) that followed (Rao, 1984), is reported to have brought about a concurrent reduction in bed bug populations (Wattal and Kalra, 1961). NMCP was undertaken by the government of India in co-operation with all state governments and the US Technical Co-operation Mission with the objective of protecting the population in the entire malarial area of the country. Low endemic and potentially malarial areas were also covered under NMEP, the program operated in collaboration with the United States Agency for International Development, the WHO and UNICEF. Virtually the entire country was thus covered, with the exception of urban areas having populations in excess of 40 000 (Rao, 1984). DDT was primarily used under the NMCP, while BHC, dieldrin, malathion, and fenitrothion were also used under NMEP (Rao, 1984), and a population of 165 million was covered by the year 1958 (Raghavendra and Subbarao, 2002).

Pesticide application was carried out over all inner walls, roofs and under eaves, doors, verandas, rafters and beams of all buildings, large or small, in which people slept or gathered at night, the undersides of furniture, and the backs of cupboards and wardrobes. Sprayable structures also included healthcare facilities, hotels, rest houses, and fishing and field huts (WHO, 2015). Considering the fact that insecticidal treatment against mosquitoes covered all areas where bed bugs are also generally located, it is not surprising that pesticide treatment turned out to be equally effective against bed bugs, as reported by Wattal and Kalra (1961).

9.5 The Modern Resurgence

The continuous use of synthetic insecticides, particularly DDT and BHC, under the NMEP is reported to have resulted in development of resistant populations amongst many mosquito vector species (Pal *et al.*, 1952; Singh *et al.*, 2014) as well as bed bugs. Development of resistance to DDT in *C. hemipterus* was first reported by Halgeri and Rao (1956) from Pune. Studies carried out at the Armed Forces Medical College, Pune, corroborated these studies and showed that *C. hemipterus* had developed a high degree of resistance to DDT and that the LC$_{50}$ could be obtained only with a concentration of 30–70% DDT (Menon *et al.*, 1958). Further tests showed that exposure of more than 5 h to 5% DDT dust was required to kill 50% of the bed bugs, whereas bed bugs in unsprayed areas required less than 10 min at the same concentration to give a 50% kill. In the case of BHC, a 96% kill of *C. hemipterus* could be obtained from the cantonment area by exposure to a higher dose of 10% dust for 10 min. In field trials, three barracks occupied by troops were separately treated with DDT, BHC, and diazinon, and an untreated barrack room served as a control. DDT and BHC were not observed to reduce the density of *C. hemipterus*, but a sudden drop was noticed in diazinon-treated barracks, with over 90% mortality. This indicated that at the time, resistance was high to the organochlorines but not to the organophosphates.

Wattal and Kalra (1960) reported that exposure of *C. hemipterus* for 10 min to 5% DDT did not cause any mortality in regularly sprayed areas in Pune in 1955, as compared to 66.6% mortality in a population 129 km away, while 50% mortality was obtained within 2 min at Pune in 1952. The same authors also reported an increase in *C. hemipterus* populations compared to that prevalent in the pre-DDT era, which could perhaps be considered the first report of bed bug resurgence.

Subsequently, both *C. hemipterus* and *C. lectularius* were reported to have developed resistance to DDT and dieldrin in other parts of the country (Kalra and Krishnamurthy, 1965; Shetty *et al.*, 1965). *Cimex hemipterus* even in the interior of Delhi, where no organized regular insecticidal spraying was carried out, were found to be resistant to DDT, dieldrin, and gamma BHC, and had increased tolerance to pyrethrins (Kalra and Krishnamurthy, 1965). The authors attributed the high degree of resistance to individual use of insecticides by householders and inadvertent transportation from sprayed areas, where selection had taken place.

In a study sponsored by the Indian Armed Forces in 1968 at different stations across the country, *C. hemipterus* from Cochin, Trivandrum, Bangalore, and Pune were found to be resistant to malathion, while those from Madras and Vishakhapatnam were susceptible (Kochar and Dixit, 1972). Bandopadhyaya and Sheigaonkar (1976) found that *C. hemipterus* was resistant to DDT in three locations in the north-east (Binaguri, Hashimara, and Gangtok) and was tolerant at Siliguri. The species had also developed intermediate resistance to malathion at Binaguri and Gangtok, but was susceptible at Siliguri and Hashimara. Although systematic spraying of barracks in the armed forces was conducted for debugging with malathion and propoxur, *C. hemipterus* infestations continued to be a major problem. A field study conducted at the Armed Forces Medical College, Pune, revealed that malathion and dichlorvos were ineffective for the control of *C. hemipterus* (Varma and Gupta, 1983a,b).

Cimex hemipterus was also reported to have developed resistance to Malathion (Kochar and Dixit, 1972), propoxur, and dichlorvos (Dhadwal, 2009). Studies by Kalra and Krishnamurthy (1965) showed that DDT- and dieldrin-resistant *C. hemipterus* demonstrated a comparative increase in tolerance to pyrethrins, but not to malathion. Among the organophosphates tested, susceptibility of bed bugs was highest to fenthion and lowest to trithion, while susceptibility to diazinon, folithion, and parathion was more or less of the same order.

It is clearly evident that the bed bug resurgence started during late 1950s and 1960s in villages and towns across the Indian subcontinent, much earlier than in Europe, USA, and Australia. It may therefore be safe to assume that resurgent populations of bed bugs that had already developed resistance spread to army cantonments, towns, and cities with personal belongings, when people joined the security forces or travelled for work.

9.6 Strategies to Combat the Bed Bug Resurgence

Bed bug infestations became more prevalent from late 1990s in urban and peri-urban areas, affecting public transport systems and other public places such as hospitals, the hospitality sector, school and college hostels, and industries employing contract labor. Mendiki *et al.* (2014) cited newspaper articles reporting bed bug infestations in railway coaches at Mangalore and Mumbai, while Bandyopadhyay *et al.* (2015) reported a serious bed bug infestation in a neonatal intensive care unit attached to the state-run Lady Hardinge Medical College, New Delhi. Bed bug outbreaks were also reported from student residential halls of the University of Dhaka (Khan and Rahman, 2012) and a tertiary public care hospital at a metropolitan city in Pakistan (Hussain *et al.*, 2014).

As per information available with the operations team of Pest Control (India) Pvt Ltd (PCI), only limited calls were received for bed bug control until the early 2000s, particularly from metropolitan cities. A wide spectrum of oil- and water-based contact insecticides were used for treatment at the time. With increasing awareness amongst the public about the side effects of chemical pesticides and the advent of gel baits and physical trapping methods of treatment of German cockroaches, only minimal quantities of chemical pesticides were used against household pests. This, combined with the increased use of plywood furniture, which provided suitable habitats, is likely to have contributed to the resurgence of bed bugs in cities. Demand for bed bug services from high-income and upper-middle-class customers, starting in 2006, indicated an upsurge in infestations, but systematic data is not available to support this.

Although Indian studies on insecticide resistance in bed bugs were initiated from the mid-1950s, concerted efforts to investigate the underlying mechanisms imparting resistance were not attempted. Elevated esterase and malathion carboxylesterase mechanisms were reported to be present in bed bug populations in Sri Lanka, and high tolerance to both DDT and pyrethroids suggested the presence of a knockdown (*kdr*) type resistance mechanism (Karunaratne *et al.*, 2007). A recent study by Dang *et al.* (2015) detected two pointed mutations, M9181 and L1014F, which were implicated as *kdr* mutations contributing resistance to the pyrethroids, in a population of *C. hemipterus* from Bangalore. Studies on resistance are warranted to indicate the most appropriate insecticides for effective bed bug management.

Insecticide-based treatments that rely on pyrethroids, organophosphates, and carbamates give highly varied results, making elimination of bed bug populations very difficult. The situation calls for the adoption of an integrated strategy involving use of chemicals in conjunction with alternate methods of control to prevent bed bug infestations from spiraling out of control. Proper tools for the early detection of infestations and knowledge of biology and ecology also play an important role. Some pest management companies such as PCI have taken the initiative in introducing non-chemical strategies such as steam, heat, controlled atmosphere treatment, and desiccant dusts for bed bug management.

References

Bandopadhyaya, D. and Sheigaonkar, V.I. (1976) A preliminary report on evaluation of susceptibility status of arthropods of medical importance in North East Assam and Kameng Frontier division of Arunachal Pradesh. *Armed Forces Medical Journal (India)*, **32**, 600–612.

Bandyopadhyay, T., Kumar, A. and Saili, A. (2015) Bed bug outbreak in a neonatal unit. *Epidemiology and Infection*, **143** (13), 2865–2870.

Bhat, H.R. (1974) A review of Indian Cimicidae (Hemiptera: Heteroptera). *Oriental Insects*, **8** (4), 545–550.

Burton, G.J. (1962) Observations on filarial larvae in bedbugs feeding on human carriers of *Wuchereria bancrofti* and *Brugia malvi* in South India. *American Journal of Tropical Medicine and Hygiene*, **11** (1), 68–75.

Burton, G.J. (1963) Natural and experimental infection of bedbugs with *Wuchereria bancrofti* in British Guiana. *American Journal of Tropical Medicine and Hygiene*, **12** (4), 541–547.

Cornwall, J.W. and Menon, T.K. (1917) On the possibility of the transmission of plague by bed-bugs. *Indian Journal of Medical Research*, **5** (1), 137–159.

Dang, K., Toi, C.S., Lilly, D.G., *et al.* (2015) Identification of putative *kdr* mutations in the tropical bed bug, *Cimex hemipterus* (Hemiptera: Cimicidae). *Pest Management Science*, **71** (7), 1015–1020.

Dhadwal, B.S. (2009) Field trial of various insecticides for the control of bedbugs. *Journal of Communicable Diseases*, **41** (1), 57–60.

Distant, W.L. (1910) *The Fauna of British India, including Ceylon and Burma. Rhynchota, Vol V, Heteroptera*, Taylor and Francis, London.

Girault, A.A. (1907) The Indian bedbug and the Kala Azar disease. *Science*, **25**, 1004.

Halgeri, A.V. and Rao, T.R. (1956) A note on resistance of bed bugs to DDT in Bombay state. *Indian Journal of Malariology*, **10** (2), 149–154.

Hussain, M., Khan, M.S., Wasim, A., *et al.* (2014) Inpatient satisfaction at tertiary care public hospitals of a metropolitan city of Pakistan. *Journal of Pakistan Medical Association*, **64** (12), 1392–97.

Kalra, R.L. and Krishnamurthy, B.S. (1965) Studies on the insecticide susceptibility and control of bed-bugs. *Bulletin of the Indian Society for Malaria and Other Communicable Diseases*, **2** (3), 223–229.

Karunaratne, S.H.P.P., Damayanthi, B.T., Fareena, M.H.J., Imbuldeniya, V. and Hemingway, J. (2007) Insecticide resistance in the tropical bedbug *Cimex hemipterus*. *Pesticide Biochemistry and Physiology*, **88** (1), 102–107.

Khan, H.R. and Rahman, M.M. (2012) Morphology and biology of the bedbug, *Cimex hemipterus* (Hemiptera: Cimicidae) in the laboratory. *Dhaka University Journal of Biological Science*, **21** (2), 125–130.

Kochar, R.K. and Dixit, R.S. (1972) Evaluation of insecticidal susceptibility/resistance of arthropods of medical importance in the field areas and important garrison stations to establish the base. *Armed Forces Medical Research Council, Pune, India, Project No.313/68*.

Kunhikannan, K. (1912) The bed bug (*Cimex rotundus*) on the common yellow bat (*Scotophilus kuhli*). *Journal of Bombay Natural History Society*, **21** (4), 1342.

Mendiki, M.J., Ganesan, K., Parashar, B.D., Sukumaran, D. and Prakash, S. (2014) Aggregation responses of *Cimex hemipterus* F. to semiochemicals identified from their excreta. *Journal of Vector Borne Diseases*, **51** (3), 224–229.

Menon, P.B., Karni, N.D.P. and Narasimhan, D. (1958) Control of bedbugs and the problem of resistance. *Indian Journal of Public Health*, **2** (4), 265–271, 325.

Mishra, R.K. (1980) Sterilization of bed-bug, *Cimex hemipterus* Fabr. with bisazir. *Zeitschrift fuer Angewandte Entomologie*, **89** (3), 247–249.

Oudhia, P. (1995) Traditional knowledge about medicinal insects, mites and spiders in Chhattisgarh, India. *Insect Environment*, **4**, 57–58.

Pal, R., Sharma, M.I.D. and Krishnamurthy, B.S. (1952) Studies on the development of resistance strains of house flies and mosquitoes. *Indian Journal of Malariology*, **6**, 303–324.

Patton, W.S. (1907) Notes on the distribution of the two species bed-bug. *Indian Medical Gazette*, **42**, 41.

Patton, W.S. (1908) *Cimex rotundus* Signoret. *Records of the Indian Museum*, **2**, 153–155.

Raghavendra, K. and Subbarao S.K. (2002). Chemical insecticides in malaria vector control in India. *ICMR Bulletin*, **32** (10), 1–7.

Rao, T.R. (1984) *The Anophelines of India*, revised edn, Malaria Research Centre, Indian Council of Medical Research, Delhi.

Sajem Betlu, A.L. (2013) Indigenous knowledge of zootherapeutic use among the Biate tribe of Dima Hasao District, Assam, Northeastern India. *Journal of Ethnobiology and Ethnomedicine*, **9** (56), 1–15.

Sharma, V.P. (1990) Ethnomedicozoological studies on the invertebrates of Rajasthan. *Uttar Pradesh Journal of Zoology*, **10** (2), 133–136.

Shetty, K.M., Nayar, S.V., Subbiah, V.K. and Panickar, K.K. (1965) A new insecticide for bed bug control. *Indian Journal of Public Health*, **9** (1), 1, 2, 22.

Siddiqui, S.S. and Raja, I.A. (2015) Molecular detection of endosymbiont bacteria *Wolbachia* in bedbug species *Cimex lectularius* from Vidarbha region of India. *International Journal of Life Sciences*, **3** (3), 200–204.

Singh, K.K., Singh, R.K.G. and Anand, L. (1998) Invertebrate animals of ethnomedicozoological importance in Manipur. *Uttar Pradesh Journal of Zoology*, **18** (2), 99–102.

Singh, R.K., Kumar, G. and Mittal, P.K. (2014) Insecticide susceptibility status of malaria vectors in India: A review. *International Journal of Mosquito Research*, **1**, 5–9.

Usinger, R.L. (1966) *Monograph of Cimicidae (Hemiptera – Heteroptera)*, Entomological Society of America, Maryland.

Varma, B.D. and Gupta, D.K.K. (1983a) Use of dichlorvos in bedbug control. *Medical Journal Armed Forces India*, **39** (2), 81–84.

Varma, B.D. and Gupta, D.K.K. (1983b) A comparative evaluation of malaria and fenitrothion in bedbug control. *Journal of Communicable Diseases*, **15** (2), 126–128.

Venkatachalam, P.S. and Belavady, B. (1962) Loss of haemoglobin iron due to excessive biting by bedbugs. A possible aetiological factor in the iron deficiency anemia of infants and adults. *Transactions of Royal Society of Tropical Medicine and Hygiene*, **56** (3), 218–221.

Verma, A.K., Prasad, S.B., Rongpi, T. and Arjun, J. (2014) Traditional healing with animals (zootherapy) by the major ethnic group of Karbi Anglong district of Assam, India. *International Journal of Pharmacy and Pharmaceutical Sciences*, **6** (8), 593–600.

Wattal, B.L. and Kalra, N.L. (1960) *Pyralis pictalis* Curt. (Pyralidae: Lepidoptera) larvae as predators of eggs of bed bug, *Cimex hemipterus* Fab. (Cimicidae: Hemiptera). *Indian Journal of Malariology*, **14**, 77–79.

Wattal, B.L. and Kalra, N.L. (1961) New methods for the maintenance of a laboratory colony of bed-bug, *Cimex hemipterus* Fabricius, with observations on its biology. *Indian Journal of Malariology*, **15** (2), 157–171.

WHO (2015) *Indoor Residual Spraying. An Operational Manual for Indoor Residual Spraying (IRS) for Malaria Transmission Control and Elimination*, 2nd edn. World Health Organization, Geneva. http://apps.who.int/iris/bitstream/10665/177242/1/9789241508940_eng.pdf (accessed 6 March 2016).

10

The Bed Bug Resurgence in the Middle East

Odelon Del Mundo Reyes

10.1 Introduction

The Middle East is considered one of the cradles of civilization and may well have been the first location where the bed bug found its niche amongst humans. The discovery of oil in the region has resulted in a major economic boom, along with an influx of workers and substantial population growth. This has probably aided the spread of bed bugs to the region and, in recent years, a resurgence has been noted. To date there has been no coordinated response to combat the rise in bed bug numbers. This chapter reviews the history of bed bugs in the Middle East and discusses the factors contributing to the modern resurgence in the region.

10.2 History of Bed Bugs in The Middle East

Usinger (1966) noted that bed bugs have been biting humans since the beginning of recorded time. The Fertile Crescent region of the Middle East and areas along the Mediterranean Sea have often been described by experts as the location of origin for the Common bed bug, *Cimex lectularius* L., although the origins of the Tropical bed bug, *Cimex hemipterus* (F.) tend to be less hypothesized.

Some believe that the early bed bug precursors first parasitized cave-dwelling bats, and then the insects also attacked the humans who co-inhabited the caves. The humans subsequently took these insects to other dwellings (Sailer, 1952; Usinger and Povolny, 1966; Potter, 2011). However, others have not discounted the possibility that bats roosting in early homes may have introduced bed bugs, which then adapted to human hosts. It apparently became a lot easier for these ectoparasites to infest humans after the formation of more established dwellings in villages and then later in crowded cities. These bed bugs eventually adapted to living with humans in their dwellings, and as human civilization spread, were carried whenever and wherever people travelled throughout the world (Usinger, 1966). Considered both as a pest and a potion by the ancients, bed bugs have been unearthed from archeological sites in Egypt dating back 3500 years (Panagiotakopulu and Buckland, 1999). Bed bugs are well described in ancient writings during the advancement of ancient Rome and Greek civilization (Potter, 2011) and history shows that these ancient societies shaped and influenced the Middle East.

In more recent times, the discovery of oil in the region, particularly in the Gulf States, has brought tremendous wealth and development in less than 50 years. Modern-westernized cities like Dubai, Abu Dhabi, Doha, Riyadh, Jeddah, Kuwait, Manama, and Muscat have replaced the old mud-brick, palm-branch villages dotting the coastlines of the region. These city centers are continually drawing in trade and commerce, tourism, and a steady influx of migrant workers from around the world who seek to take advantage of opportunities brought about by the economic boom. Ever opportunistic, taking advantage also of the growth of these new cities and

Advances in the Biology and Management of Modern Bed Bugs, First Edition.
Edited by Stephen L. Doggett, Dini M. Miller, and Chow-Yang Lee.
© 2018 John Wiley & Sons Ltd. Published 2018 by John Wiley & Sons Ltd.

the continuous flow of millions of potential hosts, are bed bugs. Similar to the experience of American, European and Australian cities (Boase, 2008; Doggett, 2006; Potter, 2011), the Middle East is also experiencing a bed bug resurgence. In Arabic, bed bugs are known as "buk" and "khatmal", especially to migrant workers from the subcontinent (India, Pakistan, Bangladesh, and Sri Lanka) who form the majority of the labor force in almost every industry.

10.3 Regional Reports of the Bed Bug Resurgence

There are few scientific reports coming out from the region with regards to the bed bug resurgence. The only materials readily available are the information sheets and alerts posted on the web sites of the health ministries and public health pest control departments of each country. These at least confirm the presence of these insects. Numerous anecdotes documenting bed bug encounters have flourished on blog sites and forums, as well as in comments posted on-line in hotel reviews for the region (such as on TripAdvisor.com). In addition, there were bed bug-related news reports emanating from leading Gulf-based newspapers.

Early fauna studies conducted by Patton (1919, 1920) reported that *Cimex* does not occur in Mesopotamia, and that "*Cimex hemiptera*" (meaning *C. hemipterus*) only occurs in the cities of Basra and Baghdad, where it is associated with Indians. After more than sixty years, Kandil (1986) mentioned that in rural families in an Iraqi village, pest organisms such as flies, bed bugs, and fleas were widespread. In a survey conducted between November 1985 and August 1987 by Abul-hab *et al.* (1989), it was reported that *C. lectularius* were collected from 44 locations (hotels, homes, apartment buildings, and government offices) in Baghdad, Iraq. In connection with these, Abul-hab and Shihab (1989) and Abul-hab (1997) reported on the presence of *C. lectularius* as an ectoparasite of bats in Iraq. The US Armed Forces Pest Medical Board (AFPMB) reported that in Baghdad, Iraq, *C. lectularius* had become established in hotels, apartments, and office buildings (AFPMB, 1999). Due to the increase in reported infestations, studies on the susceptibility of *C. lectularius* to deltamethrin, malathion, and dieldrin from a great number of Baghdad hotels, offices, and homes was conducted by Al Barzangi and colleagues (1988).

Rosen *et al.* (1987) reported the occurrence of *C. hemipterus* in poultry barns in Israel and mentioned that these parasites have invaded human habitations as well. More recently, Mumcuoglu (2008) reported his first encounter of *C. lectularius* in Israel after 25 years of practice in diagnosing ectoparasitic infestations.

A questionnaire of pest management professionals (PMPs) was conducted in 2009 by Mumcuoglu and Shalom (2011), who examined the epidemiology of bed bugs in Israel as well as management practices. Over the years 2006–2008, there was a 50–150% increase in the recorded number of bed bug infestations in comparison with the period 2001–2005. The infestations were reported in new and old hotels, new and old residences, new and old prisons, as well as in industrial areas. Infestations were mentioned as being less frequent in youth hostels, store rooms, courthouses, army barracks, and caravans for new immigrants, as well as in day schools. The majority of respondents (42.8%) thought that bed bugs were imported by travelers and, according to 26.7% of those surveyed, the increase was thought to be due to insecticide resistance. Other contributing factors to the resurgences included unawareness of the problem, the prohibition of organophosphates inside houses and workplaces, and the increasing number of foreign workers and new immigrants from developing countries. Later, Siegel-Itzkovich (2014), reporting for the *Jerusalem Post*, and citing Mumcuoglu, noted that in the five years previously, there had been a 150% increase in bed bug cases as reported by PMPs in Israel.

The US National Pest Management Association (NPMA) in conjunction with the University of Kentucky, undertook a global bed bug survey in early 2010 (Manuel, 2010; Potter *et al.*, 2010). In this survey, it was reported that 90% of PMPs from the Middle East and Africa had encountered bed bug infestations. Meek (2013) highlighted the numerous challenges brought about by the bed bug resurgence in the Middle East.

El-Azazy at al. (2013) described the resurgence of *C. lectularius* in Kuwait and the measures undertaken to combat the insect. They discussed the circumstantial evidence to suggest that the transfer of bed bugs is linked with recent immigrants and with used furniture. The authors argued that the spread of the infestation can be attributed to the increase in migrant labor and their mobility inside the country.

Haghi *et al.* (2014) reported the prevalence of the bed bug, *C. lectularius*, in human settlements in Bahnamir, Iran. Related to this, Gharari and colleagues (2016) have listed the presence of *C. lectularius* in the Iranian fauna of Cimicidae.

Balfour (2003) made a mention of *C. hemipterus* as being a biting nuisance in Dubai, United Arab Emirates (UAE), stating that in high density staff accommodation in labor camps and older inner city properties with multiple accommodation, bed bugs are extremely common. This information was further supported by Ramachandran (2012) who observed that the most common species in the UAE is *C. hemipterus*. Debusmann (2015) described a study conducted by Imdaad and https://Movesuoq.com, which noted that 80% of all pest control problems in Dubai were related to bed bugs and cockroaches. Cleland (2013) noted that during summertime in the UAE, the number of bed bug treatments increased dramatically: pest management companies (PMCs) based in Abu Dhabi declared that they normally undertake 20–24 treatments per month, which increases to 10 per week during the months of June, July, and August.

10.4 Impact of the Bed Bug Resurgence

Ahmed (2016) says that nothing else can damage the reputation of a hotel faster than the presence of bed bugs. The hospitality industry in the region is now ever more vigilant in their quest to remedy and to minimize the risk of bed bugs in their facilities. The pest control service contracts of most hotel and resort properties have a clause that requires that all guest rooms must be inspected and treated on a quarterly basis. For some hotels, a resident technician is assigned to the facility for up to 8–10 h a day, focusing on bed bug, fly, and cockroach treatments. For bed bug treatments, a 2–4 h response time is the standard and that is required at all times.

There has also been a tremendous rise in the number of PMCs operating in the region. For city centers, like Dubai and Abu Dhabi in the UAE, seven years ago the number of registered PMCs listed with the Dubai Municipality Pest Control Section and the Abu Dhabi Waste Management Center-TADWEER were less than 100 in number, but now this close to 200. The rise of PMCs is also happening in Doha (Qatar), Al Khobar (Kingdom of Saudi Arabia), Manama (Bahrain), and in Muscat (Oman). Facility management companies are adding pest management services to their portfolios, further contributing to the growth of the sector. There is an increasing demand for quality pest services for the hotel, accommodation, commercial, and residential communities. A large component of this service requirement is geared towards bed bug treatment and prevention.

Linked also to this growing demand for bed bug treatment is the dark side of accidental poisoning and deaths. The *Arab News* (Anonymous, 2009) reported two cases of accidental poisoning in Jeddah, Saudi Arabia. Similarly, Hilotin and Dhal (2011), the Gulf News (Anonymous, 2013), and Dhal (2015), described a series of incidences leading to accidental poisoning and death associated with the illegal use of aluminum phosphide, mainly used against bed bugs in residential areas in the Emirates. These incidences have triggered the government authorities to implement stricter regulations and policies on use of pesticides, including regular on-site inspections, crack-downs on illegal PMCs, as well as a massive information campaign (Anonymous, 2014; Abdullah, 2016).

10.5 Pest Management Professionals Close Encounters with Cimicids

Working as a field entomologist, trainer, and technical consultant to PMPs, hospitality and facility-management personnel, and government stakeholders in the region, the author has witnessed the development and often negative impact of bed bugs bites on the community. In this section are listed a few cases that highlight the impact of the resurgence:

> **Case 1**
>
> In a four-star hotel apartment, the executive housekeeper confided that a guest had called the police and complained that there was a bed bug in his room. Surprisingly, the police responded to the call of the foreign guest, inspected the site, called it to the attention of the hotel, ensuring that the guest was "assisted" in his bug problem!

> **Case 2**
>
> Working with a PMP in a 1000-room staff accommodation complex, I personally saw the tremendous efforts the occupants made in covering the movable ceiling boards and their metal bed bunk frames with packaging tape! The ceiling area became a bed bug reservoir. Furthermore, the off-white painted walls of the rooms were adorned with the splatter of bed bug excrement, blood stains from squashed bugs, and also black burned spots due to the use of small cigarette lighters to burn bed bugs off the wall! Adding to the challenges of the PMP, these rooms are not prepared for service, with all the personal belongings of the occupants placed underneath and on top of the bed bunks.

Among the challenges mentioned by PMPs include:

- a lack of cooperation in the preparation for treatment of staff accommodation and low-cost residential flats
- the growing use and sales of used furniture
- the difficulties in scheduling follow-up treatments in staff accommodation and residential areas
- unethical guests of hotels and apartments, who try to get free accommodation or try to damage the reputation of the facility
- the lack of appropriate pesticides
- insufficient training of technicians
- the absence of local research and development
- the competitive price war on pest management service fees (with clients focusing on the lowest price and not quality).

There are also a number of companies providing 3–6 months bed bug treatment warranties, after conducting a single treatment, which is merely a wall-to-wall surface spray.

Despite the lack of definitive evidence, some experts have stated that bed bugs, at least *C. lectularius*, originated in the Middle East, and that these parasites have subsequently spread around the world. While its importance diminished during the mid-20th century, in the Middle East as well as elsewhere, bed bugs have undergone a resurgence since around the beginning of the 21st century. There are many factors that have contributed to this resurgence within the region.

References

Abdullah, A. (2016). Beware, illegal pest control could cause even deaths. *Khaleej Times* (28 April), p. 1.

Abul-hab, J. (1997) Further survey for the ectoparasites of bats in Iraq. In: *Proceedings of the 12th International Congress of Speleology, August 10–17, 1997, La Chaux-de-Fonds, Switzerland*. Swiss Speleological Society, Switzerland.

Abul-hab, J. and Shihab, B.A. (1989) Ectoparasites of some bats from Iraq. *Bulletin of the Iraq Natural History Museum*, **8** (2), 59–64.

Abul-Hab, J., Al-Baerzangi, H., Kassal, S., Fuad, M. and Al-Obaidi, T. (1989) The common bed bug *Cimex lectularius* L. (Hemiptera, Cimicidae) in Baghdad City, Iraq. Magallat Buhut Ulum Al-Hayat = *Journal of Biological Science Research*, **20** (3), 455–462.

AFPMB (1999). *Regional Disease Vector Ecology Profile, The Middle East*, Armed Forces Pest Management Board, Washington DC.

Ahmed, T. (2016) *Understanding the Bedbug*, http://www.cleanmiddleeast.ae/news/243/understanding-the-bed-bug.html (accessed 5 August 2016).

Al-Barzangi, H.A.W., Abul-Hab, J., Al-Obaidi, T., Issa, W. and Hadi, N. (1988) Preliminary study on susceptibility of common bedbugs *Cimex lectularius* L. (Hemiptera, Cimicidae) in Baghdad to insecticides. *Bulletin of Endemic Diseases*, **29**, 17–21.

Anonymous (2009) *Pesticides Strikes in Jeddah Again*, www.arabnews.com/node/321711 (accessed 15 March 2016).

Anonymous (2013) *Timeline: Pesticide Poisoning Incidents in UAE. A Look at the Pesticide Poisoning in the Emirates*, http://m/gulfnews.com/news/uae/general/timeline-pesticide-poisoning-incidents-in-uae-1.1208815 (accessed 30 June 2016).

Anonymous (2014) *Dubai Warns 20 Pest Control Companies Over Breach of Standards*, http://www.thenational.ae/uae/dubai-warns-20-pest-control-companies-over-breach-of-standards (accessed 20 June 2016).

Balfour, J. (2003). *Arthropod Public Health Pests in the Emirates. How to Recognize them, How to Control Them*, Gulf Printing Press, Dubai.

Boase, C. (2008) Bed bugs (Hemiptera: Cimicidae): An evidence-based analysis of the current situation. In: *Proceedings of the Sixth International Conference on Urban Pests, July 13–16, 2008, Budapest, Hungary*. OOK-Press, Hungary, Veszprém.

Cleland, E. (2013) *Summertime in the UAE: A Comfort for Bed Bugs*, http://m.national.ae/news/uae-news/summertime-in-the-uae-comfort-for-bed-bugs (accessed 10 June 2016).

Dhal, S. (2015) *Say No to Killer Pesticides: Aluminum Phosphide in Homes*, http://gulfnews.com/news/uae/health/say-no-to-killer-pesticides-aluminium-phosphide-in-homes-1.1473915 (accessed 20 June 2016).

Debusmann, B. (2015) Dubai residents lets the bed bugs bite. *Khaleej Times* (20 April), p. 1.

Doggett, S. (2006) Focus: "The real millennium bug". *Syntrophy*, **7** (5), 1, 10–12.

El-Azazy, O.M., Al-Behbehani, B. and Abdou, N.E. (2013) Increasing bedbug, *Cimex lectularius*, infestations in Kuwait. *Journal of the Egyptian Society of Parasitology*, **43** (2), 415–418.

Ghahari, H., Moulet, P. and Ostovan, H. (2016) An annotated catalog of the Iranian Cimicidae and Largidae (Hemiptera: Heteroptera) and in memoriam Carl Walter Schaefer (1934–2015). *Zootaxa*, **4111** (2), 194–200.

Haghi, S.F.M., Behbodi, M., Hajati, H. and Shafaroudi, M.M. (2014) Prevalence of bed bug (*Cimex lectularius*) in human settlement of Bahnamir, Iran. *Asian Pacific Journal of Tropical Disease*, **4** (2), S786–S789.

Hilotin, J. and Dhal, S. (2011) Pesticide poisoning: "bomb" scare in the UAE. *Xpress* (18 August), p. 1.

Kandil, S.A.Y. (1986) Housing and sanitation conditions among rural families in an Iraqi village. *Alexandria Journal of Agricultural Research*, **31** (2), 47–58.

Manuel, J. (2010) *International Environmental Health: Invasion of the Bed Bug*, http://www.ncbi.nlm.nih.gov/pmc/articles/PMC2957946/ (accessed 16 June 2016).

Meek, F. (2013) Bed bug challenges Middle East region. Speech presented at the Global Bed Bug Summit in Denver, Colorado, December 11, 2013.

Mumcuoglu, K.Y. (2008) A case study of imported bed bugs (*Cimex lectularius*) infestations in Israel. *Israel Medical Association Journal*, **10** (5), 388–389.

Mumcuoglu, K.Y. and Shalom, U. (2011). Questionnaire survey of common bed bug (*Cimex lectularius*) infestations in Israel. *Israel Journal of Entomology*, **40**, 1–10.

Panagiotakopulu, E. and Buckland, P.C. (1999) *Cimex lectularius* L. the common bed bug from Pharaonic Egypt. *Antiquity*, **73** (282), 908–911.

Patton, W.S. (1919) Note on the etiology of Oriental Sore in Mesopotamia. *Bulletin de la Société de Pathologie Éxotique*, **12** (8), 500–504.

Patton, W.S. (1920) Some notes on the arthropods of medical and veterinary importance in Mesopotamia, and on their relation to disease. Part V. Some miscellaneous arthropods. *Indian Journal of Medical Research*, **8** (2), 245–256.

Potter, M.F. (2011) The history of bed bug management – with lessons from the past. *American Entomologist*, **57** (1), 14–25.

Potter, M.F, Rosenberg, B. and Henriksen, M. (2010) Bugs without borders: defining the global bed bug resurgence. *Pest World*, **Sept/Oct**, 1–12.

Ramachandran, D. (2012) Regional challenges associated with bed bug control. Speech presented at the National Pest Management Association NPMA Pest World East Conference in Dubai, UAE, April 23–24, 2012.

Rosen, S., Hadani, A., Gur Lavi, A., *et al.* (1987) The occurrence of the tropical bed bug in poultry barns in Israel. *Avian Pathology*, **16** (2), 339–342.

Sailer, R. (1952) The bed bug: an old bedfellow that's still with us. *Pest Control*, **20** (10), 22–24, 70, 72

Siegel-Itzkovich, J. (2014) *Bed Bugs in Israel Becoming Resistant to Insecticide, Says Parasite Expert*, http://m.jpost.com/Israel-News/Health/Bed-bugs-becoming-resistant-to-insecticide-378783#article (accessed 15 August 2016).

Usinger, R.L. (1966) *Monograph of Cimicidae (Hemiptera – Heteroptera)*, Entomological Society of America, College Park.

Usinger, R.L. and Povolny, D. (1966) The discovery of a possibly aboriginal population of the bedbug (*Cimex lectularius* Linnaeus, 1758). *Acta Musei Moroviae*, **51**, 237–242.

Part III

Bed Bug Impacts

11

Dermatology and Immunology[1]

Shelley Ji Eun Hwang, Stephen L. Doggett and Pablo Fernandez-Penas

11.1 Introduction

Bed bugs (both the tropical bed bug, *Cimex hemipterus* [F.] and the common bed bug, *Cimex lectularius* L.) are hematophagous arthropods that readily attack humans. The clinical consequences of bed bug exposure result in various bodily reactions, due either to the bite or from contact with the insect and its allergens (the latter aspect is discussed in Chapter 14). During the bite, a variety of proteins are injected into the skin that can cause a wide range of dermatological effects. Cutaneous reactions appear in 30–90% of individuals and can vary from erythematous macules[2], to papules[3], wheals[4] or urticarial[5] plaques, to even bullous[6] eruptions (Thomas *et al.*, 2004; Goddard and de Shazo, 2009a; Bolognia *et al.*, 2012; Cannet *et al.*, 2015). As a consequence of the bite, secondary infections may develop and victims may suffer sleep loss due to the intense itching (Cestari and Martignago, 2005; Goddard and de Shazo, 2009a; Doggett *et al.*, 2012).

Recognizing the clinical presentation and diagnostic features of bed bug bites is extremely challenging, but crucial in differentiating the bite from that of other hematophagous arthropods. This will ensure that the appropriate patient management is undertaken and that unrecognized bed bug infestations are subsequently controlled (Fallen and Gooderham, 2011; Doggett *et al.*, 2012). This chapter reviews the dermatological consequences of bed bug bites along with the immunological basis for the reaction in humans. Most bite investigations reported to date have been based around *C. lectularius* and information described herein pertains to this species, unless otherwise specifically stated.

11.2 Bed Bug Saliva

Bed bugs inject saliva into the skin during feeding. The saliva contains multiple protein fractions, some with anticoagulant properties. The saliva of *C. lectularius* contains 46 different protein components

---

1 This chapter uses general medical and specialized dermatological terminology. Although some footnotes have been added to explain some terms, it is recommended to refer to medical texts for further clarification.
2 Macule: flat lesion that only shows a change in color compared to the surrounding skin.
3 Papule: elevated, circumscribed, solid lesion, smaller than 1 cm in diameter (it is called nodule if the lesion is bigger than 1 cm).
4 Wheal: reddish, pink or pale swelling of the skin.
5 Urticarial: similar to urticaria (very pruritic cutaneous disease with multiple wheals, each one lasting less than 24 h).
6 Bulla: elevated, circumscribed lesion filled with fluid, bigger than 0.5 cm in diameter (it is called a vesicle if the lesion is smaller than 0.5 cm).

Advances in the Biology and Management of Modern Bed Bugs, First Edition.
Edited by Stephen L. Doggett, Dini M. Miller, and Chow-Yang Lee.
© 2018 John Wiley & Sons Ltd. Published 2018 by John Wiley & Sons Ltd.

(Francischetti *et al.*, 2010). Many of which contribute in delaying host hemostasis, while some contribute to host protection, and others play a role that is not fully understood yet (Doggett *et al.*, 2012). Some of the proteins are:

- nitrophorin, a vasodilator inducer (Valenzuela *et al.*, 1995)
- apyrase, a platelet function inhibiting enzyme (Valenzuela *et al.*, 1996a)
- a factor X inhibitor that interferes with coagulation cascade (Valenzuela *et al.*, 1996b).

The saliva of *C. hemipterus* also contains a smaller number of hemeproteins and proteins with anticoagulant properties than *C. lectularius*. The total protein contents of the saliva were found to be similar between these two species (Araujo *et al.*, 2009). Interestingly, despite popular belief, there is no evidence that bed bug saliva contains any anesthetic (Doggett *et al.*, 2012).

Proteins in bed bug saliva will trigger an inflammatory reaction where both innate and adaptive immune responses are involved (Goddard and de Shazo, 2009a). The host immune system recognizes these proteins as antigens, resulting in humoral and cellular responses and the release of chemokines and cytokines (Goddard *et al.*, 2013). With the recent rise in bed bug infestations, there has also been an increase in the number of patients presenting with allergic responses (Reinhardt *et al.*, 2009), and systemic reactions that have been treated with an immunosuppressant (Goddard and de Shazo, 2008).

11.3 Cutaneous Reactions

Bed bug mouthparts have exceptionally fine needle-like stylets that penetrate the skin to suck the host's blood. During feeding, bed bug saliva containing the anticoagulants is injected into the host. Once the bed bug completes its feeding, the stylets are withdrawn and the blood that oozes from the bite may stain bed sheets with small flecks of blood (Doggett *et al.*, 2012).

Usually, a barely visible punctum at the bite site can be seen, but may go unnoticed (Goddard and de Shazo, 2009a). The most common reaction for which medical attention is sought is a 2–5-mm wide lesion with a hemorrhagic punctum at the center (Steen *et al.*, 2004; Goddard and de Shazo, 2009a; Bernardeschi *et al.*, 2013). However, a range of dermatological manifestations have been reported. Bites tend to be distributed in uncovered areas of the body such as limbs and face (Steen *et al.*, 2004; Bernardeschi *et al.*, 2013). Bites tend to be edematous[7] in the early phase, forming wheals that blanch poorly and last more than 24 h, in contrast with typical urticaria (Scarupa and Economides, 2006). Pseudopodium-like extension of erythema in the peripheries of the bite site, resembling livedo or reticular pattern, has also been described (Doggett *et al.*, 2012). Typically, bed bug bites are pruritic and may appear in a linear fashion if there are multiple bed bugs present (Leverkus *et al.*, 2006; Doggett *et al.*, 2012). When there are large numbers of bed bugs, or because of scratching, the individual lesions can coalesce to form large plaques (Doggett *et al.*, 2012; Ukleja-Sokolowska *et al.*, 2013). Some authors claim that bed bug bites occur in groups of three, often referred to as 'breakfast, lunch, and dinner', but there is no objective evidence to confirm this (Doggett, *et al.*, 2012).

The onset and the severity of clinical reaction varies between individuals, and depends on the host's degree of previous exposure and the individuals' immune responses (Steen *et al.*, 2004; Leverkus *et al.*, 2006; Goddard and de Shazo, 2009a,b; Reinhardt *et al.*, 2009; Doggett *et al.*, 2012; Bernardeschi *et al.*, 2013). Individuals can exhibit immediate or delayed cutaneous reactions (Leverkus *et al.*, 2006). The onset of symptoms can vary from several hours to days, and even weeks (Leverkus *et al.*, 2006; Bernardeschi *et al.*, 2013). Patients who develop clinical signs are most likely to exhibit symptoms within 24–48 h, but a delayed reaction can appear in 9–10 days after the bite (Doggett *et al.*, 2012). The latency between the bite and cutaneous reactions decreases with repeated exposure (Reinhardt *et al.*, 2009). Bites can persist for a week or more, and rarer forms of

---

7 Edema: Swelling of the skin with similar texture to normal skin.

vesicular or bullous eruptions tend to be more painful, pruritic, and persistent (Liebold *et al.*, 2003; Doggett *et al.*, 2012; Sheele *et al.*, 2013; Laks and Wilson, 2015).

Some authors report that there may be no response to the bed bug bite (Bartley and Harlan, 1974; Gbakima *et al.*, 2002; Hwang *et al.*, 2005; Goddard and de Shazo, 2009b). However, there are conflicting findings in the literature. One study reported as many as 86% of individuals in an internally displaced-person camp in Sierra Leone, with mixed infestations of *C. lectularius* and *C. hemipterus*, developed wheals from the bites (Gbakima *et al.*, 2002). Yet, another group reported that only 30% of volunteers exposed to *C. lectularius* bites developed reactions (Ryckman, 1985; Goddard and de Shazo, 2009a). Similarly, amongst 474 individuals surveyed in the USA in 2009, 30% reported no reactions to bed bug bites (presumably from *C. lectularius*) (Potter *et al.*, 2010). Repeated exposure has demonstrated a higher incidence of cutaneous reactions (94.7% in 19 patients; Reinhardt *et al.*, 2009), and reduced the time from the bite to reactions (from weeks to seconds; Doggett *et al.*, 2012). In a small study involving 24 volunteers, 54.2% showed no skin reaction within 20 weeks of the first bite from *C. lectularius*, but upon re-exposure to bites, all except one volunteer developed a skin reaction (Reinhardt *et al.*, 2009). In a pilot study with seven volunteers from our institution (Doggett and Fernandez-Penas, unpublished results), two had no reactions, and five (three of whom had previous bed bug bite exposures) developed cutaneous reactions to *C. lectularius*. In the survey of 474 individuals mentioned above, those aged over 65 were found to have less reactivity to bed bug bites compared to the younger population (Potter *et al.*, 2010). The situation was similar in a smaller survey of 58 individuals from Kentucky, USA, where 76% (44 of 58) of individuals over the age of 65 reported no bite reactions (Potter *et al.*, 2010). A lesser immune response in the elderly could be an explanation for the reduced bite reaction (Potter *et al.*, 2010).

There are few observations or studies comparing the differences in cutaneous reactions to the bites of *C. hemipterus* and *C. lectularius*. Most published descriptions are limited to single observations, which generally do not specify the bed bug species, instar, or time from the bite (Doggett *et al.*, 2012). Usinger suggested that there may be a degree of clinical cross-reactivity in these similar species (that is, *C. hemipterus* and *C. lectularius*) compared to more distantly related species such as *Hesperocimex* and *Leptocimex* (Usinger, 1966), but this is yet to be confirmed. Comparable antigenic compounds in saliva of two species may contribute to a similar clinical presentation. However, in our single observation, no clinical reaction was observed with *C. hemipterus* bites in a highly *C. lectularius* bite-sensitive individual (Doggett *et al.*, 2012). An experimental study demonstrated that the manifestations of skin reactions to *C. hemipterus* bites take varying amounts of time to disappear in different body parts (Rahim *et al.*, 2016), although they appear to resolve more quickly than bites from *C. lectularius*. Further research is required in this area.

11.4 Dermatological Complications from Bed Bug Bites

Mostly, bed bug bites result in local cutaneous reactions that resolve within a week or two (Doggett *et al.*, 2012; Ukleja-Sokolowska *et al.*, 2013). Scratching the pruritic papules and nodules[8] predisposes patients to develop secondary infections such as impetigo, ecthyma, folliculitis, cellulitis, or lymphangitis (Thomas *et al.*, 2004; Goddard and de Shazo, 2009a; Doggett *et al.*, 2012). Another common complication is sleep deprivation due to the itch (Doggett *et al.*, 2012). One survey indicated that 29% of 474 individuals affected by bed bugs experienced sleep disturbance (Potter *et al.*, 2010). Patients wake up with itches, then scratch the lesions to relieve the itch, which then exacerbates the symptoms leading to the "itch–scratch cycle" (Thorburn and Riha, 2010; Doggett *et al.*, 2012). Sleep disturbance may have mental health impacts, due to the victim worrying that they are being bitten at night.

---

8 Nodules: papules bigger than 1 cm in diameter.

11.5 Systemic Reactions

Systemic IgE-mediated hypersensitivity reactions such as angioedema, bronchospasm, generalized urticarial or bullous eruptions, erythema multiforme, and anaphylaxis have been described in single cases or in a small series of patients following bites from *C. lectularius* (Thomas *et al.*, 2004; Leverkus *et al.*, 2006; Goddard and de Shazo, 2009a; Bernardeschi *et al.*, 2013; Sheele *et al.*, 2013; Cannet *et al.*, 2015; Laks and Wilson, 2015; Minocha *et al.*, 2017). Other symptoms and signs suggestive of multi-systemic inflammatory reaction, such as fever, general fatigue, hypereosinophilia, and polyarthralgia, may also accompany these reactions (Ukleja-Sokolowska *et al.*, 2013; Phan *et al.*, 2016).

11.6 Immunological and Pathogenic Mechanisms

The signs and symptoms developing from the bite are due to the host's response to the trauma, the subsequent inflammatory response, and immune system reaction to antigens present in the saliva. At the time of the bite there can be a painful reaction due to the disruption of pain terminals in the dermis (Reznik, 1996). However, this is usually minimal and does not trigger a response from the subject. As noted above, during feeding, bed bug saliva is injected into the host. Although the roles of innate and adaptive immune responses are largely unknown, there are some findings that can aid in the understanding of the pathogenesis of the reaction. The proteins and non-proteinaceous molecules from the saliva induce mast cells to release histamine, prostaglandins, and other mediators (such as nitric oxide, which enhances vascular permeability) (Leverkus *et al.*, 2006; Sheele *et al.*, 2013). A 17-kDa anti-coagulant (Factor X inhibitors) and a 40-kDa apyrase-like nucleotide binding enzyme delay blood clotting and inhibit platelet activation and aggregation respectively (Doggett *et al.*, 2012). In addition, macrophages and cytokines including IL-7, IL-8, IL-10, and IL-12 are also involved in cell-mediated immune response to the bites (Valenzuela *et al.*, 1996b; Valenzuela *et al.*, 1998; Goddard and de Shazo, 2009a; Fallen and Gooderham, 2011; Goddard *et al.*, 2013).

Regarding the adaptive immune response, there are reported cases of humans developing allergic IgE-mediated hypersensitivity to *C. lectularius* nitrophorin. The clinical presentations of bullous eruptions and type 1 anaphylactic reactions have been described as IgE-mediated reactions (Liebold *et al.*, 2003; Leverkus *et al.*, 2006; Ukleja-Sokolowska *et al.*; 2013). Local type III hypersensitivity reaction induced by circulating specific IgG antibodies, and immune complexes binding to receptors on leukocytes and an activating complement have been described (Abdel-Naser *et al.*, 2006; Leverkus *et al.*, 2006). Researchers from the USA developed assays measuring IgE antibodies to both *C. lectularius* extract and *C. lectularius* nitrophorin. These antibodies were common in individuals who had reported being bitten by bed bugs. However, future studies are needed to find the clinical relevance of IgE responses to bed bug allergens (Price *et al.*, 2012).

11.7 Dermatopathology

The histopathology of bed bug bites has undergone only limited study, and typically resembles other arthropod bite reactions. From the few reports published, biopsies show epidermal spongiosis, sometimes with sub-epidermal vesiculation and an inflammatory infiltrate of lymphocytes, neutrophils, eosinophils, and histiocytes in both upper and lower dermis around vessels. Edema between the collagen bundles in the papillary dermis has been reported as well (Tharakaram, 1999; Thomas *et al.*, 2004; Scarupa and Economides, 2006; Cohen *et al.*, 2010; Criado *et al.*, 2011; Doggett *et al.*, 2012). Arthropod bite reactions usually exhibit a wedge-shaped, superficial and deep perivascular and interstitial inflammatory infiltrate in the dermis. Epidermal changes of spongiosis, evolving into vesicle and epidermal necrosis, has been described (Jordaan and Schneider, 1997; Lever and

Elder, 2009). In one case with bullous lesions, the bite site progressed to exhibit leukocytoclastic vasculitis with eosinophils (de Shazo *et al.*, 2012). However, further research is needed to correlate the timing of the bite with these histopathology findings.

11.8 Differential Diagnoses of Bed Bug Bites[9]

Insect bites are difficult to distinguish from other bite reactions as they vary largely between individuals, even with the same insect species, making an accurate diagnosis and subsequent clinical management challenging. Often, the development of a wheal and a linear pattern of bites along the limb may be the suggestive signs of bed bug bites. However, the direct observation of the insect is the main criteria to diagnose the cause (Doggett *et al.*, 2012). Recently, a Western Blot test was developed as a means of serologically confirming putative bed bug bites, but it is not yet commercially or widely available (Goddard *et al.*, 2015).

When a patient is bitten by large numbers of bed bugs, multiple separate lesions can coalesce to mimic generalized maculopapular[10] exanthema. The lesions may be further modified by excoriations and lichenification[11] from persistent scratching, complicating the differential diagnoses (Doggett *et al.*, 2012). In addition, no study has described any difference between clinical presentations in different species of bed bugs, which makes it impossible for clinicians to distinguish the bites caused by one species over another (Steen *et al.*, 2004; Fallen and Gooderham, 2011; Bernardeschi *et al.*, 2013; Hoverson *et al.*, 2015).

Clinicians often misdiagnose bed bug bites as the bites of other arthropods, such as fleas (*Ctenocephalides felis* [Bouché]), mosquitoes, spiders, mites, or lice (Steen *et al.*, 2004; Fallen and Gooderham, 2011; Doggett *et al.*, 2012; Bernardeschi *et al.*, 2013; Hoverson *et al.*, 2015). The differential diagnoses could include conditions such as:

- dermatitis herpetiformis
- transient acantholytic dermatosis (Grover's disease)
- urticarial dermatoses
- autoimmune blistering diseases (Thomas *et al.*, 2004; Bernardeschi *et al.*, 2013)
- drug-induced maculopapular reactions
- miliaria
- chicken pox
- prurigo
- acral papular or papulovesciular dermatitis of Gianotti-Crosti
- acute or chronic eczema (Doggett *et al.*, 2012).

Less frequent dermatological differential diagnoses include erythema multiforme (Melnick *et al.*, 2013), Sweet's syndrome, and vasculitis (Bernardeschi *et al.*, 2013).

In cases where the patient presents with a pruritic dermatitis with macules, papules and sometimes vesicles or bullae, and they have recently travelled, or live in apartment complexes, chronic health facilities or penal institutions, bed bug bites should always be considered in the differential diagnosis (Goddard and de Shazo, 2009a; Goddard *et al.*, 2013). Most important is the identification of bed bugs or the signs of an infestation, such as eggs, cast skins, and bed bug feces. A pest management professional (PMP) specifically trained in bed bug management can be helpful in the identification process. Often without convincing history and clinical signs, some patients are misdiagnosed to have delusional parasitosis or receive inappropriate medical interventions (Doggett *et al.*, 2012; Hoverson *et al.*, 2015).

---

9 Some of the diagnoses described in this section require advanced knowledge of dermatological conditions.
10 Maculopapular: multiple macules and papules coalescing on the skin.
11 Lichenification: thickening of the skin, with more visible cutaneous fine lines and some scales, a typical reaction to frequent, chronic scratching.

Delusional parasitosis/infestation (also known as "Ekbom syndrome") is a psychiatric disorder of a somatic subtype, in which individuals have a false belief that they are infested by insects or organisms (Thakkar *et al.*, 2015). After an expert assessment, if a patient is suspected to suffer from delusional infestation-like syndrome (Freudenmann and Lepping, 2009), psychological therapy should be suggested.

Most cutaneous differential diagnoses can be ruled out by a trained dermatologist, sometimes with the help of cutaneous biopsies and immunofluorescence studies. The description of the differences of bite reactions from other dermatological conditions is beyond the scope of this chapter.

11.9 Clinical Management

The first step is bed bug eradication from the patient's living environment and prevention of reinfestation (Stibich *et al.*, 2001; Laks and Wilson, 2015). Eradication can be challenging and in most cases requires a licensed and trained PMP (Goddard and de Shazo, 2009a; see also Part 5 of this book).

In most cases, symptomatic treatment together with good hygiene can control the itch and prevent secondary infections (Steen *et al.*, 2004). When the bite lesions are inflamed and pruritic, mild-to-moderate potency topical steroids (depending on the severity of the reaction) are recommended (Leverkus *et al.*, 2006; Bernardeschi *et al.*, 2013). Topical antipruritic creams and/or systemic antihistamines can also be prescribed to manage the itch (Goddard and de Shazo, 2009a; Bernardeschi *et al.*, 2013; Laks and Wilson, 2015). In severe cases with multiple disseminated lesions or in very symptomatic patients where the quality of life is impacted, a short course of systemic prednisone (up to 0.5 mg/kg tapered over four weeks) will control the reaction. In cases of secondary infection, topical antiseptics and topical (mupirocin or fusidic acid) or systemic antibiotics may be indicated (Steen *et al.*, 2004; Thomas *et al.*, 2004; Bernardeschi *et al.*, 2013). In the rare case where a patient develops immediate hypersensitivity reaction to the bites, a prompt administration of epinephrine may be necessary (Laks and Wilson, 2015).

11.10 Conclusion

Educating physicians to become familiar with bed bug bite-induced cutaneous reactions is critical because often patients have delayed diagnosis or are completely misdiagnosed. This not only leads to recurrent bites but also allows the bed bugs to survive and the infestation to spread. Bed bug bites can cause a range of cutaneous reactions of varying severity. Therefore, establishing a prompt diagnosis and implementing the appropriate pest management is crucial for optimal patient outcome.

References

Abdel-Naser, M.B., Lotfy, R.A., Al-Sherbiny, M.M. and Sayed Ali, N.M. (2006) Patients with papular urticaria have IgG antibodies to bedbug (*Cimex lectularius*) antigens. *Parasitology Research*, **98** (6), 550–556.

Araujo, R.N., Costa, F.S., Gontijo, N.F., Goncalves, T.C.M. and Pereira, M.H. (2009) The feeding process of *Cimex lectularius* (Linnaeus 1758) and *Cimex hemipterus* (Fabricius 1803) on different bloodmeal sources. *Journal of Insect Physiology*, **55** (12), 1151–1157.

Bartley, J.D. and Harlan, H.J. (1974) Bed bug infestation: its control and management. *Military Medicine*, **139** (11), 884–886.

Bernardeschi, C., Le Cleach, L., Delaunay, P. and Chosidow, O. (2013) Bed bug infestation. *British Medical Journal*, **346**, f138.

Bolognia, J., Jorizzo, J. and Schaffer, J. (2012) *Dermatology*, 3rd edn, Elsevier Saunders, Ltd, Philadelphia.

Cannet, A., Akhoundi, M., Berenger, J.M., *et al.* (2015) A review of data on laboratory colonies of bed bugs (Cimicidae), an insect of emerging medical relevance. *Parasite*, **22** (21), 1–7.

Cestari, T.F. and Martignago, B.F. (2005) Scabies, pediculosis, bedbugs, and stinkbugs: uncommon presentations. *Clinics in Dermatology*, **23** (6), 545–554.

Cohen, P.R., Tschen, J.A., Robinson, F.W. and Gray, J.M. (2010) Recurrent episodes of painful and pruritic red skin lesions. *American Journal of Clinical Dermatology*, **11** (1), 73–78.

Criado, P.R., Belda, W., Criado, R.F.J., Silva, R.V. and Vasconcellos, C. (2011) Bedbugs (Cimicidae infestation): the worldwide renaissance of an old partner of human kind. *Brazilian Journal of Infectious Diseases*, **15** (1), 74–80.

de Shazo, R.D., Feldlaufer, M.F., Mihm, M.C. and Goddard, J. (2012) Bullous reactions to bedbug bites reflect cutaneous vasculitis. *American Journal of Medicine*, **125** (7), 688–694.

Doggett, S.L., Dwyer, D.E., Peñas, P.F. and Russell, R.C. (2012) Bed bugs: clinical relevance and control options. *Clinical Microbiology Reviews*, **25** (1), 164–192.

Fallen, R.S. and Gooderham, M. (2011) Bedbugs: an update on recognition and management. *Skin Therapy Letter*, **16** (6), 5–7.

Francischetti, I.M.B., Calvo, E., Andersen, J.F., *et al.* (2010) Insight into the sialome of the bed bug, *Cimex lectularius*. *Journal of Proteome Research*, **9** (8), 3820–3831.

Freudenmann, R.W. and Lepping, P. (2009) Delusional infestation. *Clinical Microbiology Reviews*, **22** (4), 690–732.

Gbakima, A.A., Terry, B.C., Kanja, F., *et al.* (2002) High prevalence of bedbugs *Cimex hemipterus* and *Cimex lectularius* in camps for internally displaced persons in Freetown, Sierra Leone: a pilot humanitarian investigation. *West African Journal of Medicine*, **21** (4), 268–271.

Goddard, J. and de Shazo, R. (2008) Rapid rise in bed bug populations: the need to include them in the differential diagnosis of mysterious skin rashes. *Southern Medical Journal*, **101** (8), 854–855.

Goddard, J. and de Shazo, R. (2009a) Bed bugs (*Cimex lectularius*) and clinical consequences of their bites. *The Journal of the American Medical Association*, **301** (13), 1358–1366.

Goddard, J. and De Shazo, R. (2009b) Multiple feeding by the common bed bug, *Cimex lectularius*, without sensitization. *Midsouth Entomologist*, **2**, 90–92.

Goddard, J., Hasenkampf, N., Edwards, K.T., De Shazo, R. and Embers, M.E. (2013) Bed bug saliva causes release of monocytic inflammatory mediators: plausible cause of cutaneous bite reactions. *International Archives of Allergy and Immunology*, **161** (2), 127–130.

Goddard, J., Tardo, A.C. and Embers, M E. (2015) Western blotting of human sera-can it help diagnose bed bug bites? *Skinmed*, **13** (5), 345–346.

Hoverson, K., Wohltmann, W.E., Pollack, R.J. and Schissel, D.J. (2015) Dermestid dermatitis in a 2-year-old girl: case report and review of the literature. *Pediatric Dermatology*, **32** (6), e228-e233.

Hwang, S.W., Svoboda, T.J., De Jong, L.J., Kabasele, K.J. and Gogosis, E. (2005) Bed bug infestations in an urban environment. *Emerging Infectious Diseases*, **11** (4), 533–538.

Jordaan, H.F. and Schneider, J.W. (1997) Papular urticaria: a histopathologic study of 30 patients. *The American Journal of Dermatopathology*, **19**, 119–126.

Laks, J. and Wilson, L.A. (2015) Don't let the bedbugs bite: an overlooked cause of rash in an older adult. *Journal of the American Geriatrics Society*, **63** (10), 2219–2220.

Lever, W.F. and Elder, D.E. (2009) *Lever's Histopathology of the Skin*, 10th edn. Lippincott Williams and Wilkins, Philadelphia.

Leverkus, M., Jochim, R.C., Schad, S., *et al.* (2006) Bullous allergic hypersensitivity to bed bug bites mediated by IgE against salivary nitrophorin. *Journal of Investigative Dermatology*, **126** (1), 91–96.

Liebold, K., Schliemann-Willers, S. and Wollina, U. (2003) Disseminated bullous eruption with systemic reaction caused by *Cimex lectularius*. *Journal of the European Academy of Dermatology and Venereology*, **17** (4), 461–463.

Melnick, L., Samimi, S., Elder, D., *et al.* (2013) Targetoid lesions in the emergency department. Bed bug bites (*Cimex lectularius*) with targetoid lesions on initial presentation. *The Journal of the American Medical Association Dermatology*, **149** (6), 751–756.

Minocha, R., Wang, C., Dang, K., *et al.* (2017) Systemic and erythrodermic reactions following repeated exposure to bites from the Common bed bug (*Cimex lectularius*). *Austral Entomology*, **56** (3), 345–347.

Phan, C., Brunet-Possenti, F., Marinho, E. and Petit, A. (2016). Systemic reactions caused by bed bug bites. *Clinical Infectious Diseases*, **63** (2), 284–285.

Potter, M.F., Haynes, K.F., Connelly, K., *et al.* (2010) The sensitivity spectrum: human reactions to bed bug bites. *Pest Control Technology*, **38** (2), 70–74, 100.

Price, J.B., Divjan, A., Montfort, W.R., *et al.* (2012) IgE against bed bug (*Cimex lectularius*) allergens is common among adults bitten by bed bugs. *The journal of allergy and clinical immunology*, **129** (3), 863–865.

Rahim, A., Zahran Z. and Majid A. (2016) Human skin reactions towards bites of tropical bed bug, *Cimex hemipterus* F. (Hemiptera: Cimicidae): A preliminary case study. *Asian Pacific Journal of Tropical Disease*, **6** (5), 366–371.

Reinhardt, K., Kempke, D., Naylor, R.A. and Siva-Jothy, M.T. (2009) Sensitivity to bites by the bedbug, *Cimex lectularius*. *Medical and Veterinary Entomology*, **23** (2), 163–166.

Reznik, M. (1996) Structure and functions of the cutaneous nervous system. *Pathologie Biologie*, **44**, 831–837.

Ryckman, R.E. (1985). Dermatological reactions to the bites of four species of Triatominae (Hemiptera: Reduviidae) and *Cimex lectularius* L. (Hemiptera: Cimicidae). *Bulletin of the Society of Vector Ecologists*, **10** (2), 122–125.

Scarupa, M.D. and Economides, A. (2006) Bedbug bites masquerading as urticaria. *Journal of Allergy and Clinical Immunology*, **117** (6), 1508–1509.

Sheele, J.M., Anderson, J.F., Tran, T.D., *et al.* (2013) Ivermectin causes *Cimex lectularius* (bedbug) morbidity and mortality. *The Journal of Emergency Medicine*, **45** (3), 433–440.

Steen, C.J., Carbonaro, P.A. and Schwartz, R.A. (2004) Arthropods in dermatology. *Journal of the American Academy of Dermatology*, **50** (6), 819–842.

Stibich, A.S., Carbonaro, P.A. and Schwartz, R.A. (2001) Insect bite reactions: an update. *Dermatology*, **202** (3), 193–197.

Thakkar, A., Ooi, K.G., Assaad, N. and Coroneo, M. (2015) Delusional infestation: are you being bugged? *Journal of Clinical Ophthalmology*, **9**, 967–970.

Tharakaram, S. (1999) Bullous eruption due to *Cimex lectularius*. *Clinical and Experimental Dermatology*, **24** (3), 241–242.

Thomas, I., Kihiczak, G.G. and Schwartz, R.A. (2004) Bedbug bites: a review. *International Journal of Dermatology*, **43** (6), 430–433.

Thorburn, P.T. and Riha, R.L. (2010) Skin disorders and sleep in adults: where is the evidence? *Sleep Medicine Reviews*, **14**, 351–358.

Ukleja-Sokolowska, N., Sokolowski, L., Gawronska-Ukleja, E. and Bartuzi, Z. (2013) Application of native prick test in diagnosis of bed bug allergy. *Postepy Dermatologii i Allergologii: Advances in Dermatology and Allergology*, **30** (1), 62–64.

Usinger, R. L. (1966). *Monograph of Cimicidae (Hemiptera – Heteroptera)*, Entomological Society of America, College Park.

Valenzuela, J.G., Walker, F.A. and Ribeiro, J.M.C. (1995) A salivary nitrophorin (nitric-oxide-carrying hemoprotein) in the bedbug *Cimex lectularius*. *Journal of Experimental Biology*, **198** (7), 1519–1526.

Valenzuela, J.G., Chuffe, O.M. and Ribeiro, J.M.C. (1996a) Apyrase and anti-platelet activities from the salivary glands of the bed bug *Cimex lectularius*. *Insect Biochemistry and Molecular Biology*, **26** (6), 557–562.

Valenzuela, J.G., Guimaraes, J.A. and Ribeiro, J.M. (1996b) A novel inhibitor of factor X activation from the salivary glands of the bed bug *Cimex lectularius*. *Experimental Parasitology*, **83** (2), 184–190.

Valenzuela, J.G., Charlab, R., Galperin, M.Y. and Ribeiro, J.M. (1998) Purification, cloning, and expression of an apyrase from the bed bug *Cimex lectularius*. A new type of nucleotide-binding enzyme. *The Journal of Biological Chemistry*, **273** (46), 30583–30590.

12

Bed Bugs and Infectious Diseases

Stephen L. Doggett

12.1 Introduction

The debate around bed bugs (both the Tropical bed bug, *Cimex hemipterus* [F.] and the Common bed bug, *Cimex lectularius* L.) and their ability to transmit agents of infectious disease has been highly contentious. For many people, the simple knowledge that bed bugs blood feed like other hematophagus arthropods that do transmit pathogens (notably mosquitoes and ticks), is a sufficient cause to cast suspicion onto bed bugs. Yet for decades, researchers have attempted to search for pathogens in bed bugs and to test their ability to transmit agents of diseases. Despite various human pathogens having been isolated from bed bugs (more than 50 to date) and a small number of studies having demonstrated that *C. lectularius* can transmit pathogens in the laboratory, there are presently no documented cases of bed bugs doing so in the field. Yet it must be asked whether this lack of documentation is simply the case that field studies have yet to identify local transmission, whether the correct diseases are being examined, whether bed bugs are simply incompetent to transmit pathogens, or whether there other compounding factors that ensure that they are unable to do so?

This chapter is not a detailed review of the literature covering bed bugs and infectious diseases; there has been a series of recent comprehensive publications on the topic; see Goddard (2003), Goddard and de Shazo (2009), Doggett and Russell (2009), Delauney *et al.* (2011), Doggett *et al.* (2012), and Lai *et al.* (2016), as well as detailed reviews from the past (Burton, 1963a, b). Rather, this chapter focuses on the interplay of factors that could influence the ability of bed bugs to act as vectors of human pathogens. Initially, however, a broad overview of bed bugs and infectious diseases is given, with an emphasis on the more contemporary studies.

12.2 Vectors and Transmission Pathways

Pathogens can be transmitted by arthropods via two broad means: mechanical or biological transmission (Doggett *et al.*, 2012). The former method is where the vector acts as a simple carrier of the pathogen and yet the pathogen undergoes no development within the vector. The most well-known examples of mechanical transmission are HIV and hepatitis C transmission between humans through shared syringes. Mechanical transmission is relatively rare as the pathogen has to be able to cope with extremes in environment. The other form of transmission – biological – is where the pathogen is imbibed and undergoes development within the vector, with a concomitant increase in pathogen numbers. The pathogen is subsequently transmitted to a vertebrate host via the bite of the insect, passed on via fecal contamination, transferred through crushing the vector on the skin thereby releasing the pathogen, or via ingestion of infected vectors. Biological transmission is challenging to a pathogen, as it has to overcome a number of barriers in order to complete its full replication cycle. For example with an arbovirus (arthropod-borne virus), the following sequence of events must take place:

Advances in the Biology and Management of Modern Bed Bugs, First Edition.
Edited by Stephen L. Doggett, Dini M. Miller, and Chow-Yang Lee.
© 2018 John Wiley & Sons Ltd. Published 2018 by John Wiley & Sons Ltd.

1) The vector imbibes a blood meal containing the pathogen.
2) The virus penetrates the cells of the gut wall.
3) The virus escapes out of the gut wall cell into the hemolymph.
4) The virus penetrates the cells of the salivary glands.
5) The virus escapes the salivary glands during a subsequent blood meal.
6) The virus infects cells within the new host.

This means that the surface membrane of the pathogen must be evolved to recognize, penetrate, and subsequently escape from the various vector and host cells. As such, most pathogen/vector relationships are quite specific, with pathogens being transmitted by the one species or a group of closely related species.

12.3 Bed Bugs and Infectious Diseases: an Overview

More than 50 microorganisms have been recovered from bed bugs: a range of bacteria, nematodes, and viruses, many which are known human pathogens (listed by Delauney *et al.*, 2011 and Zorrilla-Vaca, 2014). Burton (1963a) noted that some early reports implicated bed bugs as the cause of beriberi and pellagra (both vitamin deficiencies), and cancer. These papers identified by Burton were published before the etiology of the conditions had been elucidated and can now be readily dismissed. For the microorganisms detected, most studies reported on the presence of the pathogen, without undertaking histological investigations to examine if the microbe could be in a location suggestive of possible biological transmission (such as in the salivary glands), and, in many studies, if the pathogen was viable or not. Likewise, most researchers have not undertaken vector competence experiments to determine if bed bugs are capable of transmitting the pathogen (Doggett *et al.*, 2012). A brief discussion of the main infectious diseases that have been investigated in relation to possible transmission by bed bugs follows.

With the increasing incidence of hepatitis B (HBV) during the 1970s and the recognition of acquired immunodeficiency syndrome (AIDS) in the early 1980s, research on the etiology of both viruses included investigations into the potential transmission by bed bugs (reviewed by Doggett *et al.*, 2012). For AIDS, the discovery of the human immunodeficiency virus (HIV) and its largely sexual mode of transmission meant that relatively few studies were undertaken with bed bugs. While some investigations found that HIV could persist for up to four hours in bed bugs (Lyons *et al.*, 1986; Jupp and Lyons, 1987), virus replication in the insect did not occur (Webb *et al.*, 1989). HBV research was far more intensive and involved virus detection, epidemiological investigations, and transmission studies. A number of groups reported detection of viral antigens in field-collected bed bugs, and others subsequently found that HBV could persist in bed bugs for some weeks (reviewed by Doggett *et al.*, 2012). Using artificial blood-feeding devices, Jupp and McElligott (1979) found low-level transmission of HBV by bed bugs does occur, but suggested that biological transmission was unlikely. Subsequently the same group placed HBV-infected bed bugs onto chimpanzees and no transmission occurred (Jupp *et al.*, 1991). Thus, there was insufficient evidence that bed bugs acted as vectors for HIV or HBV.

In a controversial paper, two Canadian researchers proposed that methicillin-resistant *Staphylococcus aureus* Rosenbach (MRSA) and vancomycin-resistant *Enterococcus faecium* Schleifer and Kilpper-Bälz (VRE) could be transmitted by bed bugs (Lowe and Romney, 2011). Their reasoning was based on a pilot study on five bed bugs (species not stated) from a field infestation, three of which were positive for MRSA and two for VRE. However, the study failed to undertake histological examination of the bed bugs to look for bacteria in organs that could be indicative of biological replication of the pathogen. Furthermore, antibiotic-resistant bacteria are very common in the environment and a major limitation of the Lowe and Romney study was the failure to rule out or address environmental contamination. Therefore, the bacteria were probably a contaminant. Indeed, a genome for *S. aureus* was sequenced during the bed bug genome project (Benoit *et al.*, 2016). Subsequent attempts by other researchers to demonstrate MRSA transmission by *C. lectularius* found that the bacteria did not replicate in midgut and was cleared within nine days (Barbarin *et al.*, 2014). In light of the feeding behavior of bed bugs, these authors concluded that any MRSA acquired during a blood meal would be evacuated before the next feed, and thus bed bugs are unlikely to be a vector.

As noted above, pathogens tend to be transmitted by either a specific vector or closely related vectors. Thus, it is understandable that research began on the potential for bed bugs to transmit pathogens that are known to be transmitted by closely related species, namely via other hemipterans. The most infamous of all of pathogens that are transmitted by hemipterans, and infects humans, is *Trypanosoma cruzi* Chagas.

Trypanosoma cruzi is the etiological agent of Chagas disease and is vectored by triatomine bugs. The disease is most common in rural regions of Latin America in conditions of widespread poverty (CDC, 2016). In many of these regions, bed bugs (both *C. hemipterus* and *C. lectularius*) also occur. *Trypanosoma cruzi* has a simple biological transmission cycle in which development takes place in the gut of the insect and the pathogen is passed on via contact with infected feces (Schmidt and Roberts, 1981). Thus in light of the simple pathogen cycle, the relatedness of the triatomine vectors with bed bugs, and the sympatric distribution of both species, the possibility of bed bugs acting as a vector of *T. cruzi* came under consideration. Research on the possible transmission of *T. cruzi* by bed bugs began in the early 1900s. It was found that both *C. hemipterus* and *C. lectularius* could be infected with the protozoan and maintain the infection for some weeks (even for the life of the insect), and subsequently pass the pathogen onto a vertebrate host in fecal material (Brumpt, 1913; Blacklock, 1914). However, subsequent field trials found little evidence for bed bugs being naturally infected in the field (Dias, 1934; Tonn *et al.*, 1982). Other research examining the potential for bed bugs to transmit *T. cruzi* was reviewed by Doggett *et al.* (2012) and a number of later laboratory trials indicated that transmission was possible. With the current bed bug resurgence, there has been a renewed interest in this area of research. Salazar *et al.* (2015) found that in laboratory trials, feces of *C. lectularius* infected with *T. cruzi* could be transmitted to mice when applied to broken skin. Similarly, Blakely *et al.* (2014) observed transstadial transmission of *T. cruzi* in *C. lectularius* and noted that transcutaneous transmission to mice via infective feces also occurred.

There is evidence that other trypanosomes may be transmitted by various members of the Cimicidae as well as bed bugs. Gardner and Molyneux (1988a) found that *Trypanosoma incertum* Pittaluga, a parasite of the bat, *Pipistrellus pipistrellus* Schreber, could develop in laboratory-reared *Cimex pipistrelli* Jenyns and *C. lectularius*. Six uninfected bats were inoculated with the gut contents of some *C. lectularius* that were positive for the trypanosome and all bats became infected. The same authors found another related flagellate, *Trypanosoma dionisii* Bettencourt & França, could develop in *C. lectularius* when fed on infected bats (Gardner and Molyneux, 1988b). Thus the possible transmission of these pathogens to humans through exposure of infected fecal matter onto mucous membranes or compromised skin (such as wounds) as per *T. cruzi*, cannot be excluded.

Thus there is now extensive data that indicates that both *C. hemipterus* and *C. lectularius* are capable of transmitting *T. cruzi* and other trypanosomes in the laboratory, but to date, there is no evidence that they do so in the field. Admittedly, few field investigations have been undertaken, and in a Chagas-disease endemic area, differentiating human infections caused by triatomine bugs versus those from bed bugs would be difficult. Unless each insect induced a slight (and observable) genetic or phenotypic change in the parasite, which translated as some sort of unique signature in the host, it would be impossible to determine the origin of the pathogen. Furthermore, bed bugs do not usually defecate on the host in the field, which reduces the likelihood of trypanosome transmission. Clearly further research is required to examine the role of bed bugs in the field transmission of *T. cruzi* before it can be definitively stated that they are a vector of this or other trypanosomes.

As per research on the trypanosomes, the current bed bug resurgence has sparked a renewed interest in the ability of bed bugs to transmit other pathogens. Goddard *et al.* (2012) fed "bed bugs" (species not stated, but presumably *C. lectularius*) on blood infected with the spotted fever rickettsia, *Rickettsia parkeri* Lackmen *et al.* They found that the pathogen did survive in bed bugs for two weeks but could not be maintained transstadially, and thus is not likely to be transmitted. In a survey of field-collected *C. lectularius* for *Bartonella* spp. in the USA, none of the 331 insects tested were positive, although five bed bugs showed the presence of *Burkholderia multivorans* Vandamme *et al.*, an important pathogen in nosocomial infections (Saenz *et al.*,

2013). A study undertaken on *C. lectularius* in the USA on field and laboratory bed bug specimens identified a range of gram-positive bacteria from the insect, although none of the bacteria were isolated from the gut content (Cockburn *et al.*, 2013). A series of fungal isolates were obtained from field-collected *C. hemipterus* in Malaysia (Majid *et al.*, 2015), although as fungi are ubiquitous environmental contaminants, such findings have little scientific significance. In none of these studies was there any evidence to demonstrate that bed bugs could transmit the pathogen under investigation.

In contrast with the study from the USA, *Bartonella quintana* Brenner *et al.* (the agent of trench fever) was isolated from *C. hemipterus* collected in Rwandan prisons (Angelakis *et al.*, 2013), although the authors were unable to culture the pathogen from the insect. In a subsequent study, *C. lectularius* were fed three times with human blood inoculated with *B. quintana* and the bacteria persisted for 18 days, could be passed vertically to progeny, as well as transstadially, and could be detected in bed bug feces (Leulmi *et al.*, 2015). As with all other transmission studies, the role of bed bugs in the field transmission of *B. quintana* has yet to be elucidated.

In summary with regards to the role of bed bugs in the transmission of pathogens, some studies have isolated various microbes, while a few have demonstrated laboratory transmission. However, not one investigation has provided any evidence that bed bugs transmit pathogens to humans in natural infestations. Furthermore, in a very crude and conservative calculation, the total number of bed bugs produced since the start of the modern resurgence has been estimated to involve many hundreds of millions of individual bed bugs (Doggett *et al.*, 2012). This means that as there is no evidence for bed bugs being able to transmit pathogens, the risk of acquiring a pathogen from the insect has to be exceedingly small.

12.4 Why do Bed Bugs not Transmit Infectious Diseases?

As noted, presently there is no evidence that bed bugs transmit infectious pathogens to humans in the field. Their lack of ability to do so appears to be due to a range of intrinsic factors related to the insects' own physiology as well as various adaptations in biology. This ensures that there are limited opportunities to feed on multiple hosts, thereby ensuring that pathogen cycles cannot be maintained.

Lai *et al.* (2016) noted that bed bugs appear to contain a range of "neutralizing factors", which may reduce their ability to transmit pathogens. For example, the saliva of *C. lectularius* contains lysozyme and other peptides thought to have antimicrobial properties (Francischetti *et al.* 2010). Possibly related to this observation was a study that attempted to cultivate microbes from *C. lectularius*, and none were isolated from the proboscis (Reinhardt *et al.*, 2005). The ejaculate (and to a lesser extent, the hemolymph) of *C. lectularius* has been shown to have bacteriolyctic activity, which is thought to help protect sperm from microbial damage (Otti *et al.*, 2009, 2013). In female bed bugs, the spermalege has been shown to provide protection against pathogens (Reinhardt *et al.*, 2003). The authors argued that perhaps the immune system was enhanced in the spermalege but did not speculate how this was achieved. Subsequently, Moriyama *et al.* (2012) found that bed bugs express a range of genes in the spermalege that produce peptides that have known antimicrobial properties, and which also stimulate host immune responses against microbes. In a later study, one of these peptides, defensin, was demonstrated to possess antimicrobial activity against a range of human skin microflora (Kaushal *et al.*, 2016). Ulrich *et al.* (2015) found that *C. lectularius* defensive secretions inhibited the growth of the fungus, *Metarhizium anisopliae* (Metchnikoff).

These adaptations probably evolved as an immunological defense mechanism for protection against a microbe-rich environment. For example, bed bugs aggregate to prevent moisture loss (Benoit *et al.*, 2007) and higher humidity and close contact between individuals increases the risk of microbial exposure. Bed bugs also defecate in the same location as they harbor, often over each other, and digested blood provides a nutrient-rich medium for the growth of potential pathogens. Furthermore, the process of mating in bed bugs, known as "traumatic insemination" (see Chapter 16) allows for the entry of microbes (Reinhardt *et al.*, 2005). Perhaps as a consequence of the evolution of these anti-microbial defense mechanisms, such processes may have helped inhibit the development of vector/pathogen relationships.

However, while there are few unique aspects in the bed bug immune response, it is not largely different from that of other insects. For example, most blood-feeding insects have a lysozyme-like protein as part of the salivary gland proteome (Benoit *et al.*, 2016). This means it is unlikely that an immune-specific aspect is the cause of the inability of bed bugs to transmit pathogens.

As discussed in Chapter 16, biological adaptations of bed bugs – flightlessness and localization around the host's resting location – limit their opportunities to feed on multiple hosts. Key to the continual biological transmission of vector-borne diseases are sufficient host numbers and individuals not previously exposed to the pathogen (that is, immunologically naïve). Non-immune hosts need to be present such that when infected with a pathogen via the bite of a vector, they produce a sufficient pathogen titer to allow unexposed vectors to acquire the infection. It is probable that bed bugs infesting the human environment feed on insufficient unique hosts to maintain zoonotic cycles (Doggett *et al.*, 2012).

12.5 The Future Hunt for Pathogens: A Cautionary Note

Highly powerful and sensitive molecular tools such as next generation sequencing (NGS) are uncovering all manner of new organisms in hematophagus arthropods (see, for example, Coffey *et al.*, 2013; Gofton *et al.*, 2015a,b) and the environment (see, for example, Afshinnekoo *et al.*, 2015). Therefore it can be expected that exploratory investigations on field-collected bed bugs utilizing such techniques will uncover new microbes not previously identified. In fact, a transcriptomic study of *C. lectularius* by Bai *et al.* (2011) identified sequences that belonged to fungi, bacteria, viruses, and Archaea. More recently, a genomic study of *C. lectularius* by Benoit and colleagues (2016) reported sequences from a range of microbes, representing almost 90 different genera. Another genomic study, published on the same day and also undertaken on *C. lectularius* (Rosenfeld *et al.* 2016), revealed sequences that matched *Clostridium* spp.

In spite of these findings, investigations encompassing genomic techniques such as NGS need to be treated with a certain degree of caution, as studies of this type are unlikely to reveal if any newly identified organism from a bed bug is a true pathogen, a symbiont, or a commensal, if it is totally unrelated, coming from a recently acquired blood meal, or even if it is only an environmental contaminant. Thus molecular techniques do not prove the existence of pathogen/vector relationships. Despite this, certain groups are likely to latch onto and promote reports of newly discovered organisms for both personal and political gain (see for example the discussion on a recently discovered *Borrelia* and the debate on Lyme in Australia; Collignon *et al.*, 2016). Even in the bed bug world, the Lowe and Romney paper (2011) was initially hailed by some bed bug researchers – no doubt to encourage research funding – echoing the alleged threat of bed bugs being able to transmit MSRA. This lauding of the paper came despite its obvious scientific flaws (Doggett *et al.*, 2012). Science is often seen as a great race, as it is the first who makes the discovery that is most remembered to history. As a consequence, the scientific literature is besmirched by research that is now considered preliminary or poorly performed such that laboratory outcomes may not be reflective of a real world scenario, with incorrect conclusions made (for example using extremely high pathogen titers to demonstrate vector competence, the use of highly inbred bed bug strains that may have lost neutralizing factors, and the ignoring of key biological characteristics such as bed bugs tending not to defecate on the host in the field, which is key for the transmission of several pathogens). Ideally, reports of bed bugs transmitting infectious diseases should be confirmed by other laboratories when there is a lack of evidence that bed bugs do so in the field. Additionally, negative results should also be published, particularly if the studies demonstrate alternative conclusions to previous research, to ensure that future science (and monies) are appropriately directed (Weintraub, 2016).

Arguably, the search for new pathogens are mostly likely to be productive where bed bugs naturally occur in the presence of many different hosts, such as in the remnant populations of *C. lectularius* that preferentially associate with bats (Balvin *et al.*, 2012; Booth *et al.*, 2015). In such a situation, the potential spillover of a novel pathogen to the human population would be greater. While this is speculation, the most likely potential human pathogen would be an arbovirus because a number of these have been identified in cimicids infesting bats and birds (Doggett *et al.*, 2012; Adelman *et al.*, 2013). Furthermore, the arboviruses can evolve rapidly; a single genomic change can result in a highly pathogenic virus that can then enable a poor vector to become highly competent. This was the exact situation with the Chikungunya virus outbreak that began in the western islands of the Indian Ocean around 2004. A single mutation in a gene that encodes for a membrane glycoprotein resulted in a virus with increased infectivity for *Aedes albopictus* (Skuse) (Tsetsarkin *et al.*, 2007). The result was an outbreak of an unprecedented scale, encompassing parts of Asia, Europe, and the Americas, with an estimate of over four million human cases (WHO, 2016).

Yet for bed bugs, such a dramatic outbreak would be highly improbable even if a pathogen did evolve to be transmitted by the insect. As noted above, the lack of multiple hosts ensures that bed bugs are unlikely to the instigator of vector-borne pandemics. Likewise, most people will be bitten in locations where alternative hosts, such as bats or birds, do not exist, thus reducing the risk of enzootic pathogens. Furthermore, if bed bugs were involved in a disease outbreak in developed nations, it would be expected that governments would initiate health interventions (where they do not presently do so), ensuring that control of infestations took place in an appropriate and timely manner. Since bed bugs occur in relatively confined areas, a disease outbreak should be rapidly halted. Conversely, in less economically advantaged nations, health systems tend to be less developed and disease surveillance minimally undertaken (Lai *et al.*, 2016). Thus an outbreak of disease initiated by bed bugs may not be recognized as quickly and health interventions delayed or nominally undertaken.

The science of what makes a hematophagus arthropod a competent vector is a very complex field of study. As Goddard (2008) highlights in his review of the dynamics of arthropod-borne disease, not only must a potential vector be long lived, anthropophilic, abundant, and capable of transmitting a pathogen, but its distribution must also match that of the epidemiology of human disease. While the first three points are true with bed bugs, and a very limited number of researchers have shown laboratory transmission of a small number of pathogens, human disease is simply not occurring.

12.6 Conclusion

In recent years there has been a call for research on the ability of bed bugs to transmit agents of disease (Jones and Eddy, 2011). However, the lack of any epidemiological evidence for the field transmission of any pathogen to humans suggests that such calls cannot be readily justified, particularly in a world where there are multiple competing funding interests. Despite this, bed bugs should be considered a public health pest for their dermatological and mental health impacts, as well as the indirect effects on the human populace, as discussed elsewhere in this book. The factors that ensure bed bugs are not competent vectors most likely encompass various intrinsic factors as discussed, but may well be a matter of simple mathematics; there are too few unique hosts to allow the evolution of vector/pathogen relationships.

References

Adelman, Z.N., Miller, D.M. and Myles, K.M. (2013) Bed bugs and infectious disease: a case for the arboviruses. *PLoS Pathogens*, **9** (8), e1003462.

Afshinnekoo, E., Meydan, C., Chowdhury, S., *et al.* (2015) Geospatial resolution of human and bacterial diversity with city-scale metagenomics. *Cell Systems*, **1** (1), 72–87.

Angelakis, E., Socolovschi, C. and Raoult, D. (2013) *Bartonella quintana* in *Cimex hemipterus*, Rwanda. *American Journal of Tropical Medicine and Hygiene*, **89** (5), 986–987.

Bai, X., Mamidala, P., Rajarapu, S.P., Jones, S.C. and Mittapalli, O. (2011) Transcriptomics of the bed bug (*Cimex lectularius*). *PloS ONE*, **6** (1), e16336.

Balvin, O., Munclinger, P., Kratochvil, L. and Vilimova, J. (2012) Mitochondrial DNA and morphology show independent evolutionary histories of bedbug *Cimex lectularius* (Heteroptera: Cimicidae) on bats and humans. *Parasitology Research*, **111** (1), 457–469.

Barbarin, A.M., Hu, B.F., Nachamkin, I. and Levy, M.Z. (2014) Colonization of *Cimex lectularius* with Methicillin-resistant *Staphylococcus aureus*. *Environmental Microbiology*, **16** (5), 1222–1224.

Benoit, J.B., Del Grosso, N.A., Yoder, J.A. and Denlinger, D.L. (2007) Resistance to dehydration between bouts of blood feeding in the bed bug, *Cimex lectularius*, is enhanced by water conservation, aggregation, and quiescence. *American Journal of Tropical Medicine and Hygiene*, **76** (5), 987–993.

Benoit, J.B., Adelman, Z.N., Reinhardt, K., *et al.* (2016) Unique features of a global human ectoparasite identified through sequencing of the bed bug genome. *Nature Communications*, **7**, 10165.

Blacklock, B. (1914) On the multiplication and infectivity of *T. cruzi* and *Cimex lectularius*. *British Medical Journal*, **1** (2782), 912–913.

Blakely, B., Hanson, S.J. and Romero, A. (2014) Survivorship and transstadial transmission of orally-ingested *Trypanosoma cruzi* in bed bugs. Speech presented at the 2014 Entomological Society of America Annual Meeting in Portland, Oregon, November 18, 2014.

Booth, W., Balvin, O., Vargo, E.L., *et al.* (2015) Host association drives genetic divergence in the bed bug, *Cimex lectularius*. *Molecular Ecology*, **24** (3), 980–992.

Brumpt, E. (1913) Immunité partielle dan les infections à *Trypanosoma cruzi*, transmission de ce trypansome par *Cimex rotundatus*. Rôle regulateur des hôtes intermédiaires. Passage à travers la peau. *Bulletin de la Société de Pathologie Exotique*, **6**, 172–176.

Burton, G.J. (1963a) Bedbugs in relation to transmission of human diseases. *Public Health Reports*, **78** (6), 513–524.

Burton, G.J. (1963b) Bedbugs in relation to transmission of human diseases, addenda. *Public Health Reports*, **78** (11), 953.

CDC (2016) *Parasites – American Trypanosomiasis (also known as Chagas Disease)*, US Centers for Disease Control and Prevention, www.cdc.gov/parasites/chagas/ (accessed 7 November 2016).

Cockburn, C., Amoroso, M., Carpenter, M., *et al.* (2013) Gram-positive bacteria isolated from the common bed bug, *Cimex lectularius* L. *Entomologica Americana*, **119** (1), 23–29.

Coffey, L.L., Page, B., Greninger, A.L., *et al.* (2013). Enhanced arbovirus surveillance with deep sequencing: identification of novel rhabdoviruses and bunyaviruses in Australian mosquitoes. *Virology*, **448** (2014), 146–158.

Collignon, P.J., Lum, G.D. and Robson, M.B. (2016) Does Lyme disease exist in Australia? *Medical Journal of Australia*, **205** (9), 1–5.

Delaunay, P., Blanc, V., Del Giudice, P., *et al.* (2011) Bedbugs and infectious diseases. *Clinical Infectious Diseases*, **52** (2), 200–210.

Dias, E. (1934) Estudios sōbre o *Schizotrypanum cruzi*. *Memórias Do Instituto Oswaldo Cruz*, **28**, 1–111

Doggett, S.L., Dwyer, D.E., Peñas, P.F. and Russell, R.C. (2012) Bed bugs: clinical relevance and control options. *Clinical Microbiology Reviews*, **25** (1), 164–192.

Doggett, S.L. and Russell, R.C. (2009) Bed bugs. What the GP needs to know. *Australian Family Physician*, **38** (11), 880–884.

Francischetti, I.M.B., Calvo, E., Andersen, J.F., *et al.* (2010) Insight into the sialome of the bed bug, *Cimex lectularius*. *Journal of Proteome Research*, **9** (8), 3820–3831.

Gardner, R.A. and Molyneux, D.H. (1988a) *Trypanosoma (Megatrypanum) incertum* from *Pipistrellus pipistrellus*: development and transmission by cimicid bugs. *Parasitology*, **96** (3), 433–447.

Gardner, R.A. and Molyneux, D.H. (1988b) Schizotrypanum in British bats. *Parasitology*, **97** (1), 43–50.

Goddard, J. (2003) Bed bugs bounce back – but do they transmit disease? *Infections in Medicine*, **20**, 473–474.

Goddard, J. (2008) *Dynamics of Arthropod-Borne Diseases*, Humana Press, Totowa, NJ.

Goddard, J. and de Shazo, R. (2009) Bed bugs (*Cimex lectularius*) and clinical consequences of their bites. *Journal of the American Medical Association*, **301** (13), 1358–1366.

Goddard, J., Varela-Stokes, A., Smith, W. and Edwards, K.T. (2012) Artificial infection of the bed bug with *Rickettsia parkeri*. *Journal of Medical Entomology*, **49** (4), 922–926.

Gofton, A.W., Doggett, S., Ratchford, A., et al. (2015a) Bacterial profiling reveals novel "Ca. Neoehrlichia", *Ehrlichia*, and *Anaplasma* species in Australian human-biting ticks. *PLoS ONE*, **10**, e0145449.

Gofton, A.W., Oskam, C.L., Lo, N., et al. (2015b) Inhibition of the endosymbiont "*Candidatus* Midichloria mitochondrii" during 16S rRNA gene profiling reveals potential pathogens in *Ixodes* ticks from Australia. *Parasites & Vectors*, **8** (345), 1–11.

Jones, S. C. and Eddy, C. (2011) Letters to the editor. *Journal of Environmental Health*, **July/August**, 56–57.

Jupp, P.G. and McElligott, S.E. (1979) Transmission experiments with hepatitis B surface antigen and the common bedbug (*Cimex lectularius* L). *South African Medical Journal*, **56** (2), 54–57.

Jupp, P.G. and Lyons, S.F. (1987) Experimental assessment of bedbugs (*Cimex lectularius* and *Cimex hemipterus*) and mosquitoes (*Aedes aegypti formosus*) as vectors of human immunodeficiency virus. *AIDS*, **1** (3), 171–174.

Jupp, P.G., Purcell, R.H., Phillips, J.M., Shapiro, M. and Gerin, J.L. (1991) Attempts to transmit hepatitis B virus to chimpanzees by arthropods. *South African Medical Journal*, **79** (6), 320–322.

Kaushal, A., Gupta, K. and Van Hoek, M.L. (2016) Characterization of *Cimex lectularius* (bedbug) defensin peptide and its antimicrobial activity against human skin microflora. *Biochemical and Biophysical Research Communications*, **470** (4), 955–960.

Lai, O., Ho, D., Glick, S. and Jagdeo, J. (2016) Bed bugs and possible transmission of human pathogens: a systematic review. *Archives of Dermatological Research*, **308** (8), 531–538.

Leulmi, H., Bitam, I., Berenger, J.M., et al. (2015) Competence of *Cimex lectularius* bed bugs for the transmission of *Bartonella quintana*, the agent of trench fever. *PLoS Neglected Tropical Diseases*, **9** (5), e0003789.

Lowe, C.F. and Romney, M.G. (2011) Bedbugs as vectors for drug-resistant bacteria. *Emerging Infectious Diseases*, **17** (6), 1132–1134.

Lyons, S.F., Jupp, P.G. and Schoub, B.D. (1986) Survival of HIV in the common bedbug. *Lancet*, **2** (8497), 45.

Majid, A.H.A, Zahran, Z. Rahim, A.H.A., et al. (2015) Morphological and molecular characterization of fungus isolated from tropical bed bugs in Northern Peninsular Malaysia, *Cimex hemipterus* (Hemiptera: Cimicidae). *Asian Pacific Journal of Tropical Biomedicine*, **5** (9), 707–713.

Moriyama, M., Koga, R., Hosokawa, T., Nikoh, N., Futahashi, R. and Fukatsu, T. (2012) Comparative transcriptomics of the bacteriome and the spermalege of the bedbug *Cimex lectularius* (Hemiptera: Cimicidae). *Applied Entomology and Zoology*, **47** (3), 233–243.

Otti, O., Naylor, R.A., Siva-Jothy, M.T. and Reinhardt, K. (2009) Bacteriolytic activity in the ejaculate of an insect. *The American Naturalist*, **174** (2), 292–295.

Otti, O., Mctighe, A.P. and Reinhardt, K. (2013) In vitro antimicrobial sperm protection by an ejaculate-like substance. *Functional Ecology*, **27** (1), 219–226.

Reinhardt, K., Naylor, R. and Siva-Jothy, M.T. (2003) Reducing a cost of traumatic insemination: female bedbugs evolve a unique organ. *Proceedings of the Royal Society B-Biological Sciences*, **270** (1531), 2371–2375.

Reinhardt, K., Naylor, R.A. and Siva-Jothy, M.T. (2005) Potential sexual transmission of environmental microbes in a traumatically inseminating insect. *Ecological Entomology*, **30** (5), 607–611.

Rosenfeld, J.A., Reeves, D., Brugler, M.R., et al. (2016) Genome assembly and geospatial phylogenomics of the bed bug *Cimex lectularius*. *Nature Communications*, 7, 10164.

Saenz, V.L., Maggi, R.G., Breitschwerdt, E.B., et al. (2013) Survey of *Bartonella* spp. in US bed bugs detects *Burkholderia multivorans* but not *Bartonella*. *PloS ONE*, **8** (9), 1–5.

Salazar, R., Castillo-Neyra, R., Tustin, AW., et al. (2015) Bed bugs (*Cimex lectularius*) as vectors of *Trypanosoma cruzi*. *American Journal of Tropical Medicine and Hygiene*, **92** (2), 331–335.

Schmidt, G.D. and Roberts, L. (1981) *Foundations of Parasitology*, 2nd edn, C.V. Mosby, St Louis.

Tonn, R.J., Nelson, M., Espinola, H. and Cardozo, J.V. (1982) Notes on *Cimex hemipterus* and *Rhodnius prolixus* from an area of Venezuela endemic for Chagas disease. *Bulletin of the Society of Vector Ecologists*, 7, 49–50.

Tsetsarkin, K.A., Vanlandingham, D.L., McGee, C.E. and Higgs, S. (2007) A single mutation in chikungunya virus affects vector specificity and epidemic potential. *PLoS Pathogens*, **3** (12), e201.

Ulrich, K.R., Feldlaufer, M.F., Kramer, M. and St. Leger, R.J. (2015) Inhibition of the entomopathogenic fungus *Metarhizium anisopliae* sensu lato in vitro by the bed bug defensive secretions (E)-2-hexenal and (E)-2-octenal. *BioControl*, **60** (4), 517–526.

Webb, P.A., Happ, C.M., Maupin, G.O., *et al.* (1989) Potential for insect transmission of HIV: experimental exposure of *Cimex hemipterus* and *Toxorhynchites amboinensis* to human immunodeficiency virus. *Journal of Infectious Diseases*, **160** (6), 970–977.

Weintraub, P.G. (2016) The importance of publishing negative results. *Journal of Insect Science*, **16** (1), 1–2.

WHO (2016) *Chikunguna Fact Sheet*, World Health Organization, http://www.who.int/mediacentre/factsheets/fs327/en/ (accessed 1 November 2016).

Zorrilla-Vaca, A. (2014) Bedbugs and vector-borne diseases. *Clinical Infectious Diseases*, **59** (9), 1351–1352.

13

Mental Health Impacts

Stéphane Perron, Geneviève Hamelin and David Kaiser

13.1 Introduction

Bed bugs have historically been thought of as a nuisance rather than as a serious public health problem. The cutaneous manifestations of bed bug bites (pruritic papules, urticaria, secondary cellulitis) are the principal health outcomes that tend to be associated with the presence of bed bugs (Doggett *et al.*, 2012; see also Chapter 11). The presence of bed bugs is, however, increasingly recognized as a source of distress and adverse mental health impacts.

Bed bug infestations are a source of stress for a number of reasons. The mere presence of insects in their household is perceived as a stressor by most individuals. The physical discomfort associated with bed bug bites may also be a source of distress. This distress can vary according to the severity of the cutaneous reaction to the bites, the presence of children in the household, the duration of the infestation, as well as a range of other compounding factors. In addition, there is a significant financial burden associated with bed bug eradication. Preparation of the household for extermination, hiring of a pest management professional (PMP), and the replacement of items that are discarded, all incur a cost. For those with limited financial means, this may cause a significant amount of tension, especially if infestations are recurrent (Comack and Lyons, 2011). This is often the situation in apartment complexes for the socially disadvantaged. Social isolation and the stigma associated with the presence of bed bugs may be an additional burden for affected individuals.

Any stressor may have an impact on mental health if it overwhelms the capacity of the individual or household to adapt. Acute stressors are known to cause symptoms of anxiety, depression, and have impacts on sleep quality.

13.2 Methods

This chapter is a narrative synthesis based on a systematic review of all available publications regarding the relationship between bed bugs, human stress, and mental health. The articles included in this systematic review cover publications up to May 2013 (Ashcroft *et al.*, 2015). The Medline and Elsevier Embase bibliographic databases were used to scan for articles published from May 2013 to April 2016 using the same keywords as Ashcroft and colleagues. A snowballing strategy using references from relevant publications was used to identify new studies, and references were also obtained from an expert in the field of bed bugs (Stephen Doggett). Data from research published by the Montreal Public Health Department was also included (Hamelin and Perron, 2017). Case reports, surveys, and observational and experimental studies were included in this review. Only original research reports or surveys written in English or French were reviewed. All studies that were commentaries, editorials, newspaper articles, clinical guidelines, and blogs were excluded due to a lack of scientific rigor. Studies that did not assess the effects of bed bugs on human mental health or wellbeing were also excluded.

Advances in the Biology and Management of Modern Bed Bugs, First Edition.
Edited by Stephen L. Doggett, Dini M. Miller, and Chow-Yang Lee.
© 2018 John Wiley & Sons Ltd. Published 2018 by John Wiley & Sons Ltd.

Using inclusion and exclusion criteria, this chapter reviews seven studies: a case report (Burrows *et al.*, 2013); a case series (Rieder *et al.*, 2012); a survey (Potter *et al.*, 2010); a systematic analysis of blogs (Goddard and de Shazo, 2012); a case-control study (Susser *et al.*, 2012); and a randomized, controlled trial (Hamelin and Perron, 2017). A study that reported the narrative stories of disadvantaged individuals living in conditions of bed bug infestations (Comack and Lyons, 2011) was also included. The main findings of these seven studies are presented in Table 13.1.

None of the other studies reviewed contained original data regarding the mental health effects of bed bug infestations.

Table 13.1 Synthesis of the main findings of the seven studies with original data on mental health and bed bug infestations.

Study	Design	Measured outcome	Main findings
Potter *et al.* (2010)	Surveys for clients of pest control operators (474 respondents)	Open ended question on the effects of bed bugs	During infestations: • insomnia/sleeplessness (29%) • emotional distress (22%) • anxiety (20%) • stress (14%)
Comack and Lyons (2011)	In depth interviews (16 inner-city residents)	Social impacts of bed bugs	• increased stress • sleep deprivation • financial strain • strains on relationships • anxiety • depression
Goddard and de Shazo (2012)	Analysis of 135 online comments against PTSD checklist	Scoring suggestive of PTSD	One post met the criteria for PTSD
Rieder *et al.* (2012)	Case reports (n = 7)		*Real infestations*: • major depressive episodes, anxiety-spectrum disorders • chronically psychotic can suffer medical sequelae secondary to disorganization and impaired baseline functioning *Perceived bed bug infestations*: • patients with a psychotic diathesis may suffer from brief psychotic and delusional disorders
Susser *et al.* (2012)	Cross-sectional study (39 individuals with infestations, 52 without)	Anxiety, depression and sleep disturbance symptoms using validated questionnaires	Anxiety and sleep disturbance
Burrows *et al.* (2013)	Case report (n = 1)	Suicide	In individual with history of psychiatric illness, substance abuse and suicidal behavior, repeated bed bug infestations contributed to successful suicide attempt
Hamelin and Perron (2017)	Randomized controlled trial. Answers to questions on: • anxiety (n = 36) • depression (n = 35) • sleep disturbance (n = 29)	Anxiety, depression and sleep disturbance symptoms using validated questionnaires	Increase in symptoms of anxiety and depression that subsided when the bed bug infestation was controlled

13.3 Main Findings

The stress related to bed bug infestation is well characterized by a study conducted in Winnipeg, Canada. The study, conducted in the winter of 2009, consisted of in-depth interviews of 16 socially disadvantaged inner-city residents whose homes had bed bug infestations. Interviewed individuals were self-selected. Almost all of the interviewed inner-city residents reported suffering from increased stress and sleep deprivation as a result of the infestation. Interviewees also discussed the financial toll, the frustration, the social isolation, and how interpersonal conflicts in relationships were magnified by the bed bug infestation. The case reports also correlate bed bug infestation with poor school performance and workplace tension. These stories exemplify the social stigma, shame, and embarrassment caused by bed bug infestations (Comack and Lyons, 2011).

A 2009 survey of customers of PMPs documented that when individuals were exposed to bed bugs, many showed signs of distress, such as increased anxiety symptoms and sleep disturbance (Potter *et al.*, 2010). This survey was conducted in several large urban centers in the USA, and all respondents were living with a bed bug infestation. Most respondents were tenants (76%), and the rest lived in single-family homes or condominiums/townhouses (24%). There were no specific questions relating to the bed bug impacts on the residents' mental health. However, when asked the open-ended question "Have you had any other symptoms which you attribute to the presence of bed bugs in your home?", 31% of respondents mentioned symptoms related to stress or mental health. Since the question on mental health was not prompted, the findings may underestimate the impact of bed bug infestations on mental health symptoms.

Symptoms of anxiety are not pathological and may be expected when one has to deal with a bed bug infestation. For many, if not most individuals, the increase in symptoms is transient and they reduce or disappear once the infestation is controlled. However, for a proportion of individuals, this transient increase in anxiety may trigger a cascade of events that may have a more important impact on their mental health. It should be expected that individuals with pre-existing mental health conditions are more at risk of developing psychiatric problems as a consequence of experiencing a bed bug infestation. Though there is not yet epidemiological evidence to support this hypothesis, there are case reports and case series that demonstrate the effect (Rieder *et al.*, 2012, Burrows *et al.*, 2013). Living with an infestation, coupled with a fear of insects resulting in anxiety and poor sleep, may bring about acute anxious and depressive states in individuals with no previously known psychiatric illnesses (Rieder *et al.*, 2012). In one study analyzing 135 on-line postings about bed bugs, one respondent even described post-traumatic stress disorder (PTSD) symptoms as a result of experiencing an infestation (Goddard and de Shazo, 2012).

Two epidemiological studies performed in Montreal, Canada, assessed the mental health impacts of a bed bug infestation in individuals in a residential setting using validated questionnaires. In the first study, participants were recruited from two apartment complexes that had been targeted by public health interventions for unfit housing conditions between January and June 2010 (Susser *et al.*, 2012). Participants were mostly of low socioeconomic status, with 55% reporting insufficient means to make ends meet, and 60% were unemployed at the time of the study. Symptoms of depression and anxiety were evaluated using the Brief Patient Health Questionnaire Mood Scale (PHQ-9) and the Generalised Anxiety Disorder Screener (GAD-7). Sleep disturbance was measured using the 5th subscale of the Pittsburgh Sleep Quality Index. The presence of bed bugs was associated with a significantly increased odds of having anxiety symptoms (OR = 4.8) or sleep disturbance (OR = 5.0). Bed bugs were also associated with an increased odds of developing depressive symptoms (OR = 2.5), but the association was not statistically significant. Although this study documented that bed bug infestations increased anxiety symptoms and sleep disorders in a vulnerable population, the study was not designed to evaluate whether symptoms improved after the bed bugs were exterminated.

In a second study (Hamelin and Perron, 2017), sleep quality and symptoms of anxiety and depression were measured in individuals currently living with an infestation and again after the infestation had been exterminated. The same measures of outcome (GAD-7, PHQ-9, and PSQI) were used as in the cross-sectional study. The primary goal of the study was to determine if extermination time could be shortened if pre-treatment preparation was provided to vulnerable households. Participants with infested premises were recruited from within Montreal's municipal housing corporation (Office Municipal d'Habitation de Montréal), which provides subsidized housing for individuals and households with limited income. In the study sample, 9% of respondents had a university degree. In 57% of the households, at least one individual had a history of previous mental health problems. In 58% of the households, at least one individual had a physical handicap. In 42% of the households, at least one child was under 18 years old. All households had bed bugs upon recruitment and participants completed a baseline questionnaire. A follow-up questionnaire was completed three to six months after recruitment. Only those households that no longer had bed bugs at the time of the follow-up questionnaire were included in the analysis. The results of this study indicated that the presence of bed bugs was associated with elevated symptoms of anxiety and depression, and that these symptoms decreased significantly after bed bugs were exterminated. Infestation status did not have any effect on sleep disturbance; only 19% of individuals had sleep disturbance when infested with bed bugs, whereas 71% of individuals in the Susser et al. (2012) study had sleep disturbance when infested.

In both of the above studies, questions were asked regarding social isolation. In Susser et al. (2012), 10.8% of those infested with bed bugs avoided leaving their apartment whereas 38.8% avoided inviting friends or family members over to their homes (Susser, 2013). In Hamelin and Perron (2017), at the time of the baseline interview, 25% of respondents avoided going to friends' or neighbors' homes for fear of bringing bed bugs with them. Also, 47% of respondents avoided inviting friends or family to their home during an infestation. At the time of the follow up questionnaire, when most of the households were free of bed bugs, 19% still avoided going to friends' or neighbors' homes for fear of bringing bed bugs with them, and 22% still avoided inviting friends or family to their apartment. These survey findings suggest that bed bug infestations can result in social isolation of the resident even after the infestation is resolved.

Both studies conducted in Montreal strongly suggest that a bed bug infestation results in an increase in anxiety and depressive symptoms in vulnerable populations. Susser et al. (2012) also provided epidemiological evidence of a correlation between a bed bug infestation and sleep disturbance. Some of the observed differences between the two studies may be due to the very different contexts in which the investigations were conducted. In the publication by Susser and colleagues, participants were selected from privately-owned buildings in which infestations had been ongoing for more than a year and where tenants actively sought out public health intervention. The municipal housing corporation, in contrast, had a well-structured pest management plan and participants had to be actively recruited into the study.

13.4 What Can Be Inferred from the Current State of the Literature?

The available evidence indicates fairly clearly that an infestation of bed bugs in a residential setting may result in a rapid increase in anxiety-related symptoms, in addition to sleep disturbance and social isolation. These effects appear to be present only during the infestation for the majority of those affected. However, a minority of individuals, having lived with an infestation of bed bugs in their dwelling, will develop depressive episodes and anxiety disorders. The likelihood of these disorders increases amongst individuals that live with an infestation for longer periods of time (that is, those experiencing a chronic infestation) and for those with pre-existing mental health problems. In addition, individuals with pre-existing psychiatric diseases may experience relapses into substance abuse, severe mood disorders, or relapses in schizophrenic psychosis when they experience a bed bug infestation.

13.5 Limitations and Future Research

Many uncertainties remain in the understanding of the psychological effects of bed bugs in a residential setting. Although a number of studies are supportive of an association between symptoms of anxiety and depression and the presence of bed bugs, the population burden of these mental health effects has not been characterized. Studies have been performed primarily amongst individuals living in low-income households, in urban settings in North America. It is possible that these findings cannot be extrapolated to include people in other countries or of a different socioeconomic status. Although population-level statistics are difficult to obtain, it appears that low-income households appear to bear a disproportionate number of bed bug infestations. In New York, the Community Health Survey performed in 2009 determined that bed bug infestations occurred more often in low-income households, and neighbourhoods with low-income residents, than in other locations (Ralph *et al.*, 2014). Population surveys undertaken between 2010 and 2014 by Montreal Public Health in Montreal, Canada, also indicated that most bed bug infestations occur in low-income situations, in particular in those locations where residents are spending more than 30% of their income on housing (7.8%, vs. owner 0.8%; David Kaiser *et al.*, unpublished data). However, it is likely that the focus on those low-income households has underestimated the total bed bug population burden, and may have biased our full appreciation of the problem. In addition, it remains unclear to what extent symptoms of anxiety and depression will persist amongst affected individuals once the infestation has been eliminated. Although symptoms of PTSD associated with bed bug infestation have been described, the causal relationship remains unclear (Goddard and de Shazo, 2012). There is a need for studies to explore the possible link between bed bugs and PTSD. Future research should examine if the mental health effects are comparable in settings other than urban areas of North America. Research should also focus on understanding which factors are responsible for more severe and prolonged mental health impacts among those experiencing a bed bug infestation. Finally, surveillance data should be collected to characterize the total population burden of bed bug infestations in particular locations.

13.6 Conclusion

Bed bug infestations are a source of human distress for a variety of reasons, including physical discomfort associated with bed bug bites, the financial burden associated with bed bug eradication, and the social isolation and stigma associated with the presence of bed bugs in a home. Studies have also conclusively linked an association between bed bug infestation and anxiety symptoms, depressive symptoms, and sleep disturbance. Case reports have documented psychiatric consequences as a result of real or perceived bed bug infestations in individuals, with or without prior psychiatric diseases.

There is currently sufficient evidence to strongly suggest that bed bugs, beyond having an impact on individual mental health outcomes, may have a significant influence on mental health at a population level. The significant mental health impacts of bed bugs and the concentration of impacts among disadvantaged and vulnerable segments of the population warrant sufficient public resources being dedicated to prevention and control of infestations.

References

Ashcroft, R., Seko, Y., Chan, L.F., *et al.* (2015) The mental health impact of bed bug infestations: a scoping review. *International Journal of Public Health*, **60** (7), 827–837.

Burrows, S., Perron, S. and Susser, S.R. (2013) Suicide following an infestation of bed bugs. *American Journal of Case Reports*, **14**, 176.

Comack, E. and Lyons, J. (2011) *What Happens when the Bed Bugs do Bite? The Social Impacts of a Bed Bug Infestation on Winnipeg's Innercity Residents*. Canadian Centre for Policy Alternatives, Winnipeg.

Doggett, S.L., Dwyer, D.E., Peñas, P.F. and Russell, R.C. (2012) Bed bugs: clinical relevance and control options. *Clinical Microbiology Reviews*, **25** (1), 164–192.

Goddard, J. and de Shazo, R. (2012) Psychological effects of bed bug attacks (*Cimex lectularius* L.). *American Journal of Medicine*, **125** (1), 101–103.

Hamelin, G. and Perron S. (2017) *Soutien à la Préparation des Logements de Personnes Vulnérables aux Prises avec une Infestation de Punaises de Lit,* Direction régionale de santé publique, Centre Intégré Universitaire de Santé et de Services Sociaux du Centre-Sud-de-l'Île-de-Montréal, Gouvernement du Québec, Quebec.

Potter, M.F., Haynes, K.F., Connelly, K., *et al.* (2010) The sensitivity spectrum: human reactions to bed bug bites. *Pest Control Technology*, **38** (2), 70–74.

Ralph, N., Jones, H.E. and Thorpe, L.E. (2013) Self-reported bed bug infestation among New York City residents: prevalence and risk factors. *Journal of Environmental Health*, **76** (1), 38–45.

Rieder, E., Hamalian, G., Maloy, K., *et al.* (2012) Psychiatric consequences of actual versus feared and perceived bed bug infestations: a case series examining a current epidemic. *Psychosomatics*, **53** (1), 85–91.

Susser, S.R. (2013). Good night, sleep tight, don't let the bed bugs bite: Exploring the mental health fallout of urban bed bug infestation in Montréal, Québec. Masters thesis, Université de Montréal, Montréal.

Susser, S.R., Perron, S., Fournier, M., *et al.* (2012) Mental health effects from urban bed bug infestation (*Cimex lectularius* L.): a cross-sectional study. *BMJ Open*, **2** (5), e000838.

14

Miscellaneous Health Impacts
Stephen L. Doggett

14.1 Introduction

The most direct and obvious injurious effects of bed bugs are dermatological, due to allergic reactions from the bite. However, beyond mental health related issues, bed bugs have been reported as being responsible for a variety of miscellaneous health impacts. These can include factors relating to the presence of the bed bugs themselves, such as allergens triggering asthmatic reactions, severe blood loss from multiple bed bugs feeding, and the disturbance of sleep patterns. Other indirect health impacts may result from poorly implemented control programs. These can be associated with the methodologies employed by people desperate to rid themselves of a bed bug infestation, such as the dangerous mishandling of chemicals and insecticides, or as a result of the actions of inexperienced pest management professionals (PMPs).

14.2 Respiratory Issues

A variety of common household arthropods have been recorded as producing allergens that can trigger asthmatic reactions. The most well-known of these are dust mites and cockroaches (Russell, 2001; Do *et al.*, 2016). In a heavily infested dwelling, the environmental dust becomes saturated with bits and pieces of arthropod integument, which may be inhaled by people living there. Accordingly, atopic individuals living in these situations may become hypersensitive and display allergic symptoms. The first suggestion that the Common bed bug, *Cimex lectularius* L., was capable of causing asthmatic reactions was from a report of 1929, in which a 37-year-old man who suffered seasonal asthma during the warmer months of the year demonstrated a sensitivity to bed bug extracts (Sternberg, 1929). After introducing a regimen of frequent cleaning of all bedding, his attacks ceased. A second report was from Spain in 1935 (Jimenez-Diaz and Cuenca, 1935) and again involved *C. lectularius*. This patient was a 28-year-old female who also suffered from seasonal asthmatic attacks during the summer months. When exposed to extracts of *C. lectularius*, the patient produced a very strong reaction, which manifested in extreme breathing difficulty. A desensitization program was implemented through injection of bed bug extracts and eventually the patient became non-reactive to bed bugs.

A study from Egypt published in 1991 exposed 54 asthmatic patients to various extracts of *C. lectularius* (Abou Gamra *et al.*, (1991). Twenty (37%) of the participants responded positively to a skin test and this cohort was considerably higher (9%) than the control group. However, none of the asthmatic patients were reported as having breathing difficulties. A similar study from China in 1995 challenged 320 asthmatic patients with *Cimex* allergens and 202 (63%) reacted (Wanzhen and Kaisheng, 1997), whereas only 9 out of the 80 (11%) in the non-asthmatic control group reacted. Based on these results, the authors concluded that *C. lectularius* may be a contributing factor for asthma in the region. Again, none of the asthmatic patients were reported as developing breathing difficulties during the study.

Arguably, the link between asthma and bed bugs is somewhat tenuous. In the first report, the hygiene measures implemented would have reduced exposure to dust mite allergens as well, and sensitivity to these allergens was not tested for. In the studies from the 1990s, none of the patients developed breathing issues. Thus, it was only in the 1935 publication that there was evidence of a link between bed bugs and asthma. Interestingly, to date there are no published reports linking the Tropical bed bug, *Cimex hemipterus* (F.) and asthmatic reactions. Further research is needed in this area.

14.3 Blood Loss

Bed bug infestations can become massive if left uncontrolled. Populations of tens of thousands of individual insects can occur in an extreme infestation (Doggett, 2010). On average, an adult female bed bug can imbibe 7.4 µl of blood in a single meal (Usinger, 1966) and will feed multiple times. Hence, in large infestations, there can be significant blood loss for the resident. A report from India in 1962 (Venkatachalam and Belavady, 1962) implicated both *C. lectularius* and *C. hemipterus* as the cause of widespread iron deficiency anemia in infants and young children. The authors suggested that, conservatively, 100 bed bugs could imbibe 1–2 ml of blood. In this study, researchers were readily able to collect 2000 individual bed bugs from a single infested dwelling.

In Canada, a 60-year-old male presenting with symptoms of fatigue and lethargy was found to have low hemoglobin levels with no obvious underlying cause (Pritchard and Hwang, 2009). An inspection of his home revealed an infestation of thousands of bed bugs. The patient knew the bugs were present but failed to treat the infestation. After bed bug eradication, the patient's hemoglobin levels normalized. A similar case was reported from Austria, where a 67-year-old male with severe anemia, was found, on admission, to be physically covered in *C. lectularius* (Paulke-Korinek *et al.*, 2012). His prosthetic leg was harboring a massive infestation. His partner, a 60-year-old woman, was admitted on the same day, and proved to be anemic as well. In France, an 82-year-old female Alzheimer's patient had developed an iron deficiency that was determined to be the result of a massive *C. lectularius* infestation in her apartment and clothing (Sabou *et al.*, 2013).

Despite these published cases, for the most part, such incidences of severe blood loss appear uncommon. Typically, in cases of heavy infestation, the affected residents also have underlying health or social problems. In each of these reports, the host's pre-existing condition of poverty, cognitive deficiency, or drug abuse was a major factor contributing to their bed bug population growing so large that anemia resulted.

14.4 Sleep Loss

A common consequence of even minor bed bug infestations is that of loss of sleep (Doggett *et al.*, 2012). Sleep loss can result from irritation at the bite site, with patients waking during the night to scratch or, more simply, from the knowledge that bed bugs are present in their home. In the most comprehensive survey to date, 29% of people having experienced a bed bug infestation reported having insomnia (Potter *et al.*, 2010). However, it is important to note that there was no control in this study, so that the percentage of infested individuals suffering from insomnia could not be compared with that of individuals in non-infested apartments. Sleep loss can have serious short- and long-term health impacts (Sigurdson and Ayas, 2007). In fact, some of the worst disasters caused by humans have been linked to individuals suffering from sleep deprivation, including the Three Mile Island and Chernobyl disasters, and the grounding of the Exxon Valdez (Doggett *et al.*, 2012).

14.5 Chemical Exposure

People who are desperate to rid themselves of bed bugs may undertake behaviors that are harmful to themselves or their environment. These behaviors may include the application of flammable liquids, such as petroleum and ethanol, to their bodies and within their home. Such applications have led to serious fires and injury (Geist-May, 2011; Goff, 2011; Kurtzman and McKibben, 2011; Associated Press, 2015). In one of the worst examples, an inexperienced PMP attempted to used propane gas to control bed bugs. This treatment resulted in the whole apartment catching fire and sustaining damages of C$3.5 million, with six injuries, including one child left with permanent brain damage (Blais, 2013).

The overuse of insecticides has also impacted people's health. In the past, highly toxic chemicals such as arsenic, hydrogen cyanide, and mercury compounds were used to control bed bugs and resulted in human deaths (Potter, 2011). However, with the current resurgence, the misuse of even relatively low-toxicity products has resulted in some adverse health impacts. A review of acute illness from the use of bed bug insecticide products in seven states within the USA over 2003–2010, identified 111 cases of negative health reactions, including one fatality (Jacobson *et al.*, 2011). Most cases were related to the overuse of insecticides. The one death was a patient that had serious predisposing medical conditions. Between 2006 and 2011, the Texas poison centers had 267 insecticide exposure calls that were the result of bed bug remediation attempts (Forrester and Prosperie, 2013). The vast majority (83.5%) of calls involved people who had suffered with minimal effects from the exposure. Many of the adverse health impacts reported in this study were related to the use of total release aerosols, which have been shown to have little efficacy for bed bug control (Jones and Bryant, 2012). In 2013, a 66-year-old woman suffered an acute kidney injury following the overuse of pyrethroids to control a bed bug infestation (Bashir *et al.*, 2013). In 2015, a 29-year-old man used an aerosol insecticide in an attempt to eradicate a bed bug infestation, and subsequently developed a dry cough, exertional dyspnea, and chest discomfort (Halpenny *et al.*, 2015). In 2016, a 59-year-old female, after accidental exposure to a bed bug repellent, had a complete heart blockage (Singh *et al.* 2016).

Some of the most serious abuses of pesticides to control bed bugs have involved the misuse of the deadly fumigant, phosphine. A series of tourist deaths in Thailand (Anonymous, 2011a,b; Browning, 2015; Fernquest, 2015) were initially blamed on the use of chlorpyrifos (Anonymous, 2011a), but were later found to be the result of illegally used phosphine gas (Browning, 2015; Fernquest, 2015). In the Middle East, there have been a series of injuries and deaths related to uncontrolled phosphine usage in buildings (see Chapter 10). In 2015, a Canadian woman imported phosphine to kill bed bugs, which resulted in the death of her young child (Anonymous, 2015; Ellwand and Klinkenberg, 2015). This prompted the Canadian National Collaborating Centre for Environmental Health to issue an alert warning about the unrestricted use of phosphine (National Collaborating Centre for Environmental Health, 2015).

14.6 Miscellaneous Health Impacts

Bed bugs have indirectly affected human health by disrupting the normal functions of emergency and medical services. In the USA, a busy fire station had to be closed because a bed bug infestation was being treated (Doggett *et al.*, 2012). Hospital wards have been temporarily shut (Reynolds, 2008; Doggett and Russell, 2008, 2009; Doggett *et al.*, 2012; Leininger-Hogan, 2011), and patients with bed bugs have been refused medical treatment (Doggett *et al.*, 2012). One must question the ethics of those in the medical profession who refuse to treat infested patients, especially when it is well-known that bed bugs do not transmit any disease.

Other indirect health effects include the suggestion that bed bugs produce: a serum sickness (Poorten and Prose, 2005); cause "nervous disorders in sensitive people" (Scott, 1963); are responsible for cancer (Cordier, 1933); or cause nutritional deficiencies such as beriberi and pellagra in humans (Burton, 1963). Modern medical knowledge ensures that the claims of cancer and nutritional deficiencies can be dismissed, and the lack of patients with "serum sickness" and "nervous disorders" correlating with the modern bed bug resurgence casts doubt on involvement in these conditions as well. In 1916, bed bugs were also suggested to be a cause of infantile paralysis in the far north coast of New South Wales, Australia (Anonymous, 1916). However, this incident occurred in an area endemic for the Australian paralysis tick, *Ixodes holocyclus*, a known cause of human paralysis (Doggett, 2004). Therefore, the authenticity of the arthropod identification must be questioned. An engorged *C. lectularius* nymph believed to have fed on the tympanic membrane of a human ear was responsible for a case of otitis in Canada (Cimolai and Cimolai, 2012). Secondary infections following bed bug bites may occur, albeit rarely in otherwise healthy individuals, but for the elderly or diabetic, the risk is greatly increased (Doggett et al, 2012).

In Rwanda, a study from 2015 investigated the factors that impeded the local use of insecticide-impregnated bed nets intended to protect people from malaria vectors (Ingabire *et al.*, 2015). It was found that many in the community refused to sleep under nets because bed bugs would harbor in the net and bite the sleeping resident (presumably the bed bugs had developed resistance). Thus, in this region of Africa, it appears that bed bugs are maybe increasing the risk of malaria.

For the most part, other than the loss of sleep, the indirect health impacts associated with bed bugs appear relatively uncommon and are possibly overstated. More identifiable adverse health effects have resulted from the overuse or inappropriate use of insecticides. When management is undertaken by a PMP as part of an integrated control program, adverse reactions to insecticides are not reported.

References

Abou Gamra, E.M., el Shayed, F.A., Morsy, T.A., Hussein, H.M. and Shehata, E.S. (1991). The relation between *Cimex lectularius* antigen and bronchial asthma in Egypt. *Journal of the Egyptian Society of Parasitology*, **21** (3), 735–746.

Anonymous (2011a) British couple allegedly killed by bed bug spray on holiday. *International Pest Control*, **53** (3), 124.

Anonymous (2011b) Toxin linked to hotel deaths. *New Zealand Herald* (9 May), p. 1.

Anonymous (2015) *Phosphine Pesticide Used to Kill Bedbugs Causes Fort McMurray Baby's Death*, http://www.cbc.ca/news/canada/edmonton/phosphine-pesticide-used-to-kill-bedbugs-causes-fort-mcmurray-baby-s-death-1.2969189 (accessed 17 May 2016).

Anonymous (1916) Bed-bugs. *The Macleay Chronicle* (22 March), p. 2.

Associated Press (2015) *New York Man Burned Trying to Kill Bedbugs Inside Rental Car*, NY Daily News http://www.nydailynews.com/news/national/n-y-man-burned-kill-bedbugs-rental-car-article-1.2186595 (accessed 19 July 2016).

Bashir, B., Sharma, S.G., Stein, H.D., Sirota, R.A. and D'Agati, V.D. (2013) Acute kidney injury secondary to exposure to insecticides used for bedbug (*Cimex lectularis*) control. *American Journal of Kidney Diseases*, **62** (5), 974–977.

Blais, T. (2011) *Edmonton Condo Fire Caused by Calgary Exterminator Prompts Lawsuit from Family Whose Child was Left Permanently Disabled*, Edmonton Sun, http://www.edmontonsun.com/2013/07/11/edmonton-condo-fire-caused-by-calgary-exterminator-prompts-lawsuit-from-family-whose-child-was-left-permanently-disabled (accessed 17 May 2016).

Browning, S. (2015) *Phosphine Gas: Lethal to Bedbugs, Beetles and People*, https://home.greens.org.nz/node/34371 (accessed 17 May 2016).

Burton, G.J. (1963) Bedbugs in relation to transmission of human diseases. *Public Health Reports*, **78** (6), 513–524.

Cimolai, N. and Cimolai, T.L. (2012) Otitis from the common bedbug. *The Journal of Clinical and Aesthetic Dermatology*, **5** (12), 43–5.

Cordier, C. (1933) Du role des insectes dans la transmission du cancer. *Revue de Zoologie Agricole*, **32** (8), 133–135.

Do, D.C., Zhao, Y. and Gao, P. (2016). Cockroach allergen exposure and risk of asthma. *Allergy*, **71** (4), 463–474.

Doggett, S.L. (2004) Ticks: human health and tick bite prevention. *Medicine Today & Tomorrow*, **5**, 33–38.

Doggett, S.L. (2010) *A Code of Practice for the Control of Bed Bug Infestations in Australia*, 3rd edn., Department of Medical Entomology and The Australian Environmental Pest Managers Association, Sydney.

Doggett, S.L. and Russell, R.C. (2008) The resurgence of bed bugs, *Cimex* spp. *(Hemiptera: Cimicidae)* in Australia. In: *Proceedings of the Sixth International Conference on Urban Pests, July 13–16, 2008, Budapest, Hungary*. OOK-Press, Veszprém.

Doggett, S.L. and Russell, R.C. (2009) Bed bugs. What the GP needs to know. *Australian Family Physician*, **38** (11), 880–884.

Doggett, S.L., Dwyer, D.E., Peñas, P.F. and Russell, R.C. (2012) Bed bugs: clinical relevance and control options. *Clinical Microbiology Reviews*, **25** (1), 164–192.

Ellwand, O. and Klinkenberg, M. (2015) *Alberta Mother Trying to Kill Bedbugs with Imported Pesticide Accidentally Poisoned Children, Killing Baby, National Post,* http://news.nationalpost.com/news/canada/alberta-mother-trying-to-kill-bedbugs-with-imported-pesticide-accidentally-poisoned-children-killing-baby (accessed 17 May 2016).

Fernquest, J. (2015) *Mystery Tourist Deaths & Illegal Bedbug Spray, Bangkok Post*, http://www.bangkokpost.com/learning/learning-from-news/491114/mystery-tourist-deaths-illegal-bedbug-spray (accessed 17 May 2016).

Forrester, M.B. and Prosperie, S. (2013) Reporting of bedbug treatment exposures to Texas poison centres. *Public Health*, **127** (10), 961–963.

Geist-May, K. (2011) *Tenant Tries to Kill Bed Bugs, Starts Fire*, http://cincinnati.com/blogs/considerthisclermont/2011/01/24/tenant-tries-to-kill-bed-bugs-starts-fire/ (accessed 15 December 2011).

Goff, L. (2011) *FDNY Warns Don't Use Gasoline Products on Bedbugs*, http://www.qgazette.com/news/2011-03-09/Front_Page/FDNY_Warns_Dont_Use_Gasoline_Products_On_Bedbugs.html (accessed 19 July 2016).

Halpenny, D., Suh, J., Garofano, S. and Alpert, J. (2015) A 29-year-old man with nonproductive cough, exertional dyspnea, and chest discomfort. *Chest*, **148** (3), E80–E85.

Ingabire, C.M., Rulisa, A., Van Kempen, L., *et al.* (2015) Factors impeding the acceptability and use of malaria preventive measures: Implications for malaria elimination in eastern Rwanda. *Malaria Journal*, **14** (1), 1–11.

Jacobson, J.B., Wheeler, K., Hoffman, R., *et al.* (2011) Acute illnesses associated with insecticides used to control bed bugs – seven states, 2003–2010. *Morbidity and Mortality Weekly Report*, **60** (37), 1269–1274.

Jimenez-Diaz, C. and Cuenca, B.S. (1935) Asthma produced by susceptibility to unusual allergens. *The Journal of Allergy*, **6** (4), 397–403.

Jones, S.C. and Bryant, J.L. (2012) Ineffectiveness of over-the-counter total-release foggers against the bed bug (Heteroptera: Cimicidae). *Journal of Economic Entomology*, **105** (3), 957–963.

Kurtzman, L. and McKibben, P. (2011) *Bed-bug Treatment Sets House on Fire*, http://www.bucyrustelegraphforum.com/fdcp/?unique=1305594573593 (accessed 17 May 2011).

Leininger-Hogan, S. (2011) Bedbugs in the intensive care unit. A risk you cannot afford. *Critical Care Nursing Quarterly*, **34** (2), 150–153.

National Collaborating Centre for Environmental Health (2015) *Phosphine Poisoning as an Unintended Consequence of Bed Bug Treatment*, http://www.ncceh.ca/documents/guide/phosphine-poisoning-unintended-consequence-bed-bug-treatment (accessed 20 July 2016).

Paulke-Korinek, M., Szell, M., Laferl, H., Auer, H. and Wenisch, C. (2012) Bed bugs can cause severe anaemia in adults. *Parasitology Research*, **110** (6), 2577–2579.

Poorten, M.C.T. and Prose, N.S. (2005) The return of the common bedbug. *Pediatric Dermatology*, **22** (3), 183–187.

Potter, M.F. (2011) The history of bed bug mangement-with lessons from the past. *American Entomologist*, **57** (1),14–25.

Potter, M.F., Haynes, K.F., Connelly, K., *et al.* (2010) The sensitivity spectrum: human reactions to bed bug bites. *Pest Control Technology*, **38** (2), 70–74, 100.

Pritchard, J.M. and Hwang, S.W. (2009) Severe anemia from bedbugs. *Canadian Medical Association Journal*, **181** (5), 287–288.

Reynolds, E. (2008) Patients flee hospital infested by bed bugs. *The Daily Mirror* (26 September), p. 21.

Russell, R.C. (2001) The medical significance of Acari in Australia. In: *Proceedings of the 10th International Congress of Acarology, July 5–10 1998, Canberra, Australia*. CSIRO Publishing, Melbourne.

Sabou, M., Imperiale, D.G., Andres, E., et al. (2013) Bed bugs reproductive life cycle in the clothes of a patient suffering from Alzheimer's disease results in iron deficiency anemia. *Parasite*, **20**, 1–5.

Scott, H.G. (1963) *Household and Stored-Food Insects of Public Health Importance and Their Control*, US Department of Health, Education, and Welfare, Atlanta.

Sigurdson, K. and Ayas, N.T. (2007) The public health and safety consequences of sleep disorders. *Canadian Journal of Physiology and Pharmacology*, **85**, 179–183.

Singh, H., Luni, F.K., Marwaha, B., Ali, S.S. and Alo, M. (2016) Transient complete heart block secondary to bed bug insecticide: a case of pyrethroid cardiac toxicity. *Cardiology*, **153** (3), 160–163.

Sternberg, L. (1929) A case of asthma caused by the *Cimex lectularius*. *Journal of Allergy*, **129** (1), 83.

Usinger, R.L. (1966) *Monograph of Cimicidae (Hemiptera – Heteroptera)*, Entomological Society of America, College Park.

Venkatachalam, P.S. and Belavady, B. (1962) Loss of haemoglobin iron due to excessive biting by bed bugs: a possible aetiological factor in the iron deficiency anaemia of infants and children. *Transactions of the Royal Society of Tropical Medicine and Hygiene*, **56** (3), 218–221.

Wanzhen, F. and Kaisheng, Y. (1997) Relationship between bedbug antigen and bronchial asthma. *Chinese Journal of Parasitology & Parasitic Diseases*, **15** (6), 386–387.

15

Fiscal Impacts

Stephen L. Doggett, Dini M. Miller, Karen Vail and Molly S. Wilson

15.1 Introduction

There is no question that the most painful aspect of the bed bug resurgence, aside from the itching bites, has been the financial devastation that these pests have brought to homes, businesses, and organizations that have attempted to control infestations while maintaining their reputation. It has been quite evident that since the start of the modern bed bug resurgence, the world's economy has suffered billions of dollars of pest "damage" from a problem that had almost been in complete remission in the last 40 years of the 20th century. The failure of many businesses and organizations to respond proactively to bed bug infestations has resulted in increased bed bug numbers, loss of reputation, loss of clients, legal action, and an explosion in control costs. This chapter reviews the many ways in which bed bugs have had an economic impact on local, national, and international business organizations, and individuals, over the last two decades.

15.2 Types of Cost

There are a number of ways in which bed bugs impose direct financial costs on business owners and organizations, or individuals. These costs may include:

- the cost of bed bug control itself
- the loss of revenue due to hotels rooms or apartment units being taken out of service
- the loss and replacement of discarded furniture or personal items, including damage to the facility during the treatment process
- litigation.

Additionally, the presence of bed bugs can result in indirect costs to businesses by causing:

- brand damage when media reports of bed bugs identify brand-name retail stores and hotel chains
- profit losses due to people avoiding locations for fear they will encounter bed bugs
- business devaluation.

Bed bugs have not only brought new expenses to those that have become infested, but also to the pest management industry that is expected to control these infestations. The pest management industry has had to develop new bed bug contracts and services, purchase new equipment, invest time and money in additional training for technicians, and spend more on labor costs than ever before in order to satisfy clients.

Advances in the Biology and Management of Modern Bed Bugs, First Edition.
Edited by Stephen L. Doggett, Dini M. Miller, and Chow-Yang Lee.
© 2018 John Wiley & Sons Ltd. Published 2018 by John Wiley & Sons Ltd.

15.3 Costs to the Multi-Unit Housing Industry

In the USA, multi-unit housing providers have been particularly vulnerable to the economic impacts of bed bug infestations. In a 2014 survey of multi-unit housing managers in Alabama and Tennessee, 88% of respondents indicated that bed bugs affected the budget or profitability of their housing facilities due to the costs associated with treating infested units. Paying for treatment in apartments adjacent to infested units also affected 55% of respondents' budget or profitability. Other costs included (K. Vail *et al.*, unpublished results):

- additional labor/time for property management and maintenance (58% of respondents)
- an increase in the length of time apartments remained vacant (38%)
- loss of the complex's reputation (22%)
- damage to the apartment carpets or walls (20%)
- costs from employees taking bed bugs home (8%)
- loss of employees (3%)
- hotel or resident transfer costs (2%).

Bed bug impacts on operating budgets have increased every year. Survey respondents indicated decreases in operating income of 1.7% between 2008 and 2009, and decreases of 2.9% and 3.2%, respectively in the following two years (K. Vail *et al.*, unpublished results). In addition, multi-unit housing managers have also experienced *extended* apartment vacancies while bed bug treatments were conducted. In the USA, survey respondents reported a mean of 27 additional days to treat and then re-rent a previously infested apartment unit. In the same survey, more than a quarter (26%) of southern US housing managers indicated the presence of bed bugs in their facilities would have reduced the market value of their property if they were selling the property at the time. A mean estimated property value loss of 25% (range 5–50%) was indicated due to the bed bug presence (K. Vail *et al.*, unpublished results).

Similar results were reported in a 2014 survey conducted among members of the Virginia Apartment Managers Association, and US Department of Housing and Urban Development housing managers in Virginia. In a survey of 170 housing managers, 41% indicated that they had dealt with bed bug infestations within their communities. Only 38% of these indicated that they had found out about the infestations directly from the apartment residents. Another 20% of respondents indicated that the bed bugs were only found after the resident had moved out of the unit, while another 11.8% discovered the bed bugs during routine apartment maintenance. Of the respondents that had dealt with bed bug infestations, 98% indicated that bed bug remediation efforts had severe impacts on their current operating budgets (D. Miller, unpublished results).

It is interesting to note that the Virginia multi-unit housing industry did not have to deal with these costs prior to 2005 and thus the bed bug management costs had novel and unanticipated impacts on the facility operations. When Virginia apartment managers were asked to describe how bed bugs impacted their operating budgets between 2009 and 2014, 50.6% of respondents indicated that paying for commercial pest control had the most impact on their facility's budget and/or profitability. Other financial impacts included the apartment staying vacant for a longer period (25.9%), damage to the apartment itself (7.6%), loss of reputation (12.9 %), and loss of employees (0.6%). Not surprisingly, Virginia survey respondents indicated that the range of financial impacts on their operating budgets increased every year between 2009 and 2014. The survey also asked by what amount (in USD) the presence of bed bugs has decreased the net operating income for each of the years 2009 through 2012:

- Respondents indicated that, in 2009, the decrease in operating income ranged from $0 to $500.
- In 2010, respondents reported a range between $0 and $5000.
- In 2011, the range reported was $0 to $20 000.
- In 2012, the range surged to $0 to $45 000.

For each of these years, the maximum losses in revenue for a facility were reported $50 000 in 2009; $33 000 in 2010; $45 000 in 2011; and, $75 000 in 2012. Note that not all facilities were equally affected by bed bugs, but larger (more housing units) facilities did incur greater bed bug remediation costs overall (D. Miller, unpublished data).

In the open comment section of the Virginia apartment managers' survey, many respondents noted that the cost for administrative time (calling pest management companies [PMCs], dealing with maintenance orders, listening to resident complaints, etc.) also had a significant fiscal impact on the profitability of the facility. Less time and effort were spent on marketing, maintenance, and landscaping when facilities' budgets had to be redistributed to accommodate bed bug management costs (D. Miller, unpublished data).

Currently, bed bugs are one of the most expensive home-invading pests to manage due to the difficulties in achieving control. In the USA, the costs of bed bug treatments are being quoted in the range of USD 500 (Ortiz 2004) to USD 1500 (Wahlberg, 2004; Bruno, 2010). A study conducted in low-income housing in the USA compared the effectiveness and costs of two integrated pest management (IPM) programs between 2005 and 2007 (Wang *et al.*, 2009). The average cost calculated was around USD 470 per apartment. A 2012 survey of PMPs in the USA indicated that there was a median expenditure of approximately USD 1000 for bed bug control in single-family homes, though a small proportion (3%) of these services cost over USD 3000 (Potter *et al.*, 2013). By 2015, that median expenditure had risen to approximately USD 1225 (Potter *et al.*, 2015). The survey reported that in 2012, 78% of property managers spent an average of USD 5000 or less on bed bug management (Potter *et al.*, 2013), but in some cases, more than USD 50 000 had been spent treating a single apartment complex (Potter *et al.*, 2013). In Australia, costs for controlling large infestations range from over AUD 4000 (USD 2900) for individual apartments, to over AUD 40 000 (USD 29 000) to achieve complete eradication in an apartment complex (Doggett *et al.*, 2011).

In the past, because many PMCs only had limited experience with bed bugs, several companies could be hired and subsequently fired (for unsatisfactory results) before an infestation was completely eradicated. In multi-unit dwellings, the infestation often had time to spread from the original site to other apartments within the facility before being addressed, thus increasing the control costs by several magnitudes. For example, a 320-unit staff accommodation complex at a major hospital in Sydney, Australia, initially had a single bed bug infestation in May 2003. Within two years, as a result of repeated control failures, the infestation spread to 68 rooms (21%), producing costs totaling more than AUD 42 000 (USD 30 500) to eradicate bed bugs from the entire building (Doggett and Russell, 2008). These costs included (in AUD) replacement of bedding and linen (>$7220), pest control (approx. $33 000), and intellectual support (>$2000; this included meetings and surveys of the site by groups other than the pest management professionals [PMPs]). Costs not included in the total were medical expenses for treating staff, loss of productivity, and various miscellaneous expenses. As rooms were rented out on a long term lease (usually six months to a year), there was no associated loss of income with room enclosures.

Sadly this example is not unique. While it may seem to be a financial burden to employ a PMP who offers a treatment guarantee at a high cost, this is often more cost effective than hiring a less expensive PMP who offers no service warranty and subsequently fails. Ideally, multi-unit housing providers should aim to accurately determine the true expenses associated with bed bug infestations as a standard part of the risk management and review process. Unfortunately, this is rarely done.

A study of bed bug management in low-income housing in the USA found that the cost of implementing a building-wide monitoring program for early bed bug detection was only USD 12 per apartment (Wang *et al.*, 2016). A study evaluating bed bug prevention and management in a low-income apartment complex in Harrisonburg, Virginia (USA), found that a proactive desiccant dust barrier could be applied in apartment units at a cost of USD 67–110, depending on unit size (Stedfast and Miller, 2014). Another study found that implementing a building-wide IPM program in a low income housing complex, using proactive monitoring and bi-weekly treatments, reduced bed bug numbers by 98% (Cooper *et al.*, 2015). In this particular facility, annual bed bug management costs prior to the implementation of the IPM program were USD 57 215, and the facility saw no substantial reduction in the bed bug population. The cost of implementing the IPM program was USD 65 536 for the year, but based on the reduction in bed bug numbers, Cooper *et al.* (2015) suggested that that overall treatment costs would be reduced to around USD 22 000 per year in subsequent years.

Bed bugs can result in indirect or unnecessary expenses if people are unfamiliar with the appropriate control procedures. For example, a manager of a government-subsidized housing facility in Sydney, Australia, spent approximately AUD 15 000 (USD 10 800) on dry cleaning for one bed bug victim's entire wardrobe in order to decontaminate their clothing (S. Doggett, unpublished data). A more logical option would have been the purchase of a series of chest freezers and clothes dryers, which could have been used by the whole facility (for this price in Australia, five 519 l chest freezers and eleven 8-kg clothes driers could have been purchased). Another example of indirectly incurred expenses is when, in heavy infestations, furnishings and personal belongings are discarded in the hope of controlling the bed bugs. The replacement of these discarded items becomes an added expense. A further example of indirect costs comes from amateurish attempts at control. Many bed bug sufferers who cannot afford the service of a PMP have resorted to home remedies to rid themselves of an infestation. In some cases, the use of highly flammable liquids has resulted in fires, causing serious property damage (Doggett *et al.*, 2012). In one extreme case, a PMC using heat to control an infestation caused a fire in a condominium that resulted in damages of CAD 3.5 million (USD 2.57 million; Anonymous, 2012). Fortunately, such extreme examples are rare.

15.4 Cost to the Hospitality and Travel Industry

Since bed bugs have become such a popular topic in the media, many people have eschewed travel in an effort to avoid encountering bed bugs. Tourists have become aware of the regions and cities in the world that are badly afflicted with bed bugs and avoid traveling to such areas. A 2010 survey of travelers found that 12% of respondents were so concerned about bed bugs that they altered or cancelled their travel plans for fear of encountering the insect (Henriksen, 2010). Liu and Pennington-Gray (2015) noted that the media reinforces those travel concerns because international travel is often cited as one of the main contributing factors to the bed bug resurgence.

In some cases, travelers have had their holidays disturbed by bed bugs, and some travel organizations have had to reimburse guests for their expenses. For example, a motel chain in the south Pacific reimbursed guests for their accommodation and airfares as well as medical and other expenses when they were bitten by bed bugs (Wojcik, 2004). Passengers on a cruise ship were refunded USD 2800 for their accommodation, and had their airfares and additional hotel costs reimbursed after they were bitten by bed bugs while on-board a ship (Penn *et al.*, 2014).

Hotel guests often refuse to pay for their accommodation if they are bitten by bed bugs during their stay. This loss of income can range from hundreds to thousands of dollars for high-end hotels and resorts. Unfortunately, there are also many anecdotal reports of guests falsely claiming they were bitten by bed bugs in an attempt to avoid paying for their accommodation. A survey of the Australian hospitality industry from September 2004 to March 2005 reported that 40% of respondents had experienced false bed bug claims (Doggett, 2013). Hotels and resorts suffer further financial losses due to their rooms being closed for multiple days while PMPs try to achieve bed bug eradication. These losses can be quite significant for hotels and resorts during their peak seasons.

In cases where bed bugs are in fact present, the costs of remediation can be extremely high, particularly if multiple rooms are involved. Bed bugs are known to stain bedding, furnishings, and walls. In hotels, guests noting indications of past infestations may refuse to stay in their assigned room (as the senior author has done on two occasions!). Hence, all bed bug evidence should be removed once the infestation is eradicated. As an example of actual costs, 39 out of 100 rooms (39%) in a three-star hotel in Sydney, Australia, developed bed bug infestations in 2007 (Doggett and Russell, 2008). The costs associated with these infestations exceeded AUD 60 000 (2007 pricing, or USD 43 500) and included room closures, pest control, replacement of linen, room refurbishment, replacement of mattresses and furnishings, refunds to guests, and other miscellaneous expenses.

15.5 Cost to the Retail Industry

It is not only the hospitality and housing industry that has suffered the economic consequences of bed bugs. When bed bugs were discovered in well-known US retail stores, the stores had to close down for several days to "eradicate" the problem (Bland, 2010; Odell, 2010; West, 2010). Most likely, these bed bug sightings were the result of bed bugs being brought in inadvertently, rather than the presence of a breeding population. Regardless, not only did the retailer lose sales as a consequence of bed bug presence, but the "infested" stock was discarded and the brand name of the retailer was notably damaged.

15.6 Brand Damage in the Housing, Hospitality, and Retail Industries

In addition to the direct costs associated with eliminating a bed bug issue, the mere presence of a bed bug has the potential to damage a brand's reputation. For example, in the hospitality industry, a poll found that if bed bugs are found in one location of a hotel chain, then 38% of travelers would refrain from staying in any of the properties in the chain (Meek, 2007). The same poll found that 50 percent of people who are bitten by bed bugs tell five or more people of their experience. In the USA, there is a website dedicated to reporting hotels and other locations where guests have found bed bugs (http://bedbugregistry.com/). Even if only one room was infested out of a thousand, being named on this site can still be very damaging to a hotel's reputation and cause financial losses. Similarly, online travel review sites can rapidly spread bed bug reports through the community (Liu *et al.*, 2015a). Often, the hotel industry is urged to have risk management plans in place to minimize the potential damage from such adverse online reports (Liu *et al.*, 2015b). Ideally, all bed bug reports posted on travel review sites should be responded to by senior hotel management to minimize guest dissatisfaction and potential damage to the brand (Liu *et al.*, 2015c).

For hotels, attempts have been made to quantify the fiscal impacts of online bed bug reports. A 2015 survey of approximately 2100 travelers found that a single online mention of bed bugs in a hotel lowers the room value by USD 38 for corporate travelers, and USD 23 for leisure travelers (Nathe, 2015). The survey revealed that many travelers would not even book a hotel if there was an online report of a bed bug infestation. Another report indicated that travelers would be willing to pay substantially higher prices, as much as USD 100, for a bed bug-free environment (Penn *et al.*, 2014). This additional "willingness to pay" has implications for the pricing of hotel rooms.

The 2015 survey also presented travelers with a variety of potential scenarios, and asked which scenario might cause them to change hotels (Nathe, 2015). The most significant factor that would cause them to change accommodation was the presence of bed bugs. Respondents in the survey were also asked what they would do if they found bed bugs in their room. The three most frequent responses were (Nathe, 2015):

- change rooms with a request for compensation
- leave the hotel
- report the presence of bed bugs on social media.

This study suggests that even a single bed bug can have a major economic effect on a facility.

Another concern for the housing and hospitality industry has been the proliferation of bed bug infestation reports (and subsequent lawsuits) in business magazines (Arnst and Sagar, 2004; Lundine, 2003; De Marco, 2004; Ortiz, 2004; Tucker, 2003). The financial losses resulting from bed bug sightings and infestations have led to reduced profit margins, as well as the ongoing potential for litigation. These indirect effects have subsequently affected the stock prices and overall values of such businesses. Additionally, the threat of litigation causes insurance premiums to rise (Wojcik, 2004; Reinhardt *et al.* 2009).

Early in the bed bug resurgence, it was suggested that in the USA, lawsuits involving bed bugs might not be claimable against insurance (Wojcik, 2004). Since then, many companies have initiated bed bug-specific insurance policies for a range of groups that could be impacted by them (Holbrook, 2011).

Interestingly in 2014, a group of multi-unit housing managers (n = 110) in the southern USA indicated that none of them had purchased bed bug insurance (K. Vail *et al.*, unpublished results). However, bed bug insurance claims do occur. The international insurer, Allianz Global Corporate & Specialty, analyzed liability insurance claims over the period 2011 to 2016 (AGCS, 2017). The total value of all claims was USD 9.3 billion. Of these, 1880 (2%) were related to animals, with 397 claims (or 21% of the 2%), relating to bed bugs. This equated to value of USD $39.06 million and appears to be rising. The same report noted that bed bug complaints in New York hotels increased by more than 44% between 2014 and 2015.

15.7 Legal Expenses

Across the world, disputes over bed bugs have led to legal actions addressed through court proceedings. The expenses incurred and amounts awarded by the courts varies considerably between countries. For example, in Australia, settlement amounts have ranged from AUD 400–6000 (USD 295–4479) and have largely related to the reimbursement of rental costs (see Chapter 43). In the UK, legal cases have usually resulted in out-of-court settlements and payments that covered such items as doctor's fees, loss of earnings, and other compensation. Compensation payments have been in the range of GBP 750–5000 (USD 537–3581, and occasionally up to GBP 10 000 (USD 7162; see Chapter 42).

The largest legal settlement awards to date have come from the USA. Seemingly unique to this nation is the propensity for courts to award substantial penalties in the form of punitive damages. For example, in the case of *Mathias v. Accor Econ. Lodging, Inc.*, the court awarded the two plaintiffs USD 10 000 compensation, and USD 186 000 in punitive damages. Other US court cases have resulted in costs and awards ranging from several hundred dollars to over USD 2.4 million in a recent class action settlement (see Chapter 41).

15.8 Cost to Pest Management Companies

While PMCs are the agents, who receive income for bed bug remediation services, they often fail to adequately charge for their bed bug management services. When this happens, profitability for their company declines precipitously as technicians spend more time than budgeted to complete a job. It is essential when pricing a bed bug job that the salesperson consider how much time a technician (preferably, two technicians) needs to effectively treat an area of a certain size, level of clutter, and density of infestation. However, according to a survey of multi-unit housing managers, only 55% of respondents indicated that apartment size was considered in contract prices. Additionally, less than a quarter of respondents indicated that prices were based on size of the infestation (22%), number of furniture pieces (9%), or amount of household clutter (13%). Finally, 35% of respondents indicated that cost did not vary from unit to unit, despite discrepancies in unit size, clutter level, or infestation level (K. Vail *et al.*, unpublished results). PMCs do a disservice to themselves and their clients when their charges do not reflect the time and effort required to successfully service infested homes and apartment units.

The bed bug resurgence has also required the use of new protocols and novel equipment. Bed bug remediation equipment can be quite expensive for a PMC to purchase. As whole-home heat treatments become more widespread within the US, PMCs are spending between USD 50 000 and $150 000 for bed bug heating systems. Because whole-home heat treatment can take all day to complete, companies have difficulty making a return on their heat system investment unless they buy multiple heating systems so that they can treat more than one housing unit at a time. For companies that use heat chambers to treat household items, the prices for a heat chamber can range from USD 300 to more than USD 5000, depending on the size and manufacturer of the chamber (Stedfast and Miller, 2015). Steam machines with separate reservoirs and boilers (which allow for continual use) can cost anywhere from USD 500 to USD 2000. The use of bed bug detection canines requires an initial investment of around USD 10 000 per dog (2016 pricing courtesy of Stephanie Taunton, Bed Bug K9S, California) and these dogs require ongoing care and daily training. The cost to the consumer for canine inspection is around USD 300/h (Wang *et al.* 2011).

Applications of spray formulation insecticide is still the least expensive method of bed bug treatment. However, the amount of time required of the technician for detailed inspection and application, at a break-even cost of USD 1.50/min can still be a significant investment for a PMC trying to make a profit. Thus, the company has to ensure that these expenses can be recovered while still making some profit from servicing the account.

15.9 Bed Bug Management Revenues

In 2015, Curl (2016) conducted a study to predict the market forecast for the pest management industry within USA. The conclusions of this study indicated that between the years 2015 and 2016 the total number of bed bug jobs performed was 815 000. These bed bug jobs represented a USD 100 million increase in revenue over the previous year, for a total annual revenue of USD 573.2 million in 2016. Curl (2016) went on to suggest that if the current bed bug infestation trend continues, bed bug revenue might exceed USD 1 billion dollars annually within five years.

In many other countries, bed bug infestations are also continuing to increase. Therefore, there is a concomitant rise in bed bug control services. In the UK, it is predicted that there will be a compound annual growth rate of 4.7% for bed bug control services over 2016–2026, and by 2026, bed bug control will be valued at USD 12 million (Anonymous, 2016). For Germany, this rate is expected to be 5.2%, and by 2026 the market is expected to be around USD 16 million per year (Anonymous, 2016). If it is assumed that across Europe a similar rate of growth will be seen, then based on population sizes (UK: 64.1 m, Germany: 80.6 m, Europe: 743.1 m; Wikipedia, 2016), the bed bug service market for Europe will be of the order of USD 140 million. While it would not be valid to extrapolate these figures to the world bed bug market, due to differences in economic standards between developed and economically disadvantaged nations, it is clear that bed bugs represent a multi-billion dollar business for the global pest management industry.

15.10 Conclusions

It is evident that bed bugs are a major burden on the world economy in many ways, yet surprisingly these amounts have yet to be accurately determined, and can only be estimated as many billions of US dollars annually. For Australia alone, in 2011 it was conservatively estimated that the bed bugs had cost the economy approximately AUD 200 million (USD 149.3 million) since the start of the resurgence (Doggett *et al.*, 2011). With the cost of bed bugs not being fully realized, this means that there is less of a justification for grant-awarding bodies to support applications for research funding. Clearly the economic impact of bed bugs is an area of research that requires some urgency. It is also obvious that fiscal support for bed bug research should be a priority in order to help protect the financial security of the world. Ideally, precedence should be given to those projects that have some promise of leading to technological innovations or enhanced methodologies that result in a reduction in control costs. By making control less expensive, we will be on the road to defeating bed bugs…again.

References

AGCS (2017) *Global Claims, Allianz Global Corporate & Specialty*, http://www.agcs.allianz.com/assets/PDFs/Reports/AGCS-Global-Claims-Review-2017.pdf (accessed 15 May 2017).

Anonymous (2012) *Bedbug Company Fined in Downtown Condo Fire*, http://www.cbc.ca/news/canada/edmonton/bedbug-company-fined-in-downtown-condo-fire-1.1190798 (accessed 21 December 2016).

Anonymous (2016) *UK and Germany Market Study on Bed Bug Control Services: Chemical Control Service Type Segment Expected to Gain Significant Market Share by 2026*, http://www.persistencemarketresearch.com/market-research/uk-and-germany-bed-bug-control-services-market.asp (accessed 21 December 2016).

Arnst, C and Sagar, I. (2004) What's eating the guests? *Business Week* (2 February), p. 14.

Bland, S. (2010) *Bloodsucking Bed Bugs Creep Back In*, http://abcnews.go.com/Technology/victorias-secret-closes-bed-bugs-infestations-rise/story?id=11238695 (accessed 5 January 2017).

Bruno, L. (2010) *A Real Nightmare: Bed Bugs Biting All Over US*, http://www.nbcnews.com/id/38382427/ns/health-health_care/t/real-nightmare-bed-bugs-biting-all-over-us/#.UXdfAytNQ9o (accessed 21 December 2016).

Cooper, R., Wang, C. and Singh, N. (2015) Evaluation of a model community-wide bed bug management program in affordable housing. *Pest Management Science*, **72** (1), 45–56.

Curl, G.D. (2016) *U.S. Structural Pest Control Market to Reach $10 Billion in 2020*, http://www.pctonline.com/article/sc-research-pest-control-market-report/ (accessed 11 February 2017).

De Marco, D. (2004) Sleepyheads beware; Bedbugs rule the night in the U.S. *Knight Ridder Tribune Business News* (9 January), p. 1.

Doggett, S.L. (2013) The financial impacts of bed bugs. In: *Pestaway Australia Bed Bug Workshop Course Notes*, (ed. S.L. Doggett) Institute for Clinical Pathology & Medical Research, Westmead.

Doggett, S.L. and Russell, R.C. (2008) The resurgence of bed bugs, *Cimex* spp. (Hemiptera: Cimicidae) in Australia. In: *Proceedings of the Sixth International Conference on Urban Pests, July 13–16, 2008, Budapest, Hungary*. OOK-Press, Hungary, Veszprém.

Doggett, S.L., Orton, C.J., Lilly, D.G. and Russell, R.C. (2011) Bed bugs: the Australian response. *Insects*, **2** (2), 96–111.

Doggett, S.L., Dwyer, D.E., Peñas, P.F. and Russell, R.C. (2012) Bed bugs: clinical relevance and control options. *Clinical Microbiology Reviews*, **25** (1), 164–192.

Henricksen (2010) *Bed Bugs in America: New Survey Reveals Impact on Everyday Life*, http://www.pestworld.org/news-hub/press-releases/bed-bugs-in-america-new-survey-reveals-impact-on-everyday-life/ (accessed 5 January 2017).

Holbrook, E. (2011) Bed bug insurance. *Risk Management*, **July/August**, 10.

Liu, B. and Pennington-Gray, L. (2015) Bed bugs bite the hospitality industry? A framing analysis of bed bug news coverage. *Tourism Management*, **48**, 33–42.

Liu, B. J., Pennington-Gray, L., Donohoe, H. and Omodior, O. (2015a) New York City bed bug crisis as framed by tourists on TripAdvisor. *Tourism Analysis*, **20**, 243–250.

Liu, B. J., Pennington-Gray, L. and Klemmer, L. (2015b) Using social media in hotel crisis management: the case of bed bugs. *Journal of Hospitality and Tourism Technology*, **6** (2), 102–112.

Liu, B., Kim, H. and Pennington-Gray, L. (2015c) Responding to the bed bug crisis in social media. *International Journal of Hospitality Management*, **47**, 76–84.

Lundine, S. (2003) Sleep tight? Not when the bedbugs bite. *Orlando Business Journal*, **20**, 1.

Meek, F. (2007) Guests consider bed bugs. *Lodging Hospitality*, **63** (16), 94–95.

Nathe, C. (2015) *UK Research: Bed Bugs "Bite" the Wallet of Hotel Owners*, http://uknow.uky.edu/campus-news/uk-research-bed-bugs-bite-wallet-hotel-owners (accessed 21 December 2016).

Odell, A. (2010) *Bedbugs Strike Again, This Time at Victoria's Secret on Lexington*, http://nymag.com/thecut/2010/07/bedbugs_strike_again_this_time.html (accessed 5 January 2017).

Ortiz, E. (2004) Bedbugs make big comeback. *Knight Ridder Tribune Business News* (5 January), p. 1.

Penn, J., Maynard, L.J and Brown, D.O. (2014) The cost of bed bug anxiety: travelers' willingness to pay to avoid them. *Consortium Journal of Hospitality and Tourism*, **19** (1), 22–47.

Potter, M.F., Haynes, K.F., Fredericks, J. and Henriksen, M. (2013) Bed bug nation are we making any progress? *Pestworld*, **September/October**, 4–11.

Potter, M.F., Haynes, K.F. and Fredericks, J. (2015) Bed bugs across America. *Pestworld*, **November/December**, 4–14.

Reinhardt, K., Kempke, D., Naylor, R.A. and Siva-Jothy, M.T. (2009) Sensitivity to bites by the bedbug, *Cimex lectularius*. *Medical and Veterinary Entomology*, **23** (2), 163–166.

Stedfast, M., and Miller, D. (2014) Development and evaluation of a proactive bed bug (Hemiptera: Cimicidae) suppression program for low-income multi-unit housing facilities. *Journal of Integrated Pest Management*, **5** (3), 1–7.

Stedfast, M. and Miller, D. (2015) Turning up the heat: researchers at Virginia Tech put commercial bed bug heat chambers to the test. *Pest Control Technology Magazine*, **43** (6), 94, 96, 98–100, 114.

Tucker, J. (2003) "Don't let the bedbugs bite" is apt in Pueblo, Colo. *Knight Ridder Tribune Business News* (21 May), p. 1.

Wahlberg, D. (2004) Globe-trotting bedbugs find a place to crawl home. *The Atlanta Journal Constitution* (Jan 25), p. A1.

Wang, C., Gibb, T. and Bennett, G.W. (2009) Evaluation of two least toxic integrated pest management programs for managing bed bugs (Heteroptera: Cimicidae) with discussion of a bed bug intercepting device. *Journal of Medical Entomology*, **46** (3), 566–571.

Wang, C., Tsai, W.T., Cooper, R.A. and White, J. (2011) Effectiveness of bed bug monitors for detecting and trapping bed bugs in apartments. *Journal of Economic Entomology*, **104** (1), 274–278.

Wang, C., Singh, N., Zha, C. and Cooper, R. (2016) Bed bugs: prevalence in low-income communities, resident's reactions, and implementation of a low-cost inspection protocol. *Journal of Medical Entomology*, **53** (3), 639–646.

West, M.G. (2010) *Bedbugs Expand Their Horizons Beyond the Bedroom*, http://www.wsj.com/articles/SB10001424052748704525704575341462710948740 (accessed 5 January 2017).

Wikipedia (2016) *List of European Countries by Population*, https://en.wikipedia.org/wiki/List_of_European_countries_by_population (accessed 21 December 2016).

Wojcik, J. (2004) Lawsuits over bedbugs a nightmare for hotels. *Business Insurance*, **38**, 1–2.

Part IV

Bed Bug Biology

16

Bed Bug Biology

Sophie E.F. Evison, William T. Hentley, Rebecca Wilson, and Michael T. Siva-Jothy

16.1 Introduction

The features of bed bug ecology that have driven this insect's co-evolution with humans and lie at the core of understanding its control are:

- obligate blood feeding
- cryptic behavior (anachoresis) in the host's sleeping environment whilst completing their life-cycle
- their ability to cope with inbreeding.

Whilst they have many other unique biological features, these aspects of their behavior underpin their success as parasites and as pests. In this chapter, the examination of basic bed bug biology is structured around these three themes.

Bed bugs are obligate hematophagous insects (blood is their sole source of food) from the true-bug family Cimicidae. Within the 65 known species of cimicid, there are three species that regularly come into contact with, and feed on, humans: the Tropical bed bug, *Cimex hemipterus* (F.), the Common bed bug, *Cimex lectularius* L. (which is also occasionally found in the tropics), and the West African bed bug, *Leptocimex boueti* (Brumpt). There is little information on much of the biology of *C. hemipterus* and *L. boueti*. Consequently, information herein largely pertains to *C. lectularius*, unless otherwise stated.

Cimex lectularius and *C. hemipterus* develop through five nymphal instars before reaching adulthood, with each life stage requiring a blood meal in order to molt to the next instar (Usinger, 1966). Adult fecundity and spermatogenesis are dependent on blood intake, because females cannot produce eggs autogenously, and males cannot produce sperm, without the resources accrued from blood feeding. Consequently, successful growth and development at both the individual and population level means these insects need to be in close proximity to their host(s). The group has earned the name "bed bugs" because they all specialize in feeding on hosts that periodically, and predictably, spend time in spatially constrained retreats (caves, enclosed nests, burrows, and bedrooms). In the case of *C. lectularius* and *C. hemipterus*, concealment within a bed and/or bedroom provides access to a sleeping host and therefore a reliable blood resource.

16.2 Hematophagy

Hematophagy, or blood feeding, is an effective feeding strategy; blood is sterile, and relatively quick and easy to ingest. An adult *C. lectularius* takes between 5 and 20 min to feed to satiation (Usinger, 1966), but *C. hemipterus* is less efficient at feeding and may take longer (Araujo *et al.*, 2009). Blood also contains all the macronutrients the bed bug needs to develop and reproduce. Bed bugs take relatively large blood meals (compared to their body

Advances in the Biology and Management of Modern Bed Bugs, First Edition.
Edited by Stephen L. Doggett, Dini M. Miller, and Chow-Yang Lee.
© 2018 John Wiley & Sons Ltd. Published 2018 by John Wiley & Sons Ltd.

size) and store most of the meal in their anterior midgut, while a smaller portion is digested in the posterior midgut approximately 6–18 h after feeding (as shown in *C. lectularius*; Vaughan & Azad, 1993). Host-specific erythrocyte size may be a constraint on host choice since the diameter of the food canal of early instars needs to be able to accommodate the host's erythrocytes. For example, an adult *C. lectularius* has an internal rostrum diameter of 8–12 μm (Tawfik, 1968) whilst human erythrocytes are 6–8 μm in diameter. By contrast, chicken erythrocytes (a viable alternative food source for *C. lectularius* and *C. hemipterus*) are 11.2 μm in diameter, which suggests that there may be ontogenetic constraints driven by host erythrocyte size differences.

The availability of *ad libitum* blood meals is critical to both post-embryonic development, to female fecundity, and for sperm production. After feeding and mating, a female *C. lectularius* can lay eggs for around 10 days, but she requires another blood meal to continue laying (Usinger, 1966). With *ad libitum* access to a host (and males), a single female can lay 200–500 eggs in her lifetime (Bartley and Harlan, 1974), although fecundity with modern insecticide resistant strains is reported to be considerably less (see, for example, Gordon *et al.*, 2015). However, whilst obligate blood feeding has many advantages, it has a major drawback: blood lacks biotin and riboflavin, two essential micronutrients that insects cannot manufacture *de novo*. As obligate hematophages, bed bugs overcome this limitation through a mutualistic relationship with *Wolbachia* (Hosokawa *et al.*, 2010; see also Chapter 20). The *C. lectularius*-specific strain (*w*Cle) contains complete operons for both riboflavin (vitamin B2) and biotin (vitamin B7) synthesis (Nikoh *et al.*, 2014; Moriyama *et al.*, 2015). The nutritional importance of this symbiont is highlighted by the fact that it is found in both sexes, all life stages, and in laboratory and field populations of bed bugs (Sakamoto and Rasgon, 2006; Hosokawa *et al.*, 2010). Moreover, it is mainly localized in a specially evolved organ called the bacteriome, which is attached to the gonads (Usinger, 1966). Usually insect hosts with symbiotic *Wolbachia* have them spread throughout the host's tissue, rather than being localized (Dobson *et al.*, 1999). The bacteriome consists of bacteriocytic cells that contain endosymbiotic *w*Cle at densities 2000–900 000 times higher than in other bed bug organs (Hosokawa *et al.*, 2010; Nikoh *et al.*, 2014). As *w*Cle is transferred to bed bug embryos via maternal transmission (Hypša and Aksoy, 1997), the location of the bacteriome close to the ovaries suggests adaptive positioning for vertical transmission. The reason for their location next to the testes is less clear, but may function to produce components of the seminal fluid (see Reinhardt *et al.*, 2009a; Moriyama *et al.*, 2012). *Wolbachia*-cured bed bugs (such as those that have been fed with antibiotics) suffer retarded egg and nymphal development, and reduced success at imaginal eclosion, resulting in sterility and reduced fecundity. These effects are reversed when their food is supplemented with B vitamins (Hosokawa *et al.*, 2010).

16.3 Anachoresis

It is likely that hematophagy evolved only once in the cimicids given its obligate nature across the group. All hosts of cimicid species have some common features that have probably driven the host–parasite association: they have relatively high body temperature and occur in predictable, often gregarious assemblages in enclosed spaces such as caves and buildings (Reinhardt *et al.*, 2007). Host switching to humans, in species that specialize in other primary hosts, appears to be driven by starvation in times of absence of the main host (see, for example, Myers, 1928; Overal and Wingate, 1976). However, host switching may also have been facilitated by spatial and temporal host overlap/co-existence, and it is likely that human specialists, such as *C. lectularius*, *C. hemipterus*, and *L. boueti*, switched directly from bats to humans (Usinger, 1966), or via birds due to our close relationship with domesticated species (Weidner, 1958) and species that utilize human habitation, such as swallows and martins. Despite the hypothesized host switching, genetic analyses of populations of bed bugs support the maintenance of two races of *C. lectularius*: human-associated populations that exhibit elevated levels of inbreeding, and populations that remain associated with roosting bats and which exhibit higher levels of genetic diversity, suggesting a different life history in terms of mortality and dispersal (Booth *et al.* 2015).

16.4 Flightlessness

Flightlessness is another shared trait in the cimicids and has some important consequences. A core life-history strategy is concealment in harborages (or refugia) in the hosts' iterative resting place and repetitive feeding on the same host. Consequently, reduction in wing size makes anachoresis easier, and reduced investment in wing-related tissue means resources can be channeled into enhanced egg production in females and sperm production in males (bed bug males are limited in their ability to inseminate females because of the cost of producing seminal fluid – not sperm) (Reinhardt *et al.*, 2011). The reason for the high cost of seminal fluid is likely because males enhance their reproductive investment by transferring these critical micronutrients (notably B vitamins; see, for example, Reinhardt *et al.*, 2009a). Another correlate of winglessness is repeated feeding from the same host (a relative rarity in hematophagous insects in general). Adaptations such as anachoresis, flightlessness, and localization in the host's resting place results in repeated contact between the parasite and the same host. There is therefore little opportunity to vector potential disease organisms and this aspect of bed bug biology may well explain why bed bugs do not transmit disease-causing pathogens (see also Chapter 12). However, the suggestion of genetic divergence of races of bed bugs with different life history characteristics (Booth *et al.*, 2015), and differences in the biology of the three human-associated bed bugs may have important consequences for disease transmission potential.

16.5 Reproduction

Bed bugs have a unique mode of copulation: despite the fact that female bed bugs have a genital tract, males push a hardened needle-like intromittent organ (a modified paramere) through the female's abdominal cuticle and inseminate directly into her body cavity (Carayon, 1966). For this reason, copulation in bed bugs is termed 'traumatic insemination' (TI), and female cimicids have evolved a unique secondary genital system, called the paragenital system, through which sperm move (Carayon, 1966). The structure of this anatomical feature varies across cimicids, from being almost non-existent in *Primicimex cavernis* Barber, to being as, if not more, complex than "normal" internal genitalia in *Crassicimex sexualis* Ferris and Usinger (Siva-Jothy, 2006). During traumatic insemination the male deposits sperm into a unique organ, the mesospermalege, in which it remains for about 4h (Carayon, 1966). Sperm then migrate through the blood into paired sperm-storage organs, the seminal conceptacles, from where they migrate into the oviducts and fertilize eggs in the ovaries (Carayon, 1966). The trauma is anatomically localized in the female because of the presence of a groove in the female's fifth abdominal sternite (the ectospermalege; see Carayon, 1966; Stutt and Siva-Jothy, 2001). Females receive approximately five traumatic inseminations per feeding (not necessarily from the same male), with a re-mating interval of around 17 min (Stutt and Siva-Jothy, 2001). *Cimex lectularius* males direct their sexual interest at recently fed females, who have reduced resistance to male mating attempts (Siva-Jothy, 2006).

The reproductive tract, in the strict sense of the term, of female cimicids persists because it is used during oviposition. Males have never been observed to use the female's genital tract for insemination in any bed bug species. Female bed bugs therefore have both a paragenital tract that functions in mating and through which sperm migrate to their eggs, and a genital tract that functions during oviposition. This mode of copulation is costly for females as energy is expended both to repair damaged cuticle and to mount an immune response to defend against the microbes introduced during copulation (Reinhardt *et al.*, 2005), particularly when males pierce a location of the abdomen rather than the spermalege (Stutt and Siva-Jothy, 2001; Morrow and Arnqvist, 2003), which leads to elevated water stress (Benoit *et al.*, 2012), and a reduction in egg production (Polanco *et al.*, 2011). The adult female counter-adaptation to TI is the mesospermalege, which reduces the costs of TI (Morrow and Arnqvist, 2003; Reinhardt *et al.*, 2003). Reproductively active males also attempt to copulate with recently fed males and nymphs. In these cases, the recipient of the attention secretes alarm pheromone, which deters copulation (Ryne, 2009; Harraca *et al.*, 2010). This alarm response represents another adaptation to offset fitness costs. Males and nymphs lack the anatomical adaptations (the spermalege) of adult females that allow them to withstand traumatic insemination in most bed bug species. However, *Afrocimex constrictus* Ferris and Usinger males also possess a spermalege, and examination of its mating system suggests aspects of the ecology and the operational sex ratio may favor males to be less discriminating about whom they mate with. The male possession of a spermalege may therefore have evolved in this species to offset the accidental traumatic insemination of males by other males because of the increased probability of "mistaken identity", and because of the increased competition for available females (Reinhardt et al., 2007).

Considering that a female will mate with around five males per feeding cycle, and feeding cycles are iterative, there is plenty of opportunity for sperm competition. The ejaculate size of males mating subsequent to the first male is always approximately 50% of the volume of the first (Siva-Jothy and Stutt, 2003). The fact that male bed bugs can detect the presence of another male's ejaculate in their current mate's spermalege (most likely via chemodetection), and consequently reduce the volume of their ejaculate size (Siva-Jothy and Stutt, 2003), lends further support to the presence of as-yet-undocumented sperm competition effects in the bed bug. In addition, because males control the rate of mating (Reinhardt et al., 2009b), the realized mating rate is 12–20 times higher than the rate required for maximum fecundity and, in the absence of other confounding factors, mating reduces female lifespan and fecundity (Stutt and Siva-Jothy, 2001). Second-male sperm appears to take precedence in terms of siring offspring, despite a longer copulation period and ejaculate volume transferred by the first male (Stutt and Siva-Jothy, 2001). This may be due to immunological attack of sperm by the hemocytes in the spermalege (Carayon, 1966), leading to a compensatory increase in the ejaculate size of the male who mates with a virgin female. Because males traumatically puncture the female's cuticle, the likelihood of passing microbes into the female's body cavity is high; the female therefore has to cope with microbial "non-self"' as well as the males' sperm (also "non-self") during the mating process. This adds a unique dimension to sperm competition dynamics in this taxon. The seminal fluid of bed bugs can digest bacterial components through lysozyme-like activity (LLA) (Otti et al., 2009) but shows no antimicrobial peptide activity (as in *Drosophila*). Importantly, there is significant variation in this LLA titer, which may also contribute to sperm competition or precedence after multi-male copulations via differential protection of sperm from pathogens co-introduced during the TI process (Reinhardt et al., 2003, 2005). It is also apparent that components of the ejaculate increase female reproductive output as well as delaying the onset of female reproductive senescence (Reinhardt et al., 2009a), perhaps via increased protection for females from the LLA or because seminal components also provide antioxidants and micronutrients (Reinhardt et al., 2009b; Rao, 1974). Thus components of the ejaculate may provide fitness benefits to females through releasing them from trade-offs between immune defense and fecundity (Reinhardt et al., 2009a). All these details, including the unique route for the passage of sperm from testes to the ovaries and the polyandrous nature of the relationship, result in bed bugs being an important model system for understanding sexual conflict (Chapman, 2006), and particularly in understanding how sperm cells withstand the harsh internal environment of the female (Reinhardt et al., 2009b). Despite the fact that female bed bugs are wounded and infected during copulation (Reinhardt et al., 2003), it appears they still gain a net benefit from ejaculate components in terms of increased reproductive rate and delayed reproductive senescence (Reinhardt et al., 2009a).

Another unique feature of bed bug biology is that fertilization occurs in the ovaries and females lay eggs that have already begun the early stages of embryogenesis. After TI, the sperm move from the spermalege to the seminal concepticles. From here they move toward the ovaries through hemolymph-filled canals running inside the oviduct walls (the spermodes; Carayon, 1954; Davis, 1956; Steigner, 2001). Sperm eventually arrive in the ovaries, where they either fertilize mature oocytes or die. Sperm are stored in the seminal conceptacles (Carayon, 1966) and the sperm number transferred during the average copulation period is sufficient for *C. lectularius* females to lay fertile eggs for approximately 35–50 days after sexual isolation (Davis, 1956; Khalifa, 1952; Stutt and Siva-Jothy, 2001). Sperm metabolic rate is positively correlated with the production of damaging oxygen radicals, but in stored sperm these two components of sperm function are suppressed by the female for up to an average of 9.5 weeks (Reinhardt and Ribou, 2013). After this time, females become infertile unless they re-mate, which coincides with the average time (of 8–9 weeks) that it takes sexually isolated females to produce sexually mature sons in a new infestation (Saenz *et al.*, 2012; Reinhardt and Ribou, 2013): A single fertilized female can therefore generate a new colony not only by producing interbreeding offspring, but also by re-entering the breeding pool once her sons reach sexual maturity.

16.6 Egg Laying

In most species of bed bug, egg production begins 2–5 days after a blood meal, and maximum egg production ensues during the first four feeding/oviposition cycles (Johnson, 1940; How and Lee, 2010; Polanco *et al.*, 2011). Some strains of bed bug have been shown to display differential egg production patterns, but overall female fecundity does not appear to differ substantially among strains (Polanco *et al.*, 2011). However, sexually isolated females show an increased laying rate due to release from the costs of multiple TIs (Stutt and Siva Jothy, 2001), a phenomenon likely to promote successful growth of an initial infestation (Polanco *et al.*, 2011). Egg production stops rapidly if females do not feed, presumably because the proteins required for egg production are lacking or because sperm migration to the ovaries is not initiated (Davis, 1956). Laying senescence starts between days 30 and 200 of adult age, depending on species (Usinger, 1966; Ryckman *et al.*, 1981) and temperature (Janisch, 1935).

16.7 Host-seeking Behavior

Bed bug anachoresis is a major factor that hinders their effective control. Their habit of spending large amounts of time in the refuge means they are hard to detect, especially in the early phases of an infestation, and are inaccessible to most control agents and application methods. There are no empirical data on natural behavior in and around harborages, so most of what is known about leaving and returning to harborages comes from laboratory studies.

Leaving the refuge to feed is a necessity, and exposure during host-seeking and feeding is likely to carry the highest risk of mortality and so drives the bed bug's tendency to spend as little time as possible outside the harborage. Consequently, understanding the drivers and processes underpinning host-seeking and refuge-seeking behaviors are important goals to inform new pest-control strategies. When exposed to dark, heat, and carbon dioxide cues (see, for example, Rivnay, 1932; Singh *et al.*, 2012), hungry bed bugs will begin host seeking (Aak *et al.*, 2014). However, it is likely that there are other cues used to successfully find a suitable host: body odor is a common attractant in other hematophages (for example, mosquitoes; Takken and Verhulst, 2013) and vibration is also likely to be a signal for host presence. After feeding, the insects return to the refuge within around 30 min but if they are unable to feed, both sexes will continue to search for a host for many hours, before returning to the refuge approximately 2 h before daytime (Reis and Miller, 2011). In the presence of appropriate host stimuli, which modulate the movement of bed bugs (Anderson *et al.*, 2009; Harraca *et al.*, 2012), host-seeking behavior is characterized by a spiraling and tortuous meander (Rivnay, 1932). Bed bugs do not appear to wait for hosts if they are starved; rather they travel into other rooms or apartments, and infestations and can spread out from infested apartments to neighboring apartments via the hallways (Wang *et al.*, 2009, 2010). Feeding behavior in *C. lectularius* coincides with periods of minimal host activity (usually sleeping). After an adult *C. lectularius* fully engorges on host blood, it returns to its refugium. Under *ad libitum* conditions, imaginal *C. lectularius* feed once per week (Usinger, 1966). This iterative behavior is likely to be a facultative adaptation to the host's circadian rhythms and *C. lectularius* shows a clear nocturnal pattern (Romero *et al.*, 2010). Activity patterns also appear to change contingent on different cues and starvation level (Romero *et al.*, 2010).

Bugs that have recently fed generally show low levels of activity after they have returned to their refugia (Reis and Miller, 2011), but how activity changes with level of starvation appears to be gender dependent (Romero *et al.*, 2010). Male activity increases for a few days after feeding, presumably as a resource-driven trigger for mate finding, but then decreases. Conversely, female activity is low in the first few days after feeding and increases as starvation ensues (Romero *et al.*, 2010). Generally, female responses to the host are stronger and become active sooner after starvation (Aak *et al.*, 2014). These differences in behavior between the sexes map onto the diverging energy resources required for reproduction; that is, egg production begins around four days after feeding, which corresponds to the increase in female activity post-feed (Usinger, 1966; Aak *et al.*, 2014). Females also found new infestations, so their increased activity may be a bet-hedging strategy to distribute eggs to form new infestations (Hopper, 1999; Aak *et al.*, 2014).

16.8 Harborage Seeking Behavior and Aggregation

The bed bug's anachoretic behavior drives these insects into spatially restricted harborages. However, they do not distribute themselves randomly with respect to the positioning or the availability of these refugia. They tend to use harborages close to the host (presumably to reduce travel time to and from feeding) and they tend to aggregate. Harborage preference is driven by the presence of volatiles derived from faeces, exuvia, and the bed bug aggregation pheromone (Siljander *et al.*, 2008; Olson *et al.*, 2009; Domingue *et al.*, 2010; see also Chapter 17). Experiments by Naylor (2012) revealed that when given a free choice of a continuous harborage (a single three-meter strip of paper), bed bugs did not all cluster at the end closest to the feeding point, but distributed themselves in discrete clusters. This phenomenon is commonly seen in the field (Figure 16.1) and it is unclear why bed bugs distribute themselves in this way. Equally unclear are the factors and sensory drivers that bed bugs use to choose new harborages, either as a population expands within an infestation, or as it colonizes a new site after dispersing. Understanding these features of their biology will be critical in designing effective traps and monitors. Naylor (2012) also showed that habitat complexity would alter the tendency for bed bugs to disperse. Habitats with more complexity (three strips as opposed to one or two) sustained larger populations for a longer period. The habitat with a single strip showed evidence of dispersing bed bugs (moving more than 3 m away from the host) long before the phenomenon became apparent in the other treatments.

16.9 Dispersal

There are two types of dispersal used by insects: active and passive. Active dispersal is where the organism moves through its own ability. Although more commonly seen in mammals and birds, some insect species are capable of actively dispersing great distances. For example, the Monarch butterfly (*Danaus plexippus* L.) is capable of actively dispersing by flying thousands of miles (Urquhart and Urquhart, 1978). Active dispersal for flightless insects such as *C. lectularius* has a much more limited range and is likely to not be important over large distances. Passive dispersal is where the organism relies on processes such as wind or other organisms for dispersal. The complex network of human movement and trade acts as a vector for passive insect dispersal. For example, a new insect species is found in 1 out of 54 shipping containers entering US ports (Work *et al.*, 2005).

The global resurgence of *C. lectularius* has highlighted how little is currently known of its dispersal patterns and what drives dispersal. Measuring the dispersal of any insect species between populations is very difficult, but advances in molecular tools allow genetic differences and relatedness within and between populations to be easily determined. Several studies have investigated relatedness between *C. lectularius* populations at different spatial scales (see also Chapter 18). When multiple infestations occur within a building such as an apartment complex, there appears to be low genetic variability between infestations, suggesting all populations within the complex arise from a single (or few) founding female(s) (Booth *et al.*, 2012; Saenz *et al.*, 2012; Akhoundi *et al.*, 2015; Raab *et al.*, 2016). In contrast, there is a high level of

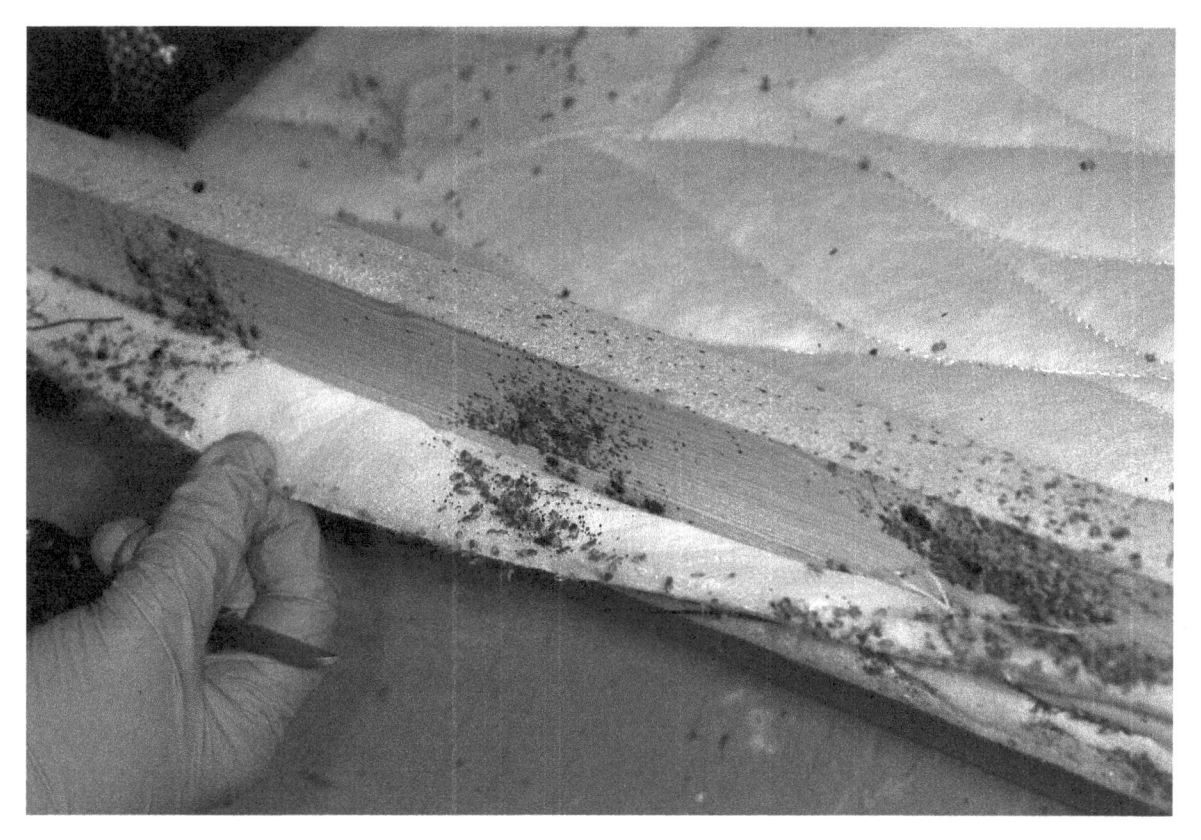

Figure 16.1 An infested bed frame showing the distinctive periodic and equidistant distribution of bed bug "clusters", despite the homogeneity of the refuge (courtesy of Richard Naylor, The Bed Bug Foundation, UK).

genetic divergence between *C. lectularius* populations infesting different buildings across cities (Akhoundi *et al.*, 2015), and both within (Saenz *et al.*, 2012; Narain *et al.*, 2015) and between countries (Fountain *et al.*, 2014). The distance between these isolated populations has no effect on their relatedness (Akhoundi *et al.*, 2015) suggesting that *C. lectularius* establish in buildings stochastically after spreading through passive dispersal.

What triggers localized dispersal in *C. lectularius* is currently unknown, but is a major issue when trying to control the spread of *C. lectularius* infestations within apartment blocks. When placed in the middle of a small (71 × 51 cm) lit arena, unfed bed bugs were less likely to disperse from the release point than those recently fed (Goddard *et al.*, 2015). This is contrary to other established theories, which suggest lack of a host – that is, hunger – is a trigger for dispersal (see, for example, Pinto *et al.*, 2007; Cooper *et al.*, 2015). Population density may be a cue triggering active dispersal; Naylor (2012) found that dispersal behavior in a laboratory population only occurred as a consequence of increasing population size.

Directly measuring *C. lectularius* dispersal in the field by traditional methods such as mark-recapture is difficult due to the size of the organism and the ethical implications for the human occupants of the infested building. One study overcame these obstacles: Cooper *et al.* (2015) collected, marked, and released *C. lectularius* in six apartments and monitored all adjoining units. They found patterns of extensive movement within and between apartments regardless of whether there was a host present or not, suggesting these insects have active mobility that may not be driven by hunger alone, but instead by endogenous rhythms of activity. Laboratory work by Romero *et al.* (2010) showed evidence that spontaneous activity (non-host dependent) is driven by a circadian rhythm. Another surprising result from this study was extensive periods of activity and survival of even early instars in the absence of hosts; unfed first and second instars were still alive after five months of no host presence. However, as the study was performed in a temperate location, the results are unlikely to be representative of all latitudes.

Single mated females have been suggested to be the dispersing phenotype (see, for example, Pfiester *et al.*, 2009) because they can "seed" a new population. However, it is unclear whether dispersal is an active or stochastic process: whether all ages and both sexes are equally likely to "disperse" (as suggested by Cooper *et al.*, 2015), or whether only mated females disperse. One hypothesis is that mated females disperse to avoid excessive TI by males. Given that observed sex ratios in natural populations tend towards 1:1 (Stutt and Siva-Jothy, 2001) it is unlikely that females have a higher propensity to disperse compared to males. Whatever the dispersal mechanism, newly established populations will have gone through a bottle-neck; it is therefore likely that bed bugs are adapted to repeated cycles of inbreeding. Population genetic data supports the notion that new infestations go through a bottle-neck (see, for example, Fountain *et al.*, 2014), but is less clear whether bed bugs are adapted to inbreeding or suffer from outbreeding depression. Fountain *et al.* (2015) created inbred and outbred crosses and showed that outbred lines had significantly higher survival rates under starvation conditions compared to inbred offspring. However, by the third generation of outbreeding, there was no longer a survival benefit. Outbreeding may initially result in fitness advantages under stressful conditions, but these are lost if outbreeding continues. At present it is unclear how, or if, bed bugs cope with the deleterious effects of repeated cycles of inbreeding.

In this chapter, aspects of bed bug biology are reviewed under the themes that drive their "pest" status: their feeding, their anachoretic behavior, and their ability to cope with population bottle-necks. These traits have made them incredibly effective human ectoparasites and underpin the issues that make them so difficult to control.

References

Aak, A., Rukke, B.A., Soleng, A. and Rosnes, M.K. (2014) Questing activity in bed bug populations: male and female responses to host signals. *Physiological Entomology*, **39** (3), 199–207.

Akhoundi, M., Kengnem P., Cannet, A., *et al.* (2015) Spatial genetic structure and restricted gene flow in bed bugs (*Cimex lectularius*) populations in France. *Infection, Genetics and Evolution*, **34**, 236–243.

Anderson, J.F., Ferrandino, F.J., McKnight, S., *et al.* (2009) A carbon dioxide, heat and chemical lure trap for the bed bug, *Cimex lectularius*. *Medical and Veterinary Entomology*, **23** (2), 99–105.

Araujo, R.N., Costa, F.S., Gontijo, N.F., *et al.* (2009) The feeding process of *Cimex lectularius* (Linnaeus 1758) and *Cimex hemipterus* (Fabricius 1803) on different bloodmeal sources. *Journal of Insect Physiology*, **55** (12), 1151–1157.

Bartley, J.D. and Harlan, H.J. (1974) Bed bug infestation: its control and management. *Military Medicine*, **139** (11), 884–886.

Benoit, J.B., Jajack, A.J. and Yoder, J.A. (2012) Multiple traumatic insemination events reduce the ability of bed bug females to maintain water balance. *Journal of Comparative Physiology*, **182** (2), 151–172.

Booth, W., Saenz, V.L., Santangelo, R.G., *et al.* (2012) Molecular markers reveal infestation dynamics of the bed bug (Hemiptera: Cimicidae) within apartment buildings. *Journal of Medical Entomology*, **49** (3), 535–546.

Booth, W., Balvin, O., Vargo, E.L., *et al.* (2015) Host association drives genetic divergence in the bed bug, *Cimex lectularius*. *Molecular Ecology*, **24** (3), 980–992.

Carayon, J. (1954) Un type nouveau d'appareil glandulaire propre aux males de certains Hémiptères Anthocoridae. *Bulletin du Muséum National d'Histoire Naturelle*, **26**, 602–606.

Carayon, J. (1966) Paragenital System, in *Monograph of the Cimicidae (Hemiptera – Heteroptera)* (ed. R. Usinger), Entomological Society of America, College Park, pp. 81–167.

Chapman, T. (2006) Sexual conflict. *Nature*, **439**, 537.

Cooper, R.A., Wang, C. and Singh, N. (2015) Mark-release-recapture reveals extensive movement of bed bugs (*Cimex lectularius* L.) within and between apartments. *PLoS ONE*, **10** (9), e013642.

Davis, N.T. (1956) The morphology and functional anatomy of the male and female reproductive systems of *Cimex lectularius* L. (Heteroptera, Cimicidae). *Annals of the Entomological Society of America*, **49** (5), 466–493.

Dobson, S.L., Bourtzis, K., Braig, H.R., et al. (1999) *Wolbachia* infections are distributed throughout insect somatic and germ line tissues. *Insect Biochemistry and Molecular Biology*, **29** (2), 153–160.

Domingue, M.J., Kramer, M. and Feldlaufer, M.F. (2010) Sexual dimorphism of arrestment and gregariousness in the bed bug (*Cimex lectularius*) in response to cuticular extracts from nymphal exuviae. *Physiological Entomology*, **35** (3), 203–213.

Fountain, T., Duvaux, L., Horsburgh, G., Reinhardt, K. and Butlin, R.K. (2014) Human-facilitated metapopulation dynamics in an emerging pest species, *Cimex lectularius*. *Molecular Ecology*, **23** (5), 1071–1084.

Fountain, T., Butlin, R. K., Reinhardt, K. and Otti, O. (2015) Outbreeding effects in an inbreeding insect, *Cimex lectularius*. *Ecology and Evolution*, **5** (2), 409–418.

Goddard, J., Caprio, M. and Goddard, J. (2015) Diffusion rates and dispersal patterns of unfed *versus* recently fed bed bugs (*Cimex lectularius* L.). *Insects*, **6** (4), 792–804.

Gordon, J.R., Potter, M.F. and Haynes, K.F. (2015) Insecticide resistance in the bed bug comes with a cost. *Scientific Reports*, **5** (10807), 1–7.

Harraca, V., Ryne, C. and Ignell, R. (2010) Nymphs of the common bed bug (*Cimex lectularius*) produce anti-aphrodisiac defence against conspecific males. *BMC Biology*, **8**, 121.

Harraca, V., Ryne, C., Birgersson, G. and Ignell, R. (2012) Smelling your way to food: can bed bugs use our odour? *Journal of Experimental Biology*, **215** (4), 623–629.

Hopper, K.R. (1999) Risk-spreading and bet-hedging in insect population biology. *Annual Review of Entomology*, **44**, 535–560.

Hosokawa, T., Koga, R., Kikuchi, Y., Meng, X. and Fukatsu, T. (2010) *Wolbachia* as a bacteriocyte-associated nutritional mutualist. *Proceedings of the National Academy of Sciences of the United States of America*, **107** (2), 769–774.

How, Y.F. and Lee, C.Y. (2010) Fecundity, nymphal development and longevity of field-collected tropical bedbugs, *Cimex hemipterus*. *Medical and Veterinary Entomology*, **24** (2), 108–116.

Hypša, V. and Aksoy, S. (1997) Phylogenetic characterization of two transovarially transmitted endosymbionts of the bedbug *Cimex lectularius* (Heteroptera: Cimicidae). *Insect Molecular Biology*, **6** (3), 301–304.

Janisch, E. (1935) Über die Vermehrung der Bettwanze *Cimex lectularius* in verschiedenen Temperaturen (Beobachtungen bei der Aufzucht von Bettwanzen II). *Parasitology Research*, **7** (4), 408–439.

Johnson, C.G. (1940) Development, hatching and mortality of the eggs of *Cimex lectularius* L. (Hemiptera) in relation to climate, with observations on the effects of reconditioning to temperature. *Parasitology*, **32** (2), 345–461.

Khalifa, A. (1952) A contribution to the study of reproduction in the bed-bug (*Cimex lectularius* L.). *Bulletin de la Societe Fouad ler d'Entomologie*, **36**, 311–336.

Moriyama, M., Hosokawa, T., Futahashi, R., et al. (2012) Comparative transcriptomics of the bacteriome and the spermalege of the bedbug *Cimex lectularius* (Hemiptera: Cimicidae). *Applied Entomology and Zoology*, **47** (3), 233–243.

Moriyama, M., Nikoh, N., Hosokawa, T. and Fukatsu, T. (2015) Riboflavin provisioning underlies *Wolbachia*'s fitness contribution to its insect host. *mBio*, **6** (6), e01732–15.

Morrow, E.H. and Arnqvist, G. (2003) Costly traumatic insemination and a female counter-adaptation in bed bugs. *Proceedings of the Royal Society of London Series B*, **270** (1531), 2377–2381.

Myers, L.E. (1928) The American swallow bug, *Oeciacus vicarius* Horvath (Hemiptera, Cimicidae). *Parasitology*, **20** (2), 159–172.

Narain, R.B. and Kamble, S.T. (2015) Effects of Ibuprofen and caffeine concentrations on the common bed bug (*Cimex lectularius* L.) feeding and fecundity. *Entomology, Ornithology & Herpetology*, **4** (2), 152.

Naylor, R. (2012) Ecology and dispersal of the bedbug. PhD thesis, University of Sheffield, UK.

Nikoh, N., Hosokawa, T., Moriyama, M., *et al.* (2014) Evolutionary origin of insect-*Wolbachia* nutritional mutualism. *Proceedings of the National Academy of Sciences of the United States of America*, **111** (28), 10257–10262.

Olson, J.F., Moon, R.D. and Kells, S.A. (2009) Off-host aggregation behavior and sensory basis of arrestment by *Cimex lectularius* (Heteroptera: Cimicidae) *Journal of Insect Physiology*, **55** (6), 580–587.

Otti, O., Naylor, R., Siva-Jothy, M.T. and Reinhardt, K. (2009) Bacteriolytic activity in the ejaculate of an insect. *American Naturalist*, **174** (2), 292–295.

Overal, W.L. and Wingate, L.R. (1976) The biology of the batbug *Stricticimex antennatus* (Hemiptera: Cimicidae) in South Africa. *Annals of the Natural History Museum of Pietermaritzburg*, **22** (3), 821–828.

Pfiester, M., Koehler, P. and Pereira, R. (2009) Effect of population structure and size on aggregation behavior of *Cimex lectularius* (Hemiptera: Cimicidae). *Journal of Medical Entomology*, **46** (5), 1015–1020.

Pinto, L.J., Cooper, R. and Kraft, S.K. (2007) *Bed Bug Hand-Book: The Complete Guide to Bed Bugs and Their Control*, Pinto & Associates. Mechanicsville, MD.

Polanco, A.M., Miller, D.M. and Brewster, C.C. (2011) Reproductive potential of field-collected populations of *Cimex lectularius* L. and the cost of traumatic insemination. *Insects*, **2** (2), 326–335.

Raab, W.R., Moore, J.E., Vargo, E.L., *et al.* (2016) New introductions, spread of existing matrilines, and high rates of resistance result in chronic infestation of bed bugs (*Cimex lectularius* L.) in lower-income housing. *PLoS ONE*, **11** (2), e0117805.

Rao, H.V. (1974) Free amino acids in the seminal fluid and haemolymph of *Cimex lectularious* [sic] L. and their possible role in sperm metabolism. *Indian Journal of Experimental Biology*, **12** (4), 346–348.

Reinhardt, K. and Ribou, A.C. (2013) Females become infertile as the stored sperm's oxygen radicals increase. *Scientific Reports*, **3** (2888), 1–5.

Reinhardt, K., Naylor, R. and Siva-Jothy, M.T. (2003) Reducing a cost of traumatic insemination: Female bedbugs evolve a unique organ. *Proceedings of the Royal Society of London Series B*, **270** (1531), 2371–2375.

Reinhardt, K., Naylor, R. and Siva-Jothy, M.T. (2005) Potential sexual transmission of environmental microbes in a traumatically inseminating insect. *Ecological Entomology*, **30** (5), 607–611.

Reinhardt, K., Harney, E., Naylor, R., Gorb, S. and Siva-Jothy, M.T. (2007) Female-limited genitalia polymorphism in a traumatically inseminating insect. *American Naturalist*, **170** (6), 931–935.

Reinhardt, K., Naylor, R.A. and Siva-Jothy, M.T. (2009a) Ejaculate components delay reproductive senescence while elevating female reproductive rate in an insect. *Proceedings of the National Academy of Sciences of the United States of America*, **106** (51), 21743–21747.

Reinhardt, K., Naylor, R. and Siva-Jothy, M.T. (2009b) Situation exploitation: higher male mating success when female resistance is reduced by feeding. *Evolution*, **63** (1), 29–39.

Reinhardt, K., Naylor, R. and Siva-Jothy, M.T. (2011) Male mating rate is constrained by seminal fluid availability in bedbugs, *Cimex lectularius*. *PLoS ONE*, **6** (7), e22082.

Reis, M.D. and Miller, D.M. (2011). Host searching and aggregation activity of recently fed and unfed bed bugs (*Cimex lectularius* L.). *Insects*, **2** (2), 186–194.

Rivnay, E. (1932) Studies in the tropisms of the bed bug *Cimex lectularius* L. *Parasitology*, **24** (1), 121–136.

Romero, A., Potter, M.F. and Haynes, K.F. (2010) Circadian rhythm of spontaneous locomotor activity in the bed bug, *Cimex lectularius* L. *Journal of Insect Physiology*, **56** (11), 1516–1522.

Ryckman, R.E., Bentley, D.G. and Archbold, E.F. (1981) The Cimicidae of the Americas and oceanic islands, a checklist and bibliography. *Bulletin of the Society of Vector Ecology*, **6**, 93–142.

Ryne, C. (2009) Homosexual interactions in bed bugs: alarm pheromones as male recognition signals. *Animal Behaviour*, **78** (6), 1471–1475.

Saenz, V.L., Booth, W., Schal, C. and Vargo, E.L. (2012) Genetic analysis of bed bug populations reveals small propagule size within individual infestations but high genetic diversity across infestations from the eastern United States. *Journal of Medical Entomology*, **49** (4), 865–875.

Sakamoto, J.M. and Rasgon, J.L. (2006) Geographic distribution of *Wolbachia* infections in *Cimex lectularius* (Heteroptera: Cimicidae). *Journal of Medical Entomology*, **43** (4), 696–700.

Siljander, E., Gries, R., Khaskin, G. and Gries, G. (2008) Identification of the airborne aggregation pheromone of the common bed bug, *Cimex lectularius*. *Journal of Chemical Ecology*, **34** (6), 708–718.

Singh, N., Wang, C.L., Cooper, R. and Liu, C. (2012) Interactions among carbon dioxide, heat, and chemical lures in attracting the bed bug, *Cimex lectularius* L. (Hemiptera: Cimicidae). *Psyche: A Journal of Entomology*, **2012**, 1–9.

Siva-Jothy, M.T. (2006) Trauma, disease and collateral damage: conflict in cimicids. *Philosophical Transactions of the Royal Society of London Series B*, **361** (1466), 269–275.

Siva-Jothy, M.T. and Stutt, A. (2003) A matter of taste: direct detection of mating status in the bed bug. *Proceedings of the Royal Society of London Series B*, **270** (1515), 649–652.

Steigner, A. (2001) Licht- und elektronenmikroskopische Untersuchungen am inneren Genitalapparat weiblicher Bettwanzen (*Cimex lectularius* Linnaeus 1758, Cimicidae, Heteroptera). PhD thesis. University of Saarbrücken, Germany.

Stutt, A.D. and Siva-Jothy, M.T. (2001) Traumatic insemination and sexual conflict in the bed bug *Cimex lectularius*. *Proceedings of the National Academy of Sciences of the USA*, **98** (10), 5683–5687.

Takken, W. and Verhulst, N.O. (2013) Host preferences of blood-feeding mosquitoes. *Annual Review of Entomology*, **58**, 433–453.

Tawfik, M.S. (1968) Feeding mechanisms and the forces involved in some blood-sucking insects. *Quaest Entomology*, **4** (3), 92–111.

Urquhart, F.A. and Urquhart, N.R. (1978). Autumnal migration routes of the eastern population of the monarch butterfly (*Danaus p. plexippus* L.; Danaidae; Lepidoptera) in North America to the overwintering site in the Neovolcanic Plateau of Mexico. *Canadian Journal of Zoology*, **56**, 1759–1764.

Usinger, R.L. (1966) *Monograph of Cimicidae (Hemiptera - Heteroptera)*, Entomological Society of America, College Park.

Vaughan, J.A. and Azad, A.F. (1993) Patterns of erythrocyte digestion by bloodsucking insects: constraints on vector competence. *Journal of Medical Entomology*, **30** (1), 214–216.

Wang, C., Gibb, T., Bennett, G.W. and McKnight, S. (2009) Bed bug (Heteroptera: Cimicidae) attraction to pitfall traps baited with carbon dioxide, heat, and a chemical lure. *Journal of Economic Entomology*, **102** (4), 1580–1585.

Wang, C., Saltzmann, K., Chin, E. Bennett, G.W. and Gibb, T. (2010) Characteristics of *Cimex lectularius* (Hemiptera: Cimicidae), infestation and dispersal in a high-rise apartment building. *Journal of Medical Entomology*, **103** (1), 172–177.

Weidner, H. (1958) Die entstehung der Hausinsekten. *Zeitschrift für Angewandte Entomologie*, **42**, 429–447.

Work, T.T., McCullough, D.G., Cavey, J.F. and Komsa, R. (2005) Arrival rate of nonindigenous insect species into the United States through foreign trade. *Biological Invasions*, **7**, 323–332.

17

Chemical Ecology

Gerhard Gries

17.1 Introduction

This chapter summarizes current knowledge about the chemical ecology of the Common bed bug, *Cimex lectularius* L., and the Tropical bed bug, *Cimex hemipterus* (F.), and is organized in three sections. Section 17.2 highlights the sensory basis (olfaction and contact chemoreception) for the reception of semiochemicals (semio (Latin) = message; semiochemicals = message-bearing chemicals), which precedes any behavioral response. Section 17.3 describes the types of pheromones that *C. lectularius* and *C. hemipterus* produce for communication, and the analytical methods and behavioral bioassays applied to identify them. Section 17.4 presents the foraging cues that *C. lectularius* exploits to locate hosts. This section is concise because the topic is also covered in Chapter 16. As *C. lectularius* has been more intensely studied, it is the focal species in this chapter and will be referred to as "bed bug". If *C. hemipterus* is mentioned in any context, it is only if pertinent information is available.

17.2 Olfaction and Contact Chemoreception

17.2.1 General Introduction

Sensory reception of odorants in hematophagous insects proceeds in several steps (Gullan and Cranston, 2000; Logan and Birkett, 2007; Guidobaldi *et al.*, 2014). Odorants enter the olfactory sensilla on the insects' antennae through pores. Odorant-binding proteins then transport odor molecules through the sensillum lymph to specific odorant receptors located in dendrites of olfactory receptor neurons (ORNs) (Guidobaldi *et al.*, 2014). The variable chain of the olfactory receptor binds the odor molecule and confers specificity, whereas the somewhat universal constant chain of the receptor (Jones *et al.*, 2005) functions as an odorant receptor co-receptor (Orco) that is critical for olfaction but does not directly bind to odorants. Upon binding an odor molecule, the odorant receptor/Orco complex mediates cation influx and signal transduction (Guidobaldi *et al.*, 2014), generating receptor and action potentials (Hallem *et al.*, 2006; Grant and Dickens, 2011; Leal, 2013) that are interpreted by the central nervous system, and followed by behavioral responses. In single sensillum recordings, these neuronal responses to odorants can be characterized in terms of number of action potentials (spikes), the spike amplitude, and the temporal dynamics in spike occurrence following exposure to odorants (Liu and Liu, 2015).

In bed bugs, olfaction and contact chemoreception are of paramount importance in the context of foraging and intraspecific communication (aggregation, alarm behavior, and mate recognition). The sensory receptors are located on the antennae and have been investigated in morphological, histological, electrophysiological, molecular, and behavioral studies, sometimes with partially or completely antennectomized insects.

Advances in the Biology and Management of Modern Bed Bugs, First Edition.
Edited by Stephen L. Doggett, Dini M. Miller, and Chow-Yang Lee.
© 2018 John Wiley & Sons Ltd. Published 2018 by John Wiley & Sons Ltd.

17.2.2 Olfactory Sensilla on Bed Bug Antennae and their Responses to Odorants

The antenna of bed bugs consists of a barrel-shaped scapus, an elongated pedicellus, and a two-segmented flagellum (Levinson *et al.*, 1974a). The 44 olfactory sensilla are located on the terminal antennal flagellum (Steinbrecht and Müller, 1976). They comprise 9 grooved peg sensilla (C sensilla) housing 4–5 olfactory cells, 6 smooth peg sensilla (D sensilla) housing 8–19 olfactory cells, and 29 hairs (E sensilla) housing 1–3 olfactory cells (Steinbrecht and Müller, 1976; Harraca *et al.*, 2010). The remaining surface of the antenna bears bristles that have mechanoreceptive or contact chemoreceptive functions (Steinbrecht and Müller, 1976).

Morphological studies of *C. hemipterus* and *C. lectularius* antennae have come to inconsistent conclusions regarding the number, distribution, and types of sensilla borne on the antennae. Singh *et al.* (1996) inferred that *C. hemipterus* antennae have more olfactory and mechanoreceptive sensilla than *C. lectularius* antennae, and that *C. hemipterus* females have 1.5 times more C- and D-type sensilla than males. Other studies (Prakash *et al.*, 1996; Liedtke *et al.*, 2011) found no such sexual dimorphism and concluded (Liedtke *et al.*, 2011) that the total number, distribution, and types of sensilla born on *C. hemipterus* antennae correlate well with findings reported for *C. lectularius* (Steinbrecht and Müller, 1976; Harraca *et al.*, 2010).

The morphology of bed bug sensilla is correlated with the electrophysiological responses of the ORNs that they house (Harraca *et al.*, 2010). Subjecting sensilla to odorants that elicited behavioral responses from bed bugs or other hematophagous insects revealed that ORNs in all nine grooved peg (C) sensilla respond to ammonia in a dose-dependent manner, whereas ORNs in the six smooth peg (D) sensilla responded to α-pinene, indole, ethyl butyrate, 6-methyl-5-hepten-2-one, (*E*)-2-hexanal, (*E*)-2-octenal, and dimethyltrisulfide (DMTS) (Harraca *et al.*, 2010). D sensilla were grouped into three functional pairs (Dα, Dβ, Dγ) based on their response spectra to test odorants (Harraca *et al.*, 2010). Interestingly, the two aldehydes and DMTS were later determined to be essential components of the bed bug aggregation pheromone (Gries *et al.*, 2015).

As bed bugs sense human host odors (Aboul-Nasr and Erakey, 1968), neuronal responses of bed bugs to 104 human odorants were thoroughly investigated by Liu and Liu (2015). The study revealed that D-type receptors detect human odorants, and that aldehydes (C3–C10) and various types of alcohols, in particular, trigger strong receptor responses. This suggests that these two groups of odorants may play a role as semiochemicals during foraging. In contrast, carboxylic acids, the largest cohort of human odorants that attract other blood-feeding insects (Lehane, 2005), elicited only weak receptor responses (Liu and Liu, 2015).

In a study focused on potential repellent semiochemicals for bed bugs, Liu *et al.* (2014) investigated the responses of olfactory sensilla to synthetic or botanic odor sources that are repellent to other insects and ticks. In their study, Dγ sensilla produced high firing rates (spikes/sec) when exposed to terpenes and terpenoids including eucalyptol and camphor. Specific sensilla of *Culex quinquefasciatus* Say mosquitoes also respond to eucalyptol and camphor, supporting the notion that olfactory sensilla may have converged function across blood-feeding insects (Liu *et al.*, 2013). Whether and to what extent terpenes or terpenoids are equally repellent to various blood-feeding insects is yet to be investigated.

17.2.3 Molecular Mechanisms Underlying Olfaction

The molecular mechanisms underpinning the electrophysiological responses of bed bug ORNs to odorants have rarely been examined. Employing RNA sequencing of bed bug sensory organs, Hansen *et al.* (2014) identified partial cDNAs of at least six odor-binding proteins and of at least 16 odorant receptors. They also characterized the bed bugs' Orco, a 451 amino acid protein, and reported its expression in olfactory receptor neurons and spermatozoa. In a nearly parallel study, Liu and Liu (2015) succeeded in annotating two putative odorant receptors and their co-receptors, expressing them in *Xenopus* (African clawed frog) oocytes, and in recording action potentials of such oocytes in response to human odorants. The Orco amino sequence of bed bugs most closely aligns with that of the kissing bug *Rhodnius prolixus* (Stahl), and resembles the Orco of plant bugs, such as *Apolygus lucorum* (Meyer-Dür) and *Lygus hesperus* (Knight), more closely than that of other insects, including mosquitoes (Hansen *et al.*, 2014; Liu and Liu, 2015).

Much less is known about contact chemoreceptors on the antennae of bed bugs. In a study aimed at understanding the sensory basis for the arrestment behavior (aggregation by reduced or suspended locomotory activity) of bed bugs (Olson *et al.*, 2009), the propensity of bed bugs to arrest on bed bug-exposed paper, which contains the aggregation pheromone including the arrestment-inducing component histamine (Gries *et al.*, 2015), was reduced following the removal of the antennal pedicel (Olson *et al.*, 2009). It is thus conceivable that the contact chemoreceptor for histamine is located on the pedicel of the antennae.

Bed bugs have fewer odor-binding proteins and odorant receptors than kissing bugs (*R. prolixus*) and mosquitoes (*Anopheles gambiae* Say, *Aedes aegypti* L.) (Bohbot *et al.*, 2007; Hansen *et al.* 2014), correlating well with the low number of olfactory sensilla (44) present on bed bug antennae (Steinbrecht and Müller, 1976). The fact that bed bugs have 50-fold fewer olfactory sensilla than the hematophagous hemipteran *Triatoma infestans* (Klug) (Lehane, 2005), which unlike bed bugs feeds on diverse vertebrate hosts, underlines the close association of bed bugs with their human host (Hansen *et al.*, 2014). With the bed bug genome now sequenced (Benoit *et al.*, 2016), further studies on bed bug olfaction and contact chemoreception can be expected to proceed at an accelerated pace.

17.3 Pheromones

17.3.1 Alarm Pheromone

In a state of alarm, bed bug nymphs and adults discharge the content of their scent glands, thus triggering increased activity and ultimately dispersal from the site of alarm pheromone release (Levinson and Bar Ilan, 1971; Levinson *et al.*, 1974b). Scent glands contain (*E*)-2-hexenal and (*E*)-2-octenal in large proportions (77–92%), with smaller proportions (8–27%) of 2-butanone, acetylaldehyde and, in nymphs, 4-oxo-(*E*)-2-hexenal and 4-oxo-(*E*)-2-octenal (Feldlaufer *et al.*, 2010). The blend of (*E*)-2-hexenal and (*E*)-2-octenal was as bioactive as the entire scent gland content (Levinson *et al.*, 1974b), indicating that these two aldehydes represent the complete alarm pheromone of bed bugs. The threshold concentration at which the airborne alarm pheromone elicits alarm behavior is much higher than the sensory detection threshold, suggesting that the alarm pheromone components at low concentration may play a role in a context other than alarm behavior (see aggregation pheromone).

The selectivity of the alarm pheromone receptor was tested by exposing E1- and E2-type sensilla to (*E*)-2-hexenal and (*E*)-2-octenal, the corresponding alkenes – (*E*)-hexene and (*E*)-2-octene – and to three saturated aldehydes (hexanal, pentanal, butanal) (Levinson *et al.*, 1974b). Measuring the resulting receptor potentials (voltages), it became apparent that molecular prerequisites of a possible ligand to stimulate the alarm pheromone receptor are a chain length of at least six carbon atoms and a terminal carbonyl group, whereas a double bond at C2 is immaterial. Levinson *et al.* (1974b) conceded that ORNs in sensilla adjacent to the E1- and E2-types may have contributed to the recorded receptor potentials, and Harraca *et al.* (2010) demonstrated that ORNs responding to the alarm pheromone are housed exclusively in D sensilla.

Studying the alarm pheromone of *C. hemipterus*, Liedtke *et al.* (2011) reported that distressed males, females, and nymphs emit (*E*)-2-hexenal and (*E*)-2-octenal at adult- and nymph-specific ratios and that *C. hemipterus* nymphs, like *C. lectularius* nymphs (Feldlaufer *et al.*, 2010), also emit two oxo-aldehydes, namely 4-oxo-(*E*)-2-hexenal and 4-oxo-(*E*)-2-octenal. At high concentration, whole-body extract of nymphs (but not of males or females) repels nymphs as well as male and female adults (Liedtke *et al.*, 2011).

17.3.2 Aggregation Pheromone

Bed bugs aggregate within harborages (Usinger, 1966; Figure 17.1) and return to them after each blood meal (Reis and Miller, 2011). As early as 1955, the aggregation behavior of bed bugs was attributed to a "nest odor" (Marx, 1955), which is now recognized as aggregation pheromone. Various studies investigated the effects of

Figure 17.1 Following a recent blood meal, replete *Cimex lectularius* aggregate in folds of bed linen.

bed bug strain, developmental stage, sex, and mating and feeding status on the responses of bed bugs to sources of the aggregation pheromone (Weeks *et al.*, 2011a). For pheromone analysis, it was most critical to know that bed bugs sense aggregation pheromone both by olfaction (Siljander *et al.*, 2008; Weeks et. al., 2011b) and contact chemoreception (Siljander *et al.*, 2007), indicating that the pheromone blend comprises volatile and non-volatile components.

An early attempt to identify the volatile components of the aggregation pheromone had limited success. While the research afforded an eight-component synthetic blend that attracted bed bugs in the laboratory (Siljander *et al.*, 2008), the blend was ineffective in the field (Gries *et al.*, unpublished results), indicating that one or more components of the aggregation pheromone were still missing. A recent study identified all essential components of the aggregation pheromone (Gries *et al.*, 2015) and is described here in detail not only to convey the complexity of the pheromone analyses but also because the synthetic aggregation pheromone holds promise as a bait for effective bed bug traps (Bennett *et al.*, 2016).

Pheromone analyses focused on bed bug feces and exuviae (Figure 17.2), which are both present in bed bug shelters and have been associated with arrestment behavior (Domingue *et al.*, 2010; Reis, 2010). Two-choice, still-air, single-insect olfactometer experiments established that physical contact with exuviae is required to induce the arrestment behavior, indicating that the arrestant component is sensed by contact chemoreception rather than olfaction. To isolate the arrestant for molecular characterization, more than 18 000 shed bed bug exuviae as well as filter paper stained with bed bug feces were extracted with organic solvents and bioassayed (Gries *et al.*, 2015). Only methanol extracts of exuviae and feces-stained paper induced the arrestment behavior of bed bugs, indicating that the arrestant component has unusual pheromonal properties and thus would not be detectable in routine pheromone analyses by gas chromatography-mass spectrometry (GC-MS). Analyzing bioactive methanol extracts by two-dimensional nuclear magnetic resonance spectroscopy rather than by GC-MS, and testing identified candidate pheromone components – amino acids, histamine, N-acetylglucosamine, dimethylaminoethanol, 3-hydroxy-kynurenine-*O*-sulfate – in laboratory bioassays, demonstrated that only histamine (Figure 17.3) induced arrestment behavior of bed bugs (Gries *et al.*, 2015). However, histamine has low volatility (boiling point >200 °C), lacks attractive properties, and when presented physically inaccessible to bed bugs elicits no behavioral responses.

The inability of histamine to attract bed bugs prompted the search for additional attractive pheromone components. Focusing on bed bug feces, a proven source of the arrestant pheromone, feces-stained filter paper was heated to 90 °C to enhance the release of volatiles. Analyzing these volatiles by GC-MS resulted in the identification of a complex blend of 15 compounds with either oxygen or sulfur hetero-atoms. With

Chemical Ecology | 167

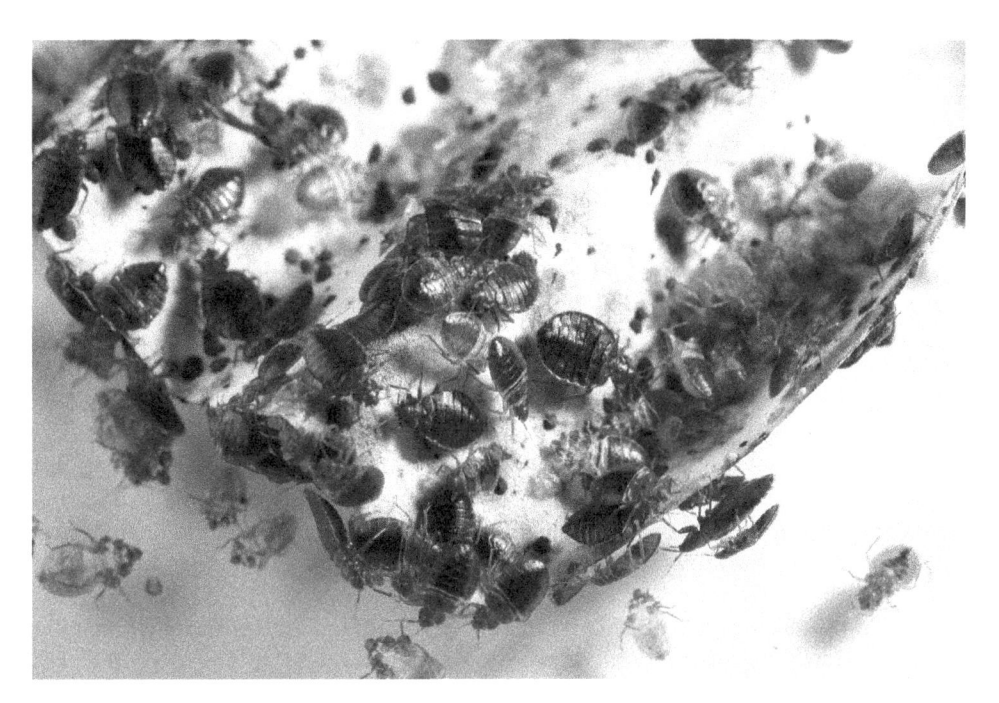

Figure 17.2 Exuviae and feces (dark spots on filter paper) of *Cimex lectularius*, sources of the aggregation pheromone.

(a)
(*E*)-2-hexenal
(*E*)-2-octenal

(b)
dimethyldisulfide
dimethyltrisulfide
(*E*)-2-hexenal
(*E*)-2-octenal
2-hexanone
histamine

(c)
(*E*)-2-hexenal
(*E*)-2-octenal

(d)
(*E*)-2-hexenal
4-oxo-(*E*)-2-hexenal
(*E*)-2-octenal
4-oxo-(*E*)-2-octenal

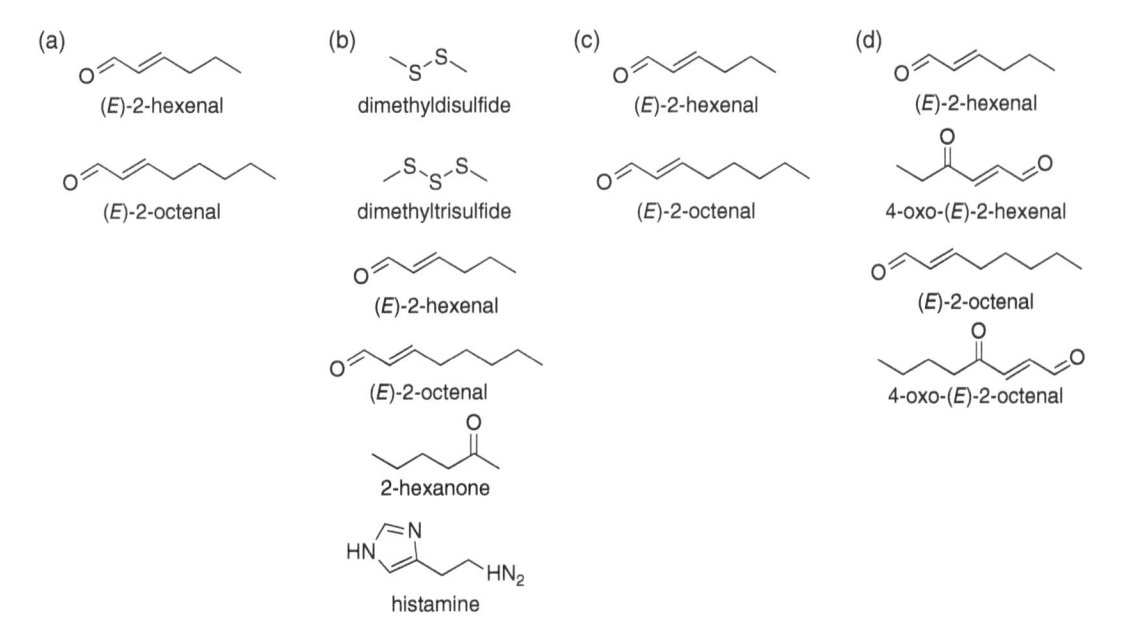

Figure 17.3 Types of *Cimex lectularius* pheromones: (a) alarm pheromone produced by nymphs and adults in a state of alarm causing dispersal; (b) aggregation pheromone mediating aggregation of all instar nymphs as well as male and female adults; (c) anti-mating pheromone released by males mounted by males; and (d) anti-aphrodisiac pheromone produced by nymphs mounted by males. Notes: (1) some of the same components re-occur in context-specific functions; (2) in *Cimex hemipterus*, whole-body extracts of nymphs [containing the components in (d)] – but not of males or females – repel nymphs as well as male and female adults. Sources: Levinson and Bar Ilan (1971), Levinson *et al*. (1974b), Feldlaufer *et al*. (2010), Harraca *et al*. (2010), Ryne (2009), Liedtke *et al*. 2011, Gries *et al*. (2015).

evidence that the 15-component blend attracted bed bugs in olfactometer experiments, 35 follow-up experiments determined the essential volatile pheromone components (VPCs) of the blend (Figure 17.3) (Gries *et al*., 2015):

- two sulfides: dimethyldisulfide (DMDS) and dimethyltrisulfide (DMTS)
- two aldehydes: (*E*)-2-hexenal, (*E*)-2-octenal
- one ketone: 2-hexanone.

The effectiveness of these VPCs and histamine on attraction and arrestment behavior of bed bugs was ultimately tested in bed bug-infested premises (Gries *et al*., 2015). Experimental replicates consisted of paired corrugated cardboard shelter traps placed against a wall or furniture. Within each pair, one trap was baited with the complete pheromone blend (VPCs and histamine) or a partial blend (VPCs or histamine), whereas the other trap was unbaited. In these experiments, the complete synthetic blend was highly effective, attracting and retaining up to 6.7 times more bed bugs than unbaited control traps. Traps baited with partial blends, in contrast, were as ineffective as unbaited control traps. These results demonstrated a remarkable synergism between the VPCs that attract bed bugs and histamine that causes their arrestment upon contact. Capturing all five nymphal instars (fed and non-fed) as well as male and female adults (fed and non-fed) in pheromone-baited traps is evidence that the pheromone attracts and retains bed bugs irrespective of their developmental stage, gender, or feeding status.

The aggregation pheromone is unusual in that it contains oxygen-, sulfur-, and nitrogen-containing components, which as a blend induce both attraction and arrestment of bed bugs. DMDS and DMTS are typically associated with microbial decomposition of organic materials (Khoga *et al*., 2002) but are clearly also components of the aggregation pheromone. This is backed by data showing that synthetic pheromone blends without these two sulfides are ineffective (Gries *et al*., 2015) and that specific olfactory sensilla on the bed bugs' antennae are "tuned" to DMDS and DMTS (Harraca *et al*., 2010). That the alarm pheromone components (*E*)-2-hexenal and (*E*)-2-octenal are also part of the aggregation pheromone supports the idea that pheromone components can express attractive or repellent properties in a dose-dependent manner, as has also been shown for bark beetles (Miller *et al*., 2005). The two aldehydes at high concentration alarm bed bugs (Levinson and Bar Ilan, 1971; Levinson *et al*., 1974b) but at low concentration attract them (Ulrich *et al*., 2016).

In search for the aggregation pheromone of *C. hemipterus*, Mendki *et al*. (2014) identified 33 volatile constituents in *C. hemipterus* feces, laboratory bioassayed synthetic components both individually and in groups of identical chemical functionality (for example, aldehydes or esters), and concluded that hexanoic acid, (*E*)-2-hexenoic acid, hexanal, and (*E*)-2-hexenal elicit positive behavioral responses in nymphs and adults. Whether these compounds are effective attractants in field settings and truly serve as pheromone components has yet to be tested.

17.3.3 Sex-attractant Pheromone

Sex-attractant pheromones (produced by one sex and attracting a prospective mate) do not seem to exist in bed bugs. Of several possible explanations (Reinhardt *et al*., 2008), the most plausible may be that virgin (but not mated) females respond to aggregation pheromone (Siljander *et al*., 2008) and together with nymphs and males form large-density aggregations (Reinhardt and Siva-Jothy, 2007) where they have ample mating opportunities even without attracting males.

17.3.4 Anti-mating and Anti-aphrodisiac Pheromones

With no evidence for a bed bug sex attractant pheromone and just movement of any bed bug-sized or shaped object prompting approach and mounting by males (Rivnay, 1933; Schaefer, 2000), mate recognition appears to rely primarily on cues perceived after mounting. They include morphological, tactile, and behavioral traits of the mounted individual (Rivnay, 1933; Davis, 1956) as well as the pheromone it releases in response to being mounted (Ryne, 2009). Surprisingly, cuticular hydrocarbons as potential contact pheromones do not differ between male and female bed bugs (Feldlaufer and Blomquist, 2011). However, males and females respond differently in response to being mounted. Unlike females, males mounted by males emit alarm pheromone, thus signaling their sex and reducing the duration of homosexual mountings (Ryne, 2009). In this male–male context the alarm pheromone serves as an anti-mating signal, benefiting both parties: the mounted male, who is less likely to be wounded by hypodermic insemination attempts of the mounting male, and the mounting male, who otherwise would miss real mating opportunities and waste spermatozoa and seminal fluid (Ryne, 2009), constraining future mating rates (Reinhardt et al., 2011).

Parallel "interests" of males (optimal reproductive fitness) and nymphs (least possible injury) have likely driven the evolution of communication also between nymphs and males, because both signaler (nymph) and receiver (male) benefit from the information transfer (Haynes et al., 2010). Late instar nymphs overlap in size with adult females and thus are potential mate targets for males. Harraca et al. (2010) hypothesized, and showed experimentally, that nymphs emit deterrent semiochemicals that act as a status signal, reducing sexual harassment of nymphs and saving energy for males. Harassed by males, nymphs emit from their unique dorsal abdominal scent gland a four-component blend of (E)-2-hexenal, (E)-2-octenal, 4-oxo-(E)-2-hexenal, and 4-oxo-(E)-2-octenal, the former three components detected by olfactory receptor neurons on the males' antennae (Harraca et al., 2010). Nymphs with their scent gland experimentally blocked were mated by males as often as females, and exposure of male–female pairs to either 4-oxo-(E)-2-hexenal, or (E)-2-hexenal and (E)-2-octenal, decreased the rate of mating to a level not different from that of male–nymph pairs (Harraca et al., 2010). These data together demonstrate that the alarm pheromone of bed bugs not only informs conspecifics of potential danger, but as an anti-mating or anti-aphrodisiac pheromone also plays a communicative role between males, and nymphs and males, reducing the risk of injury to males and nymphs, and enhancing the reproductive fitness of males.

17.4 Host Seeking

The reliance of bed bugs on their (human) host for development, reproduction, and survival (Usinger, 1966; Reinhardt and Siva-Jothy, 2007) provides selection pressure for the evolution of "host-guided" foraging behavior. Marx (1955) concluded that bed bugs sense their hosts from a distance of at least 1.5 m and he surmised that host encounters are not random but mediated by host cues, primarily host breath (containing CO_2 and volatile odorants) and host body heat.

Carbon dioxide on its own, or in combination with human odorants and heat, is the most important host foraging cue for bed bugs (Marx, 1955; Anderson et al., 2009; Wang et al., 2009; Aak et al., 2014). Humans and experimental CO_2 are comparable in their ability to trigger foraging by bed bugs (Aak et al., 2014). Human breath not only prompts foraging responses but also guides bed bugs to the site of release (Suchy and Lewis, 2011).

Even though bed bugs can sense heat and resolve differentials between host body and room temperatures of just 1–2 °C (Weeks et al., 2011a), the contributing effect of heat to the human-host cue complex during host foraging is limited. The thermal gradient between any warm object such as a warm-blooded host and its surrounding is steep (Baierlein, 1999), and bed bugs fail to detect it over a distance of merely a few centimeters (Aboul-Nasr and Erakey, 1967). This may explain the inconsistent effect of heat in bed bug trapping experiments (Anderson et al., 2009; Wang et al., 2009).

Least understood is the role of host odorants on foraging behavior of bed bugs. As trap baits, synthetic blends of human odorants such as

- propionic acid, butyric acid, valeric acid, L-lactic acid, and (RS)-1-octen-3-ol (Anderson et al., 2009)
- L-lactic acid and (RS)-1-octen-3-ol (Wang et al., 2009)

afforded inconclusive effects on bed bug captures, probably because these odorants – while attractive to kissing bugs (Barrozo and Lazzari, 2004a,b) – are not perceived by bed bug olfactory sensilla (Harraca *et al.*, 2010; Liu and Liu, 2015). Based on evidence that bed bugs distinguish between host odors (Aboul-Nasr and Erakey, 1968) and sense skin emanations (Rivnay, 1932), Harraca *et al.* (2012) collected and tested odorants from human volunteers, concluding that human odor, on its own, has only a weak effect on bed bug behavior. At close range to a host, the odor of sebaceous gland secretions from the skin (sebum) and ear (cerumen) was found to be attractive to bed bugs (Rivnay, 1932; Aboul-Nasr and Erakey, 1968; Weeks *et al.*, 2011a), whereas human sweat elicited attractive, indifferent, or repellent responses (Rivnay, 1932; Aboul-Nasr and Erakey, 1968; Levin, 1975). Once in contact with the host, the warm skin triggers probing by bed bugs followed by engorgement (Araujo *et al.*, 2011), with adenosine triphosphate serving as an effective phagostimulant (Romero and Schal, 2014).

To conclusively assess a potential semiochemical function of odorants in human breath and skin emanations, future studies need to thoroughly test a select group of odorants (see, for example, Liu and Liu, 2015) in combination with specific human host cues, primarily carbon dioxide and heat.

References

Aak, A., Rukke, A.S. and Rosnes, M.K. (2014) Questing activity in bed bug populations: male and female responses to host signals. *Physiological Entomology*, **39** (3), 199–207.

Aboul-Nasr, A.E. and Erakey, M.A.S. (1967) On the behaviour and sensory physiology of the bed bug: 1. Temperature reactions (Hemiptera: Cimicidae). *Bulletin of the Entomological Society of Egypt*, **51**, 43–54.

Aboul-Nasr, A.E. and Erakey, M.A.S. (1968) The effect of contact and gravity reactions upon the bed bug, *Cimex lectularius* L. *Bulletin of the Entomological Society of Egypt*, **52**, 363–370.

Anderson, J.F., Ferrandino, F.J., McKnight, S., Nolen, J. and Miller, J. (2009) A carbon dioxide, heat and chemical lure trap for the bedbug, *Cimex lectularius*. *Medical and Veterinary Entomology*, **23** (2), 99–105.

Araujo, R.N., Gontijo, N.F., Guarneri, A.A., *et al.* (2011) Electromyogram of the cibarial pump and the feeding process in hematophagous Hemiptera. In: *Advances in Applied Electromyography* (ed. J. Mizrahi), InTech, Rijeka, pp. 137–158.

Baierlein, R. (1999) *Thermal Physics*, Cambridge University Press, Cambridge.

Barrozo, R.B. and Lazzari, C.R. (2004a) The response of the blood-sucking bug *Triatoma infestans* to carbon dioxide and other host odours. *Chemical Senses*, **29** (4), 319–329.

Barrozo, R.B. and Lazzari, C.R. (2004b) Orientation behaviour of the bloodsucking bug *Triatoma infestans* to short chain fatty acids: synergistic effect of L-lactic acid and carbon dioxide. *Chemical Senses*, **29** (9), 833–841.

Bennett, G.W., Gondhalekar, A.D., Wang, C., Buczkowskia, G. and Gibb, T.J. (2016) Using research and education to implement practical bed bug control programs in multifamily housing. *Pest Management Science*, **72** (1), 8–14.

Benoit, J.B., Adelman, Z.N., Reinhardt. K., *et al.* (2016) Unique features of a global human ectoparasite identified through sequencing of the bed bug genome. *Nature Communications*, **7**, 10165.

Bohbot, J., Pitts, R.J., Kwon, H.W., *et al.* (2007) Molecular characterization of the *Aedes aegypti* odorant receptor gene family. *Insect Molecular Biology*, **16** (5), 525–537.

Davis, N.T. (1956) The morphology and functional anatomy of the male and female reproductive systems of *Cimex lectularius* L. (Heteroptera, Cimicidae). *Annals of the Entomological Society of America*, **49** (5), 466–493.

Domingue, M.J., Kramer, M. and Feldlaufer, M.F. (2010) Sexual dimorphism of arrestment and gregariousness in the bed bug (*Cimex lectularius*) in response to cuticular extracts from nymphal exuviae. *Physiological Entomology*, **35** (3), 203–213.

Feldlaufer, M.F. and Blomquist, G.J. (2011) Cuticular hydrocarbons from the bed bug *Cimex lectularius* L. *Biochemical Systematics and Ecology*, **39** (4–5), 283–285.

Feldlaufer, M.F., Domingue, M.J., Chauhan, K.R. and Aldrich, J.R. (2010) 4-Oxo-aldehydes from the dorsal abdominal glands of the bed bug (Hemiptera: Cimicidae). *Journal of Medical Entomology*, **47** (2), 140–143.

Grant, A.J. and Dickens, J.C. (2011) Functional characterization of the octenol receptor neuron on the maxillary palps of the yellow fever mosquito, *Aedes aegypti. PLoS ONE*, **6**, e21785.

Gries, R., Britton, R., Holmes, M., Zhai, H., Draper, J. and Gries, G. (2015) Bed bug aggregation pheromone finally identified. *Angewandte Chemie International Edition*, **54** (4), 1135–1138.

Guidobaldi, F., May-Concha, I.J. and Guerenstein, P.G. (2014) Morphology and physiology of the olfactory system of blood-feeding insects. *Journal of Physiology Paris*, **108** (2–3), 96–111.

Gullan, P.J. and Cranston P.S. (2000) Sensory systems and behaviour. In: *The Insects: an Outline of Entomology* (eds P.J. Gullan and P.S. Cranston), Blackwell Science Ltd, Oxford, pp. 83–110.

Hallem, E.A., Dahanukar, A. and Carlson, J.R. (2006) Insect odor and taste receptors. *Annual Review of Entomology*, **51**, 113–35.

Hansen, I.A., Rodriguez, S.D., Drake, L.L., *et al*. (2014) The odorant receptor co-receptor from the bed bug, *Cimex lectularius* L. *PLoS ONE*, **9** (11), e113692.

Harraca, V., Ignell, R., Löfstedt, C. and Ryne, C. (2010) Characterization of the antennal olfactory system of the bed bug (*Cimex lectularius*). *Chemical Senses*, **35** (3), 195–204.

Harraca, V., Ryne, C., Birgersson, G. and Ignell, R. (2012) Smelling your way to food: can bed bugs use our odour? *Journal of Experimental Biology*, **215** (4), 623–629.

Haynes, K.F., Goodman, M.H. and Potter, M.F. (2010) Bed bug deterrence. *BMC Biology*, **8**, 117.

Jones, W.D., Nguyen, T.A., Kloss, B., Lee, K.J. and Vosshall, L.B. (2005) Functional conservation of an insect odorant receptor gene across 250 million years of evolution. *Current Biology*, **15** (4), 119–121.

Khoga, J.M., Toth, E., Marialigeti, K. and Borossay, J. (2002) Fly-attracting volatiles produced by *Rhodococcus fascians* and *Mycobacterium aurum* isolated from myiatic lesions of sheep. *Journal of Microbiological Methods*, **48** (2–3), 281–287.

Leal, W.S. (2013) Odorant reception in insects: roles of receptors, binding proteins, and degrading enzymes. *Annual Review of Entomology*, **58**, 373–391.

Lehane, M.J. (2005) *Biology of Blood-Sucking in Insects*, Cambridge University Press, Cambridge.

Levin, N.A (1975) Olfactory activity of *Cimex lectularius* in relation to the sex and age composition and density of the populations. *Ekologiya (Moscow)*, **6**, 99–101.

Levinson, H.Z. and Bar Ilan, A.R. (1971) Assembling and alerting scents produced by the bedbug *Cimex lectularius* L. *Experientia*, **27** (1), 102–103.

Levinson, H.Z., Levinson, A.R., Muller, B. and Steinbrecht, R.A. (1974a) Structure of sensilla, olfactory perception, and behaviour of bedbug, *Cimex lectularius*, in response to its alarm pheromone. *Journal of Insect Physiology*, **20** (7), 1231–1248.

Levinson, H.Z., Levinson, A.R. and Maschwitz, U. (1974b) Action and composition of alarm pheromone of bedbug *Cimex lectularius*, L. *Naturwissenschaften*, **61** (12), 684–685.

Liedtke, C. H., Abjornsson, K., Harraca, V., *et al*. (2011) Alarm pheromones and chemical communication in nymphs of the tropical bed bug *Cimex hemipterus* (Hemiptera: Cimicidae). *PLoS ONE*, **6** (3), 1–7.

Liu, F. and Liu, N. (2015) Human odorant reception in the common bed bug, *Cimex lectularius. Scientific Reports*, **5**, 15558.

Liu, F., Chen, L., Appel, A.G. and Liu, N. (2013) Olfactory responses of the antennal trichoid sensilla to chemical repellents in the mosquito, *Culex quinquefasciatus. Journal of Insect Physiology*, **59** (11), 1169–1177.

Liu, F., Haynes, K.F., Apple, A.G. and Liu, N. (2014) Antennal olfactory sensilla responses to insect chemical repellents in the common bed bug, *Cimex lectularius. Journal of Chemical Ecology*, **40** (6), 522–533.

Logan, J. and Birkett, M. (2007) Semiochemicals for biting fly control: their identification and exploitation. *Pest Management Science*, **63** (7), 647–657.

Marx, R.U. (1955) Über die Wirtsfindung und die Bedeutung des artspezifischen Duftstoffes bei *Cimex lectularius* Linne. *Zeitschrift für Parasitenkunde*, **17**, 41–72.

Mendki, M.J., Ganesan, K, Parashar, B.D., Sukumaran, D. and Prakashet, S. (2014) Aggregation responses of *Cimex hemipterus* F. to semiochemicals identified from their excreta. *Journal of Vector Borne Diseases*, **51** (3), 224–229.

Miller D.R., Lindgren B.S. and Borden, J.H. (2005) Dose-dependent pheromone responses of mountain pine beetle in stands of lodgepole pine. *Environmental Entomology*, **34** (5), 1019–1027.

Olson, J.F., Moon, R.D. and Kells, S.A. (2009) Off-host aggregation behavior and sensory basis of arrestment by *Cimex lectularius* (Heteroptera: Cimicidae). *Journal of Insect Physiology*, **55** (6), 580–587.

Prakash, S., Chauhan, R.S., Parashar, B.D. and Rao, K.M. (1996) Morphology and distribution of antennal sensilla in the postembryonic developmental stages of *Cimex hemipterus* Fabricius. *Italian Journal of Zoology*, **63** (2), 131–133.

Reinhardt, K. and Siva-Jothy, M.T. (2007) Biology of the bed bugs (Cimicidae). *Annual Review of Entomology*, **52**, 351–374.

Reinhardt, K., Naylor, R.A. and Siva-Jothy, M.T. (2008) Situation exploitation: higher male mating success when female resistance is reduced by feeding. *Evolution*, **63** (1), 29–39.

Reinhardt, K., Naylor, R. and Siva-Jothy, M.T. (2011) Male mating rate is constrained by seminal fluid availability in bedbugs, *Cimex lectularius*. *PLoS ONE*, **6** (7), e22082.

Reis, M.D. (2010) An evaluation of bed bug (*Cimex lectularius* L.) host location and aggregation behavior. Masters thesis, Virginia Polytechnic Institute and State University.

Reis, M.D. and Miller, D.M. (2011) Host searching and aggregation activity of recently fed and unfed bed bugs (*Cimex lectularius* L.). *Insects*, **2** (2), 186–194.

Rivnay, E. (1932) Studies in tropisms of the bed bug *Cimex lectularius* L. *Parasitology*, **24** (1), 121–136.

Rivnay, E. (1933) The tropisms effecting copulation in the bed bug. *Psyche*, **40** (4), 115–20.

Romero, A. and Schal, C. (2014) Blood constituents as phagostimulants for the bed bug *Cimex lectularius* L. *Journal of Experimental Biology*, **217** (4), 552–557.

Ryne, C. (2009) Homosexual interactions in bed bugs: alarm pheromones as male recognition signals. *Animal Behaviour*, **78** (6), 1471–1475.

Schaefer, C.W. (2000) Bed bugs (Cimicidae). In *Heteroptera of Economic Importance* (eds C.W. Schaefer and A.R. Panizzi), CRC Press, Boca Raton, pp. 519–538.

Siljander, E., Penman, D., Harlan, H. and Gries, G. (2007) Evidence for male- and juvenile-specific contact pheromones of the common bed bug *Cimex lectularius*. *Entomologia Experimentalis et Applicata*, **125** (2), 215–219.

Siljander, E., Gries, R., Khaskin, G. and Gries, G. (2008) Identification of the airborne aggregation pheromone of the common bed bug, *Cimex lectularius*. *Journal of Chemical Ecology*, **34** (6), 708–718.

Singh, R.N., Singh K., Prakash S., Mendki, M.J. and Rao, K.M. (1996) Sensory organs on the body parts of the bed-bug *Cimex hemipterus* Fabricius (Hemiptera: Cimicidae) and the anatomy of its central nervous system. *International Journal of Insect Morphology & Embryology*, **25** (1–2), 183–204.

Steinbrecht, R.A. and Müller, B. (1976) Fine structure of antennal receptors of bed bug, *Cimex lectularius* L. *Tissue & Cell*, **8** (4), 615–636.

Suchy, J.T. and Lewis, V.R. (2011) Host-seeking behavior in the bed bug, *Cimex lectularius*. *Insects*, **2** (1), 22–35.

Ulrich, K.R., Kramer, M. and Feldlaufer, M.F. (2016) Ability of bed bug (Hemiptera: Cimicidae) defensive secretions (E)-2-hexenal and (E)-2-octenal to attract adults of the common bed bug *Cimex lectularius*. *Physiological Entomology*, **41** (2), 103–110.

Usinger, R. (1966) *Monograph of Cimicidae (Hemiptera – Heteroptera)*, Entomological Society of America, College Park.

Wang, C., Gibb, T., Bennett, G.W. and McKnight, S. (2009) Bed bug (Heteroptera: Cimicidae) attraction to pitfall traps baited with carbon dioxide, heat and chemical lure. *Journal of Economic Entomology*, **102** (4), 1580–1585.

Weeks, E.N.I., Birkett, M.A., Cameron, M.M., Pickett, J.A. and Logan, J.G. (2011a) Pest semiochemicals of the common bed bug, *Cimex lectularius* L. (Hemiptera: Cimicidae), and their potential for use in monitoring and control. *Pest Management Science*, **67** (1), 10–20.

Weeks, E.N.I., Logan, J.G., Gezan, S.A., *et al.* (2011b) A bioassay for studying the behavioural responses of the common bed bug, *Cimex lectularius* (Hemiptera: Cimicidae) to bed bug-derived volatiles. *Bulletin of Entomological Research*, **101** (1), 1–8.

18

Population Genetics

Warren Booth, Coby Schal and Edward L. Vargo

18.1 Introduction

In their review of bed bug biology, Reinhardt and Siva-Jothy (2007) identified population and dispersal ecology as aspects of bed bug biology for which data were sparse. In recent years, however, substantial progress has been made, with notable advances, particularly in our understanding of fine-scale dispersal and population genetics of the Common bed bug, *Cimex lectularius* L. (Szalanski *et al.*, 2008; Pfiester *et al.*, 2009; Booth *et al.*, 2012, 2015; Saenz *et al.*, 2012; Fountain *et al.*, 2014; Akhoundi *et al.*, 2015; Cooper *et al.*, 2015; Narain *et al.*, 2015; Raab *et al.*, 2016); unfortunately, such information is virtually non-existent for the Tropical bed bug, *Cimex hemipterus* (F.). Through a combination of ecological and genetic approaches, insight has been gained into three aspects of bed bug population ecology key to the development of effective management strategies:

- the origins of new infestations
- pathways to invasion
- propagule pressure.

With the array of molecular markers now available, questions previously out-of-reach using purely behavioral or ecological approaches can now be addressed with relative ease (Avise, 2004). To date, studies of bed bug population genetics have focused on two classes of molecular marker: mitochondrial DNA (mtDNA) and nuclear microsatellites. However, with advances in technology and lower costs, next-generation sequencing methods are likely to become prominent in future studies.

It should be noted that this chapter deals exclusively with *C. lectularius*. To date, only a single study has assessed the use of genetic markers to study *C. hemipterus* (Majid and Kee, 2015). With its likely introduction and spread into sub-tropical and temperate regions, genetic studies of this species are warranted.

18.2 The Evolution of Modern Bed Bugs

Cimex lectularius has a long-term association with humans, with evidence suggesting the human-associated lineage evolved from the ectoparasites of cave-dwelling bats (Usinger, 1966; Balvín *et al.*, 2015). Data derived from three classes of molecular markers lend support to this hypothesis (Booth *et al.*, 2015), with one or more lineages subsequently switching to hominids approximately 245 000 years before present (Balvín *et al.*, 2012a). Across the Old World, populations of *C. lectularius* still exist in association with their putative ancestral hosts (Povolný and Usinger, 1966; Usinger, 1966; Balvín *et al.*, 2012a), and recent evidence indicates genetic and reproductive isolation between lineages associated with bats and humans within homes (Balvín *et al.*, 2012a; Booth *et al.*, 2015). Hybridization experiments between the two host-associated

Advances in the Biology and Management of Modern Bed Bugs, First Edition.
Edited by Stephen L. Doggett, Dini M. Miller, and Chow-Yang Lee.
© 2018 John Wiley & Sons Ltd. Published 2018 by John Wiley & Sons Ltd.

lineages failed to produce viable offspring, so Wawrocka *et al.* (2015) concluded that post-mating barriers separate these two lineages. Usinger (1966) reported that *C. hemipterus* has been found to utilize birds, humans, and bats as hosts. With a tropicopolitan distribution, the ancestral origin of *C. hemipterus* is currently unknown, but molecular genetic studies of populations on these alternative hosts may clarify the evolutionary history of this species.

18.3 Genetic Variation Within Populations

At the infestation level, basic population genetic data (such as number of alleles/haplotypes, levels of expected and observed heterozygosity, and haplotype/nucleotide diversity) provide information relevant to understanding genetic diversity, and thus potentially the number of introductions into a given structure. Careful selection of the most appropriate markers should be made in relation to the question at hand because different markers do not necessarily yield the same information (Avise, 2004).

With some exceptions, notably *C. lectularius* (see Section 18.5), mtDNA has proven useful when inferring propagule pressure of invasive populations, given that the presence of multiple mitochondrial lineages suggests either several distinct introduction events, or the introduction of a single, genetically diverse propagule (Fitzpatrick *et al.*, 2012). Following the sequencing of 22 spatially distinct infestations in the USA, Canada, and Australia at the 16S rRNA gene, Szalanski *et al.* (2008) observed multiple haplotypes within many structures. This finding led the authors to suggest that gene flow among populations was common, and that within these countries bed bugs utilized alternative host species that served as reservoirs prior to their recent resurgence.

Microsatellite DNA – Mendelian inherited, short tandem repeat sequences dispersed throughout the nuclear genome – revealed a contrasting picture. A scenario with four or fewer alleles at each locus within an infestation is possible only when the founding propagule consisted of a single, singly-mated female, a single sexual pair, or a group of highly related individuals (Scenario A). In contrast, observing more than four alleles at two or more loci suggests a single introduction from a genetically diverse source population, or multiple introductions from more than one source population (Scenario B). Note, however, that genetic diversity informs us about the number of *genetically* distinct introductions but not the number of potential *introduction events* from the same source population. Moreover, extensive use of insecticides could selectively eradicate haplotypes that are associated with susceptibility to insecticide, and thus obscure the history of unique introductions, as suggested by Dang *et al.* (2015) in regards to *C. lectularius* in Australia. To date, microsatellite studies have revealed limited within-infestation genetic diversity in human-associated populations (Table 18.1, Figure 18.1a) (Booth *et al.*, 2012, 2015; Saenz *et al.*, 2012; Fountain *et al.*, 2014; Akhoundi *et al.*, 2015; Narain *et al.*, 2015; Raab *et al.*, 2016). As such, with four or fewer alleles observed within most infestations (but see Raab *et al.*, 2016), introductions appear to follow Scenario A, in contrast to the conclusions of Szalanski *et al.*, (2008). Conversely, multiple infestations of the bat-associated lineage follow Scenario B (Booth *et al.*, 2015), a finding not unexpected given their limited exposure to insecticides (see Section 18.5.3), higher prevalence of bed bugs in bat roosts, and greater host density permitting more frequent passive dispersal events (Balvín *et al.*, 2012b; Bartonička and Růžičková, 2013). Therefore, populations of bat-associated *C. lectularius* are likely to be more stable than human-associated populations and likely open to more frequent introductions.

With infestations founded by only one or a few individuals, inbreeding occurs rapidly, elevating genome-wide homozygosity and thus exposing recessive deleterious alleles. This may result in one of two potential outcomes: population extinction or, conversely, the purging of genetic load. Across seven studies for which within-infestation relatedness (r) could be calculated, the average value within human-associated populations was 0.751 (range 0.636–0.822), whereas within the bat-associated lineage, the average value was lower ($r = 0.590$) (Table 18.1). To place the significance of these values in context, within an outbred population r is

Table 18.1 Within and among *Cimex lectularius* infestation population genetic summary statistics.

Sample location	No. of loci	n	Average N_A	He	Ho	Average r	Reference
Within infestation							
	24	322	2.21	0.229	0.207	0.733	Booth *et al.*, 2012
	9	206	1.9	0.265	0.225	0.780	Saenz *et al.*, 2012
	8	80	2.83	0.433	0.406	0.636	Narain *et al.*, 2015
	6	183	1.81	0.453	NA	NA	Akhoundi *et al.*, 2015
	19	156	1.85	0.231	0.194	0.822	Fountain *et al.*, 2014 [A,*]
	21	63	1.71	0.288	0.273	0.728	Fountain *et al.*, 2014 [B,*]
	20	130	2.42	0.388	0.303	0.590	Booth *et al.*, 2015 [C]
	20	525	2.21	0.166	0.129	0.805	Booth *et al.*, 2015 [D]
Among infestations							
	24	322	5.75	0.592	0.214	NA	Booth *et al.*, 2012
	9	206	10.3	0.779	0.222	NA	Saenz *et al.*, 2012
	8	80	12	0.800	0.406	NA	Narain *et al.*, 2015
	6	183	NA	0.307	NA	NA	Akhoundi *et al.*, 2015
	19	156	4.89	0.566	0.194	NA	Fountain *et al.*, 2014 [A,*]
	21	63	5.67	0.680	0.266	NA	Fountain *et al.*, 2014 [B,*]
	20	187	10.5	0.721	0.306	NA	Booth *et al.*, 2015 [C]
	20	525	7.5	0.603	0.130	NA	Booth *et al.*, 2015 [D]

n, sample size; N_A, number of alleles; H_e, expected heterozygosity; H_O, observed heterozygosity; r, relatedness;
* data calculated from DRYAD dataset (doi:10.5061/dryad.cg10d);
[A] five populations across UK and Australia;
[B] within London, UK;
[C] bat-associated lineage;
[D] human-associated lineage.

expected to be ~0 between unrelated individuals, whereas between full-sibling or parent–offspring pairs r should be ~0.5. The lower value seen in the bat-associated lineage reflects increased allelic diversity due to more frequent immigration events. However, r is still >0.5, suggesting mating among siblings or parent–offspring pairs is common. As expected, where population size is small and consanguineous matings common, within-population heterozygosity is low (Table 18.1, Figure 18.1b). These values sharply contrast with within-population levels of heterozygosity of another significant human-commensal pest, the German cockroach (*Blattella germanica* L.) (Figure 18.1b), highlighting the unique nature of population foundation by the bed bug. Remarkably, heterotic effects experienced by bed bugs resulting from outbreeding appear short-lived (Fountain *et al.*, 2015). Therefore, inbreeding may actually promote local adaptation through the purging of deleterious alleles and the fixation of beneficial gene-complexes.

18.4 Genetic Variation Among Populations

Regardless of host lineage, dispersal outside of contiguous structures relies exclusively on passive events tied to the respective host. With the increasing frequency and relative ease of both national and international travel, *C. lectularius* now has a distribution spanning the world's temperate regions. The bat-associated

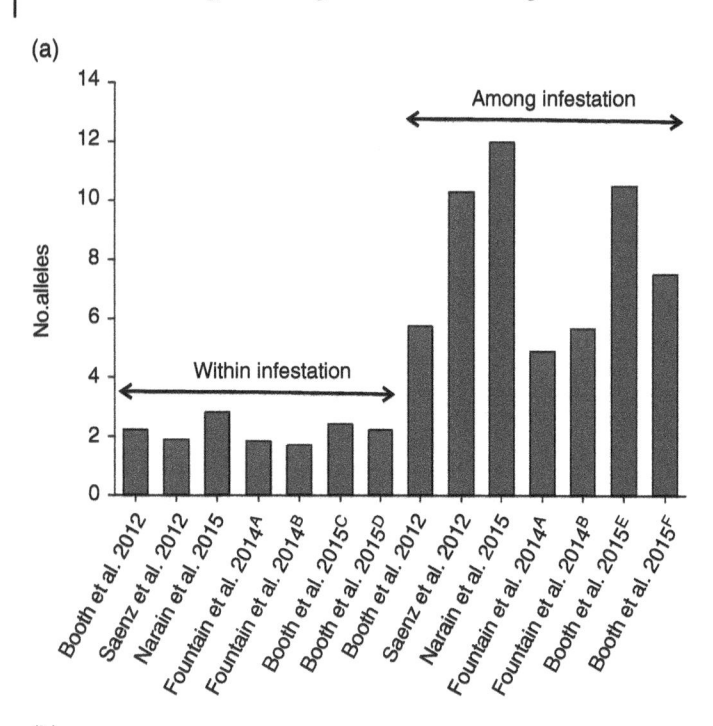

Figure 18.1 Genetic diversity in *Cimex lectularius* populations: (a) Within and among infestation mean number of alleles per locus; (b) Comparison of heterozygosities between the Common bed bug (*C. lectularius*) and the German cockroach (*B. germanica*). Hashed bar, expected heterozygosity; open bar, observed heterozygosity; [A]within-infestation, [B]within city, [C]bat-associated lineage, [D]human-associated lineage. [E]within USA, [F]within Eurasia.

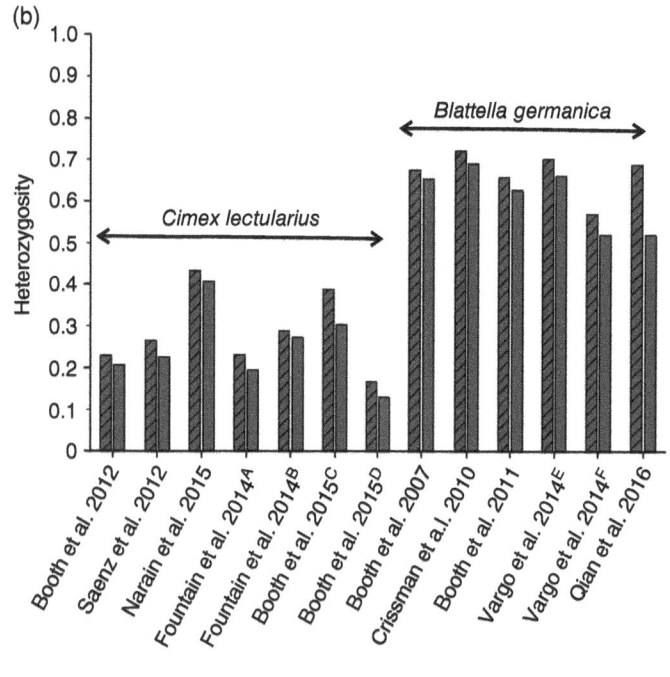

lineage, however, has been recorded only in the Old World (Povolny and Usinger, 1966; Balvín *et al.*, 2012b), and with the limited likelihood of transatlantic bat migrations (Constantine, 2003), is unlikely to spread further by bats.

Nevertheless, bats are often found on ships and shipping containers, and their association with human structures could result in eastward and westward dispersal of the bat-associated bed bug lineage with humans. Passive dispersal in association with high population turnover, driven through pest control, has resulted in human-associated populations existing in highly structured metapopulations, consisting of many island-like populations each founded by a genetically depauperate propagule with limited gene flow among them (Saenz *et al.*, 2012; Fountain *et al.*, 2014; Booth *et al.*, 2015).

As multiple distinct metapopulations likely exist within even limited geographic ranges (say, within a city) (Rosenfeld *et al.*, 2016), each founded from genetically distinct sources, high levels of structuring in association with insufficient sampling of multiple populations within a given metapopulation results in exceptionally high values in measures of genetic differentiation, such as F_{ST}. Comparisons with other species are difficult to make as few species exist in such human-centered metapopulations where dispersal is restricted to host-associated movements. The German cockroach and the human louse (*Pediculus humanus* L.) may be the only species to which comparisons are possible. Therefore, when we compare overall F_{ST} values, for example, for *B. germanica*:

- Crissman *et al.*, 2010: 0.048
- Booth *et al.*, 2011: 0.171
- Vargo *et al.*, 2014: 0.026–0.391

with those found in bed bugs (Table 18.1; Range: 0.472–0.718), the latter represent extreme levels of differentiation. Even in *P. humanus*, F_{ST} values in this upper range are only recorded between ecotypes; in other words, head versus body lice (F_{ST} = 0.409, Ascunce *et al.*, 2013).

When mtDNA genes are considered, diversity appears moderate to high, regardless of scale (regional or continental). For example, haplotype diversity at Cytochrome Oxidase I (COI) is comparable within two US states (0.698 ± 0.052 in Oklahoma and Kansas, Booth and Grant Robison, The University of Tulsa, Department of Biological Science, Tulsa, unpublished results) and across Europe (0.693 ± 0.057; Balvín *et al.*, 2012). Variation within the bat-associated lineage shows slightly elevated values (0.797 ± 0.020; Booth *et al.*, 2015). A similar pattern of lower diversity within the human-associated lineage versus the bat-associated lineage in Europe is observed at the rRNA 16S gene (Balvín *et al.*, 2012a; Booth *et al.*, 2015). As bats likely represent the ancestral host, greater genetic diversity in the bat-associated lineage is not surprising (Booth *et al.*, 2015).

When allelic diversity across populations is compared over scales ranging from single multi-unit apartment buildings (Booth *et al.*, 2012; Raab *et al.*, 2016) to cities or even continents (Saenz *et al.*, 2012; Fountain *et al.*, 2014; Booth *et al.*, 2015; Akhoundi *et al.*, 2015; Narain *et al.*, 2015), the species-level gene pool is diverse (Figure 18.1a). Following an apparently global population decline in the 1950s, and the resurgence of two species concurrently, it is unlikely that the recent resurgence was driven by a single diverse source population, but instead more likely that multiple propagules from a genetically diverse region in which populations of both species persisted were reintroduced globally. Given that populations of both species have been present in high densities in Africa prior to the global resurgence (Newberry and Jansen, 1986; Newberry *et al.*, 1987; Newberry and Mchunu, 1989; Newberry, 1990), it is possible that this continent represents the source of today's resurgence. Future genetic studies should therefore prioritize the collection of samples from across Africa as part of a global investigation. The likelihood that resurgence did not result from multiple local sources is also supported by the absence of detectable patterns of isolation by distance over large geographic scales (Table 18.2, but see Saenz *et al.*, 2012 which potentially resulted from samples collected along a frequently travelled interstate route). Locally, a metagenomic single nucleotide polymorphism (SNP) analysis, from environmental swabs collected across New York City, suggests higher relatedness among populations in close proximity (say, linked by subway lines), whereas differentiation occurred among boroughs (Rosenfeld *et al.*, 2016).

18.5 Mitochondrial Heteroplasmy

Mitochondrial DNA has long been utilized as an informative marker in studies of phylogeography, population genetics, and invasion biology. This has been due to the inherent properties considered characteristic of mtDNA, specifically a relatively high mutation rate, uniparental inheritance, homoplasmy, and a lack of recombination (Avise, 2000). However, in recent years it has come to light that these characteristics may not

Table 18.2 Among *Cimex lectularius* infestation measures of genetic differentiation.

Sample location	Number of populations	Approx. distance (km)	F_{ST}	Pairwise F_{ST} range	Isolation by distance	Reference
Raleigh, NC and Jersey City, NJ, USA	3	680	0.472	0.176–0.597	NA	Booth *et al.*, 2012*
Eastern USA	21	2,000	0.68	0.325–0.983	$P < 0.001$	Saenz *et al.*, 2012
Nebraska and Kansas City, USA	10	265	0.481	0.193–0.742	NA	Narain *et al.*, 2015
France	14	900	0.556	0.004–0.862	NS	Akhoundi *et al.*, 2015
UK	13	43.4	0.592	−0.035–0.764	NS	Fountain *et al.*, 2014**
UK & Australia	5	17,000	0.709	0.492–0.834	NA	Fountain *et al.*, 2014**
Europe – Bat lineage	14	2,300	0.468	0.031–0.659	NS	Booth *et al.*, 2015
Europe – Human lineage	55	2,300	0.718	0.078–0.982	NS	Booth *et al.*, 2015

Approx. distance (km) is approximate distance between most distant infestations;
NA, data not available;
NS, not significant,
* calculated across five STRUCTURE defined populations,
** data calculated from DRYAD dataset (doi:10.5061/dryad.cg10d).

be universal across species, and in cases where these "laws" are violated, the utility of mtDNA as a marker of genetic structure is compromised (White *et al.*, 2008). With advances in DNA sequencing technologies and the ability to accurately identify low-frequency variants, reports of heteroplasmy (having two or more distinct mitochondrial lineages present within a cell) are increasing. At the population or individual level heteroplasmy is often rare (White *et al.*, 2008); but in the bed bug, it appears to be geographically widespread and pervasive (Robison *et al.*, 2015).

18.5.1 Variation in Heteroplasmy Across Host Lineages and Among Populations

MtDNA heteroplasmy has been identified across two continents (North America and Europe) and in both host lineages (Booth *et al.*, 2015; Robison *et al.*, 2015). Among the human-associated lineage it has been found to range from 17% of infestations (south central USA; Robison *et al.*, 2015) to 28.8% (Europe; Booth *et al.*, 2015), whereas in the bat-associated lineage it was found in 21.6% of populations (Booth *et al.*, 2015). Robison *et al.* (2015) further assessed the frequency of heteroplasmic variants within individual bed bugs, and found uncharacteristic patterns in comparison to previous studies of heteroplasmy in other systems. Assessing heteroplasmic individuals from five infestations collected in Oklahoma and Missouri, each was found to possess two mtDNA variants in an approximate 2:1 ratio. Within infestations, some nymphs were also found to be differentially homoplasmic for the variants of heteroplasmic adult females, suggesting that transitions occur from heteroplasmy to homoplasmy. Other individuals have been found to harbor as many as five distinct mtDNA variants, but three variants appeared in low frequency (accounting for 8% of the total variants) (Robison and Booth, unpublished results). It should be noted that the methods for heteroplasmy detection employed here, namely Sanger sequencing of one to two genes, likely underestimates the true rate of heteroplasmy, as variants in other genes may be present. Furthermore, the ability to detect low-frequency variants may be limited. As such, future whole mitogenome studies utilizing next-generation sequencing technology may reveal a rate higher than currently reported. Preliminary studies of "families" generated from differentially homoplasmic laboratory colonies support paternal leakage as the mechanism underlying heteroplasmy (Booth, Vargo and Schal, unpublished results).

18.5.2 Implications of Heteroplasmy

The presence of heteroplasmy has the potential to significantly impact mtDNA-based analyses through the generation of phylogenetic uncertainty resulting from the inability to identify individual haplotypes in a heteroplasmic sequence (White *et al.*, 2008). While there are methods to separate haplotypes and thus derive distinct haplotypes from heterozygous genotypes, this is difficult in practice without PCR product cloning and re-sequencing. Analysis of populations may therefore require the exclusion of heteroplasmic individuals (see, for example, Booth *et al.*, 2015), the elimination of ambiguous nucleotide sites, or the coding of ambiguous sites as a fifth character state.

In *C. lectularius*, microsatellite studies suggest that the majority of infestations follow Scenario A (see Section 18.3), predicting the occurrence of a single mtDNA haplotype per infestation. However, when samples from a multi-unit apartment building, previously inferred to have resulted from a single introduction (Booth *et al.*, 2012), were sequenced at the COI gene, five mtDNA haplotypes were identified and heteroplasmy was found to be common (Robison and Booth, unpublished results). The strongly supported inference through highly polymorphic nuclear markers of a single introduction, coupled with the understanding that heteroplasmy is pervasive in this species, questions the use of mtDNA for inferring bed bug invasion history and the dating of host-associated lineage divergence.

18.5.3 Insecticide Resistance, *kdr*, and Geographic Variation

Although several mechanisms of insecticide resistance have been documented in *C. lectularius* (see, for example, Adelman *et al.*, 2011; Koganemaru *et al.*, 2013; see also Chapter 29), knockdown resistance (*kdr*) has been investigated most extensively at the population level. One or more mutations occur in the voltage-sensitive sodium channel α-subunit gene, resulting in the modification of the amino acid sequence and lower sensitivity of the target site to pyrethroid insecticides. In *C. lectularius*, three mutations have been found. Two are geographically widespread: V419L (valine to leucine) and L925I (leucine to isoleucine) (Yoon *et al.*, 2008; Zhu *et al.*, 2010; Vargo *et al.*, 2011; Booth *et al.*, 2015; Dang *et al.*, 2015; Palenchar *et al.*, 2015; Raab *et al.*, 2016; Holleman and Booth, The University of Tulsa, Department of Biological Science, Tulsa, unpublished results). These were detected in infestations across the USA, Europe, Middle East, and Australia. A third mutation, I936F (isoleucine to phenylalanine), was identified in Australia and Israel (Dang *et al.*, 2015; Palenchar *et al.*, 2015). Note that the former amino acid in each is the wild type and is associated with susceptibility, whereas the latter is associated with resistance.

Independent of their contribution to insecticide resistance, it may be possible to view these mutations as informative markers in population genetic studies. For example, within the USA, haplotype B (419 susceptible, 925 resistant) and haplotype C (419 resistant, 925 resistant) are prevalent (32.3% and 57.1%, respectively) (Figure 18.2). In contrast, within Europe, haplotype B dominates (90%); A was found in 4%, C in 2.0%, and the remaining 4% represented populations heterozygous for haplotypes A and B. Within a complex of two high-rise apartment buildings in Paris, France, samples collected from 26 apartments exhibited only haplotype B (Durand *et al.*, 2012). Samples collected in Israel were comparable to those in Europe (92% haplotype B, 8% haplotype C). In contrast, haplotype A (419 susceptible, 925 susceptible) is the only haplotype detected in the bat-associated lineage (Booth *et al.*, 2015), providing evidence that it represents the ancestral haplotype. This is consistent with microsatellite and mtDNA data, indicating that contemporary gene-flow between host-associated lineages is absent or minimal, and that it is unlikely that bat-associated populations provide propagules for human-associated infestations. Furthermore, the lack of broad homogeneity of the haplotypes across sampled regions suggests that movement between the Old World and New World populations may be limited. Within the USA alone, while samples at present are from only 25 of 50 states and sample size across various states is often small and spatially uneven (Figure 18.2), it is clear that regions such as the east and west coasts exhibit high diversity, with all haplotypes present despite relatively limited sampling. Within the south-central USA, in contrast, only haplotypes B and C have been detected, despite larger sample sizes (Figure 18.2). From this asymmetrical pattern of diversity, we may speculate that patterns of dispersal differ markedly across geographic regions.

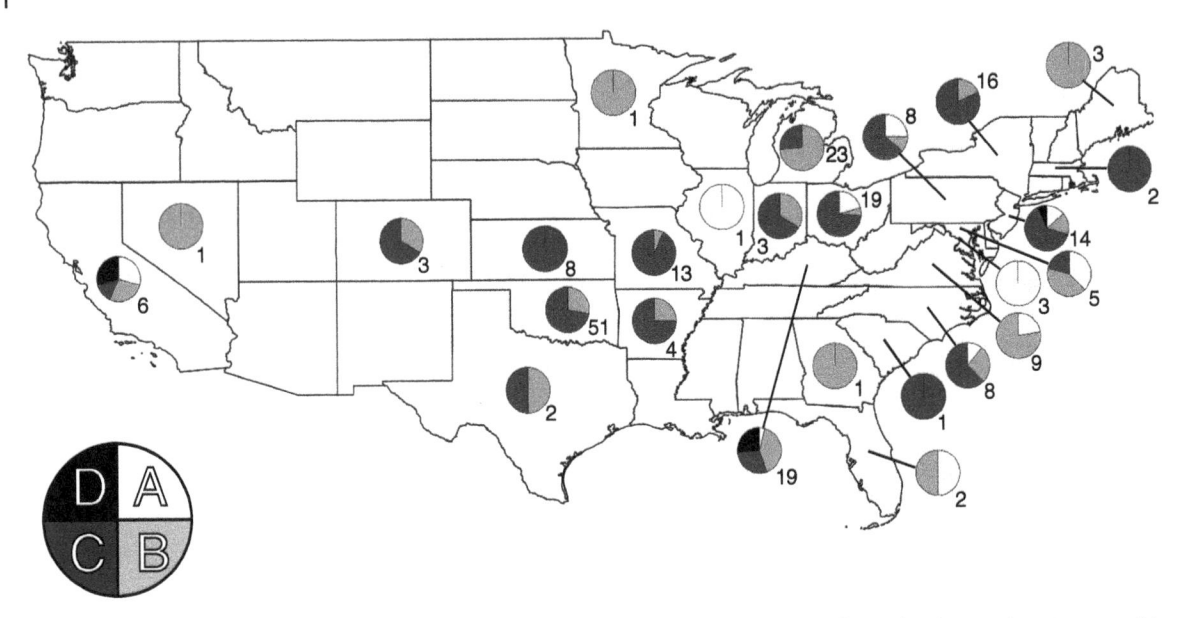

180 | *Advances in the Biology and Management of Modern Bed Bugs*

Figure 18.2 *kdr* haplotype distribution in *Cimex lectularius* across the USA. Letter (A–D) refers to haplotype. A: 419 susceptible, 925 susceptible, B: 419 susceptible, 925 resistant, C: 419 resistant, 925 resistant, D: 419 resistant, 925 susceptible. Number below pie is total number of infestations screened. Data combined from Zhu *et al.*, (2008); Vargo *et al.*, (2011); Holleman and Booth (unpublished results).

18.6 Future Directions in Bed Bug Population Genetics

It is clear that while significant advances have been made in our understanding of the population genetics of *C. lectularius*, these represent the tip of the iceberg. With the recent publication of its genome (Benoit *et al.*, 2016; Rosenfeld *et al.*, 2016), the bed bug looks poised to become an important model organism for the study of a suite of evolutionary processes. Nevertheless, a number of areas still lack clarity. These include defining genetic associations among globally distributed populations, and understanding the temporal and spatial patterns of *kdr* mutation frequency, from the apartment level to the global scale. Studies using genome-wide markers such as SNPs to identify sources of resurgent populations are critical and may shed light on the evolution of resistance in this species. Taking advantage of the sequenced genome to identify genes underlying host-associated lineage differentiation represents an exciting area for future investigation. Likewise, identifying the mechanisms that underlie heteroplasmy may revolutionize our understanding of eukaryote mitochondrial inheritance.

References

Adelman, Z.N., Kilcullen, K.A., Koganemaru, R., *et al.* (2011) Deep sequencing of pythreroid-resistant bed bugs reveals multiple mechanisms of resistance within a single population. *PLoS ONE*, **6** (10), e26228.

Akhoundi, M., Kengne, P., Cannet, A., *et al.* (2015) Spatial genetic structure and restricted gene flow in bed bugs (*Cimex lectularius*) populations in France. *Infection, Genetics, & Evolution*, **34**, 236–243.

Ascunce, M.S., Toups, M.A., Kassu, G., *et al.* (2013) Nuclear genetic diversity in human lice (*Pediculus humanus*) reveals continental differences and high inbreeding among worldwide populations. *PLoS ONE*, **8** (2), e57619.

Avise, J.C. (2000) *Phylogeography: The History and Formation of Species*, Harvard University Press, Cambridge, MA.

Avise, J.C. (2004) *Molecular Markers, Natural History and Evolution*, Chapman and Hall, New York.

Balvín, O., Munclinger, P., Kratochvíl, L. and Vilímova, J. (2012a) Mitochondrial DNA and morphology show independent evolutionary histories of bedbug *Cimex lectularius* (Heteroptera: Cimicidae) on bats and humans. *Parisitology Research*, **111** (1), 457–169.

Balvín, O., Ševčík, M., Jahelková, H., Bartonička, T., Orlova, M. and Vilímová, J. (2012b) Transport of bugs of the genus *Cimex* (Heteroptera: Cimicidae) by bats in western Palaearctic. *Vespertilio*, **16**, 43–54.

Balvín, O., Roth, S. and Vilímova, J. (2015) Molecular evidence places the swallow bug genus *Oeciacus* Stal within the bat and bed bug genus *Cimex* Linneaus (Heteroptera: Cimicidae). *Systematic Entomology*, **40** (3), 652–665.

Bartonička, T. and Růžičková, L. (2013) Recolonization of bat roost by bat bugs (*Cimex pipistrelli*): could parasite load be a cause of bat roost switching? *Parasitology Research*, **112** (4), 1615–1622.

Benoit, J.B., Adelman, Z.N., Reinhardt, K., et al. (2016) Unique features of a global human ectoparasite identified through sequencing of the bed bug genome. *Nature Communications*, **7**, 10165.

Booth, W., Santangelo, R.G., Vargo, E.L., et al. (2011) Population genetic structure in German cockroaches (*Blattella germanica*): differentiated islands in an agricultural landscape. *Journal of Heredity*, **102**, 175–183.

Booth, W., Saenz, V.L., Santangelo, R.G., et al. (2012) Molecular markers reveal infestation dynamics of the bed bug (Hemiptera: Cimicidae) within apartment buildings. *Journal of Medical Entomology*, **49** (3), 535–546.

Booth, W., Balvín, C., Vargo, E.L., et al. (2015) Host association drives significant genetic differentiation in the common bed bug, *Cimex lectularius*. *Molecular Ecology*, **24**, 980–992.

Cooper, R., Wang, C. and Singh. N. (2015) Mark-release-recapture reveals extensive movement in bed bugs (*Cimex lectularius* L.) within and among apartments. *PLoS ONE*, **10** (9), e0136462.

Constantine, D.G. (2003) Geographic translocation of bats: known and potential problems. *Emerging Infectious Disease*, **9** (1), 17–21.

Crissman, J.R., Booth, W., Santangelo, R.G., et al. (2010) Population genetic structure of the German cockroach (Blattodea: Blattellidae) in apartment buildings. *Journal of Medical Entomology*, **47** (4), 553–564.

Dang, K., Toi, C.S., Lilly, D.G., et al. (2015) Detection of knockdown resistance (*kdr*) mutations in the common bed bug, *Cimex lectularius* (Hemiptera: Cimicidae) in Australia. *Pest Management Science*, **71** (7), 914–922.

Durand, R., Cannet, A., Berdjane, Z., et al. (2012) Infestation by pyrethoids resistant bed bugs in the suburb of Paris, France. *Parasite*, **19** (4), 381–387.

Fitzpatrick, B.M., Fordyce, J.A., Niemiller, M.L. and Reynolds, R.G. (2012) What can DNA tell us about biological invasions? *Biological Invasions*, **14** (2), 245–253.

Fountain, T., Davaux, L., Horsburgh, G., et al. (2014) Human-facilitated metapopulation dynamics in an emerging pest species, *Cimex lectularius*. *Molecular Ecology*, **23** (5), 1071–1084.

Fountain, T., Butlin, R.K., Reinhardt, K. and Otti, O. (2015) Outbreeding effects in an inbreeding insect, *Cimex lectularius*. *Ecology and Evolution*, **5** (2), 409–418.

Koganemaru, R., Miller, D.M. and Adelman, Z.N. (2013) Robust cuticular penetration resistance in the common bed bug (*Cimex lectularius* L.) correlates with increased steady-state transcript levels of CPR-type cuticle protein genes. *Pesticide Biochemistry and Physiology*, **106** (3), 190–197.

Majid, A.H.A. and Kee, K.L. (2015) Genotyping of tropical bed bugs, *Cimex hemipterus* F. from selected urban areas in Malaysia inferred from microsatellite marker (Hemiptera: Cimicidae). *Entomology, Ornithology and Herpetology*, **4**, 155.

Narain, R.B., Lalithambika, S. and Kamble, S.T. (2015) Genetic variability and geographic diversity of the common bed bug (Hemiptera: Cimicidae) populations from the Midwest using microsatellite markers. *Journal of Medical Entomology*, **52** (4), 566–572.

Newberry, K. (1990) The tropical bedbug *Cimex hemipterus* near the southernmost extent of its range. *Transactions of the Royal Society of Tropical Medicine & Hygiene*, **84** (5), 745–747.

Newberry, K. and Jansen, E.J. (1986) The common bedbug *Cimex lectularius* in African huts. *Transactions of the Royal Society of Tropical Medicine & Hygiene*, **80** (4), 653–658.

Newberry, K. and Mchunu, Z.M. (1989) Changes in the relative frequency of occurrence of infestations of two sympatric species of bedbug in northern Natal and KwaZulu, South Africa. Transactions *of the Royal Society of Tropical Medicine & Hygiene*, **83** (2), 262–264.

Newberry, K., Jansen, E.J. and Thibaud, G.R. (1987). The occurrence of the bedbugs *Cimex hemipterus* and *Cimex lectularius* in northern Natal and KwaZulu, South Africa. *Transactions of the Royal Society of Tropical Medicine & Hygiene*, **81** (3), 431–433.

Palenchar, D.J., Gellatly, K.Y., Yoon, K.S., et al. (2015) Quantitative sequencing for the determination of kdr-type resistance allele (V419L, L925I, I936F) frequencies in common bed bug (Hemiptera: Cimicidae) populations collected from Israel. *Journal of Medical Entomology*, **52** (3), 1018–1027.

Pfiester, M., Koehler, P.G. and Pereira, R.M. (2009) Effect of population structure and size on aggregation behavior of *Cimex lectularius* (Hemiptera: Cimicidae). *Journal of Medical Entomology*, **46** (5), 1015–1020.

Povolný, D. and Usinger, R.L. (1966) The discovery of a possibly aboriginal population of the bed bug (*Cimex lectularius* Linnaeus, 1958). *Acta Mus Moraviae, Sci Biol*, **51**, 237–242.

Raab, R.W., Moore, J.E., Vargo, E.L., et al. (2016) New introductions, spread of existing matrilines, and high rates of pyrethroid resistance result in chronic infestations of bed bugs (*Cimex lectularius* L.) in lower–income housing. *PLoS ONE*, **11** (2), e0117805.

Reinhardt, K. and Siva-Jothy, M.T. (2007) Biology of the bed bugs (Cimicidae). *Annual Review of Entomology*, **52**, 351–374.

Robison, G.A., Balvín, O., Schal, C., et al. (2015) Extensive mitochondrial heteroplasmy in natural populations of a resurging human pest, the bed bug (Hemiptera: Cimicidae). *Journal of Medical Entomology*, **52** (4), 734–738.

Rosenfeld, J.A., Reeves, D., Brugler, M.R., et al. (2016) Genome assembly and geospatial phylogenomics of the bed bug *Cimex lectularius*. *Nature Communications*, **7**, 10164.

Saenz, V.L., Booth, W., Schal, C. and Vargo, E.L. (2012) Genetic analysis of bed bug populations reveals small propagule size within individual infestations but high genetic diversity across infestations from the eastern United States. *Journal of Medical Entomology*, **49** (4), 865–875.

Szalanski, A.L., Austin, J.W., Mckern, J.A., Steelman, C.D. and Gold, R.E. (2008) Mitochondrial and ribosomal internal transcribed spacer 1 diversity of *Cimex lectularius* (Hemiptera: Cimicidae). *Journal of Medical Entomology*, **45** (2), 229–236.

Usinger, R.L. (1966) *Monograph of Cimicidae (Hemiptera – Heteroptera)*, Entomological Society of America, College Park.

Vargo, E.L., Booth, W., Saenz, V., et al. (2011) Genetic analysis of bed bug infestations and populations. In: *Proceedings of the Seventh International Conference on Urban Pests, 2011, August, 7–10, São Paulo, Instituto Biológico, São Paulo*.

Vargo, E.L., Crissman, J.R., Booth, W., et al. (2014) Hierarchical genetic analysis of German cockroach (*Blattella germanica*) populations from within buildings to across continents. *PLoS ONE*, **9** (7), e102321.

Wawrocka, K., Balvín, O. and Bartonička, T. (2015) Reproductive barrier between two lineages of bed bug (*Cimex lectularius*) (Heteroptera: Cimicidae). *Parasitology Research*, **114** (8), 3019–3025.

White, D.J., Wolff, J.N., Pierson, M. and Gemmell, N.J. (2008) Revealing the hidden complexities of mtDNA inheritance. *Molecular Ecology*, **17** (23), 4925–4942.

Yoon, K.S., Kwon, D.H., Strycharz, J.P., et al. (2008) Biochemical and molecular analysis of deltamethrin resistance in the common bed bug (Hemiptera: Cimicidae). *Journal of Medical Entomology*, **45** (6), 1092–1101.

Zhu, F., Wigginton, J., Romero, A., et al. (2010) Widespread distribution of knockdown resistance mutations in the bed bug, *Cimex lectularius* (Hemiptera: Cimicidae), populations in the United States. *Archives of Insect Biochemistry and Physiology*, **73** (4), 245–257.

19

Physiology

Joshua B. Benoit

19.1 Introduction

This chapter provides a synopsis of the current knowledge associated with the physiology of the Common bed bug, *Cimex lectularius* L., with comparisons to the Tropical bed bug, *Cimex hemipterus* (F.), when available and applicable. The chapter is organized based on two periods during the life cycle of *Cimex*: the long-term period spent off-host, and the immediate post-feeding intervals of digestion and reproduction, with a specific focus on factors that have allowed bed bugs to develop and persist as human pests. General bed bug biology has been covered in other chapters of Part 4 of this book and will be discussed in this chapter only as necessary. This contribution will specifically provide information on:

- how dehydration and starvation resistance promotes bed bug off-host survival between blood meals
- bed bug thermal tolerance
- features underpinning blood digestion
- specific aspects of female reproductive physiology.

Other areas of bed bug physiology have been omitted from this chapter as they are the focus of other sections, including traumatic insemination, host-seeking behavior, and dispersal (Chapter 16), chemical ecology (Chapter 17), symbiosis (Chapter 20), and insecticide resistance (Chapter 29). These studies highlight the unique physiological traits that this pest has evolved to optimize its close association with humans, as well as its ability to tolerate, if necessary, prolonged host absence. The key aspects associated with bed bug stress tolerance and starvation resistance are shown in Table 19.1.

19.2 Stress Tolerance and Starvation Resistance

To establish and proliferate in an environment, small arthropods must be able to adequately respond to change in their local habitat (Chown and Nicolson, 2004; Benoit and Denlinger, 2010; Benoit, 2011; Benoit and Lopez-Martinez, 2011). As bed bugs commonly reside within the indoor biome, we would expect minimal shifts in environmental conditions as seasonal changes are buffered (Martin *et al.*, 2015). Even though this is the case seasonally, significant variations in local temperature and relative humidity (RH) can occur on a daily basis or even shorter timescales (Martin *et al.*, 2015). For example, running bathroom showerheads or sink faucets (increased heat, water, and local RH), household appliances (increased heat), and leaking pipes (increased water and local RH), can greatly alter local environmental conditions. One specific environmental stress that will likely be exacerbated within the indoor biome is dehydration (Brimblecombe and Lankester, 2012; Martin *et al.*, 2015). Indoor RH is lower than the comparable outdoor RH, with the exception of indoor/

Advances in the Biology and Management of Modern Bed Bugs, First Edition.
Edited by Stephen L. Doggett, Dini M. Miller, and Chow-Yang Lee.
© 2018 John Wiley & Sons Ltd. Published 2018 by John Wiley & Sons Ltd.

Table 19.1 Stress tolerance in the Common bed bug, *Cimex lectularius*.

Stress		Egg	Early instars	Adults/late instars
Dehydration tolerance[1]		25 to 30%	35 to 40%	30 to 35%
Water loss rate[1]		0.03 to 0.05%/h	0.3 to 0.4%/h	0.1 to 0.4%/h
Heat tolerance[2]	Short term[a]	36 to 39°C	44 to 46°C	44 to 46°C
	Long term[b]	32 to 35°C	35 to 37°C	36 to 39°C
Cold tolerance[2]	Short term	−25 to −20°C	−25 to −20°C	-25 to -20°C
	Long term	−8 to −6°C	−10 to −15°C	-10 to -15°C
Starvation resistance[3,c]	Susceptible	—	27 to 70 days	70 to 130 days
	Resistant	—	10 to 30 days	40 to 75 days

a, 1–2 h; b, 1 week; c starvation resistance likely equivalent to dehydration resistance. Sources: [1]Usinger, 1966; Benoit, 2011; Benoit *et al.*, 2007, 2009a,b; [2]Johnson, 1941; Usinger, 1966; Benoit *et al.*, 2009a; Naylor and Boase, 2010; Olson *et al.*, 2013; Rukke *et al.*, 2015. [3]Polanco *et al.* (2011b). Diagrams based on Benoit and Attardo (2013); early instar, 1st and 2nd instar nymphs; late instar, 3rd–5th instar nymphs. Dehydration tolerance in percent water loss tolerated. Susceptible and resistant relate to pesticide exposure.

outdoor dynamics in dry regions such as Phoenix, Arizona, USA. In addition, the indoor biome is warmer than outdoor conditions in areas with colder winters (Martin *et al.*, 2015), which can allow specific pests to survive in regions that would typically be inhabitable due to overwintering conditions below the cold tolerance of a specific species. Stress tolerance along with some aspects of basic physiology in bed bugs has been previously reviewed by the author of this chapter (Benoit, 2011). However, with the increasing research effort on *C. lectularius* physiology there has been additional and significant progress in this area.

For hematophagous insects, prevention of dehydration between each blood meal is critical. The insect must remain hydrated until the next host can be located (Needham and Teel, 1991; Benoit and Denlinger, 2010; Benoit *et al.*, 2014). Many blood feeding arthropods will ingest water or absorb water vapor from the atmosphere, but bed bugs do not ingest non-blood liquid when dehydrated (Johnson, 1941; Usinger, 1966; Benoit *et al.*, 2007; Benoit *et al.*, 2009a). As a result, bed bugs must rely on water retained from their previous blood meal until the next host is located. This infrequent "drinking" makes the bed bug an excellent candidate for control using a desiccant dust when hosts are absent. Desiccant dust exposure exacerbates water loss to induce mortality if water cannot be ingested (Benoit *et al.*, 2009b; Anderson and Cowles, 2012; Aak *et al.* 2016; Akhtar and Isman, 2016). This process is further described in Section 30.3.3. When the RH is high and/or a host is available at regular intervals, bed bug dehydration is not likely to occur. Bed bugs, specifically adults and late instar nymphs, have water-loss rates equivalent to those observed in arthropods adapted to survive in desert-like habitats (Hadley, 1994; Benoit *et al.*, 2007; Benoit, 2010; Benoit and Denlinger, 2010). Dehydration tolerance (DT, amount of water loss tolerated before succumbing to dehydration) is 30–40%, depending on the life stage, which is comparable to most other insects (Hadley, 1994; Benoit *et al.*, 2007; Benoit and Denlinger, 2010). The egg, however, has a lower DT (Table 19.1, ~25%), which is a characteristic seen in other arthropods (Yoder *et al.*, 2004; Urbanski *et al.*, 2010). The low water loss rate is likely due to the low permeability of the cuticle to water, and a slightly lower respiration rate compared to other arthropods (DeVries *et al.*, 2013, 2015, 2016). The first instar bed bug is most susceptible to dehydration, surviving only a few days at low RH (<10% RH). This is likely due to their small size (high surface-to-volume ratio, which promotes water loss; Wharton, 1985), and a relatively high metabolic rate compared to other life stages (DeVries et al., 2015). Older life stages, such as fifth instar nymphs and adults, are much more resistant to dehydration and will survive well over two weeks at extremely low RH (Mellanby, 1939b; Johnson, 1941; Benoit et al., 2007; Benoit et al., 2009b; How and Lee, 2010). When kept under conditions similar to those found in most human dwellings (40–70% RH, 20–22°C; Martin et al., 2015), an unfed bed bug infestation will likely persist for at least 4–6 months if adults and late instars are present. After this period, a blood meal will be necessary to prevent death (Omori, 1941; Benoit et al., 2007; Benoit et al., 2009a,b). At lower temperatures (~15°C) and higher RH (~80%), *C. lectularius* infestations will likely persist for well over a year between blood meals. The extreme ability to resist dehydration indicates that bed bugs might succumb to dehydration or starvation depending on the local environmental conditions (discussed further later in this section). Clustering in harborages further reduces water loss by increasing the local RH (Benoit et al., 2007). Very little is known about the exact mechanisms underlying dehydration resistance and the molecular response to dehydration in bed bugs. Metabolism declines as starvation, and water loss, progress (DeVries et al., 2015), suggesting that water loss through respiration is likely reduced as the time since the last blood meal increases. Alterations to the composition and thickness of the cuticle (proteins and cuticle lipids) or distribution/sizes of the spiracle will impact water loss and have been documented in other insect systems (Benoit et al., 2005; Reidenbach et al., 2014; Mamai et al., 2016). Increases in cuticle thickness and cuticle protein expression in specific bed bug lineages have been linked to pesticide resistance (Koganemaru et al., 2013; Zhu et al., 2013; Lilly et al., 2016), which may in turn alter water loss. Yet, this has not been studied. Traumatic insemination will compromise the cuticle if mating occurs outside of the spermalage (Benoit et al., 2012), which is possible since bed bugs mate ~5–6 times after each meal (Stutt and Siva-Jothy, 2001; Siva-Jothy and Stutt, 2003). If mating occurs at the spermalege, water loss will be minimized due to increased resilin, an elastometric protein, associated with this structure, which will likely reduce cuticle damage compared to piercing of a more rigid area (Michels et al., 2015; Benoit et al., 2016).

Information on the molecular response to dehydration in bed bugs has been minimal, with only a single study examining the expression of heat-shock proteins, specifically chaperone proteins, which prevent or repair protein damage during dehydration (Benoit et al., 2009a). This study suggests that these chaperone proteins perform critical functions during dehydration stress. Interestingly, long exposure to high RH reduces bed bug survival (Benoit et al., 2009a), suggesting that RH near saturation (100% RH) is detrimental to bed bugs (Omori, 1941; Usinger, 1966; Benoit et al. 2009a; How and Lee, 2010). The exact mechanisms of this reduced survival are unknown, but may be the result of increased microbial infection. Studies in *C. hemipterus* suggest a dehydration resistance similar to that in *C. lectularius*, where survival drops significantly during continual exposure to RH below 40% (Usinger, 1966; Benoit et al. 2007; Benoit, 2011; How and Lee, 2010). The innate ability of bed bugs to survive prolonged dehydration exposure is likely a contributing factor in their ability to survive until a host is encountered when transported to a new location. *Cimex* ability to survive extended periods with no water likely developed before or during their use of cave-dwelling bats or birds as their primary host, preceding the switch to humans, since access to bats or birds was more likely periodic rather than continual (Usinger, 1966).

Temperature tolerance has been examined in bed bugs primarily to determine their thermal death points for use in temperature-based remediation (Olson et al., 2013; Puckett et al., 2013; DeVries et al., 2016; Loudon, 2017). Depending on the method utilized, the upper lethal temperature for short-term exposure (less than 1–2 h), was ~44–46°C depending on the developmental stage (Usinger, 1966; Benoit et al., 2009a; Benoit, 2011). Prolonged exposure (over one week), to temperatures above 37°C also resulted in mortality (Rukke et al., 2015). Interestingly, exposure to high, sublethal temperatures (33–35°C) is known to have negative impacts on female fecundity and adult blood feeding (Rukke et al., 2015). In addition, the feeding, development, and survival of offspring produced from heat-treated mothers is reduced, indicating a maternal effect (Rukke et al., 2015). This transgenerational effect is likely due to failed transmission of the obligate mutualistic

Wolbachia to the progeny (Rukke *et al.*, 2015). Previous studies have indicated that prolonged heat stress will impact the vertical transfer of *Wolbachia* (Li *et al.*, 2014), which acts as a critical source of B vitamins for bed bugs (Hosokawa *et al.*, 2010; Nikoh *et al.*, 2014; also Chapter 20). For bed bug eggs, continual exposure to temperatures as low as 34°C will begin to prevent hatching. Extended periods above 37°C will result in near complete suppression of egg viability (Johnson, 1941; Omori, 1941; Rukke *et al.*, 2015).

Similar to heat tolerance, the duration of cold exposure has a significant impact on bed bug survival (Usinger, 1966; Benoit *et al.*, 2009a; Naylor and Boase, 2010; Benoit, 2011; Olsen *et al.*, 2013). Bed bug lethal temperatures have been reported to be as high as −14°C (Johnson, 1941; Usinger, 1966; Benoit *et al.*, 2009a; Naylor and Boase, 2010) to as low as −25°C (Olsen *et al.*, 2013) for short exposures. Prolonged exposure for one to three weeks to less extreme cold (−7 to −5°C) will begin to induce mortality, prevent nymphal molting, and suppress progeny production in surviving individuals (Rukke *et al.*, 2017). Supercooling points, the temperature where the bed bug freezes, are noted to be between −20 and −25°C for most developmental stages. Whether or not the individual has fed beforehand has no impact on the supercooling point (Benoit *et al.*, 2009a; Olsen *et al.*, 2013). Differences in cold tolerance observed between research studies can likely be attributed to the blood source utilized to maintain the bed bug colonies, age of the bed bugs, and the rates of cooling and warming during treatment (Usinger, 1966; Benoit *et al.*, 2009a; Naylor and Boase, 2010; Benoit, 2011; Olsen *et al.*, 2013). Finally, even though direct mortality does not occur, continual exposure of bed bugs to temperatures below 10°C prevents developmental progression (Usinger, 1966; Johnson, 1941). Heat tolerance of *C. hemipterus* has been documented to be slightly higher than *C. lectularius* (Omori, 1941; Usinger, 1966; How and Lee, 2010), but no significant studies have assessed cold tolerance of *C. hemipterus*. Outside of heat shock protein expression following cold and heat exposure (Benoit *et al.*, 2009a; Benoit *et al.*, 2011), little is known about the exact underlying biochemical, molecular, and physiological mechanisms associated with temperature tolerance in bed bugs. Information on the use of temperature as a control is included in Chapter 28.

Due to the ability to survive long periods between blood feeding, starvation resistance in bed bugs has been examined on several occasions (Mellanby, 1932; Johnson, 1941; Usinger, 1966; Polanco *et al.*, 2011b; DeVries *et al.*, 2015). As mentioned previously, disentangling the effects of dehydration from those of starvation is difficult in bed bugs due to their obligate use of blood as their water source. Starved bed bug survivorship is higher when the RH is near 75% rather than below 30%, suggesting that starvation (as opposed to dehydration) may play an increasing role in bed bug mortality as the RH increases. Recent studies monitoring the metabolic rates throughout starvation have identified distinct periods that occur after feeding/molting (DeVries *et al.*, 2015). In general, there was a reduction in the bed bug metabolic rate for all developmental stages as starvation progressed. Processing of the blood meal and molting yielded the highest peaks in metabolism for adults and immatures, respectively (DeVries *et al.*, 2015). Both oxygen consumption and the percentage of mass loss declined after blood feeding/molting for 10–20 days and remained stable (DeVries *et al.*, 2015), which is a likely a critical factor allowing for prolonged starvation resistance. Between adults and juvenile stages, the only major difference in metabolism was an initial plateau at 8–9 days in the juvenile stages, which correlated with the 7–9-day period since the last meal. Starvation resistance studies of two pyrethroid-susceptible and two pyrethroid-resistant populations revealed that resistant populations have a lower starvation threshold (Polanco *et al.*, 2011a,b), which is likely due to increased expression of detoxification or other resistance mechanisms that have metabolic costs (Adelman *et al.*, 2011; Mamidala *et al.*, 2012; Zhu *et al.*, 2013; Koganemaru *et al.*, 2013). Mating has also been found to have a negative effect on starvation resistance (Mellanby, 1932; Omori, 1941; How and Lee, 2010; Polanco *et al.*, 2011a,b), which is likely due to the stress of repeated mating or physiological investment into reproductive efforts. The starvation threshold for *C. hemipterus* is comparable to *C. lectularius* (Mellanby, 1932; Usinger, 1966; How and Lee, 2010; Polanco *et al.*, 2011a,b). The results of these starvation studies suggest that bed bug colonies, if late instars and adults are present, can persist for up to at least 4–6 months without a host depending on environmental conditions, and the specific bed bug population.

19.3 Blood Feeding

After extended periods of starvation/dehydration, the ingestion of blood represents a drastic shift in bed bug physiology. The blood meal is critical for developmental progression and mating in the adults because it is their only external source of nutrition. However, feeding will produce thermal stress in the bed bug due to the heat of the blood meal (Benoit *et al.*, 2011; Lahondère and Lazzari, 2012), and osmotic stress due to the rapid ingestion of a high volume of fluid (Benoit and Denlinger, 2010). RNA-seq analyses of bed bugs have revealed that ingestion of a blood meal induces substantial transcriptional shifts (Rosenfeld *et al.*, 2016). This is unsurprising because similar large-scale shifts in gene expression have been identified in other hematophagous insects after feeding (Bonizzoni *et al.*, 2011). This differential expression of genes is likely to allow for the proper digestion and storage of nutrients from the blood meal, as well as rapid removal of excessive water, and the buffering of thermal effects from the warm blood meal (Benoit and Denlinger, 2010; Benoit *et al.*, 2011; Bonizzoni *et al.*, 2011; Benoit *et al.*, 2014; Vannini *et al.*, 2014). Genomic analyses revealed that Cathepsin D, an aspartic peptidase (Sojka *et al.*, 2011), is expanded in *Rhodnius prolixus* (Stahl) and *C. lectularius* when compared to other non-blood feeding hemipterans (Benoit *et al.*, 2016). This expansion of Cathepsin D might be critical for the digestion of blood, but this has not been examined. Blood provides the necessary lipid, protein, and carbohydrate sources for bed bug reproduction and nymphal development, but does not contain sufficient levels of specific micronutrients. These lacking micronutrients – B vitamins – are provided by bed bugs' obligate *Wolbachia* symbionts (Hosokawa *et al.*, 2010; Nikoh *et al.*, 2014; Moriyama *et al.*, 2015), which are discussed thoroughly in Chapter 20. There are many aspects of bed bug blood feeding and waste removal that have yet to be studied. Currently, only a single study has examined bed bug Malpighian tubules, the organ associated with diuresis, and that study focused on potassium transporters (Mamidala *et al.*, 2013). Therefore little is known about bed bug digestion or diuresis in comparison to other blood feeding insects (Esquivel *et al.*, 2014, 2016).

19.4 Reproduction and Development

There have been a multitude of studies on early periods of bed bug reproduction due to the uniqueness of bed bug mating (traumatic insemination) and male–female conflict in relation to the frequency and number of mating events (see Chapter 16). Outside of the direct effects of mating on females, specific components of the spermalege, and sperm/seminal fluid components (Stutt and Siva-Jothy, 2001; Siva-Jothy, 2006; Reinhardt *et al.*, 2009; Tartarnic *et al.*, 2014), there is little known about the physiology of reproduction in bed bugs. Currently, there has only been a single study that has identified the expression of *vitellogenin* as being critical to bed bug egg production (Moriyama *et al.*, 2016). *Vitellogenin* is highly expressed in the fat body, suggesting that this protein will be transported from the fat body to ovaries for oogenesis. RNA interference of *vitellogenin* resulted in adult females that were swollen, with atrophied fat bodies and ovaries. These maternal effects also resulted in few developing oocytes, and consequently a reduction in egg production. Even in the eggs that were produced, there was a low rate of hatching, suggesting that insufficient *vitellogenin* had been allocated to the eggs. With the exception of the Moriyama *et al.* (2016) study, other mechanisms underlying female bed bug reproductive physiology have yet to be examined. Studies that have been conducted to examine the impact of juvenile hormone (JH or analogs) on bed bug reproduction and development (Takahashi and Ohtaki, 1975; Goodman *et al.*, 2013) indicate that high levels of JH may suppress reproduction. Mechanisms that underlie reproduction in other blood feeding or closely-related hemipterans could provide a framework for research devoted to bed bug reproductive physiology.

19.5 Summary and Future Directions

Early studies by Mellanby (1932, 1935, 1939a,b) and Johnson (1941) examined bed bug physiology prior to the 1950s, when bed bugs were still a major worldwide pest and focus of entomological research. After their near eradication, basic research projects on bed bugs became rare. The bed bug resurgence has prompted renewed research interest in this pest, including in-depth studies on their physiology, examination of the mechanisms that underlie reproduction, host location, off-host stress tolerance, pesticide resistance, and physiological variations between populations. The recent sequencing of the bed bug genome (Benoit *et al.*, 2016; Rosenfeld *et al.*, 2016) along with next-generation analyses (RNA-seq, proteomics, metabolomics) and effective gene interference technologies (dsRNA injection) will allow for more focused studies utilizing molecular techniques to understand physiological properties of bed bugs. There are many topics worth examining, such as using combined genomics to identify single nucleotide polymorphisms that alter the expression of genes associated with pesticide resistance (Zhu *et al.*, 2013), or those genes responsible for mating incapability between populations that utilize different hosts (Balvín *et al.*, 2012; Booth *et al.*, 2015).

References

Aak, A., Roligheten, E., Rukke, B.A. and Birkemoe, T. (2016) Desiccant dust and the use of CO_2 gas as a mobility stimulant for bed bug: a potential control solution? *Journal of Pest Science*, **90** (1), 249–259.

Adelman, Z.N., Kilcullen, K.A., Koganemaru, R., *et al.* (2011) Deep sequencing of pyrethroid-resistant bed bugs reveals multiple mechanisms of resistance within a single population. *PLoS ONE*, **6** (10), e26228.

Akhtar, Y. and Isman, M.B. (2016) Efficacy of diatomaceous earth and a DE-aerosol formulation against the common bed bug *Cimex lectularius* Linnaeus in the laboratory. *Journal of Pest Science*, **89** (4), 1013–1021.

Anderson, J.F. and Cowles, R.S. (2012) Susceptibility of *Cimex lectularius* (Hemiptera: Cimicidae) to pyrethroid insecticides and to insecticidal dusts with or without pyrethroid insecticides. *Journal of Economic Entomology*, **105** (5), 1789–1795.

Balvín, O., Munclinger, P., Kratochvíl, L. and Vilímová, J. (2012) Mitochondrial DNA and morphology show independent evolutionary histories of bedbug *Cimex lectularius* (Heteroptera: Cimicidae) on bats and humans. *Parasitology Research*, **111** (1), 457–469.

Benoit, J.B. (2010) Water management by dormant insects: comparisons between dehydration resistance during summer aestivation and winter diapause. In: *Aestivation*, (eds C.A. Navas and J.E. Carvalho), Springer-Verlag, Berlin Heidelberg.

Benoit, J.B. (2011) Stress tolerance of bed bug: a review of factors that cause trauma to *Cimex lectularius* and *C. hemipterus*. *Insects*, **2** (2), 151–172.

Benoit, J.B. and Attardo, G.M. (2013) Mechanisms that contribute to the establishment and persistence of bed bug infestations. *Terrestrial Arthropod Reviews*, **6** (3), 227–246.

Benoit, J.B. and Denlinger D.L. (2010) Meeting the challenges of on-host and off-host water balance in blood-feeding arthropods. *Journal of Insect Physiology*, **56** (10), 1366–1376.

Benoit, J.B. and Lopez-Martinez, G. (2011) Role of conventional and unconventional stress proteins during the response of insects to traumatic environmental conditions. In: *Hemolymph Proteins and Functional Peptides: Recent Advances in Insects and Other Arthropods*, (eds M. Tufail and M. Takeda), Bentham Science Publishers, Oak Park.

Benoit, J.B., Yoder, J.A., Rellinger, E.J., Ark, J.T. and Keeney, G.D. (2005) Prolonged maintenance of water balance by adult females of the American spider beetle, *Mezium affine* Boieldieu, in the absence of food and water resources. *Journal of Insect Physiology*, **51** (5), 565–573.

Benoit, J.B., Del Grosso, N.A., Yoder, J.A. and Denlinger, D.L. (2007) Resistance to dehydration between bouts of blood feeding in the bed bug, *Cimex lectularius*, is enhanced by water conservation, aggregation, and quiescence. *American Journal of Tropical Medicine and Hygiene*, **76** (5), 987–993.

Benoit, J.B., Lopez-Martinez, G., Teets, N.M., *et al.* (2009a) Responses of the bed bug, *Cimex lectularius*, to temperature extremes and dehydration: levels of tolerance, rapid cold hardening and expression of heat shock proteins. *Medical and Veterinary Entomology*, **23** (4), 418–425.

Benoit, J.B., Phillips, S.A., Croxall, T.J., *et al.* (2009b) Addition of alarm pheromone components improves the effectiveness of desiccant dusts against *Cimex lectularius*. *Journal of Medical Entomology*, **46** (3), 572–579.

Benoit, J.B., Lopez-Martinez, G., Patrick, K.R., *et al.* (2011) Drinking a hot blood meal elicits a protective heat shock response in mosquitoes. *Proceedings of the National Academy of Sciences of the USA*, **108** (19), 8026–8029.

Benoit, J.B., Jajack, A.J. and Yoder, J.A. (2012) Multiple traumatic insemination events reduce the ability of bed bug females to maintain water balance. *Journal of Comparative Physiology B*, **182** (2), 189–198.

Benoit, J.B., Hansen, I.A., Szuter, E.M., *et al.* (2014) Emerging roles of aquaporins in relation to the physiology of blood-feeding arthropods. *Journal of Comparative Physiology B*, **184** (7), 811–825.

Benoit, J.B., Adelman, Z.N., Reinhardt, K., *et al.* (2016) Unique features of a global human ectoparasite identified through sequencing of the bed bug genome. *Nature Communications*, **7** (10165), 1–10.

Bonizzoni, M., Dunn, W.A., Campbell, C.L., *et al.* (2011) RNA-seq analyses of blood-induced changes in gene expression in the mosquito vector species, *Aedes aegypti*. *BMC Genomics*, **12** (82), 1–13.

Booth, W., Balvín, O., Vargo, E.L., *et al.* (2015) Host association drives genetic divergence in the bed bug, *Cimex lectularius*. *Molecular Ecology*, **24** (5), 980–992.

Brimblecombe, P. and Lankester, P. (2013) Long-term changes in climate and insect damage in historic houses. *Studies in Conservation*, **58** (1), 13–22.

Chown, S. and Nicolson, S.W. (2004) *Insect Physiological Ecology: Mechanisms and Patterns*, Oxford University Press, Oxford.

DeVries, Z.C., Kells, S.A. and Appel, A.G. (2013) Standard metabolic rate of the bed bug, *Cimex lectularius*: Effects of temperature, mass, and life stage. *Journal of Insect Physiology*, **59** (11), 1133–1139.

DeVries, Z.C., Kells, S.A. and Appel, A.G. (2015) Effects of starvation and molting on the metabolic rate of the bed bug (*Cimex lectularius* L.). *Physiological and Biochemical Zoology*, **88** (1), 53–65.

DeVries, Z.C., Kells, S.A. and Appel, A.G. (2016) Estimating the critical thermal maximum (CT max) of bed bugs, *Cimex lectularius*: Comparing thermolimit respirometry with traditional visual methods. *Comparative Biochemistry and Physiology Part A*, **197**, 52–57.

Esquivel, C.J., Cassone, B.J. and Piermarini, P.M. (2014) Transcriptomic evidence for a dramatic functional transition of the malpighian tubules after a blood meal in the Asian tiger mosquito *Aedes albopictus*. *PLoS Neglected Tropical Disease*, **8** (6), e2929.

Esquivel, C.J., Cassone, B.J. and Piermarini, P.M. (2016) A de novo transcriptome of the Malpighian tubules in non-blood-fed and blood-fed Asian tiger mosquitoes *Aedes albopictus*: insights into diuresis, detoxification, and blood meal processing. *PeerJ*, **4**, e1784.

Goodman, M.H., Potter, M.F. and Haynes, K.F. (2013) Effects of juvenile hormone analog formulations on development and reproduction in the bed bug *Cimex lectularius* (Hemiptera: Cimicidae). *Pest Management Science*, **69** (2), 240–244.

Hadley, N.F. (1994) *Water Relations of Terrestrial Arthropods*, Academic Press, New York.

Hosokawa, T., Koga, R., Kikuchi, Y., *et al.* (2010) *Wolbachia* as a bacteriocyte-associated nutritional mutualist. *Proceedings of the National Academy of Sciences of the USA*, **107** (2), 769–774.

How, Y.F. and Lee, C.Y. (2010) Effects of temperature and humidity on the survival and water loss of *Cimex hemipterus* (Hemiptera: Cimicidae). *Journal of Medical Entomology*, **47** (6), 987–995.

Johnson, C.G. (1941) The ecology of the bed-bug, *Cimex lectularius* L, in Britain – Report on research, 1935–40. *Journal of Hygiene*, **41** (4), 345–461.

Koganemaru, R., Miller, D.M. and Adelman, Z.N. (2013) Robust cuticular penetration resistance in the common bed bug (*Cimex lectularius* L.) correlates with increased steady-state transcript levels of CPR-type cuticle protein genes. *Pesticide Biochemistry and Physiology*, **106** (3), 190–197.

Lahondère, C. and Lazzari, C.R. (2012) Mosquitoes cool down during blood feeding to avoid overheating. *Current Biology*, **22** (1), 40–45.

Li, Y.Y., Floate, K., Fields, P. and Pang, B.P. (2014) Review of treatment methods to remove *Wolbachia* bacteria from arthropods. *Symbiosis*, **62** (1), 1–15.

Lilly, D.G., Latham, S.L., Webb, C.E. and Doggett, S.L. (2016) Cuticle thickening in a pyrethroid-resistant strain of the common bed bug, *Cimex lectularius* L.(Hemiptera: Cimicidae). *PLoS ONE*, **11** (4), e0153302.

Loudon, C. (2017) Rapid killing of bed bugs (*Cimex lectularius* L.) on surfaces using heat: application to luggage. *Pest Management Science*, **73** (1), 64–70.

Mamidala, P., Wijeratne, A.J., Wijeratne, S., *et al.* (2012) RNA-Seq and molecular docking reveal multi-level pesticide resistance in the bed bug. *BMC Genomics*, **13** (6), 1–16.

Mamidala, P., Mittapelly, P., Jones, S.C., *et al.* (2013) Molecular characterization of genes encoding inward rectifier potassium (Kir) channels in the bed bug (*Cimex lectularius*). *Comparative Biochemistry and Physiology B*, **164** (4), 275–279.

Mamai, W., Mouline, K., Parvy, J.P., *et al.* (2016) Morphological changes in the spiracles of *Anopheles gambiae* s.l. (Diptera) as a response to the dry season conditions in Burkina Faso (West Africa). *Parasites & Vectors*, **9** (11), 1–9.

Martin, L.J., Adams, R.I., Bateman A., *et al.* (2015) Evolution of the indoor biome. *Trends in Ecology & Evolution*, **30** (4), 223–232.

Mellanby, K. (1932) Effects of temperature and humidity on the metabolism of the fasting bed-bug (*Cimex lectularius*), Hemiptera. *Parasitology*, **24** (3), 419–428.

Mellanby, K. (1935) A comparison of the physiology of the two species of bed bugs which attack man. *Parasitology*, **27** (1), 111–122.

Mellanby, K. (1939a) Fertilization and egg production in the bed bug (*Cimex lectularius* L.). *Parasitology*, **31** (2), 193–199.

Mellanby, K. (1939b) The physiology and activity of the bed bug (*Cimex lectularius* L.) in a natural infestation. *Parasitology*, **31** (2), 200–211.

Michels, J., Gorb, S.N. and Reinhardt, K. (2015) Reduction of female copulatory damage by resilin represents evidence for tolerance in sexual conflict. *Journal of The Royal Society Interface*, **12** (104), 20141107.

Moriyama, M., Nikoh, N., Hosokawa, T. and Fukatsu, T. (2015) Riboflavin provisioning underlies *Wolbachia's* fitness contribution to its insect host. *MBio*, **6** (6), e01732.

Moriyama, M., Hosokawa, T., Tanahashi, M., *et al.* (2016) Suppression of bedbug's reproduction by RNA Interference of *vitellogenin*. *PLoS ONE*, **11** (4), e0153984.

Naylor, R.A. and Boase, C.J. (2010) Practical solutions for treating laundry infested with *Cimex lectularius* (Hemiptera: Cimicidae). *Journal of Economic Entomology*, **103** (1), 136–139.

Needham, G.R. and Teel, P.D. (1991) Off-host physiological ecology of ixodid ticks. *Annual Review of Entomology*, **36**, 659–681.

Nikoh, N., Hosokawa, T., Moriyama, M., *et al.* (2014) Evolutionary origin of insect–*Wolbachia* nutritional mutualism. *Proceedings of the National Academy of Sciences*, **111** (28), 10257–10262.

Olson, J.F., Eaton, M., Kells, S.A., *et al.* (2013) Cold tolerance of bed bugs and practical recommendations for control. *Journal of Economic Entomology*, **106** (6), 2433–2441.

Omori, N. (1941) Comparative studies on the ecology and physiology of common and tropical bed bug, with special reference to the reactions to temperature and moisture. *Journal of Medical Association of Taiwan*, **60**, 555–729.

Polanco, A.M., Brewster, C.C. and Miller, D.M. (2011a) Population growth potential of the bed bug *Cimex lectularius* L.: a life table analysis. *Insects*, **2** (2), 173–185.

Polanco, A.M., Miller, D.M. and Brewster, C.C. (2011b) Survivorship during starvation for *Cimex lectularius* L. *Insects*, **2** (2), 232–242.

Puckett, R.T., McDonald, D.L. and Gold, R.E. (2013) Comparison of multiple steam treatment durations for control of bed bugs (*Cimex lectularius* L.). *Pest Management Science*, **69** (9), 1061–1065.

Reidenbach, K.R., Cheng, C., Liu, F., *et al.* (2014) Cuticular differences associated with aridity acclimation in African malaria vectors carrying alternative arrangements of inversion 2La. *Parasites & Vectors*, **7** (176), 1–13.

Reinhardt, K., Naylor, R.A. and Siva-Jothy, M.T. (2009) Situation exploitation: higher male mating success when female resistance is reduced by feeding. *Evolution*, **63** (1), 29–39.

Rosenfeld, J.A., Reeves, D., Brugler, M.R., *et al.* (2016) Genome assembly and geospatial phylogenomics of the bed bug *Cimex lectularius*. *Nature Communications*, **7** (10164), 1–10.

Rukke, B.A., Aak, A. and Edgar, K.S. (2015) Mortality, temporary sterilization, and maternal effects of sublethal heat in bed bugs. *PLoS ONE*, **10** (5), e0127555.

Rukke, B.A., Hage, M. and Aak, A. (2017) Mortality, fecundity, and development among bed bugs (*Cimex lectularius*) exposed to prolonged, intermediate cold stress. *Pest Management Science*, **73** (5), 838–843.

Siva-Jothy, M.T. (2006) Trauma, disease and collateral damage: conflict in cimicids. *Philosophical Transactions of the Royal Society B-Biological Sciences*, **361** (1466), 269–275.

Siva-Jothy, M.T. and Stutt, A.D. (2003) A matter of taste: direct detection of female mating status in the bedbug. *Proceedings of the Royal Society of London B: Biological Sciences*, **270** (1515), 649–652.

Sojka, D., Francischetti, I.M., Calvo, E. and Kotsyfakis, M. (2011) Cysteine proteases from bloodfeeding arthropod ectoparasites. In: *Cysteine Proteases of Pathogenic Organisms* (eds M.W. Robinson and J.P. Dalton), Springer, Berlin.

Stutt, A.D. and Siva-Jothy, M.T. (2001) Traumatic insemination and sexual conflict in the bed bug *Cimex lectularius*. *Proceeding of the National Academy of Sciences USA*, **98** (10), 5683–5687.

Takahashi, M. and Ohtaki, T. (1975) Ovicidal effects of two juvenile hormone analogs, methoprene and hydroprene, on the human body louse and the bed bug. *Medical Entomology and Zoology*, **26** (4), 237–239.

Tatarnic, N.J., Cassis, G. and Siva-Jothy, M.T. (2014) Traumatic insemination in terrestrial arthropods. *Annual Review of Entomology*, **59**, 245.

Urbanski, J., Benoit, J.B., Michaud, M.R., *et al.* (2010) The molecular physiology of increased egg desiccation resistance during diapause in the invasive mosquito, *Aedes albopictus*. *Proceedings B*, **277**, 2683–2692.

Usinger, R.L. (1966) *Monograph of Cimicidae (Hemiptera - Heteroptera)*. Entomological Society of America, College Park.

Vannini, L., Dunn, W.A., Reed, T.W. and Willis, J.H. (2014) Changes in transcript abundance for cuticular proteins and other genes three hours after a blood meal in *Anopheles gambiae*. *Insect Biochemistry and Molecular Biology*, **44**, 33–43.

Wharton, G.A. (1985) Water balance of insects. In: *Comprehensive Insect Physiology, Biochemistry and Pharmacology* (eds G.A. Kerkut and L.I. Gilbert), Pergamon Press, Oxford.

Yoder, J.A., Benoit, J.B. and Opaluch, A.M. (2004) Water relations in eggs of the lone star tick, *Amblyomma americanum*, with experimental work on the capacity for water vapor absorption. *Experimental and Applied Acarology*, **33**, 235–242.

Zhu, F., Gujar, H., Gordon, J.R., *et al.* (2013). Bed bugs evolved unique adaptive strategy to resist pyrethroid insecticides. *Scientific Reports*, **3** (1456), 1–7.

20

Symbionts

Mark Goodman

20.1 Introduction

The term "symbiosis" was first used by Anton de Bary in 1878 to describe dissimilar organisms living together (Oulhen *et al.*, 2016). Endosymbiosis is a special case of symbiosis in which one organism lives within another. These relationships occur on a continuum, ranging from mutualistic interactions, in which both host and symbiont benefit from the relationship, to parasitism, in which the symbiont benefits to the detriment of the host (Moran *et al.*, 2008). A plethora of diverse interactions of symbionts with their hosts have been described in recent years. For example, Kikuchi *et al.* (2012) determined that endosymbionts confer insecticide resistance in stink bugs. Tsuchida *et al.* (2010) found that endosymbionts dictated host color in aphids, and Kuriwada *et al.* (2010) documented that endosymbionts influenced the size and development time of a weevil. Endosymbionts are also known to provide a defense against aphid parasitoids and pathogens (Brownlie and Johnson, 2009; Scarborough *et al.*, 2005). In many insects, symbionts such as *Wolbachia* can provision nutrients for the host (Brownlie *et al.*, 2009; Nikoh *et al.*, 2014) or cause reproductive anomalies including cytoplasmic incompatibility and male-killing (Werren *et al.*, 2008). Multiple endosymbiotic organisms may exist within a single host, resulting in more complex interactions (Wu *et al.*, 2006; Moran *et al.*, 2008).

The wide-ranging impact symbionts often have on a host's overall biology and success indicate that a basic understanding of the identity and impacts of bed bug endosymbionts is crucial to understanding the insect itself. The first published studies of symbionts associated with bed bugs came in the 1920s (Arkwright *et al.*, 1921; Buchner, 1921, 1923; Hase, 1929; Hertig and Wolbach, 1924) and addressed only the Common bed bug, *Cimex lectularius* L. These researchers and others in the following decades, meticulously characterized the bacteria detected within *C. lectularius* using light and scanning electron microscopy, histology, and chemical diagnoses (Dasgupta and Ray, 1956; Krieg, 1959; Chang and Musgrave, 1973; Louis *et al.*, 1973). Initially there was some confusion among researchers over the identity of the symbionts, with different names being provided based on physical descriptions and where they were found in the host. Recent technological advances have allowed us to more accurately characterize the bed bug endosymbionts that are the focus of this chapter.

20.2 Identity of Endosymbionts

Many different microorganisms have been found living inside bed bugs (Khan, 1974; Strand, 1977; Meriweather *et al.*, 2013). However, only two are considered symbionts and these are found consistently across and within populations, and these two have now been identified using PCR and gene sequencing (Hypsa and Aksoy, 1997; Rasgon and Scott, 2004; Hosokawa *et al.*, 2010).

Advances in the Biology and Management of Modern Bed Bugs, First Edition.
Edited by Stephen L. Doggett, Dini M. Miller, and Chow-Yang Lee.
© 2018 John Wiley & Sons Ltd. Published 2018 by John Wiley & Sons Ltd.

The best known of the two symbiont types is *Wolbachia*, which has been identified not only in *C. lectularius*, but also in the Tropical bed bug, *Cimex hemipterus* (F.), *Cimex adjuctus* Barber, *Cimex incrassatus* Usinger and Ueshima, *Oeciacus hirundinis* (Lamarck), *Oeciacus vicarius* Horvath, *Afrocimex constrictus* Ferris and Usinger, and *Haematosiphon inodorus* (Duges) (Hase, 1929; Pfeiffer, 1931; Rasgon and Scott, 2004; Sakamoto et al., 2006; Meriweather et al., 2013; Siddiqui and Raja, 2015). In these species, including both *C. hemipterus* and *C. lectularius*, the *Wolbachia* symbiont belongs to the F Clade which has been associated primarily with termites, weevils, and nematode worms. This clade has been shown in some instances to provide nutritional benefits to the host (Lo et al., 2002). It is probable that *Wolbachia* is found in most species across the family Cimicidae, and there is no record of any natural population of *C. lectularius* lacking this symbiont. Within the bed bug host, *Wolbachia* is found primarily in specific organs. These include the mycetomes (specialized, paired organs for holding bacteria, found in both male and female bed bugs) and the reproductive organs, including the ovaries, testes, eggs, and mesospermalege (Hosokawa et al., 2010).

The second symbiont is an unnamed gamma-proteobacterium, designated here as BLS following the precedent in the literature ("BEV-like symbiont", referring to its similarity to the bacterial symbiont of *Euscelidius variegatus* (Kirschbaum), a leafhopper) (Campbell and Purcell, 1993; Hypsa and Aksoy, 1997; Sakamoto and Rasgon, 2006a). While the presence of BLS has not been determined in most other cimicid species (including *C. hemipterus*), it can be found in most populations of *C. lectularius* around the world. BLS occurs in the same organs as *Wolbachia*, although generally in lower titers (Sakamoto and Rasgon, 2006b; Sakomoto et al., 2006; Hosokawa et al., 2010; Meriweather et al., 2013; Akhoundi et al., 2016). In addition to the mycetomes and reproductive organs, BLS can also be found in high concentrations in the Malpighian tubules and hemolymph, where it is highly motile and can be easily seen moving when viewed under a negative phase-contrast microscope (Arkwright et al., 1921; Louis et al., 1973; Goodman, 2016).

20.3 Impact of Symbionts on Bed Bug Biology

It has long been accepted that to thrive, those arthropods that feed exclusively on nutritionally deficient materials such as blood and xylem must have symbionts to provide supplemental nutrients, including vitamins and amino acids (Koch, 1960). This has been demonstrated in aphids, tsetse flies, lice, and other insects (Nakabachi and Ishikawa, 1999; Brownlie et al., 2009; Rio et al., 2016). This theory has led to the suggestion that at least one of these symbionts must provide a nutritional benefit to bed bugs. Early studies in which *C. lectularius* were fed on nutrient-deficient rats suggested that B vitamins might be supplied to the bed bugs via symbionts (De Meillon and Goldberg, 1947; De Meillon et al., 1947). However, this host–symbiont interaction remained unproven until Hosokawa et al. (2010) published a study which demonstrated that *Wolbachia* played a major role in providing nutrients to *C. lectularius*, without which bed bugs experienced decreased fecundity and slowed development. In populations of *C. lectularius* in which the symbionts were eliminated, and especially in the offspring of aposymbiotic bugs, there was slower development, a lower fecundity for both male and female bugs, and about a 10% decrease in adult size (Heaton, 2013; Goodman, 2016). These detrimental effects were probably due in part to the loss of *Wolbachia*. These strains recover to nearly normal development and fecundity when provided with supplemental B vitamins (Hosokawa et al., 2010). Specifically, it has been demonstrated that the B vitamins biotin and riboflavin are both produced by this *Wolbachia* symbiont in *C. lectularius* (Nikoh et al., 2014). When both symbionts are eliminated, low fecundity and slow development make it unlikely that bed bug populations will survive and reproduce. This is a significant finding, because most *Wolbachia* infections in other arthropod hosts demonstrate reproductive manipulation, but not benefits to the host (Werren et al., 2008; Rio et al., 2016). Perhaps the discovery of nutritional provisioning by *Wolbachia* should not be surprising, since the C, D and F clades of *Wolbachia* provide similar benefits to nematodes and other arthropod hosts (Bandi et al., 2001; Nikoh et al., 2014). Similar investigations are yet to be undertaken on *C. hemipterus*.

The role of BLS is less vital, and losing it does not appear to be detrimental to the bed bugs so long as *Wolbachia* is present. However, female aposymbiotic *C. lectularius*, which have BLS restored, are more fecund than those which have no symbionts at all (Goodman, 2016), indicating that BLS may also be beneficial to the host.

20.4 Transmission of Symbionts

Researchers have effectively documented the transovarial transmission of both symbionts in *C. lectularius*. Even the movement of symbionts into the egg at time points prior to fertilization has been measured (Buchner, 1923; Hosokawa *et al.*, 2010). Since all natural populations of *C. lectularius* have *Wolbachia* present (Sakamoto and Rasgon, 2006b; Meriweather *et al.*, 2013) it seems clear that even if *Wolbachia* transmission is not 100% efficient, its obligate nature keeps it nearly universal within populations. There is no evidence of *Wolbachia* being transmitted horizontally during mating or other normal activities (Goodman, 2016).

Transmission of BLS, however, is not limited to transovarial. Experiments with aposymbiotic strains of *C. lectularius* indicate that transmission from the male to female occurs during traumatic insemination (Goodman, 2016). This horizontal transmission may be important in keeping the symbiont established within populations despite it not being obligate for survival. Since male bed bugs mate repeatedly with many different female partners (Reinhart and Siva-Jothy, 2007), it is likely that if BLS is introduced to any given population, it spreads readily.

20.5 Symbionts and Bed Bug Management

Resistance to insecticides has been attributed to symbiont presence in stink bugs (Kikuchi *et al.*, 2012). While resistance to pyrethroid insecticides has been widely reported in bed bugs (Gordon *et al.*, 2014), both *Wolbachia* and BLS symbionts are found in pyrethroid-resistant and susceptible populations of *C. lectularius* (Goodman, 2016). Thus resistance due to symbiont presence is unlikely.

Since aposymbiotic strains of bed bugs have low fitness, symbiont elimination may appear to be an appealing control option. But while symbionts are vital to bed bug survival, it is unlikely that they can be utilized as part of an effective control strategy. Consistent elimination of the symbionts requires feeding bed bugs antibiotics through an artificial feeding system (Hosokawa *et al.*, 2010). If one can feed them antibiotics, it is simpler to deliver an insecticide to kill the bug immediately, and not go to great effort simply to reduce their fecundity.

While clear applications for symbionts in control of bed bugs may not yet be recognized, we still know relatively little about the relationships between these organisms, and the field of study is rapidly expanding. It may be that the most notable interactions between bed bugs and their symbionts have yet to be discovered.

References

Akhoundi, M., Cannet, A., Loubatier, C., *et al.* (2016) Molecular characterization of *Wolbachia* infection in bed bugs (*Cimex lectularius*) collected from several localities in France. *Parasite*, **23** (31), doi: 10.1051/parasite/2016031.

Arkwright, J.A., Atkin, E.E. and Bacot, A. (1921) An hereditary Rickettsia-like parasite of the bed bug (*Cimex lectularius*). *Parasitology*, **13** (1), 27–38.

Bandi, C., Trees, A.J. and Brattig, N.W. (2001) *Wolbachia* in filarial nematodes: evolutionary aspects and implications for the pathogenesis and treatment of filarial diseases. *Veterinary Parasitology*, **98** (1–3), 215–238.

Brownlie, J.C. and Johnson, K.N. (2009) Symbiont-mediated protection in insect hosts. *Trends in Microbiology*, **17** (8), 348–354.

Brownlie, J.C., Cass, B.N., Riegler, M., et al. (2009) Evidence for metabolic provisioning by a common invertebrate endosymbiont, *Wolbachia pipientis*, during periods of nutritional stress. *PLoS Pathogens*, **5** (4), e1000368.

Buchner, P. (1921) Uber ein neues symbiontisches organ der bettwanze. *Biologisches Zentralblatt Leipzig*, **41**, 570–574.

Buchner, P. (1923) Studien an intracellularen symbionten. IV. Die Bakteriensymbiose der Bettwanze. *Archiv fur Protistenkunde*, **46**, 225–263.

Campbell, B.C. and Purcell, A.H. (1993) Phylogenetic affiliation of BEV, a bacterial parasite of the leafhopper *Euscelidius variegatus*, on the basis of 16S rDNA sequences. *Current Microbiology*, **26** (1), 37–41.

Chang, K.P. and Musgrave, A.J. (1973) Morphology, histochemistry, and ultrastructure of mycetome and its rickettsial symbiotes in *Cimex lectularius* L. *Canadian Journal of Microbiology*, **19** (9), 1075–1081.

Dasgupta, B. and Ray, H.N. (1956) Observations on the "corpus luteum" and the "NR bodies" in the female gonads of bed-bug. *Proceedings of the Zoological Society of Calcutta*, **9** (2), 55–63.

De Meillon, B. and Goldberg, L. (1947) Preliminary studies on the nutritional requirements of the bed bug (*Cimex lectularius* L.) and the tick *Ornithodorus moubata* Murray. *Journal of Experimental Biology*, **24** (1–2), 41–61.

De Meillon, B., Thorpe, J.M. and Hardy, F. (1947) The development of *Cimex lectularius* and *Ornithodorus moubata* on riboflavin deficient rats. *The South African Journal of Medical Sciences*, **12**, 111–116.

Goodman, M.H. (2016) Endosymbiotic bacteria in the bed bug, *Cimex lectularius* L. (Hemiptera: Cimicidae). PhD thesis, University of Kentucky.

Gordon, J.R., Goodman, M.H., Potter, M.F. and Haynes, K.F. (2014) Population variation in and selection for resistance to pyrethroid-neonicotinoid insecticides in the bed bug. *Scientific Reports*, **4** (3836), 1–7.

Hase, A. (1929) Weitere versuche zur kenntnis der bettwanzen *Cimex lectularius* L. und *Cimex rotundatus* Sign. (Hex.-Rhynch.). *Parasitology Research*, **2** (3), 368–418.

Heaton, L.L. (2013) *Wolbachia* in bed bugs *Cimex lectularius*. PhD thesis, University of Sheffield.

Hertig, M. and Wolbach, S.B. (1924) Studies on Rickettsia-like microorganisms in insects. *The Journal of Medical Research*, **44** (3), 329–374.

Hosokawa, T., Koga, R., Kikuchi, Y., Meng, X.Y. and Fukatsu, T. (2010). *Wolbachia* as a bacteriocyte-associated nutritional mutualist. *Proceedings of the National Academy of Sciences of the United States of America*, **107** (2), 769–774.

Hypsa, V. and Aksoy, S. (1997) Phylogenetic characterization of two transovarially transmitted endosymbionts of the bedbug *Cimex lectularius* (Heteroptera: Cimicidae). *Insect Molecular Biology*, **6** (3), 301–304.

Khan, A.M. (1974) Isolation of intracellular bacterium from *Cimex hemipterus* Walker. *Indian Journal of Entomology*, **36** (3), 240–241.

Kikuchi, Y., Hayatsu, M., Hosokawa, T., et al. (2012) Symbiont-mediated insecticide resistance. *Proceedings of the National Academy of Sciences*, **109** (22), 8618–8622.

Koch, A. (1960) Intracellular symbiosis in insects. *Annual Review of Microbiology*, **14**, 121–140.

Krieg, A. (1959) Über die natur von "NR-bodies" bei Rickettsien-infektionen von insekten. *Naturwissenshaften*, **46**, 231–232.

Kuriwada, T., Hosokawa, T., Kumano, N., et al. (2010) Biological role of *Nardonella* endosymbiont in its weevil host. *PLoS ONE*, **5** (10), e13101.

Lo, N., Casiraghi, M., Salati, E., Bazzocchi, C. and Bandi, C. (2002) How many *Wolbachia* supergroups exist? *Molecular Biology and Evolution*, **19** (3), 341–346.

Louis, C., Laporte, M., Carayon, J. and Vago, C. (1973) Mobilité, ciliature et caractères ultrastructuraux des micro-organismes symbiotiques endo et exocellulaires de *Cimex lectularius* L. (Hemiptera Cimicidae). *Comptes rendus de l'Académie des Sciences Serie D*, **277** (6), 607–611.

Meriweather, M., Matthews, S., Rio, R. and Baucom, R.S. (2013). A 454 survey reveals the community composition and core microbiome of the common bed bug (*Cimex lectularius*) across an urban landscape. *PloS ONE*, **8** (4), e61465.

Moran, N.A., McCutcheon, J.P. and Nakabachi A, (2008) Genomics and evolution of heritable bacterial symbionts. *Annual Reviews of Genetics*, **42**, 165–190.

Nakabachi, A. and Ishikawa, H. (1999) Provision of riboflavin to the host aphid, *Acyrthosiphon pisum*, by endosymbiotic bacteria, *Buchnera*. *Journal of Insect Physiology*, **45** (1), 1–6.

Nikoh, N., Hosokawa, T., Moriyama, M., et al. (2014) Evolutionary origin of insect–*Wolbachia* nutritional mutualism. *Proceedings of the National Academy of Sciences*, **111** (28), 10257–10262.

Oulhen, N., Schulz, B.J. and Carrier T.J. (2016) English translation of Heinrich Anton de Bary's 1878 speech, 'Die Erscheinung der Symbiose' ('De la symbiose'). *Symbiosis*, **69**, 131–139.

Pfeiffer, H. (1931) Beitrage zu der bakteriensymbiose der bettwanze (*Cimex lectularius*) und der Schwalbenwanze (*Oeciacus hirundinis*). *Zentralblatt fur Bakteriologie, Parasitenkunde, Infektionskrankheiten und Hygiene*, **123**, 151–171.

Rasgon, J.L. and Scott, T.W. (2004) Phylogenetic characterization of *Wolbachia* symbionts infecting *Cimex lectularius* L. and *Oeciacus vicarius* Horvath (Hemiptera: Cimicidae). *Journal of Medical Entomology*, **41** (6), 1175–1178.

Reinhardt, K. and Siva-Jothy, M.T. (2007) Biology of the bed bugs (Cimicidae). *Annual Review of Entomology*, **52**, 351–374.

Rio, R.V.M., Attardo, G.M. and Weiss, B.L. (2016) Grandeur alliances: symbiont metabolic integration and obligate arthropod hematophagy. *Trends in Parasitology*, **32** (9), 739–749.

Sakamoto, J.M. and Rasgon, J.L. (2006a) Endosymbiotic bacteria of bed bugs: evolution, ecology and genetics. *American Entomologist*, **52** (2), 119–122.

Sakamoto, J. M. and Rasgon, J.L. (2006b) Geographic distribution of *Wolbachia* infections in *Cimex lectularius* (Heteroptera: Cimicidae). *Journal of Medical Entomology*, **43** (4), 696–700.

Sakamoto, J. M., Feinstein, J. and Rasgon, J.L. (2006) *Wolbachia* infections in the Cimicidae: museum specimens as an untapped resource for endosymbiont surveys. *Applied Environmental Microbiology*, **72** (5), 3161–3167.

Scarborough, C.L., Ferrari, J. and Godfray, H.C.J. (2005) Aphid protected from pathogen by endosymbiont. *Science*, **310** (5755), 1781–1781.

Siddiqui, S.S. and Raja, I.A. (2015) Molecular detection of endosymbiont bacteria *Wolbachia* in bedbug species *Cimex lectularius* from Vidarbha region of India. *International Journal of Life Sciences*, **3** (3), 200–204.

Strand, M.A. (1977). Pathogens of Cimicidae (bedbugs). *Bulletin of the World Health Organization*, **55** (Suppl 1), 313–315.

Tsuchida, T., Koga, R., Horikawa, M., et al. (2010) Symbiotic bacterium modifies aphid body color. *Science*, **330** (6007), 1102–1104.

Werren, J.H., Baldo, L. and Clark, M.E. (2008) *Wolbachia*: master manipulators of invertebrate biology. *Nature Reviews Microbiology*, **6** (10), 741–751.

Wu, D., Daugherty, S.C., Van Aken, S.E., et al. (2006) Metabolic complementarity and genomics of the dual bacterial symbiosis of sharpshooters. *PLoS Biology*, **4** (6), e188.

21

Bed Bug Laboratory Maintenance

Mark F. Feldlaufer, Linda-Lou O'Connor and Kevin R. Ulrich

21.1 Introduction

The worldwide resurgence of bed bug infestations has resulted in an increased research effort by academic, industry, and government laboratories, to develop strategies to mitigate this blood-sucking insect pest. Since all cimicids, including the Common bed bug, *Cimex lectularius* L., and the Tropical bed bug, *Cimex hemipterus* (F.), are obligate blood-feeding insects, the blood-feeding of these species on a routine basis is required to maintain colonies. The laboratory maintenance and rearing requirements of bed bugs were the subject of several recent reviews (Aak and Rukke, 2014; Feldlaufer *et al.*, 2014; Cannet *et al.*, 2015). This current review will focus on general colony maintenance, different feeding techniques available to rear *C. lectularius* and its tropical relative *C. hemipterus*, the insect's need for blood plasma in any feeding regime, and the development of an "artificial" blood source.

21.2 General Colony Maintenance

21.2.1 Containers and Harborages

Since bed bugs are found in a wide variety of human habitations, hiding in whatever cracks and crevices are available, the containers and harborages used for their laboratory rearing are similarly diverse. For laboratory maintenance, bed bugs are usually housed in containers of varying size and material (glass or plastic) that contain some sort of harborage material – filter paper, blotting paper, cardboard, manila, and so on – where the bed bugs can hide, mate, and oviposit. Oftentimes, the person or persons responsible for rearing use containers and harborages that are both readily available and cost-effective. Whatever container is chosen, they generally have mesh at one or both ends to facilitate blood-feeding. The mesh is of a size that allows individuals to feed, yet prevents even the smallest nymphs from escaping. A US standard mesh size of "100 mesh" (0.150 mm openings) generally will fulfill this requirement. An example of a suitable glass rearing (mason) jar and other components associated with bed bug rearing is shown in Figure 21.1, although any container that does not allow bed bugs to escape is suitable.

Since *C. lectularius* does not move effectively or quickly on smooth surfaces (Loudon and Boudaie, 2009), manipulating either individual bed bugs and/or colonies on Teflon (DuPont, Wilmington, Delaware, US) or Fluon (GC Chemicals Europe, Lancashire, United Kingdom) treated surfaces can thwart their unwanted movement and dispersal (Feldlaufer *et al.*, 2014). The situation with *C. hemipterus*, however, appears quite different because Teflon coated surfaces do not appear to limit mobility (Kim *et al.*, 2017).

Figure 21.1 Mason Jar and components used to house bed bugs. Unassembled and assembled cap with mesh screening, and folded, filter paper harborage.

21.3 Feeding Techniques

All cimicids suck blood and blood is required for normal growth, development, and reproduction. Members of the Cimicidae feed primarily on birds, bats, and humans (Usinger, 1966). While both *C. lectularius* and *C. hemipterus* will feed on poultry (Rosen *et al.*, 1987; Axtell and Arends, 1990; Steelman *et al.*, 2008), it is their sucking of human blood that has vaulted these insects into their position as a pest of major public concern.

A number of *in-vivo* and *in-vitro* methods are currently used to feed *C. lectularius* and *C. hemipterus*, and the methods can be found in the literature after 2000 (Table 21.1). Prior to this time, several *in-vivo* and *in-vitro* methods were used to feed bed bugs that were to be used either for specific studies or for colonization. References are made to these previous studies in this review article because newer techniques have often evolved from older ones (see Girault, 1910; Rivnay, 1930; Rendtorff, 1938; De Meillon and Goldberg, 1946, 1947; Davis, 1956; Adkins and Arant, 1959; Peterson, 1959; Wattal and Kalra, 1961; Gilbert, 1964; Bell and Schaefer, 1966; Burden, 1966; Hall *et al.*, 1979; Ogston and Yanovski, 1982). Some of these earlier citations contain references that the authors were either unable to acquire in their full text, or were in languages other than English.

21.3.1 *In-vivo* and *In-vitro* Blood Sources

A review of the literature since 2000 indicates that *in-vivo* feeding of bed bugs has involved a variety of blood sources, although it appears that feeding upon humans is the most common (Table 21.1). *In-vivo* feeding upon humans may require a human ethics or Institutional Review Board (IRB) approval, depending on the country and institution in which the studies are being conducted. A distinct disadvantage of using humans is that some individuals can become sensitized to bed bug bites and systemic allergic reactions in humans have been reported following the feeding of bed bugs (Minocha *et al.*, 2017). When vertebrates other than humans are used as an *in-vivo* source of blood, some sort of animal ethics approval (for example IACUC; Institutional Animal Care and Use Committee in the USA) is usually required to ensure the host animals are housed and handled in a humane manner. In many countries, these animal ethics committees are also responsible for reviewing research protocols that involve live vertebrate animals. In some countries, the inspection of facilities is conducted by groups external to the organization involved.

Table 21.1 Blood sources used for *in-vivo* and *in-vitro* feeding of *C. lectularius* and *C. hemipterus*[1].

	Reference(s)	
Blood source	In-vivo	In-vitro
Cattle	—	Montes *et al.* (2002) (h), Romero (s) (d)[3]
Chicken	Pfeister *et al.* (2008)	Montes *et al.* (2002) (h), Romero *et al.* (2007) (h), Siljander *et al.* (2007) (h), Benoit *et al.* (2009), Ryne (2009) (d), Zhu *et al.* (2010) (h), Adelman *et al.* (2011) (c), Reis and Miller (2011) (c), Jones and Bryant (2012) (h)
Human	Moore and Miller (2006), Siljander *et al.* (2007), Wintle and Reinhardt (2008), How and Lee (2011), Kilpinen *et al.* (2012), Cockburn *et al.* (2013), Rahim and Majid (2014), Naylor (s)[2]	Moore and Miller (2006), Yoon *et al.* (2008), Araujo *et al.* (2009) (c), Olson *et al.* (2009), Feldlaufer *et al.* (2010) (c), Seong *et al.* (2010), Kells and Goblirsch (2011) (h), Barbarin *et al.* (2013), Tawatsin *et al.* (2013), Aak and Rukke (2014) (h), Leulmi *et al.* (2015) (c), Rahim *et al.* (2015), Balvin (s)
Mouse	Araujo *et al.* (2009), Lee (s)	—
Pig	—	Vander Pan (s) (d)
Sheep	—	Montes *et al.* (2002) (d, h), Harraca *et al.* (2010) (d), Weeks *et al.* (2011) (h), Naylor (s)
Pigeon	Araujo *et al.* (2009)	—
Rabbit	Stutt and Siva-Jothy (2001), Reinhardt *et al.* (2003), Anderson *et al.* (2009), Vander Pan (s)	Romero *et al.* (2009) (c), Hosokawa *et al.* (2010), Zhu *et al.* (2012) (c), Chin-Heady *et al.* (2013) (c), Jones *et al.* (2013), Singh *et al.* (2013) (d), Wang *et al.* (2013) (d), Choe and Campbell (2014) (c), Romero and Schal (2014) (d), Campbell and Miller (2015) (d), Hinson (2015) (d), Lee (s), Pereira (s)
Guinea pig	Wang (s)	—
Hamster	Tawatsin *et al.* (2011)	—
Rat	Aak and Rukke (2014), Lilly and Doggett (s)	—

[1] since 2000;
[2] (s) = survey: (respondent may be part of a larger laboratory);
[3] for *in vitro* feeding. Anticoagulant (when mentioned): d, defibrinated; h, heparin; c, citrated.

In-vitro feeding methods have also used a wide variety of blood sources to feed laboratory colonies of bed bugs (Table 21.1), and it appears that chicken blood, human blood, and rabbit blood are currently used the most. While *C. lectularius* and *C. hemipterus* naturally feed upon humans (Johnson, 1941), they obviously can be reared on different kinds of vertebrate blood, with varying degrees of success. The literature contains a large number of publications evaluating the nutritional value (based on bed bug fecundity, longevity, and so on) of blood from different vertebrate sources (see Usinger, 1966, and references therein). For *in-vitro* feeding of bed bugs, whole blood must contain an anticoagulant, such as heparin, citrate, or blood in which the fibrin has been removed (defibrinated). While certain studies have addressed the suitability of different blood sources and different anticoagulants, often with seemingly contradictory results (Montes *et al.*, 2002; Barbarin *et al.*, 2013; Aak and Rukke, 2014; Rahim *et al.*, 2015), it appears that a variety of blood sources and anticoagulants are currently used (Table 21.1). There is even a reference in the literature where quail (*Coturnix coturnix* L.) was used to maintain *C. hemipterus* (Azevedo *et al.*, 2009).

21.3.2 *In-vitro* feeding units

For *in-vitro* feeding, the blood supply is normally heated, to mimic the temperature of the host (Hall *et al.*, 1979; Montes *et al.*, 2002). Several commercial feeding units fashioned from glass are shown in Figure 21.2. The amount of blood needed to fill these units can vary from less than 3 ml to more than 40 ml, and this can often be a consideration when choosing a feeding unit. The advantage of any unit that holds a minimal amount of blood (3 ml, or less) can be negated by the fact that larger colonies of bed bugs may quickly deplete the blood supply, necessitating replenishment. Conversely, units that hold larger amounts of blood are not without their issues because blood can be expensive and/or time-consuming to collect. The glass units depicted in Figure 21.2 are essentially "jacketed," in that they allow for recirculating, heated water to be pumped into an outer jacket, which holds the blood at a suitable temperature in the inner chamber. A commercially-available feeding apparatus is available from Hemotek Ltd (Blackburn, United Kingdom). This uses electricity to maintain heat in a heating unit to which the blood-reservoir unit is attached. The Hemotek feeding system is shown in Figure 21.3. Individually-designed feeding units have also been successfully used (Chin-Heady *et al.*, 2013; Aak and Rukke, 2014).

A suitable membrane must also be employed to allow for bed bug feeding and to contain the blood. Currently, most laboratories use Parafilm M (Bemis, Oshkosh, WI, US), (or Nescofilm, Karlan Research Products Corp., Cottonwood, AZ, USA, if available), although other readily available membranes can be substituted (Wawrocka and Bartonička, 2013). Overall, the type of feeding unit employed for the *in-vitro* feeding of bed bugs, along with the actual blood source, anticoagulant, and feeding membrane used is largely dependent on the preference of the researchers, the materials' availability, and the cost.

In a laboratory study evaluating the *in-vitro* rearing of *C. hemipterus*, no significant differences were found in the numbers of live bed bugs produced (after seven weeks), when the colonies were fed on whole blood, red blood cells (RBC), or RBC admixed with plasma (Rahim *et al.*, 2015). However, these same authors reported significant mortality when bed bugs were fed *in vitro* – regardless of blood composition – when compared to *C. hemipterus* that fed directly upon a volunteer's arm. Rahim and Majid, (2014) a year earlier, also reported significant mortality in two other strains of *C. hemipterus* fed *in vitro*, when compared to those bed bugs fed on a volunteer's arm.

Regardless of whether an *in-vivo* or *in-vitro* method is chosen to feed bed bugs, the toxicity of heme proteins, produced as a result of hemoglobin digestion, might be an issue. How blood-sucking insects and other arthropods deal with heme is dealt with in a review by Graça-Souza and colleagues (2006). Despite the toxicity of unbound heme, bed bug salivary heme proteins –termed nitrophorins – are responsible for delivering nitrous oxide to the bite site, where the compound is responsible for vasodilation and other actions involved in the feeding process (Valenzuela *et al.*, 1995; Badgandi, 2009; Francischetti *et al.*, 2010).

Figure 21.2 Commercially-available glass feeders used for feeding bed bugs *in vitro*.

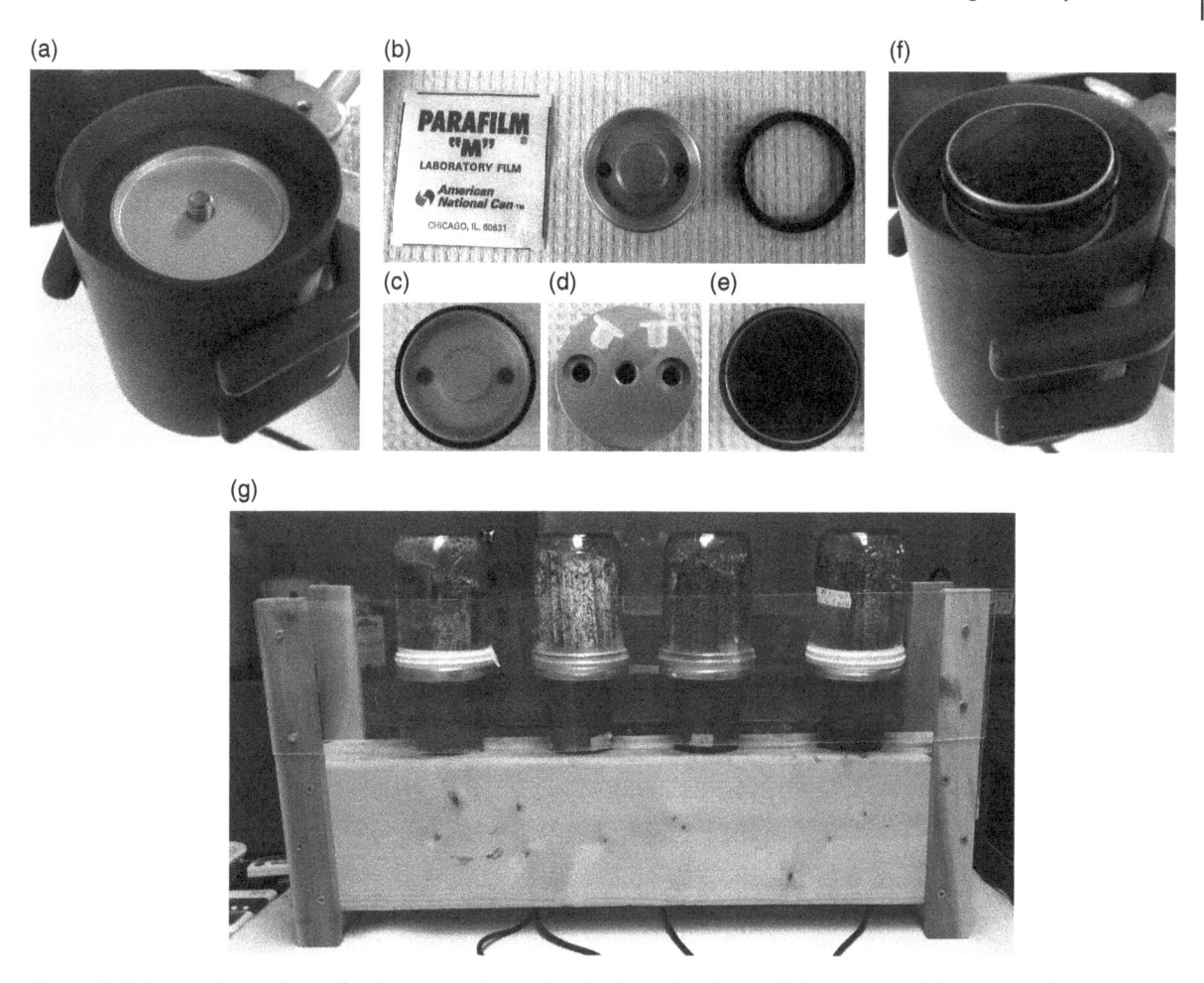

Figure 21.3 Hemotek membrane feeding system for *in-vitro* feeding of bed bugs: (a) heating unit; (b) Parafilm, left; blood reservoir, center; and O-ring to attach Parafilm to blood reservoir, right; (c) assembled blood reservoir; (d) reverse side of blood reservoir (after attachment of Parafilm and O-ring) showing the two fill holes, left and right, and stoppers. Middle opening accepts the screw from heating unit. To fill the unit, after tilting the reservoir about 45°, approximately 3.5 ml blood is introduced through the bottom hole; the unit is then tilted 45° in the opposite direction and the left hole is plugged; after tilting 45° in the original direction, the right hole is plugged, resulting in a filled blood reservoir; (e) filled blood reservoir; (f) blood reservoir screwed into the heating unit; and (g) wooden stand to support the heating units and bed bug colonies (inverted for feeding).

21.4 Need for Plasma

Despite the seemingly widespread use of whole blood product to feed bed bugs, some researchers cite the *in-vitro* feeding method of Takano-Lee and colleagues (2003), which was originally developed for the feeding of head lice. This method utilized human RBC admixed with blood plasma in a ratio of 1.25:1.00 (RBC:plasma). Since this method requires the acquisition of two components (RBC and plasma), the senior author of this review chapter investigated the need for plasma (Mark Feldlaufer, unpublished results). Five adult male and five adult female *C. lectularius* (Harlan strain) were fed weekly on either RBC admixed with blood plasma (1.25:1; Takano-Lee *et al.* 2003) or RBC alone. After 10 weeks, adults and nymphs were counted in both groups. In the group fed RBC plus plasma, more than 6 times the number of adults were recorded (236 vs 39; $\chi^2 = 14.12$, 1 d.f., $P < 0.0001$) than in the group that was fed RBC, alone. In addition, about 2.6 times the number of immature stages (698 vs. 266; $\chi^2 = 193.59$, 1 d.f., $P < 0.0001$) were found in the groups fed RBC plus plasma, versus the groups that were fed RBC alone. It should be noted that the bed bugs used to initiate this study had been fed on RBC plus plasma, so it may have taken a period of time for the absence of the plasma to become observable in bed bug population numbers (reproduction). Based on these current observations, one could speculate that this downward population trend might continue, because the RBC alone does not adequately support colony growth. Of course, an interesting study that was not conducted might determine if plasma, alone, would satisfy the nutritional requirement of bed bugs, since plasma contains many of the electrolytes, lipids, carbohydrates, and vitamins and so on that are found in RBC, albeit in different concentrations (Altman, 1961). There are reports (see Galun, 1971) that while blood plasma alone may initiate bed bug probing, RBC are needed by other blood-sucking arthropods, such as mosquitoes and tsetse flies, and fleas, for them to be able to fully engorge.

21.5 Development of an Artificial Blood Source

The development of an "artificial" blood, which contains all the necessary nutrients and phagostimulants suitable for feeding bed bugs, would be ideal because it would eliminate the need for purchasing blood, blood products, or extracting it from the animal hosts. The nutritional requirements of bed bugs are reviewed by Usinger (1966). The factors that affect feeding in blood-sucking insects and other arthropods have been addressed in Friend and Smith (1997), Galun (1971), and Lehane (2005). Although bed bugs were not the focus in these publications, bed bugs nevertheless offer useful insights into the blood-feeding process of all insects and arthropods. Evidently, the stimulation required to initiate probing and subsequent engorgement in a blood-sucking insect or arthropod is complex. Galun noted, as reported in Lehane (2005), that components of the blood acted as phagostimulants, but that different blood sources cause differences in probing and engorgement even within closely-related taxonomic groups. So, while an artificial blood source developed for one *Cimex* sp. may be suitable for another *Cimex* sp., it may be that an artificial blood source developed for another blood-sucking arthropod, such as a mosquito or tick, would not be suitable for bed bugs.

Phagostimulants associated with *C. lectularius* probing and engorgement have recently been investigated (Romero and Schal, 2014). These authors concluded that adenine nucleotides such as adenosine triphosphate (ATP) were important phagostimulants, although relatively high engorgement rates were also achieved using phosphate-buffered saline and even certain solutions of sodium chloride. *Cimex* sp. acceptance and ingestion of an artificial blood source would of course be essential in any bed-bug-rearing regime to ensure colony growth and reproduction. However, Romero and Schal (2014) correctly point out that even partial feeding by *Cimex* sp. would be useful in a bed bug control program that utilized toxic baits, since the engorgement would not be necessary to kill the insects.

Disclaimer: Mention of trade names or commercial products in this article is solely for the purpose of providing specific information or examples, and does not imply recommendation or endorsement by any of the authors or the US Department of Agriculture.

References

Aak, A. and Rukke, B.A. (2014) Bed bugs, their blood sources and life history parameters: a comparison of artificial and natural feeding. *Medical and Veterinary Entomology*, **28** (1), 50–59.

Adelman, Z.N., Kilcullen, K.A., Koganemaru, R., *et al.* (2011) Deep sequencing of pyrethroid-resistant bed bugs reveals multiple mechanisms of resistance within a single population. *PLoS ONE*, **6** (10), e26228.

Adkins, T.R. and Arant, F.S. (1959) A technique for the maintenance of a laboratory colony of *Cimex lectularius* L. on rabbits. *Journal of Economic Entomology*, **52** (4), 685–686.

Altman, P.L. (1961) *Blood and Other Body Fluids*, Federation of American Societies for Experimental Biology, Washington D.C.

Anderson, J. F., Ferrandino, F.J., McKnight, S., Nolen, J. and Miller, J. (2009) A carbon dioxide, heat and chemical lure trap for the bedbug, *Cimex lectularius*. *Medical and Veterinary Entomology*, **23** (2), 99–105.

Araujo, R. N., Costa, F.S., Gontijo, N.F., Gonçalves, T.C. and Pereira, M.H. (2009) The feeding process of *Cimex lectularius* (Linnaeus 1758) and *Cimex hemipterus* (Fabricius 1803) on different bloodmeal sources. *Journal of Insect Physiology*, **55** (12), 1151–1157.

Axtell, R.C. and Arends, J.J. (1990) Ecology and management of arthropod pests of poultry. *Annual Review of Entomology*, **35**, 101–126.

Azevedo, D.O., Neves, C.A., Dos Santos Mallet, J.R., *et al.* (2009) Notes on midgut ultrastructure of *Cimex hemipterus* (Hemiptera: Cimicidae). *Journal of Medical Entomology*, **46** (3), 435–441.

Badgandi, H.B. (2009) Biology facilitated by heme proteins as seen in *Cimex* nitrophorin and ecdysone inducible protein 75. PhD thesis, University of Arizona, Tucson.

Barbarin, A.M., Gebhardtsvauer, R. and Rajotte, E.G. (2013) Evaluation of blood regimen on the survival of *Cimex lectularius* L. using life table parameters. *Insects*, **4** (2), 273–286.

Bell, W. and Schaefer, C.W. (1966) Longevity and egg production of female bed bugs, *Cimex lectularius*, fed various blood fractions and other substances. *Annals of the Entomological Society of America*, **59** (1), 53–56.

Benoit, J.B., Phillips, S.A., Croxall, T.J., *et al.* (2009) Addition of alarm pheromone components improves the effectiveness of desiccant dusts against *Cimex lectularius*. *Journal of Medical Entomology*, **46** (3), 572–579.

Burden, G.S. (1966) Bed bugs. In: C. Smith (ed.), *Insect Colonization and Mass Production*. Academic Press, New York, pp. 175–182.

Campbell, B.E. and Miller, D.M. (2015) Insecticide resistance in eggs and first instars of the bed bug, *Cimex lectularius* (Hemiptera: Cimicidae). *Insects*, **6** (1), 122–132.

Cannet, A., Akhoundi, M., Berenger, J.M., Michel, G., Marty, P. and Delaunay, P. (2015) A review of data on laboratory colonies of bed bugs (Cimicidae), an insect of emerging medical relevance. *Parasite*, **22** (21), 1–7.

Chin-Heady, E., DeMark, J.J., Nolting, S., *et al.* (2013) A quantitative analysis of a modified feeding method for rearing *Cimex lectularius* (Hemiptera: Cimicidae) in the laboratory. *Pest Management Science*, **69** (10), 1115–1120.

Choe, D.H. and Campbell, K. (2014) Effect of feeding status on mortality response of adult bed bugs (Hemiptera: Cimicidae) to some insecticide products. *Journal of Economic Entomology*, **107** (3), 1206–1215.

Cockburn, C., Amoroso, M., Carpenter, M., *et al.* (2013) Gram-positive bacteria isolated from the common bed bug, *Cimex lectularius* L. *Entomologica Americana*, **119** (1), 23–29.

Davis, N.T. (1956) The morphology and functional anatomy of the male and female reproductive systems of *Cimex lectularius* L. (Heteroptera: Cimicidae). *Annals of the Entomological Society of America*, **49** (5), 466–493.

De Meillon, B. and Goldberg, L. (1946) Nutritional studies on blood-sucking arthropods. *Nature*, **158** (4008), 269–270.

De Meillon, B. and Goldberg, L. (1947) Preliminary studies on the nutritional requirements of the bedbug (*Cimex lectularius* L.) and the tick *Ornithodorus moubata* Murray. *Journal of Experimental Biology*, **24** (1–2), 61–63.

Feldlaufer, M. F., Domingue, M.J., Chauhan, K.R. and Aldrich, J.R. (2010) 4-oxo- aldehydes from the dorsal abdominal glands of the bed bug (Hemiptera: Cimicidae). *Journal of Medical Entomology*, **47** (2), 140–143.

Feldlaufer, M.F., Harlan, H.H. and Miller, D.M. (2014) Laboratory rearing of bed bugs. In: K. Maramorosch and F. Mahmood (eds), *Rearing Animal and Plant Pathogen Vectors*, pp. 118–130. CRC Press, Boca Raton.

Francischetti, I.M.B., Calvo, E., Andersen, J.F., *et al.* (2010) Insight into the sialome of the bed bug, *Cimex lectularius*. *Journal of Proteome Research*, **9** (8), 3820–3831.

Friend, W.G. and Smith, J.J.B. (1997) Factors affecting feeding by bloodsucking insects. *Annual Review of Entomology*, **22**, 309–331.

Galun, R. (1971) Recent developments in the biochemistry and feeding behavior of haematophagous arthropods as applied to their mass rearing. In: *Sterility Principles for Insect Control or Eradication*, pp. 273–282. International Atomic Energy Agency, Vienna.

Gilbert, I.H. (1964) Laboratory rearing of cockroaches, bed-bugs, human lice and fleas. *Bulletin of the World Health Organization*, **31** (4), 561–563.

Girault, A.A. (1910) Preliminary studies on the biology of the bed bug, *Cimex lectularius* Linn. *Journal of Experimental Biology*, **5** (3), 88–91.

Graça-Souza, A.V., Maya-Monteiro, C., Paiva-Silva, G., et al. (2006) Adaptations against heme toxicity in blood-feeding arthropods. *Insect Biochemistry and Molecular Biology*, **36** (4), 322–335.

Hall, R.D., Turner, Jr., E.C. and Gross, W.B. (1979) A simple apparatus for providing blood diets at constant temperature and with different corticosterone levels to individual bed bug colonies (Hemiptera: Cimicidae). *Journal of Medical Entomology*, **16** (3), 259–261.

Harraca, V., Ignell, R., Lofstedt, C. and Ryne, C. (2010) Characterization of the antennal olfactory system of the bed bug (*Cimex lectularius*). *Chemical Senses*, **35** (3), 195–204.

Hinson, K. (2015) Biology and control of the bed bug *Cimex lectularius* L. PhD thesis, Clemson University, Clemson, South Carolina.

Hosokawa, T., Koga, R., Kikuchi, Y., Meng, X.Y. and Fukatsu, T. (2010) *Wolbachia* as a bacteriocyte-associated nutritional mutualist. *Proceedings of the National Academy of Sciences*, **107** (2), 769–774.

How, Y.-F. and Lee, C.Y. (2011) Surface contact toxicity and synergism of several insecticides against different stages of the tropical bed bug, *Cimex hemipterus* (Hemiptera: Cimicidae). *Pest Management Science*, **67** (6), 734–740.

Johnson, C.G. (1941) The ecology of the bed-bug, *Cimex lectularius* L., in Britain. *Journal of Hygiene*, **41** (4), 345–461.

Jones, S.C. and Bryant, J.L. (2012) Ineffectiveness of over-the-counter total-release foggers against the bed bug (Heteroptera: Cimicidae). *Journal of Economic Entomology*, **105** (3), 957–963.

Jones, S.C., Bryant, J.L. and Harrison, S.A. (2013) Behavioral responses of the bed bug to permethrin-impregnated ActiveGuard^{TM} fabric. *Insects*, **4** (2), 230–240.

Kells, S.A. and Goblirsch, M.J. (2011) Temperature and time requirements for controlling bed bugs (*Cimex lectularius*) under commercial heat treatment conditions. *Insects*, **2** (3), 412–422.

Kilpinen, O., Liu, D. and Adamsen, P.S. (2012) Real-time measurement of volatile chemicals released by bed bugs during mating activities. *PLoS ONE*, **7** (12), e50981.

Kim, D.Y., Billen, J., Doggett, S.L. and Lee, C.Y. (2017) Differences in climbing ability of *Cimex lectularius* and *Cimex hemipterus* (Hemiptera: Cimicidae). *Journal of Economic Entomology*, **110** (3), 1179–1186.

Lehane, M.J. (2005) *The Biology of Blood-Sucking in Insects*, Cambridge University Press, New York.

Leulmi, H., Bitam, I., Berenger, J.M., et al. (2015) Competence of *Cimex lectularius* bed bugs for the transmission of *Bartonella quintana*, the agent of trench fever. *PLoS Neglected Tropical Diseases*, **9** (3), e0003789.

Loudon, C. and Boudaie, J. (2009) Walking with grappling hooks: bed bug locomotion on different surfaces. In: *The 57th Annual Meeting of the Entomological Society of America*, pp. 95. Entomological Society of America, Lanham, Maryland.

Minocha, R., Wang, C., Dang, K., et al. (2017) Systemic and erythrodermic reactions following repeated exposure to bites from the Common bed bug *Cimex lectularius* (Hemiptera: Cimicidae). *Austral Entomology*, **56** (3), 345–347.

Montes, C., Cuadrillero, C. and Vilella, D. (2002) Maintenance of a laboratory colony of *Cimex lectularius* (Hemiptera: *Cimicidae*) using an artificial feeding technique. *Journal of Medical Entomology*, **39** (4), 675–679.

Moore, D.J. and Miller, D.M. (2006) Laboratory evaluations of insecticide product efficacy for control of *Cimex lectularius*. *Journal of Economic Entomology*, **99** (6), 2080–2086.

Ogston, C.W. and Yanovski, A.D. (1982) An improved artificial feeder of bloodsucking insects. *Journal of Medical Entomology*, **19** (1), 42–44.

Olson, J.F., Moon, R.D. and Kells, S.A. (2009) Off-host aggregation behavior and sensory basis of arrestment by *Cimex lectularius* (Heteroptera: Cimicidae). *Journal of Insect Physiology*, **55** (6), 580–587.

Peterson, A. (1959) Complete rearing bedbug. In: A. Peterson (ed), *Entomological Techniques*. Edward Bros, Inc, Ann Arbor, pp. 117–118.

Pfiester, M., Koehler, P.G. and Pereira, R.M. (2008). Ability of bed bug-detecting canines to locate live bed bugs and viable bed bug eggs. *Journal of Economic Entomology*, **101** (4), 1389–1396.

Rahim, A.H.A. and Majid, A.H.A. (2014) Laboratory rearing of tropical bed bugs (*Cimex hemipterus*, Fabricius) using artificial feeding system. In: *Proceeding of the 13th Symposium of the Malaysian Society of Applied Biology*, Cherating, Pahang, Malaysia.

Rahim, A.H.A., Majid, A.H.A. and Ahmad, A.H. (2015) Laboratory rearing of *Cimex hemipterus* F. (Hemiptera: Cimicidae) feeding on different types of human blood compositions by using modified artificial feeding system. *Asian Pacific Journal of Tropical Disease*, **5** (12), 930–934.

Reinhardt, K., Naylor, R. and Siva-Jothy, M.T. (2003) Reducing a cost of traumatic insemination: female bedbugs evolve a unique organ. *Proceeding of the Royal Society London (Series B-Biological Sciences)*, **270** (1531), 2371–2375.

Reis, M.D. and Miller, D.M. (2011) Host searching and aggregation activity of recently fed and unfed bed bugs (*Cimex lectularius* L.). *Insects*, **2** (2), 186–194.

Rendtorff, R.C. (1938) A method for rearing the bedbug, *Cimex lectularius* L., for studies in toxicology and medical entomology. *Journal of Economic Entomology*, **31** (6), 781.

Rivnay, E. (1930) Techniques in artificial feeding of the bed bug, *Cimex lectularius* L. *Journal of Parasitology*, **16** (4), 246–249.

Romero, A. and Schal, C. (2014) Blood constituents as phagostimulants for the bed bug *Cimex lectularius* L. *Journal of Experimental Biology*, **217** (4), 552–557.

Romero, A., Potter, M.F., Potter, D.A. and Haynes, K.F. (2007) Insecticide resistance in the bed bug: A factor in the pest's sudden resurgence? *Journal of Medical Entomology*, **44** (2), 175–178.

Romero, A., Potter, M.F. and Haynes, K.F. (2009) Behavioral responses of the bed bug to insecticide residues. *Journal of Medical Entomology*, **46** (1), 51–57.

Rosen, S., Hadani, A., Gur Lavi, A., Berman, E., Bendheim, U. and Hisham, A.Y. (1987) The occurrence of the tropical bedbug (*Cimex hemipterus*, Fabricius) in poultry barns in Israel. *Avian Pathology*, **16** (2), 339–342.

Ryne, C. (2009) Homosexual interactions in bed bugs: alarm pheromones as male recognition signals *Animal Behaviour*, **78** (6), 1471–1475.

Seong, K.M., Lee, D.Y., Yoon, K.S., *et al.* (2010) Establishment of quantitative sequencing and filter contact vial bioassay for monitoring pyrethroid resistance in the common bed bug, *Cimex lectularius*. *Journal of Medical Entomology*, **47** (4), 592–599.

Siljander, E., Penman, D., Harlan, H. and Gries, G. (2007) Evidence for male- and juvenile-specific contact pheromones of the common bed bug *Cimex lectularius*. *Entomologia Experimentalis et Applicata*, **125** (2), 215–219.

Singh, N., Wang, C. and Cooper, R. (2013) Effectiveness of a reduced-risk insecticide based bed bug management program in low-income housing. *Insects*, **4** (4), 731–742.

Steelman, C.D., Szalanski, A.L., Trout, R., *et al.* (2008) Susceptibility of the bed bug *Cimex lectularius* L. (Heteroptera: Cimicidie) collected in poultry production facilities to selected insecticides. *Journal of Agriculture and Urban Entomology*, **25** (1), 41–51.

Stutt, A.D. and Siva-Jothy, M.T. (2001) Traumatic insemination and sexual conflict in the bed bug *Cimex lectularius*. *Proceedings of the National Academy of Sciences (USA)*, **98** (10), 5683–5687.

Takano-Lee, M., Velten, R.K., Edman, J.D., Mullens, B.A., and Clark, J.M. (2003) An automated feeding apparatus for in vitro maintenance of the human head louse, *Pediculus capitas* (Anoplura: Pediculidae). *Journal of Medical Entomology*, **40** (6), 795–799.

Tawatsin, A., Thavara, U., Chompoosri, J., *et al.* (2011) Insecticide resistance in bedbugs in Thailand and Laboratory evaluation of insecticides for the control of *Cimex hemipterus* and *Cimex lectularius* (Hemiptera: Cimicidae). *Journal of Medical Entomology*, **48** (5), 1023–1030.

Tawatsin, A., Lorlertthum, K., Phumee, A., *et al.* (2013) Discrimination between tropical bed bug *Cimex hemipterus* and common bed bug *Cimex lectularius* (Hemiptera: Cimicidae) by PCR-RFLP. *Thai Journal of Veterinary Medicine*, **43** (3), 421–427.

Usinger, R.L. (1966) *Monograph of Cimicidae (Hemiptera – Heteroptera)*, Entomological Society of America, College Park.

Valenzuela, J.G., Walker, F.A. and Ribeiro, J.M. (1995) A salivary nitrophorin (nitric-oxide-carrying hemoprotein) in the bedbug *Cimex lectularius*. *Journal of Experimental Biology*, **198** (7), 1519–1526.

Wang, C., Singh, N., Cooper, R., Liu, C. and Buczkowski, G. (2013) Evaluation of an insecticide dust band treatment method for controlling bed bugs. *Journal of Economic Entomology*, **106** (1), 347–352.

Wattal, B.L. and Kalra, N.L. (1961) New methods for the maintenance of a laboratory colony of bed-bug, *Cimex hemipterus* Fabricius, with observations on its biology. *Indian Journal of Malariology*, **15** (2), 157–171.

Wawrocka, K. and Bartonička, T. (2013) Two different lineages of bedbug (*Cimex lectularius*) reflected in host specificity. *Parasitology Research*, **112** (11), 3897–3904.

Weeks, E.N., Logan, J.G., Gezan, S.A., et al. (2011) A bioassay for studying behavioural responses of the common bed bug, *Cimex lectularius* (Hemiptera: Cimicidae) to bed bug-derived volatiles. *Bulletin of Entomological Research*, **101** (1), 1–8.

Wintle, K. and Reinhardt, K. (2008) Temporary feeding inhibition caused by artificial abdominal distension in the bedbug, *Cimex lectularius*. *Journal of Insect Physiology*, **54** (7), 1200–1204.

Yoon, K.S., Kwon, D.H., Strycharz, J.P., et al. (2008) Biochemical and molecular analysis of deltamethrin resistance in the common bed bug (Hemiptera: Cimicidae). *Journal of Medical Entomology*, **45** (6), 1092–1101.

Zhu, F., Wigginton, J., Romero, A., et al. (2010) Widespread distribution of knockdown resistance mutations in the bed bug, *Cimex lectularius* (Hemiptera: Cimicidae), populations in the United States. *Archives of Insect Biochemistry and Physiology*, **73** (4), 245–257.

Zhu, F., Sams, S., Mournal, T., et al. (2012) RNA interference of NADPH-cytochrome P450 reductase results in reduced insecticide resistance in the bed bug, *Cimex lectularius*. *PLoS ONE*, **7** (2), e31037.

Part V

Bed Bug Management

22

Bed Bug Industry Standards: Australia

Stephen L. Doggett

22.1 Introduction

As in other nations, the bed bug resurgence in Australia caught all those that have to deal with bed bugs unprepared. Early on with the resurgence, treatment failures by pest management professionals (PMPs) were common (Doggett, 2005a) and suspicion of insecticide resistance was high (May, 2005). This suspicion was soon confirmed locally with the demonstration of extremely high levels of resistance (Lilly *et al.*, 2009c; Lilly *et al.*, 2015) and later, multiple mechanisms contributing to the resistance (Dang *et al.*, 2014, 2015; Lilly *et al.*, 2016a,b). Clearly bed bugs were, for all practical purposes, a "new" pest and few in the pest management industry had the knowledge or experience to successfully control active infestations. As discussed in more detail elsewhere in this book, the response to bed bug management in Australia was to develop a three-pronged strategy (see Chapter 7). The cornerstone of this approach was the world's first industry standard for the control of modern bed bugs. In 2005, the draft first edition of *A Code of Practice for the Control of Bed Bug Infestations in Australia* was launched. The history, governance, aims, key components, and evolution of the code, forms the basis of this chapter.

22.2 Why was the Code Required?

The need for the code became evident not only due to the increasing number of treatment failures but also with the suggestion that control failures resulted in bed bug dispersal. This was a particular problem in apartment complexes where multiple infestations could result from a poorly treated initial infestation (Doggett and Russell, 2008). Thus, poor pest management appeared to be a major contributing factor to the degree of the resurgence (Doggett, 2005a; Doggett and Russell, 2008). Furthermore, a survey in 2006 of Australian PMPs revealed that unregistered insecticides were commonly being employed and methodologies with little scientific basis used, due to product failures with the then currently registered products (Doggett and Russell, 2008). It was clear that Australian PMPs were ill-equipped to deal with this "new" pest and that urgent measures were needed to provide best practice guidance.

22.3 The History and Aims of the Code

In Australia, the peak industry body for pest management professionals is the Australian Environmental Pest Managers Association (AEPMA). In 2005, AEPMA formed the Bed Bug Working Party (BBWP) to develop an industry standard. This group consisted of representatives from the pest management industry, insecticide manufacturers, universities, and government.

Advances in the Biology and Management of Modern Bed Bugs, First Edition.
Edited by Stephen L. Doggett, Dini M. Miller, and Chow-Yang Lee.
© 2018 John Wiley & Sons Ltd. Published 2018 by John Wiley & Sons Ltd.

The primary aim of the code was to promote the use of best practice for the control of active bed bug infestations *and* the management of potential infestations. The groups targeted included anyone tasked with the eradication or management of bed bugs infestations in any situation (Doggett *et al.*, 2011). Best practice was promoted through the recommendation of employing integrated pest management (IPM) procedures. Products and strategies were called "Best Practice" if they were found to be effective at controlling bed bugs. This evidence had to be obtained through independent research (preferably peer-reviewed publications) or where there was evidence of efficacy through common usage and was assessed by the BBWP. The latter clause was necessary, as in 2005 there were very few recent publications dealing with bed bug control. It was envisaged that by adopting best practice, the fiscal and health impacts of bed bugs would be reduced, ultimately leading to a reversal of the resurgence.

Through the development of the code, knowledge gaps were quickly identified. This led to research that helped fill those gaps. Much of this research later appeared in industry and scientific publications, which led to the registration of more efficacious products, and in turn led to subsequent editions of the code (Doggett *et al.*, 2006, 2008; Doggett and Russell, 2007, 2008, 2009, 2010; Lilly *et al.*, 2009a,b) and a new edition envisaged for 2018 (Doggett, unpublished results). With the growing interest in bed bugs across the world and the rebirth of research, it was necessary to update the code on a regular basis in order to maintain relevance. In fact, the code was even designated with something quite unique amongst guidelines: a "use by" date. So far there have been four draft editions and four final editions of the code (Doggett, 2005b, 2006, 2007a,b, 2009, 2010, 2011b, 2013). As scientific research can be complex to the lay person, the code aims to distil relevant information from publications into a consumable form for PMPs and other parties.

Another key objective of the code was to achieve broad acceptance across all industries. This was accomplished through a process of document development that was both transparent and inclusive. With each draft edition, applicable industry groups were widely consulted and feedback sought. Every comment was then reviewed and considered in the development of the finalized edition. All comments were then placed on the code's web site (www.bedbug.org.au). If a particular suggestion was not incorporated, the reasons for doing so were stated. Furthermore, to ensure probity and that undue influence could not be asserted by vested interests, guidelines were developed for the management and establishment of working parties (AEPMA, 2009). These guides are now in use for all AEPMA working parties.

To encourage all to adopt the code, the AEPMA BBWP agreed to make it available for free. It was also offered to other industry associations for their use as there is little benefit in the long term of only one group addressing a global issue. With this offer, the code was adopted by the New Zealand and Italian pest management associations, and later formed the basis of the first edition of the European code of practice (BBF, 2011).

Another aim of the code was that of *protection*. Since the start of the resurgence the marketplace around the world has been deluged with all sorts of bed bug control devices. Quite often, efficacy data does not exist for these products, while some efficacy data appears highly questionable due to known insect resistance (notably data on the pyrethroids when applied as a residual). Thus, if a product is included in the code, it must not only have been demonstrated to be efficacious in independent tests, but its limitations must also have been included. It is important to include such limitations, as it is just important to know how a product does not work, as much as how it does. Thus, the code protects consumers from bad products. Additionally, by promoting best practice in terms of management, the client is provided protection from PMPs whose treatments fail as a result of not adhering to the code. Finally, protection is also provided for the PMPs against difficult clients, as the code details the processes for bed bug management and why eradication is so costly and difficult to achieve. The code also protects PMPs by encouraging them to undertake best practice, which in turn reduces the risk of treatment failures, and decreases the possibility of adversely impacting their companies' reputations.

22.4 The Key Elements of the Code

The main components of the first edition of the code included:

- aims
- required philosophies (the only acceptable outcome for clients is complete eradication)

- training requirements for PMPs and housekeeping staff
- customer relation (for the PMP this covers aspects of customer confidentiality and warranties; for the accommodation sector, it covers the sensitive handling of guest complaints)
- work health and safety (to protect the health and environment of all)
- an overview of bed bug control
- planning and preparation prior to treatment (that is, the inspection process)
- treatment procedures, post-treatment evaluation, and measurements of success
- bed bug preventative measures (Doggett, 2006).

The second edition of the code was more developed regarding insecticide efficacy and had a new section dealing with extreme levels of infestation (Doggett, 2007a). The third edition provided several new sections, including how to select a PMP, bed bug identification, high risk factors, the need to have formalized bed bug management plans, and a component on situational control (Doggett, 2010). The third edition expanded the section on reducing the risk of bed bugs by explaining the four broad phases of a bed bug infestation (introduction, establishment, growth, and spread) and what strategies might be implemented against each of these phases. This third edition was the first that recommended that accommodation providers should have a bed bug management policy.

The fourth (and current) edition of the code expanded on bed bug management plans, recommending use of proactive plans to minimize the risk of bed bugs, and reactive plans for the eradication of an active infestation (Doggett, 2013). What had been called "prevention" was renamed "reducing bed bug risks", as it had been recognized that it is not possible with current technologies to actually prevent bed bugs. In all editions, a list of Australian insecticides registered for use against bed bugs at the time of publication was included, as well as various appendices. The appendices contained definitions, references, key people in the BBWP, suppliers, and service checklists for the client and PMP. The code is extremely detailed, being some 90 pages long. However, it was always aimed to be a reference document. From it, more streamlined articles that conform to the code could be produced, targeting particular sectors. One of these is *A Bed Bug Management Policy & Procedural Guide*, which is aimed directly at those who provide accommodation for others (Doggett, 2011a). Like the code, the main emphasis of the guide is best practice in bed bug management.

22.5 The Benefits of the Code

A code for bed bug management offers numerous advantages. An industry standard is relatively cheap to produce and offers immediate tangible benefits over other strategies such as research, which is costly and takes time to complete (although research is used to refine best practice and to evolve the code over time). For insecticide manufacturers, the benefit of promoting IPM is that it prolongs the market life of insecticides by reducing the risk of resistance developing. A measure of the success of the code is that there appear to be fewer control failures being reported (S. Doggett, unpublished data). PMPs have stated that the code has assisted in client education and communication, helped streamline the treatment process, and increased the chance of eradication success. The code has also formed a template for other pest management codes of practice in Australia, including a code for pest management in the food industry (AEPMA, 2014a), for pre-purchase timber pest inspections (AEPMA, 2014b), and others in development.

Currently, the BBWP is taking steps to devise a training curriculum based on the code, to ensure that it is incorporated into the competencies required for future training of Australian PMPs.

Thus in Australia, the code has been the main stratagem for combating the bed bug resurgence. The code has gained general acceptance through its transparent and inclusive process of development, and integrated methods of pest management promulgated. It is also a testament to the success of the code that the Australian Pesticides and Veterinary Medicines Authority, the government body responsible for registration of insecticides, has allowed the inclusion of a recommendation for the use of the code on the labels of new insecticidal products registered for bed bug control. Perhaps most importantly, a survey of Australian PMPs in 2016 (Doggett, 2016) has indicated that bed bugs infestations are no longer on the increase, which has been the ultimate aim of the code and the associated strategies implemented.

References

AEPMA (2009) *Guidelines for the Establishment and Management of AEPMA Code-Of-Practice Working Parties (V2.1.3)*, Australian Environmental Pest Managers Association, Sydney.

AEPMA (2014a) *A Code of Practice for Pest Management in the Food Industry in Australia & New Zealand*, Australian Environmental Pest Managers Association, Queensland.

AEPMA (2014b) *A Code of Practice for Prior to Purchase Specialist Timber Pest Inspections*, Australian Environmental Pest Managers Association, Queensland.

BBF (2011) *European Code of Practice, Bed Bug Management*, Bed Bug Foundation, London.

Dang, K., Toi, C.S., Lilly, D.G., et al. (2014) Detection of Knockdown Resistance (*kdr*) in Cimex lectularius and Cimex hemipterus (Hemiptera: Cimicidae). In: *Proceedings of the Eighth International Conference on Urban Pests, 2014, July 20–23, Zurich, Switzerland*. OOK-Press, Veszprém.

Dang, K., Toi, C.S., Lilly, D.G., et al. (2015) Identification of putative *kdr* mutations in the tropical bed bug, Cimex hemipterus (Hemiptera: Cimicidae). *Pest Management Science*, **71** (7), 1015–1020.

Doggett, S.L. (2005a) *Bed Bug Ecology and Control*. Proceedings of Pests of Disease and Unease, April 22, 2005, Westmead Hospital, Australia. Westmead Hospital, Sydney.

Doggett, S.L. (2005b) *A Code of Practice for the Control of Bed Bug Infestations in Australia (draft)*, Department of Medical Entomology and The Australian Environmental Pest Managers Association, Sydney.

Doggett, S.L. (2006) *A Code of Practice for the Control of Bed Bug Infestations in Australia*, Department of Medical Entomology and The Australian Environmental Pest Managers Association, Sydney.

Doggett, S.L. (2007a) *A Code of Practice for the Control of Bed Bug Infestations in Australia, 2nd ed.*, Department of Medical Entomology and The Australian Environmental Pest Managers Association, Sydney.

Doggett, S.L. (2007b) *A Code of Practice for the Control of Bed Bug Infestations in Australia. 2nd ed. (draft)*, Department of Medical Entomology and The Australian Environmental Pest Managers Association, Sydney.

Doggett, S.L. (2009) *A Code of Practice for the Control of Bed Bug Infestations in Australia. 3rd ed. (draft)*, Department of Medical Entomology and The Australian Environmental Pest Managers Association, Sydney.

Doggett, S.L. (2010) *A Code of Practice for the Control of Bed Bug Infestations in Australia, 3rd ed.*, Department of Medical Entomology and The Australian Environmental Pest Managers Association, Sydney.

Doggett, S.L. (2011a) *A Bed Bug Management Policy & Procedural Guide for Accommodation Providers*, Westmead Hospital, Sydney.

Doggett, S.L. (2011b) *A Code of Practice for the Control of Bed Bug Infestations in Australia. 4th ed. (draft)*, Department of Medical Entomology and The Australian Environmental Pest Managers Association, Sydney.

Doggett, S.L. (2013) *A Code of Practice for the Control of Bed Bug Infestations in Australia, 4th ed.*, Department of Medical Entomology and The Australian Environmental Pest Managers Association, Sydney.

Doggett, S.L. (2016) Bed bugs in Australia 2016: are we biting back? Results of the national bed bug survey. *Professional Pest Manager*, **Aug/Sep**, 28–30.

Doggett, S.L. and Russell, R.C. (2007) Bed Bugs – Latest Trends and Development. *Proceedings of the Australian Environmental Pest Managers Association National Conference, June 4–6, 2007, Coffs Harbour, Australia*. Australian Environmental Pest Managers Association, Sydney.

Doggett, S.L. and Russell, R.C. (2008) The Resurgence of Bed Bugs, Cimex spp. (Hemiptera: Cimicidae) in Australia. In: *Proceedings of the Sixth International Conference on Urban Pests, July 13–16, 2008, Budapest, Hungary*. OOK-Press, Veszprém.

Doggett, S.L. and Russell, R.C. (2009) Emerging Challenges in Bed Bug Management. *In: Proceedings of the Emerging Pest Management Challenges Symposium, September 22, 2009, Westmead Hospital, Australia.* Westmead Hospital, Sydney.

Doggett, S.L. and Russell, R.C. (2010) *Laboratory Investigations of the 'BB Secure Ring' and its Ability to Act as a Barrier to the Common Bed Bug, Cimex lectularius*, Westmead Hospital, Sydney.

Doggett, S.L., Geary, M.J. and Russell, R.C. (2006) Encasing mattresses in black plastic will not provide thermal control of bed bugs, *Cimex* spp. (Hemiptera: Cimicidae). *Journal of Economic Entomology*, **99** (6), 2132–2135.

Doggett, S.L., Geary, M.J., Lilly, D. and Russell, R.C. (2008) The efficacy of diatomaceous earth against the common bed bug, *Cimex lectularius.* Westmead Hospital, Sydney.

Doggett, S.L., Orton, C.J., Lilly, D.G. and Russell, R.C. (2011) Bed bugs: the Australian response. *Insects*, **2** (2), 96–111.

Lilly, D.G., Doggett, S.L., Orton, C.J. and Russell, R.C. (2009a) Bed bug product efficacy under the spotlight - part 1. *Professional Pest Manager*, **February/March**, 14–16.

Lilly, D.G., Doggett, S.L., Orton, C.J. and Russell, R.C. (2009b) Bed bug product efficacy under the spotlight - part 2. *Professional Pest Manager*, **April/May**, 14–15, 18.

Lilly, D.G., Doggett, S.L., Zalucki, M.P., Orton, C.J. and Russell, R.C. (2009c) Bed bugs that bite back: confirmation of insecticide resistance in Australia in the common bed bug *Cimex lectularius. Professional Pest Manager*, **August/September**, 22–24, 26.

Lilly, D.G., Zalucki, M.P., Orton, C.J., *et al.* (2015) Confirmation of insecticide resistance in *Cimex lectularius* (Hemiptera: Cimicidae) in Australia. *Austral Entomology*, **54** (1), 96–99.

Lilly, D.G., Dang, K., Webb, C.E. and Doggett, S.L. (2016a) Evidence for metabolic pyrethroid resistance in the common bed bug (Hemiptera: Cimicidae). *Journal of Economic Entomology*, **109** (3), 1364–1368.

Lilly, D.G., Latham, S.L., Webb, C.E. and Doggett, S.L. (2016b) Cuticle thickening in a pyrethroid-resistant strain of the common bed bug, *Cimex lectularius* L. (Hemiptera: Cimicidae). *PloS ONE*, **11** (4), e0153302.

May, P. (2005) Bed bug resistance to insecticides - real or imagined? *Professional Pest Manager*, **February/March**, 25–29.

23

Bed Bug Industry Standards: Europe

Richard Naylor

23.1 Introduction

The resurgence of the Common bed bug, *Cimex lectularius* L., across Europe was so rapid and unexpected that many in the pest control industry were left playing catch-up to develop reliable control strategies. The academic community played an important role in identifying insecticide resistance as one of the primary causes of the resurgence in the UK (Boase *et al.*, 2006), the USA (Romero *et al.*, 2007), and Australia (Lilly *et al.*, 2009). Despite the fact that much of this information had been published in industry magazines, the dissemination of information to the European pest management industry has been slow, and as a result many pest management professionals (PMPs) continue to use products with ever-diminishing efficacy (Boase and Naylor, 2014). This highlighted the need for the development of industry standards that promoted best practice in bed bug management.

23.2 Why was the Code Required?

As the pest control industry became aware of the increasing bed bug problem and declining efficacy of the traditionally used insecticides, the pest control product market was inundated with new bed bug products, including amateur- and professional-use insecticides, and non-chemical treatment solutions such as heaters, steam machines, monitors, traps and barriers. For the most part, independent efficacy data on these products was lacking, but even where it did exist, it was frequently missed by the pest control industry, who tend not to read academic journals where the independent efficacy data is published. One of the most important roles of an industry code of practice is therefore to bridge the gap in information flow from the academic community to the PMPs and other affected parties.

Trade magazines have also played a crucial role in publishing articles about relevant findings from the academic literature. However, these are printed alongside articles and adverts from manufacturers and suppliers, which are far from independent. The trade magazines also have a potential conflict of interest because most rely heavily on the sale of advertising space to pest control product manufacturers.

23.3 The History and Aims of the Code

The first "European Code of Practice for Bed Bug Management" (ECoP) was published in 2011, although this was to some extent preceded by a set of five "Good Practice Guides" released by the Greater London Pest Liaison Group in 2009 (Boase, 2009). The ECoP was based largely on the third edition of *A Code of Practice for the Control of Bed Bug Infestations in Australia* (Doggett, 2009).

The Bed Bug Foundation (BBF), which was responsible for producing the ECoP, was formed in 2010 as a not-for-profit charity for the provision of information and education on bed bug management. The chairman of the BBF, Oliver Madge, was joined by a working party of industry professionals from across Europe, as well as Stephen Doggett, from the Australian Bed Bug Code of Practice Working Party, to form the "Senate" of the BBF (Anonymous, 2011a).

After the first draft of the document was completed in late 2010, there was a period of public consultation and feedback, so that the wider pest management community could review and respond to the content of the document. This feedback was reviewed by the BBF Senate and the document was revised accordingly. This process was repeated in 2012/13 with the publication of Version 2 of the ECoP (ECoPv2) (Anonymous, 2012; Anonymous, 2013; Bed Bug Foundation, 2013).

Version 1 of the ECoP followed the example of the Australian code, providing information for all parties affected by bed bugs. As such, it included advice for travelers, hotel staff, hotel management, landlords, home owners, and PMPs alike. The ECoPv2 was much more targeted towards the pest management industry. This shift in focus reflected the widely held view within the UK pest management industry that the treatment of bed bug infestations should be conducted by a PMP, because amateur treatments with ineffective over-the-counter insecticides was frequently noted to cause dispersal of bed bugs throughout the property, making treatment more difficult (Anonymous, 2011b). Furthermore, by tailoring the ECoPv2 to the PMP, it was possible to provide more details on the range of treatment options available, without increasing the overall length of the document. At only 33 pages, the ECoPv2 is just over a third of the length of its Australian counterpart (Doggett, 2013), making it as accessible as possible.

Although the ECoPv2 is tailored to provide guidance specifically for PMPs, it is still made freely available to all, and it is intended that all affected parties should read it. This allows customers to see what to expect from their PMP. It also empowers customers to ask the PMP to justify the use of a particular product or control method, particularly if this method is discouraged within the code.

The ECoPv2 (the current version at the time of writing) has several key aims:

Product information: To provide general information on the range of products that are currently marketed for bed bugs, as well as highlighting the advantages and shortfalls of such products. This will help the private PMP and pest management companies (PMCs) to select appropriate treatment options from a marketplace that is cluttered with products that currently lack independent efficacy testing.

Research findings: To bridge the gap between academic journals and non-academic interested parties, by providing a synthesis of the relevant academic information in a concise and accessible form.

Standards: To set a high, but achievable standard for bed bug control, and to encourage pest controllers to meet this standard.

Due to the stigma surrounding bed bugs and the false association with poor hygiene, people tend not to discuss the treatment of infestations even with close friends. As a result of this, people rarely seek recommendations prior to engaging the services of a PMP for bed bug control. Consequently, competition within the pest control industry tends to select for PMPs who provide treatments for the lowest possible price, rather than those who provide an effective service. Spraying liquid insecticides remains the cheapest option, particularly in terms of the time required to administer the treatment. This means that many service providers wishing to offer more effective, but expensive and labor-intensive services find that they are unable to compete with companies who insist on using the traditional insecticidal approach. The ECoPv2 pushes back against this trend by:

- highlighting the issue of insecticide resistance
- strongly encouraging PMCs to offer warranties for the treatments they provide.

By making it the norm for PMCs to offer warranties, and by teaching potential customers to ask for them, it makes it more difficult for companies offering cheap but ineffective insecticide treatments to compete, because repeated return visits cost them money and make ineffective treatment options unviable.

In addition to a comprehensive overview of the range of treatment options available for bed bug management, the ECoPv2 provides detailed information on the biology of bed bugs, including reproductive rates and development times, and the significance of temperature on the above. This is important to consider when planning post-treatment inspections and follow-up treatments, as it is generally advisable to allow time for any viable eggs to hatch before engaging in further treatments or declaring the "all clear".

The ECoPv2 pays particular attention to the importance of correct specimen identification. Detailed photographs and diagrams are included to assist in the identification of live bed bug stages, fecal material, eggs, and cast skins. Misidentification is well known to result in unwarranted insecticide treatments and unnecessary expenses for the customer. To address this issue, the BBF offers a free specimen identification service for any suspected bed bugs encountered around the home or work place. The most commonly encountered "bed bug imposters" are carpet beetles (Family: Dermestidae), which account for approximately 13% of specimens and photographs submitted to the BBF for identification (Richard Naylor, unpublished results). Details of how to recognize carpet beetles, along with book lice (Order: Psocoptera), head lice (*Pediculus humanus capitis* L.), and a range of other commonly encountered species is provided in Appendix 1 of the ECoPv2.

Bed bugs are a problem throughout Europe (see Chapter 5), and because legislation relating to pest management is increasingly occurring at the European level (see Chapter 42), the tools we have to deal with the problem are largely ubiquitous within Europe. It therefore seems appropriate for an industry code to have a pan-European perspective. However, the main challenge to this is the issue of translation.

To date all translations have been provided by volunteers who have recognized the value of the ECoPv2 and wished to make it available in their own language. The range of languages currently available (English, French, Dutch, German, Italian, and Swedish) therefore bears testament to the widespread acceptance of the ECoPv2. However, this acceptance has taken almost three years to achieve, which is a significant lag time for a document that requires regular updating.

In 2016, the Deutscher Schädlingsbekämpfer Verband (the largest German trade association for the pest control industry) adopted the ECoPv2 (with minor amendments) as their own code of practice for bed bug management. The German translation was published back-to-back with the English version, and given away for free to everybody who attended their 2016 annual Pest Protect conference in Stuttgart, Germany.

23.4 The Benefits of the Code

It is difficult to determine the impact that the ECoP/v2, or any other industry standard, has played in curtailing the resurgence of *C. lectularius*. There is no doubt that the PMP who follows the advice laid out in the ECoP will be more successful at eradicating bed bugs. However, PMPs who take the time to read the ECoP also tend to take a more general interest in providing a quality service, and therefore read other literature, experiment with different products and strategies, offer guarantees, and make return visits to determine treatment success. The real challenge is to reach those PMPs who take no interest in improving their quality of service and persist in the industry by providing low-cost treatments that are unguaranteed and ineffective.

References

Anonymous (2011a) European code of practice launched for the management of bed bugs. *International Pest Control*, **53** (1), 42.

Anonymous (2011b) DIY not good for your health. *Pest*, **17**, 19.

Anonymous (2012) Australia 4 – Europe 2…but it's not over yet! *International Pest Control*, **54**, 91.

Anonymous (2013) Revised bed bug code released. *Pest*, **25**, 24.

Bed Bug Foundation (2013) *European Code of Practice for Bedbug Management, version 2*, http://bedbugfoundation.org/en/ecop/ (accessed 15 April 2016).

Boase, C. (2009) Bedbugs – information upsurge. *International Pest Control*, **54** (4), 181–184.

Boase, C. and Naylor, R. (2014) Bed bug management. In: *Urban Insect Pests – Sustainable Management Strategies*, (ed P. Dhang), CABI, Oxfordshire, UK, pp. 8–22.

Boase, C.J., Small, G. and Naylor, R. (2006) Interim report on insecticide susceptibility status of UK bedbugs. *Professional Pest Controller*, **Summer**, 6–7.

Doggett, S.L. (2009) *A Code of Practice for the Control of Bed Bug Infestations in Australia*, 3rd edn. *(draft)*, Department of Medical Entomology and The Australian Environmental Pest Managers Association, Westmead Hospital, Sydney, Australia.

Doggett, S.L. (2013) *A Code of Practice for the Control of Bed Bug Infestations in Australia* 4th edn, Department of Medical Entomology and The Australian Environmental Pest Managers Association, Westmead Hospital, Sydney, Australia.

Lilly, D.G., Doggett, S.L., Orton, C.J. and Russell, R.C. (2009) Bed bug product efficacy under the spotlight – part 1. *Professional Pest Manager*, **February/March**, 14–16.

Romero, A., Potter, M.F. and Haynes, K.F. (2007) Insecticide-resistant bed bugs: implications for the industry. *Pest Control Technology*, **35** (7), 42, 44, 46, 48, 50.

24

Bed Bug Industry Standards: USA

Jim Fredericks

24.1 Introduction

The bed bug resurgence in the early 2000s left many established pest management companies (PMCs) in North America in need of educational resources to begin developing treatment protocols. The National Pest Management Association's *Best Management Practices for Bed Bugs* (NPMA BMP) were developed in response to the needs of PMCs seeking guidance (NPMA, 2011a). Subsequently, both new and established businesses began delivering bed bug-related products and services.

24.2 History and Development of the NPMA Best Management Practices for Bed Bugs

The NPMA is the leading industry association for pest management professionals (PMPs) in the USA. In 2010, the group convened the Blue Ribbon Bed Bug Task Force in response to public policy demands and the industry need for educational resources, standards, and research. This task force represented a broad-based stakeholder group (PMPs, industry suppliers, academics, and state and federal government regulators) whose goal was the development of a comprehensive response to the bed bug resurgence in the USA. This response included the development of educational materials, standards, public policy support, and research. One of the needs identified by the task force was to identify acceptable, though not prescriptive, protocols. In other words, the task force needed to identify the boundaries of good practice. Following the first meeting of the task force at the 2010 NPMA PestWorld convention in Honolulu, Hawaii, several workgroups were appointed, including one tasked with writing comprehensive bed bug best management practices, and minimum standards for canine scent detection certification testing. The BMP was first adopted by the NPMA Board of Directors in January 2011 (NPMA, 2011a). In October 2013, a second edition was adopted, followed by a third edition in July of 2016 (NPMA, 2013; NPMA, 2016).

24.3 Target Audience

The NMPA BMP was initially developed with PMPs as the primary audience. Both English and Spanish language versions were published in early 2011 (NPMA, 2011a). A consumer-friendly version was adapted from the professional document to create consumer demand for PMCs adhering to the BMPs (NPMA, 2011b). Additionally, the consumer version could be used by PMPs to educate clients about their role in the integrated pest management (IPM) process.

Advances in the Biology and Management of Modern Bed Bugs, First Edition.
Edited by Stephen L. Doggett, Dini M. Miller, and Chow-Yang Lee.
© 2018 John Wiley & Sons Ltd. Published 2018 by John Wiley & Sons Ltd.

24.4 Key Elements of the NPMA Best Management Practices

The NPMA BMP comprises 15 sections and 2 appendices. Each section is arranged in outline format and is designed to provide guidance without being highly prescriptive. This allows the PMP appropriate levels of flexibility to design and provide a bed bug management program that suits each situation. A description of each section is included below:

Business practices PMCs are encouraged to maintain high levels of professionalism and business ethics by obtaining appropriate licenses and certifications, and practicing fairness and honesty in all advertising and dealings with clients. For example, companies are encouraged to provide a description of control tactics that will be employed and all costs associated with the service. Prior to any service being rendered, PMPs are also encouraged to describe the client's role in preparing for service, and noting any potential obstacles to successful control so that the client has realistic expectations.

Service agreements The NPMA BMP recommends that PMCs utilize a written service agreement designed specifically for bed bug management. This provides protection for the company and discloses important information to the client. Due to the variation in laws and regulations between US states, PMCs are encouraged to have any contracts or service agreements reviewed by an attorney who has experience in evaluating the critical factors associated with bed bug contracts and services.

Record keeping Recordkeeping and documentation are critical components of an IPM program. Detailed records can be used to enhance communication with clients and to protect companies in case of disputes. In some specialized services, such as heat or fumigation treatments, additional data including fumigant concentrations, or time to bed bug lethal temperature should be recorded.

Technician and sales staff training When the NPMA BMP was first created, few credible educational resources focusing on bed bug biology, behavior, and control were available. To ensure that representatives of PMCs were well trained and able to communicate facts to clients, a minimum training component was included in the BMP. It was recommended that company representatives who might encounter or be asked about bed bugs receive basic training in biology, habits, elements of control, signs of infestation, and the inspection process. Technicians and sales people involved directly in delivering bed bug services require advanced training.

Client education Clients are an integral part of the IPM process and should be educated as part of the bed bug management program. The NPMA BMP encourages PMCs to make handouts or educational training sessions available for their clients to ensure that cooperation is achieved, and client expectations remain reasonable.

Disposal of beds, furniture, and possessions PMCs are often asked if beds and other possessions should be discarded as part of the control process. The cost of replacing infested items after disposal can be cost prohibitive, and discarded items are often repurposed by other residents. This repurposing can spread bed bugs to additional locations. Therefore, this practice is not encouraged in the BMP.

Client cooperation and treatment preparations Client cooperation is critical to the success of an IPM program. PMCs have the responsibility for providing property managers, residents, and other responsible parties with reasonable instructions for the preparation of their home in advance of the service visit.

Bed bug detection PMCs need to determine if treatment is necessary before providing bed bug management services. The necessity of service can be dependent on the needs of the client. If an infestation cannot be identified during the inspection phase of the service, continued monitoring is recommended in most cases. However, some clients (apartment owners and hoteliers), may insist upon prophylactic treatment based upon complaints from their residents. Recommendations for comprehensive inspection procedures are included in the BMP.

Bed bug scent detection canine teams In the USA, canine bed bug scent detection teams are employed by some companies to detect infestations, especially when bed bug populations are small. The BMP states that all canine scent detection teams must be certified by an independent third party. Certification testing demonstrates the ability of the canine and handler's competence and specify their ability to distinguish live bed bugs and viable eggs from non-target odors (dead bed bugs or cast skins) in real-world environments. Certification testing must be performed in accordance with the minimum standards for canine scent detection testing as described in the BMP.

Integrated pest management and methods of control Due to the varied nature of bed bug infestations, the BMP does not describe a standardized protocol for bed bug IPM. Instead, the BMP identifies the various legitimate control methods that can be combined with education, monitoring, and physical, mechanical, or cultural control techniques as part of a comprehensive bed bug management program.

Insecticides In addition to non-chemical control methods described in the IPM portion of the BMP, insecticides are addressed in a standalone section. This section describes the safe and effective use of these tools. Insecticide recommendations refer the PMP to the US Environmental Protection Agency approved label instructions.

Surrounding units Because of the propensity for bed bugs to move from infested locations to adjacent units in multifamily housing, the BMP recommends that adjacent areas be included in the inspection and treatment plan.

Post-treatment evaluation As part of a comprehensive bed bug IPM program, a post-treatment evaluation of the bed bug treatment methods is recommended. Post-treatment evaluation may take the form of monitoring, client interviews, inspections, or other tactics. It is noted that the presence of a bed bug following a treatment does not always indicate a service failure. In heavy infestations, it is not unusual for one or two bed bugs to avoid being treated, but these can usually be dealt with using monitors alone. In addition, there is always the possibility that bed bugs are being reintroduced, or that they are immigrating in from nearby infested locations. It is for these reasons that post-treatment evaluation is critical for any bed bug IPM program.

Health and safety of technicians In addition to standard occupational safety and health considerations, PMPs are encouraged to guard against accidentally carrying bed bugs from one infested unit to another in equipment or clothing. Additionally, precautions are recommended regarding the handling of bed linens soiled with blood, and exposure to client-applied insecticide products.

Health and safety of clients In addition to strict adherence to pesticide label instructions, the BMP instructs PMPs to communicate with clients regarding the application of professional-use pesticides in their home. In this way, the PMP can address any concerns about potential exposure.

24.5 Marketing and Adoption of the NPMA Best Management Practices

Upon adoption, the NPMA bed bug BMP was promoted broadly, both within the industry and to consumers. The consumer public awareness campaign was managed by the Professional Pest Management Alliance (PPMA), an organization associated with the NPMA that acts as its public relations and consumer education arm. The BMP was promoted as a tool to help consumers identify good bed bug service practitioners. Consumers were encouraged to seek out PMCs that were familiar with and adhering to the BMP. Training related to the BMP included online programs, promotion at conferences, conventions, and state association meetings. Within the industry, the BMPs were promoted widely through webinars, electronic newsletters, trade publications (Fredericks, 2011; McKenna, 2011; NPMA, 2011b), and urban entomology conferences.

24.6 Acceptance of the BMP

Acceptance of the BMP has been high. In 2011, the NPMA and the University of Kentucky conducted an online survey to assess the state of the bed bug industry in North America (Potter *et al.*, 2011). As part of the self-administered survey, respondents were asked about the BMP, which had been published earlier that year.

More than two-thirds (68%) of the respondents indicated that they were currently adhering to, or planning to adopt, the BMP as part of their operating procedure. Twenty-nine percent indicated that they were not yet aware of the BMP, while only 3% reported that they did not plan to follow the guidelines. In a 2013 survey (Potter et al., 2013), the majority of respondents (61%) said their company was either currently following, or planned to follow, the guidelines. Fewer firms indicated they either were not familiar with the BMP (15%, down from 29%), or that they had not made a decision regarding adoption of the guidelines (23%). A further survey, in 2015, again reported that the majority of respondents were following or planning to follow the BMP (Potter et al., 2015).

The development and adoption of the NPMA BMP has resulted in a clearer understanding of the bed bug management methods and procedures that can result in an effective service. PMCs have been able to use the BMP as a guidance document to develop and hone their operating procedures for bed bug control. In addition, consumers have a ready resource available to them for evaluating the service offerings from PMPs.

References

Fredericks, J. (2011) An evolving discussion. *Pest Management Professional*, **79** (2), 57.

McKenna, L. (2011) *Beacon Call: NPMA's Best Management Practices Represent a Comprehensive Response to The Bed Bug Pandemic*, http://www.pctonline.com/article/pct-0311-bed-bugs-pandemic-response-npma/ (accessed 21 May 2016).

NPMA (2011a) NPMA library update: Best management practices for bed bugs. *PestWorld*, **March/April**, I–XVI.

NPMA (2011b) *Best Management Practices for Bed Bugs: A Summary for Consumers*, US National Pest Management Association http://www.bedbugbmps.org/PDF/BMP-bedbug_BedBugBMPsURL.pdf (accessed 11 October 2016).

NPMA (2013). *Best Management Practices for Bed Bugs*, 2nd edn, US National Pest Management Association, http://www.bedbugbmps.org/best-practices.html (accessed 2 July 2016).

NPMA (2016) *NPMA Best Management Practices for Bed Bugs. National Pest Management Association, Fairfax*, US National Pest Management Association, http://npmapestworld.org/default/assets/File/techresources/NPMA%20Bed%20Bug%20BMPs%20approved%202016_07_28.pdf (accessed 11 October 2016).

Potter, M.F., Haynes, K.F., Rosenberg, B. and Henriksen, M. (2011) 2011 bugs without borders survey. *Pestworld*, **Nov/Dec**, 4–15.

Potter, M.F., Haynes, K.F., Rosenberg, B. and Henriksen, M. (2013) Bed bug nation. Are we making any progress? *PestWorld*, **Sep/Oct**, 5–9.

Potter, M.F., Haynes, K.F. and Fredericks, J. (2015) Bed bugs across America. Results from the 2015 bed bug survey. *PestWorld*, **Nov/Dec**, 5–14.

25

A Pest Control Company Perspective

Joelle F. Olson, Mark W. Williams and David G. Lilly

25.1 Introduction

The global resurgence of bed bugs is a well-known phenomenon that has impacted virtually all sectors of society. Two of the biggest challenges associated with the resurgence have been the lack of public education and the rapid response required by the pest management industry. In addition to responding to the resurgence, the pest management industry has also been forced to quickly adopt new information on biology, behavior, and control. Differences in insecticide availability, local regulations, and public tolerance have made it even more difficult to provide consistent services for bed bugs around the world. This chapter focuses on the multitude of challenges faced by the pest management industry in the control of bed bug infestations.

25.2 The Resurgence

25.2.1 Rapid and Sustainable Growth

At the turn of the 21st century, a resurgence in bed bugs occurred simultaneously throughout much of the developed world, including Europe, Asia, Australia, the Middle East, and North and South America (see Part 2 of this book). The magnitude of the resurgence was of particular significance for the pest management professional (PMP) given that the growth in bed bug-related service requests occurred not only very rapidly, but has now been sustained for the better part of two decades. Although, the pest management industry has financially benefited from this growth in bed bug infestations, the resurgence has also brought forth many challenges.

25.2.2 Lack of Industry Preparedness

The lack of knowledge, specialized equipment, and experience in controlling bed bugs meant that many PMPs were ill-prepared to offer efficacious bed bug services at the onset of the resurgence. Despite the need for further education and experience, the industry has been accused of being slow to offer leadership on best practices (Doggett *et al.*, 2012). This has arguably led to a diversity of control strategies being used by individual pest management companies (PMCs) around the world and even by the same company in disparate locations (Boase, 2008). In 2005, the Australian Environmental Pest Managers Association (AEPMA) was the first to offer guidance to the pest management industry by developing an industry standard for bed bug management (Doggett, 2005). However, it took another six years for Europe (BBF, 2011) and the USA (Anonymous, 2011; NPMA, 2011) to develop similar standards for their respective regions. In the USA, a federal bed bug working group was formed in 2009, but did not publish recommendations until six years later (Federal Bed Bug Workgroup, 2015). A number of pest management associations have developed bed bug eradication guides, but most regions are

Advances in the Biology and Management of Modern Bed Bugs, First Edition.
Edited by Stephen L. Doggett, Dini M. Miller, and Chow-Yang Lee.
© 2018 John Wiley & Sons Ltd. Published 2018 by John Wiley & Sons Ltd.

Figure 25.1 A cluttered apartment, which is infested with bed bugs (*Source:* Shutterstock, under licence).

still operating without one, and thus poor control practices persist in some areas. Government officials, academics, manufacturers, and PMPs have yet to develop a coordinated strategy for the pest management industry to ensure that bed bug products work as effectively and efficiently as the manufacturers claim.

25.2.3 Lack of Public Awareness

Although bed bugs have been in association with humans for more than five millennia, the resurgence that many parts of the world are currently experiencing is relatively recent. Consequently, there continues to be a lack of awareness and basic understanding about this pest amongst the general public. News reports by the media have arguably increased public awareness of bed bugs, but often sensationalize the issue and usually leverage fear and exaggeration to grab the attention of the audience (Barclay, 2010). In addition, online resources often provide inaccurate or out-of-date information. The combination of fear and misinformation has led some to attempt drastic eradication methods, which has likely contributed to the spread of bed bugs in many areas (Wang *et al.*, 2010; Lilly *et al.*, 2011). Poor information transmitted online and by the media has made it even more challenging for PMPs to properly educate the public and the industry.

Fortunately, recent public forums held in many countries have created a platform for university researchers, government officials, and PMPs to collaborate and share experiences for the combined goal of improving bed bug control efforts and properly educating each other, and the public (EPA, 2009a,b; Fenner, 2011). Continuation of collaborative efforts between professional groups will help towards improving bed bug control efforts in the future.

25.3 Responsibility and Liability

25.3.1 Educating the Client

The lack of quality information on bed bug biology and control has forced many PMPs to develop their own educational materials for their clients (Olson, Williams, and Lilly, unpublished results). At minimum, a series of guidelines and checklists should be provided prior to executing any type of bed bug service (Doggett, 2012b; Cooper, 2015). Checklists provide clear instructions, while guidelines can set expectations for services. The delivery of quality information should be seen as a necessary part of good customer communication, which will hopefully reduce the risk of treatment failure and assist in protecting the company brand.

25.3.2 Liability for Services

The perception of bed bugs among the public is highly unfavorable (Goddard and deShazo, 2009; Eddy and Jones, 2011a,b; NPMA, 2011). While it is not possible for hotels or other properties to prevent bed bugs from being introduced, guests or tenants may become very upset if they encounter bed bugs during their stay. Regardless of fault, there have been several lawsuits filed against hotel managers by guests that have encountered bed bugs during their stay (Bello, 2003; Jones, 2011). Most of these cases have settled out of court or were only proven favorable for the guest when there was sufficient evidence to suggest that the property manager was negligent such as failing to follow the directions recommended by their pest management provider or by renting a room that has an active bed bug infestation (Keeling, 2008). Again, this demonstrates the importance of developing a partnership with the client, providing clear instructions, and setting initial expectations before executing any type of bed bug service.

PMPs are not excluded from potential bed bug legal situations. However, there are ways to minimize risk (Pinto, 2013). Every PMP should work with a legal representative to develop and maintain contracts for services rendered. Legally sound pest service contracts have helped protect PMCs from liability (Harbison, 2009; Everitt, 2014). Secondly, it is important for PMPs to maintain proper documentation for all pest services rendered. The service report should include:

- the date of the service
- the cost
- detailed findings during inspection and treatment
- type of treatment performed
- list of insecticidal products applied
- dates of follow-up inspections
- date when the infestation was declared eradicated
- recommendations to the client
- any variations from the initial proposal to the client.

Any limitations in achieving a potentially successful outcome (in other words, the eradication of the infestation) must also be noted. This may include limited access to critical areas for inspection and treatment within the premises, not being able to inspect adjoining premises in multiple-occupancy dwellings, and a lack of client cooperation. Finally, PMCs must ensure all their PMPs are properly trained and that all procedures, protocols, and documentation are regularly reviewed in order to maintain compliance with any ongoing change in best practice and local regulations.

25.3.3 Financial Burden

Hotel managers, apartment managers, and facility operators often struggle with determination of responsibility when bed bugs are discovered at their properties. Bed bugs can be introduced at any time and disperse or spread from room to room and neighboring units, making determination of responsibility difficult (Cooper, 2015). The cost for service and room downtime can be extremely burdensome for the property manager. Some parts of the USA have started to pass local laws specifically addressing the person or group responsible for payment of the treatment of the apartment or dwelling. While some laws have been considered an invasion of privacy, such as those requiring full disclosure of the previous activity to new tenants (Stewart, 2016), the laws have improved processes for PMPs and ensured payment for services in some cases. However, it should be noted that these types of regulation are only relevant to the USA at this time.

25.4 Inspection and Control Methods

25.4.1 Inspection and Monitoring Tools

One of the most important aspects of any bed bug service is the inspection process, which must be extremely thorough to detect all bed bug life stages and harborage areas (Diggs, 2009). There are a variety of tools available to assist in the bed bug inspection process, including traps and monitors, and canine detection teams (Wang and Cooper, 2011; see also Chapter 27). However, the efficacy and accuracy of these options has been the subject of debate (Doggett, 2012a).

There is an abundance of monitors on the market that claim to be effective at detecting bed bugs (Gangloff-Kaufmann *et al.*, 2006; Anonymous, 2009, 2012, 2013a,b, 2014; Doggett and Russell, 2011). Many of these devices have been proven ineffective, but recent studies have demonstrated potential for detection of even light infestations in multi-unit housing and residential settings (Wang *et al.*, 2011; Cooper *et al.*, 2015a,b). Regardless of whether or not the monitor or trap will catch live bed bug activity, one critical component to their success is the client partnership and willingness to inspect and maintain the trap between service visits. Without client support, these types of devices can be misplaced, discarded, ignored, or not maintained according to manufacturer recommendations, thus rendering them useless.

Similar to monitors, controversy exists about the accuracy of canine inspection teams (Wang *et al.*, 2011, see also Chapter 27). Initial research indicated that canines could achieve 95% accuracy in identifying bed bug infestations (Pfiester *et al.*, 2008), but this work was undertaken in a highly controlled environment lacking the distractions that occur under field conditions. Recent studies have found that the accuracy of canine detection teams in the field is far more variable, in the range of 10–100% with a mean accuracy rate of 44% (Cooper *et al.*, 2014). While poor performance has been linked to inadequate training, the variability in success makes it difficult for PMPs to build confidence in this type of service regardless of whether the client prefers the use of canines.

25.4.2 Chemical Applications and Resistance

Chemical applications have been used to control insects for hundreds of years and the history of chemical control for bed bugs is discussed in Chapters 2 and 30. Historically, PMPs used a range of products such as hydrogen cyanide and sulfur dioxide to eliminate bed bugs (Potter, 2008, 2011). The introduction of the organochlorines, such as DDT, and the organophosphates virtually eliminated the pest problem from many areas of the world. However, when the bed bug resurgence first began in the late 1990s, there were very few options for chemical applications in many countries because manufacturers had stopped including bed bugs as a target pest on their product labels. In addition, there was little information on which products were even effective. This made it very difficult for the pest management industry to confidently treat for bed bugs.

The lack of novel insecticide chemistries for bed bug management is perhaps one of the most disappointing aspects of the resurgence. There has been a significant reduction in the number of insecticide classes available for bed bug control (Anonymous, 2010). While the resurgence has encouraged insecticide manufacturers to bring forth new products, most contain actives from the same insecticide class, namely the pyrethroids. For example, in Australia in 2006, 104 products were registered for use against bed bugs, of which 81% were pyrethroids (Doggett, 2006). By 2013, there were 143 products registered, however, 93% were pyrethroids (Doggett, 2012b), which meant the actual number of non-pyrethroid products decreased, from 20 in 2006 to 10 in 2013. The loss of chemical class diversity has also been observed in the UK and the USA, but is not as well documented. In addition to limited product availability, reports pertaining to insecticide resistance increased during the mid-2000s (see Chapter 29), leaving PMPs concerned about their ability to achieve successful bed bug eradication.

The tendency for manufacturers to test only against susceptible strains, and not modern, relevant, resistant strains continues to frustrate the pest management industry (Doggett and Lilly, 2015). Consequently, manufacturers have been accused of prioritizing convenience over efficacy (Doggett *et al.*, 2012; Jones and Bryant, 2012; Abejuela-Matt, 2014). In the USA, an additional concern is that actives that are considered to have low human toxicity can be exempted from the insecticide registration process (without any efficacy evaluation). Subsequent research has found a number of these compounds are largely ineffective in controlling bed bugs and several manufacturers have been fined for false claims of efficacy (see Chapter 31). Furthermore, researchers from different laboratories have produced highly variable data, even ranging from zero to complete control with the same insecticide (Doggett *et al.*, 2012). These differences can partially be explained by the use of diverse test methods and bed bug strains, but it still leaves the PMP confused about which products to use and may result in lowered confidence in the service provided.

Recently, efficacy guidelines were drafted by the US EPA. These require manufacturers to provide data to support product claims when applied according to the label directions (EPA, 2013). Superficially, this type of involvement by government officials would appear to be welcomed by the pest management industry. However, disparate bed bug populations possess an enormous range of resistance mechanisms that confer different levels of resistance. Therefore, results obtained in one laboratory with a particular bed bug strain may not be reflected in the field for the same insecticidal product. Consequently, a PMP cannot solely rely on efficacy data and manufacturer claims in assessing whether an insecticide treatment will be successful in eradicating every bed bug infestation. For this reason, follow up inspections to assess the success of treatment should always be undertaken.

25.4.3 Non-chemical Applications

The limited availability of products and reports of resistance have left PMPs scrambling for alternative solutions to control bed bugs. The use of non-chemical control methods, such as vacuuming and extremes in temperature, have been explored (albeit in a limited research capacity) and are routinely recommended for bed bug control (Kells, 2006; Zehnder *et al.*, 2014). Each of these methods has its own benefits and limitations (Doggett, 2007; Michigan Department of Community Health, 2013; Quarles, 2015; Wang *et al.*, 2015), and are reviewed in Chapter 28.

The two biggest challenges associated with non-chemical control methods for bed bugs are cost of equipment and labor. For example, heating equipment, whether for the entire structure or for smaller-scale containment heating, can be very expensive and may require a minimum of two individuals to operate (Olson, Williams, and Lilly, unpublished results). Smaller non-chemical options, such as vacuuming and steam, may require only one operator but significant amounts of time, and attention to detail is required in order to achieve success. This can be challenging for PMPs when trying to adhere to a standard service schedule, as bed bug service requests are typically an unscheduled event.

25.4.4 Training and Maintaining Service Consistency

Maintaining consistency in any pest service offering is critical (Boase and Naylor, 2014; Troyano, 2016). This can be challenging within the same country, but is especially difficult for global companies that operate in locations around the world. Product regulations and availability vary significantly across the globe, making it difficult for larger PMCs to consistently train and offer services to corporate clients located in multiple regions. Many regions have to rely on the products and equipment available to them, leaving some clients questioning why a service may be different in one location versus another.

Bed bugs are considered to be one of the hardest insects to eradicate and treatment failures are not uncommon. Therefore, it is important to spend considerable time educating new technicians on the basics of biology and control as well as providing them with in-field training. Some companies construct mock rooms and technicians are tasked with finding all bed bug harborages (Troyano, 2016). Even with experienced PMPs, it is important to stay educated and current on best practice and service standards.

Probably one of the biggest challenges in achieving successful bed bug eradication is the quantity of personal belongings in residential facilities (Figure 25.1). Clutter can significantly contribute to the bed bug problem by providing numerous harborage areas. This is especially problematic for multi-unit housing, where large infestations may go unnoticed for extended periods of time, creating a risk to nearby tenants. Individuals with hoarding tendencies pose an extreme risk for property managers and their homes are difficult to treat (Molluso, 2017). It should not be the responsibility of the PMP to remove the clutter in order to properly treat the area. However, in some cases the tenant or client is unable to properly prepare the area for treatment. Therefore, some PMCs have offered to provide a preparation service at an additional charge to the client. If the personal items or clutter are not discarded or properly treated, the potential for failure is extremely high. This further illustrates the importance of proper client education and partnership.

25.5 Conclusion

In summary, the bed bug resurgence has been both profitable and challenging for the pest management industry. PMPs have been forced to take on much of the responsibility in educating the public, developing communication tools, establishing treatment protocols, and providing expertise. Current control options have their limitations and there is no silver bullet to ensure successful eradication of this pest. Therefore, it is important for PMPs to stay educated, share best practices, and develop a partnership with the client in order to be successful in the field.

References

Abejuela-Matt, V.L. (2014) Bedbugs biting back? A multifactorial consideration of bedbug resurgence. *Journal of Patient-Centered Research and Reviews*, **1** (2), 93.

Anonymous (2009) First monitor for bedbugs arrives. *Pest*, **March/April**, 12.

Anonymous (2011) European code of practice launched for the management of bed bugs. *International Pest Control*, **53** (1), 42.

Anonymous (2010) Europe's bedbug challenge continues. *Pest*, **March/April**, 22–22.

Anonymous (2012) Navigating the monitor maze. What's available and at what price? *Pest*, **July/August**, 19–21.

Anonymous (2013a) Bed bug battle continues. *Pest*, **January/February**, 22.

Anonymous (2013b) Bed bug product explosion! *Pest*, **July/August**, 28–29.

Anonymous (2014) Advances in bed bug detection devices. *Pest*, **July/August**, 25–27.

Barclay, E. (2010) *Don't Let the Bedbugs Fright*, http://www.slate.com/articles/health_and_science/science/2010/09/dont_let_the_bedbugs_fright.html (accessed 24 November 2016).

BBF (2011) *European Code of Practice, Bed Bug Management*, Bed Bug Foundation, London.

Bello, P. (2003) *Over 201 Things to Know About Bed Bugs*, http://www.acacamps.org/resource-library/camping-magazine/over-201-things-know-about-bed-bugs (accessed 24 November 2016).

Boase, C. (2008) Bed bugs (Hemiptera: Cimicidae): an evidence-based analysis of the current situation. In: *Proceedings of the Sixth International Conference on Urban Pests, July 13–16, 2008, Budapest, Hungary*. OOK-Press, Hungary, Veszprém.

Boase, C. and Naylor, R. (2014). Bed bug management. In: *Urban Insect Pests: Sustainable Management Strategies*, (ed. P. Dhang), CABI, Oxfordshire, UK.

Cooper, R. (2015) Customer education is key to cooperation. *Pest Control Technology*, **43** (3), 94–97.

Cooper, R., Wang, C. and Singh, N. (2014) Accuracy of trained canines for detecting bed bugs (Hemiptera: Cimicidae). *Journal of Economic Entomology*, **107** (6), 2171–2181.

Cooper, R., Wang, C. and Singh, N. (2015a) Effectiveness of various interventions, including mass trapping with passive pitfall traps on low level bed bug populations in apartments. *Journal of Economic Entomology*, **109** (2), 762–769.

Cooper, R., Wang, C. and Singh, N. (2015b) Mark-release-recapture reveals extensive movement of bed bugs (*Cimex lectularius* L.) within and between apartments. *PLoS ONE*, **10** (9), e0136462.

Diggs, R. (2009) Thorough inspection works best. *Pest Management Professional*, **77** (6), 43, 45–48.

Doggett, S.L. (2005) *A Code of Practice for the Control of Bed Bug Infestations in Australia (draft)*, Department of Medical Entomology and The Australian Environmental Pest Managers Association, Sydney.

Doggett, S.L. (2006) *A Code of Practice for the Control of Bed Bug Infestations in Australia*, Department of Medical Entomology and The Australian Environmental Pest Managers Association, Sydney.

Doggett, S.L. (2007) *A Code of Practice for the Control of Bed Bug Infestations in Australia*, 2nd edn. *(draft)*. Department of Medical Entomology and The Australian Environmental Pest Managers Association, Sydney.

Doggett, S.L. (2012a) Caveat emptor! Bed bug product buyers beware. *Pest*, **Jul/Aug**, 16–17.

Doggett, S.L. (2012b) *A Code of Practice for the Control of Bed Bug Infestations in Australia*, 4th edn. Department of Medical Entomology and The Australian Environmental Pest Managers Association, Sydney.

Doggett, S. L., and Lilly, D. (2015) Bed bugs on the label…but what does this really mean? *Pest*, **February/March**, 34–35.

Doggett, S.L., and Russell, R.C. (2011) Battling bedbugs the latest in weaponry. *Pest*, **July/August**, 20–22.

Doggett, S.L., Dwyer, D.E., Peñas, P.F. and Russell, R.C. (2012) Bed bugs: clinical relevance and control options. *Clinical Microbiology Reviews*, **25** (1), 164–192.

Eddy, C., and Jones, S.C. (2011a) Bed bugs, public health, and social justice: Part 1, an opinion survey. *Journal of Environmental Health*, **73** (8), 1–14.

Eddy, C., and Jones, S.C. (2011b) Bed bugs, public health, and social justice: Part 2, an opinion survey. *Journal of Environmental Health*, **73** (8), 15–17.

EPA (2009a) *EPA's National Bed Bug Summit: Participant Recommendations. US Environmental Protection Agency*, http://ipm.ifas.ufl.edu/pdfs/EPA_NATL_BEDBUG_SUMMIT.pdf (accessed 24 November 2016).

EPA (2009b) *EPA's National Bed Bug Summit: Recommendations From The Ten Workgroups. US Environmental Protection Agency*, http://ipm.ifas.ufl.edu/pdfs/EPA_NATL_BEDBUG_SUMMIT.pdf (accessed 24 November 2016).

EPA (2013) *Effectiveness of Bed Bugs Pesticides. US Environmental Protection Agency*, https://www.epa.gov/bedbugs/effectiveness-bed-bug-pesticides#regchanges (accessed 24 November 2016).

Everitt, J. (2014) *What Happened? I Still Have Bed Bugs*, http://www.pctonline.com/article/pct1214-ineffective-bed-bug-treatments/ (accessed 24 November 2016).

Federal Bed Bug Workgroup (2015) *Collaborative Strategy on Bed Bugs*, https://www.epa.gov/sites/production/files/2015-02/documents/fed-strategy-bedbug-2015.pdf (accessed 24 November 2016).

Fenner, J. (2011) EPA hosts 2nd annual bed bug summit. *Pest Control Technology*, **39** (3), 118, 120, 122, 124, 126–127.

Gangloff-Kaufmann, J., Hollingsworth, C., Hahn, J., *et al.* (2006) Bed bugs in America: a pest management industry survey. *American Entomologist*, **52** (2), 105–106.

Goddard, J., and de Shazo, R. (2009) Bed bugs (*Cimex lectularius*) and clinical consequences of their bites. *Journal of the American Medical Association*, **301** (13), 1358–1366.

Harbison, B. (2009) *Cooper, Hardigree Provide Bed Bug Seminar Attendees With Plenty of Food For Thought*, http://www.pctonline.com/News/news.asp?ld=7233 (accessed 21 November 2010).

Jones, J.M. (2011) *Women Sue Over Bed Bug Infestation at Glendale Motel*, http://latimesblogs.latimes.com/lanow/2011/03/bed-bugs-lawsuit-glendale.html (accessed 24 November 2016).

Jones, S.C., and Bryant, J.L. (2012) Ineffectiveness of over-the-counter total-release foggers against the bed bug (Heteroptera: Cimicidae). *Journal of Economic Entomology*, **105** (3), 957–963.

Keeling, B. (2008) *Downtown Bedbug Attack Costs Ramada $71,000*, http://sfist.com/2008/10/24/downtown_bedbug_attack_costs_ramada.php (accessed 24 November 2016).

Kells, S.A. (2006) Nonchemical control of bed bugs. *American Entomologist*, **52** (2), 109–110.

Lilly, D., Jones, G., Doggett, S.L., Orton, C.J. and Russell, R.C. (2011) Use your Bed Bug Code of Practice to get more money from jobs by educating your customers. Speech presented at the Australian Environmental Pest Managers Association National Conference, Homebush, Australia, June 2, 2011.

Michigan Department of Community Health (2013) *Getting The Bed Bugs Out: A Guide to Controlling Bed Bugs in Your Home*, https://www.michigan.gov/documents/emergingdiseases/Getting_the_Bed_Bugs_Out_Guide_442175_7.pdf (accessed 24 November 2016).

Molluso, J. (2017) Hoarding and bed bugs do not mix. *Pest Management Professional*, **85** (2), 56–58.

NPMA (2011) *Best Management Practices for Bed Bugs*. US National Pest Management Association, http://www.pestworld.org/media/560191/bed_bug_bmps_for_consumers_final.pdf (accessed 24 November 2016).

Pfiester, M., Koehler, P. and Pereira, R. (2008) Ability of bed bug detecting canines to locate live bed bug an dviable bed bug eggs. *Journal of Economic Entomology*, **101** (4), 1389–1396.

Pinto, L. (2013) Minimizing your risk from bed bug lawsuits. *Pest Control Technology*, **41** (4) 80,82, 84–86, 88, 89.

Potter, M.F. (2008) Bed bug supplement. The history of bed bug management. *Pest Control Technology*, **36** (8), S1, S3–6, S8–10, S12.

Potter, M.F. (2011) The history of bed bug management-with lessons from the past. *American Entomologist*, **57** (1), 14–25.

Quarles, W. (2015) New IPM methods for bed bugs. *The IPM Practitioner*, **34** (7/8), 1–9.

Stewart, M. (2016) *Landlord Responsibility For Bed Bugs*, http://www.nolo.com/legal-encyclopedia/landlord-responsibility-bed-bugs.html (accessed 24 November 2016).

Troyano, N. (2016). Fifty shades of bed bug grey: out of the book and into the bedroom! Speech presented at the 2016 XXV International Congress of Entomology in Orlando, Florida, September 30, 2016.

Wang, C. and Cooper, C. (2011) Bed bug supplement: Detection tools and techniques. *Pest Control Technology*, **39** (8), 72, 74, 76, 78–79, 112.

Wang, C., Saltzmann, K., Chin, E., Bennett, G. and Gibb, T. (2010) Characteristics of *Cimex lectularius* (Hemiptera: Cimicidae) infestations and dispersal in high-rise apartment buildings. *Journal of Economic Entomology*, **103** (1), 172–177.

Wang, C., Tsai, W., Cooper, R. and White, J. (2011) Effectiveness of bed bug monitor for detecting and trapping bed bugs in apartments. *Journal of Economic Entomology*, **104** (1), 274–278.

Wang, C., Singh, N. and Cooper, R. (2015) Bed bug control – the resident factor. *Pest*, **August/September**, 21–23.

Zehnder, C., Schmidt, M. and Hasenböhler, A. (2014) Thermal treatments for bed bugs. In: *Proceedings of the Eighth International Conference on Urban Pests, 2014, July 20–23, Zurich, Switzerland*. OOK-Press, Veszprém.

26

Prevention

Molly S. Wilson

26.1 Introduction

There are various bed bug control methods and products currently available to consumers and pest management professionals (PMPs) throughout the world. However, being proactive and minimizing the risk of bed bugs becoming established should be the first step when it comes to the management of bed bugs (Cooper *et al.*, 2016a). While it is not possible with current technologies to completely prevent bed bugs, various strategies can be implemented to reduce the risk of an infestation. The intentions of bed bug preventative measures are:

- to decrease the potential for bed bug introductions
- to increase the probability of early bed bug detection
- to limit the potential for bed bugs to spread to other locations.

Many methods of bed bug prevention have been evaluated for their efficacy, ease of application, low cost, and practicality of use in different facilities (Stedfast and Miller, 2014). Bed bug preventative methods can be used independently, but the use of multiple strategies will decrease the risk of an infestation becoming established and becoming a serious issue (Wang and Cooper, 2011). *The Code of Practice for the Control of Bed Bugs in Australia* (Doggett, 2013) recommends minimizing potential bed bug harborage areas as part of an overall bed bug risk reduction program. In this code, it is suggested that cracks and crevices should be sealed, bed bug-friendly furniture and fixtures should be eliminated (for example, avoiding wicker furniture, installing hard tiles instead of carpet), and certain construction elements should be avoided (such as open brickwork). The US Environmental Protection Agency bed bug guidelines, for the US Department of Housing and Urban Development, specifically suggest having a regular housekeeping regimen and reducing clutter in homes (HUD, 2012). As an aid for early detection of infestations, methodologies such as the use of monitors or scent detection canines can be employed (Wang and Cooper, 2011). All these methodologies can be used to prevent bed bug infestations from becoming established, but none specifically address preventing bed bugs from being initially introduced into a building.

This chapter will discuss the bed bug preventative methodologies that have been recommended by the Virginia Tech Bed Bug and Urban Pest Information Center, Virginia, USA, in response to requests for bed bug training from different organizations, such as social services, school districts, family shelters, correctional halfway houses, drug rehabilitation centers, elderly assisted living, healthcare facilities, and public housing authorities, as well as other recommendations often promoted. These types of organizations face unique challenges, which include:

- servicing transient populations that regularly bring personal items into their facilities
- dormitory-style living environments connected by community rooms
- patrons' daily excursion activities to and from potentially infested locations.

Advances in the Biology and Management of Modern Bed Bugs, First Edition.
Edited by Stephen L. Doggett, Dini M. Miller, and Chow-Yang Lee.
© 2018 John Wiley & Sons Ltd. Published 2018 by John Wiley & Sons Ltd.

All of these factors introduce the potential for bed bugs to be brought in each day and spread from room to room within a facility.

In light of the challenges that these facilities face on a daily basis, various approaches have been suggested for preventing bed bug entry and spread within a building. The recommended methods can be broadly categorized into: education, early detection, harborage reduction, prophylactic insecticide applications, and heat treatments.

At Virginia Tech, education about these methods is delivered to personnel and clients, and includes the dissemination of accurate bed bug information and materials. The installation of passive monitoring devices is suggested for early bed bug detection, while the installation of mattress and box spring (bed base) encasements reduces potential harborages. Proactive treatments include the interior perimeter application of desiccant dusts, and heat treatments for a wide range of belongings. Ultimately, all accommodations should have and adhere to a policy for the management of bed bug infestations.

26.2 Education

Bed bug education needs to be delivered to all personnel, residents, visitors, patrons, and patients associated with facilities that are vulnerable to bed bug introductions or chronic infestations. Educational topics for high risk facilities must include:

- how to identify bed bugs and the signs of their presence
- bed bug biology and behavior in terms of how bed bugs are transported on personal belongings
- how to inspect oneself and one's personal belongings for bed bugs
- behaviors that could increase the risk of acquiring bed bugs
- what to do if a bed bug is encountered (and who should be informed).

Bed bug educational topics can be presented in a variety of ways, including one-on-one interactions, educational videos, group seminars, and print or electronic media. However, it is recommended that all participants are provided with printed materials that include photographs, which can be used as a reference for identifying live bed bugs and their signs. For more information related to bed bug education, refer to Chapter 32.

26.3 Monitors

Monitoring devices are important preventative tools for many reasons: the early detection of low-level infestations, to detect bed bug reintroductions, and to determine if bed bugs are still present after treatment (Cooper, 2011; see also Chapter 26). As bed bugs are so prevalent in the indoor environment, it is important to be constantly on the look-out for their presence, especially in the high-risk environments mentioned above. Monitoring devices can be used to keep a room or building under constant bed bug surveillance, as bed bugs can enter a facility at any time, with any person.

Many of the more effective monitoring devices are of the "pitfall" style, which are specialized plastic dishes that contain at least two interior wells. These interior wells have slick plastic walls that prevent bed bugs from climbing out of them, which traps the insect, allowing for later inspection. Passive (no attractant) bed bug monitoring devices are intended to intercept and trap bed bugs that may be using bed or furniture legs as routes to access a host, and thus also act as a barrier to impede access to the host. Monitors can be installed under the legs of all beds, sleeping furniture, couches, and chairs. They can also be placed directly on the floor in sleeping areas, offices, community rooms, and waiting rooms, for bed bug monitoring purposes (Wang *et al.*, 2009a).

Monitoring devices offer a constant level of surveillance that humans alone are unable to provide. However, monitors must be checked regularly and maintained, otherwise dirt and debris can render them less effective. It is highly recommended that a specific individual be assigned the job of checking the monitors on a calendar-based schedule (once a day, once a week, once a month, every three months, and so on). The frequency of monitor inspection should be principally determined by the infestation history of the facility, but other factors may need to be taken into consideration, for example if infestations are more associated with certain people. Failure to regularly inspect the monitors will allow bed bug introductions to become established infestations.

Several investigations have evaluated the use and success of monitoring devices. One study used passive monitoring devices to characterize the infestation level and dispersal of bed bugs in a high-rise apartment building (Wang *et al.*, 2010). Another study determined that interception devices are effective tools for detecting low-level bed bug infestations (Wang *et al.*, 2009a). A third study documented the effectiveness of a bed bug management program by using interception devices (and subsequent bed bug catch data) to monitor the reduction of the bed bug populations in apartment units over time (Wang *et al.*, 2009b). A recent study has compared the efficacy of monitors with scent detection canines and visual inspections conducted by PMPs for detecting new infestations (Cooper *et al.*, 2016b). This study found that monitors were much more reliable at detecting bed bug introductions than either dogs, PMPs, or resident complaints.

26.4 Mattress Encasements

Mattress and box spring encasements are principally used to prevent bed bugs from infesting beds. However, the purposes of the encasements are manifold. They can trap bed bugs already on or in the mattress and box spring, which means the mattress does not have to be discarded. Encasements are constructed with few seams and thus provide few places for new bed bugs to harbor, compared with the multitude of locations where bed bugs can aggregate on mattresses and box springs (Cooper, 2007). Finally, the color of encasements is also important. White encasements make it much easier to observe bed bugs and their signs (that is, fecal spotting) when present. If mattresses are encased prior to being infested, bed bugs are unable to access and contaminate the mattresses or box springs. If installed after an infestation has been discovered, an encasement can seal bed bugs inside, preventing them from biting the host, and causing them to eventually starve to death. Mattress encasements are particularly beneficial to individuals who cannot afford to replace their bed sets, or in dormitory-style locations where replacing multiple bed sets would be cost prohibitive (i.e. homeless shelters, fire stations, hospitals, and college dormitories).

Many mattress encasements are available to consumers, but not all are effective (Cooper, 2011). When installed, an appropriate mattress encasement will prevent even the smallest bed bug instar from escaping, or entering the mattress. This makes encasements particularly useful when an individual is relocating from one dwelling to another. The most important features of a "bed bug-proof" encasement are a zipper with tightly interlocking teeth, and a secure closure at the zipper's ends, as young bed bug instars can escape through larger zipper teeth or at the zipper closure point (Bell *et al.*, 2007; Cooper, 2007). Mattress encasements must be made of durable (tear-resistant) material that will also prevent the bed bugs from feeding through the encasement (bite-proof) when trapped inside (Miller, 2009a; Goldberg, 2015). The encasement should be of the appropriate size for the mattress or box spring so that it fits tightly and does not have any folds or loosely hanging areas, especially around the zipper, that could provide bed bug harborages (Miller, 2009a).

A report detailing the use of mattress encasements concluded that full encasements (that completely seal up the mattress) provided the greatest benefit by trapping bed bugs within them and reducing bed bug hiding places (Cooper, 2007). A 2012 product evaluation of one mattress encasement brand found that when the encasement was used according to label instructions (in other words that is was fully zipped), adult and late instar bed bugs were unable to escape the encasement during a four-week test period (Getty *et al.*, 2012). While encasement efficacy has been evaluated in a few studies, it is important to note that most of the published literature regarding mattress encasements consists of product reviews, which evaluate only the cost, material, sizes, and bed-bug-proof qualities of the encasements.

Mattress encasements are used by 84% of pest management professionals in the USA (Potter *et al.*, 2015). As more encasements become available to consumers throughout the world, it becomes increasingly important to be aware of the qualities that make an encasement effective and appropriate. Mattress encasements serve to reduce the costs associated with replacing an infested bed, provide fewer harborages, and makes detections easier. However, they do not prevent bed bug introductions or remediate bed bug infestations.

26.5 Desiccant Dusts

Desiccant dusts can be applied as a physical barrier between connected rooms, offices, or apartment units, to prevent bed bugs from successfully moving to and infesting a new location (Wang *et al.*, 2013). Desiccant dusts, such as insecticide grade diatomaceous earth and amorphous silica gel dust, are highly effective for killing bed bugs (Romero *et al.*, 2009). These products can be applied prophylactically as part of a bed bug preventative program. Desiccant dusts can be applied to bed frames, box springs, upholstered furniture, electrical outlets, and under light switch plate covers (note that there may be use restrictions in different countries and thus the product label should always be consulted). A perimeter barrier of dust (around a specific apartment, room, or office space) can be applied to intercept bed bugs as they attempt to move to adjacent locations through the walls in an infested building. The presence of a desiccant dust barrier forces the bed bugs either to avoid the application by staying in the same area, or to cross the barrier and pick up a lethal dose of the dust (Romero *et al.*, 2009).

The manner in which a perimeter desiccant dust barrier is applied varies depending on building construction. Different building construction features offer bed bugs a variety of passages between rooms and floors, each of which will need to be treated with the desiccant dust to complete the perimeter barrier and avoid creating any "safe spaces" for bed bugs to aggregate within the building. Common locations that need to be treated are between the tacking strip (carpet gripper) and the baseboard (skirting) of carpeted rooms, underneath baseboards, or in wall voids. Desiccant dust applications should not be applied in areas of high foot traffic or with significant air movement, because the application will not remain in place.

The implementation of a desiccant dust perimeter barrier does not take a lot of time (typically half an hour for a one-bedroom apartment) when professional-grade power dusters are used. This is true even if there has been limited preparation of the home or unit prior to application. Factors that may influence the application time of the perimeter dust barrier include the size of the home or unit, the clutter level, the presence of the homeowner or resident, and the type of desiccant dust applicator used (power duster versus bulb duster) (Stedfast and Miller, 2014). It is important to note that in an apartment complex, a desiccant dust perimeter barrier could be applied during a scheduled maintenance visit or when the unit is vacant. Because desiccant dusts are a physical insecticide and have no risk of evaporating, these applications can remain effective for as long as they remain in place.

26.6 Heat

Heat exposure is a very effective way of treating potentially infested items such as clothing, furniture, electronics, and other belongings before they are moved into new environments. The recognized thermal death point (full mortality in 1 min) for all bed bug life stages and eggs is 50°C (122°F) (Kells and Goblirsch, 2011). For more information about heat treatments, see Chapter 28.

26.6.1 Clothes Dryers

The high heat setting on most household clothes dryers can reach the thermal death point for bed bugs and will kill all life stages as long as the dryer is loosely packed, with room for air to flow (Potter *et al.*, 2007; Miller, 2009b; Naylor and Boase, 2010). Therefore, in those nations where clothes dryers are common, they can be used to treat most potentially infested fabric items, such as clothes, bedding, curtains, table linens, and fabric décor. For example, individuals who may have encountered bed bugs while travelling can prevent bed bugs from establishing in their home by placing luggage contents into the clothes dryer. Also, many organizations have clients that may transport bed bug-infested belongings into their facilities. Therefore, it is advisable that those belongings be heat treated in a dryer prior to storing them. Finally, some group housing facilities host individuals, such as emergency response teams, poultry house workers, or home healthcare providers, that are known to work in bed bug infested environments. In these facilities, multiple clothes dryers can be placed in an area where workers can change their clothing and heat treat them after visiting potentially infested premises.

26.6.2 Heat Chambers

Heat chambers and portable heating units can be used to treat potentially infested furniture, electronics, and other household items. Researchers at the University of Florida developed an inexpensive "do-it-yourself" heat chamber that was constructed using common materials that could be purchased at hardware stores (Pereira *et al.*, 2009). Using thermocouples, the researchers monitored the temperature within the heat chamber, which could reach the thermal death point for bed bugs within six hours. While heat chambers require more time to reach lethal temperatures than clothes dryers, more and larger items like furniture, and items that cannot be put in the dryer, like books and electronics, can be treated (Miller, 2009a).

In the USA, and on the internet, a variety of commercially-produced heat chambers are available in a wide range of shapes, sizes, and prices (Stedfast and Miller, 2015). Individuals can use these heat chambers to treat their luggage (including suitcases) upon returning from a trip. Second-hand stores can use these heat chambers to treat potentially infested donations, such as furniture, appliances, books, shoes, and toys. Shelters can use heat chambers to heat treat potentially infested bunk beds and incoming personal belongings day after day. Finally, multi-unit housing facilities can employ heat chambers to treat the belongings and furniture of incoming residents prior to move-in as part of a bed bug prevention protocol. It is important to note that while whole home or whole structure heat treatments are used to kill bed bugs that are aggregating in inaccessible locations and cannot be readily found, they are not a practical preventative method because they are often prohibitively expensive and time-consuming to employ.

26.7 Bed Bug Management Policy

Those facilities and organizations that provide beds for others should have a bed bug management policy as part of their routine risk-management processes (Doggett, 2011). The policy should include aspects such as who is responsible for pest management and its costs, staff training on how to recognize bed bugs and how to use the policy, the procedures for documenting infestations, the processes involved when an infestation is identified, methods of minimizing bed bugs, such as ongoing inspections, and any occupational health and safety requirements. Ideally such a policy should be developed in consultation with a PMP who is proficient in bed bug management, and reviewed on a regular basis to maintain currency. A well-developed bed bug management policy, which is strictly adhered to, should help minimize bed bug impacts as well as potential litigation.

References

Bell, J., Chen, Y.J. and Yeh, C.Y. (2007) Mattress encasement for preventing bed bug escapement via a zipper opening. US Patent 7,552,489 B2, filed 31 May 2007 and issued 30 June 2009.

Cooper, R. (2007) Just encase. *Pest Control Magazine*, **75** (4), 64–66, 68, 70, 72–75.

Cooper, R. (2011) Ectoparasites, Part three: Bed bugs and kissing bugs, in *Mallis Handbook of Pest Control*, 10th edn (eds S.A. Hedges and D. Moreland), Mallis Handbook Company LLC, Richfield, Ohio, pp. 587–632.

Cooper, R., Wang, C. and Singh, N. (2016a) Evaluation of a model community-wide bed bug management program in affordable housing. *Pest Management Science*, **72** (1), 45–56.

Cooper, R., Wang, C. and Singh, N. (2016b) Effects of various interventions, including mass trapping with passive pitfall traps, on low-level bed bug populations in apartments. *Journal of Economic Entomology*, **109** (2), 762–769.

Doggett, S.L. (2011) *A Bed Bug Management Policy and Procedural Guide*, Department of Medical Entomology, Westmead Hospital, Sydney.

Doggett, S.L. (2013) *A Code of Practice for the Control of Bed Bug Infestations in Australia*, 4th edn. Westmead Hospital, Sydney, Australia, Department of Medical Entomology and The Australian Environmental Pest Managers Association.

Getty, G., Moore, S., Tabuchi, R. and Lewis, V. (2012) A trial evaluation of the London Luxury bed bug mattress protector. *Pest Control Technology Magazine*, **28** (4).

Goldberg, G. (2015) Understanding essential quality characteristics of mattress and box spring bed bug encasements. *International Pest Control*, **57** (3), 156–157.

HUD (2012) *Guidelines on Bedbug Control and Prevention in Public Housing*, US Department of Housing and Urban Development, Washington, DC.

Kells, S.A. and Goblirsch, M.J. (2011) Temperature and time requirements for controlling bed bugs (*Cimex lectularius*) under commercial heat treatment conditions. *Insects*, **2** (4), 412–422.

Miller, D.M. (2009a) *Non-chemical Bed Bug Management*, Virginia Department of Agriculture and Consumer Services, Richmond, Virginia.

Miller, D.M. (2009b) Living with bed bugs. *Pest Management Professional*, **77** (7), 37, 39–41, 43.

Naylor, R.A. and Boase, C.J. (2010) Practical solutions for treating laundry infested with *Cimex lectularius* (Hemiptera: Cimicidae). *Journal of Economic Entomology*, **103** (1), 136–139.

Pereira, R.M., Koehler, P.G., Pfiester, M. and Walker, W. (2009) Lethal effects of heat and use of localized heat treatment for control of bed bug infestations. *Journal of Economic Entomology*, **102** (3), 1182–1188.

Potter, M.F., Romero, A., Haynes, K.F. and Hardebeck, E. (2007) Killing them softly: Battling bed bugs in sensitive places. *Pest Control Technology*, **35** (1), 24, 32.

Potter, M.F., Haynes, K.F. and Fredericks, J. (2015) Bed bugs across America. *PestWorld Magazine*, **November/December**, 10.

Romero, A., Potter, M.F. and Haynes, K.F. (2009) Are dusts the bed bug bullet? *Pest Management Professional*, **77** (5), 22–25.

Stedfast, M.L. and Miller, D.M. (2014) Development and evaluation of a proactive bed bug (Hemiptera: Cimicidae) suppression program for low-income multi-unit housing facilities. *Journal of Integrated Pest Management*, **5** (3), 1–7.

Stedfast, M.L. and Miller, D.M. (2015) *Tuning Up the Heat*, http://www.pctonline.com/article/pct0615-commercial-portable-heat-chambers-bed-bugs/ (accessed 20 January 2016).

Wang, C. and Cooper, R. (2011) Environmentally sound bed bug management solutions. In: *Urban Pest Management: An Environmental Perspective* (ed. P. Dhang), CABI International, Cambridge, Massachusetts, pp. 44–63.

Wang, C., Gibb, T., Bennett, G.W. and McKnight, S. (2009a) Bed bug (Heteroptera: Cimicidae) attraction to pitfall traps baited with carbon dioxide, heat, and chemical lure. *Journal of Economic Entomology*, **102** (4), 1580–1585.

Wang, C., Gibb, T. and Bennett, G.W. (2009b) Evaluation of two least toxic integrated pest management programs for managing bed bugs (Heteroptera: Cimicidae) with discussion of a bed bug intercepting device. *Journal of Medical Entomology*, **46** (3), 566–571.

Wang, C., Saltzmann, K., Chin, E., Bennett, G.W. and Gibb, T. (2010) Characteristics of *Cimex lectularius* (Hemiptera: Cimicidae) infestation and dispersal in a high-rise apartment building. *Journal of Economic Entomology*, **103** (1), 172–177.

Wang, C., Singh, N., Cooper, R., Liu, C. and Buczkowski, G. (2013) Evaluation of an insecticide dust band treatment method for controlling bed bugs. *Journal of Economic Entomology*, **106** (1), 347–352.

27

Detection and Monitoring

Richard Cooper and Changlu Wang

27.1 Importance of Detection and Monitoring

In spite of their intimate association with human hosts, the detection of bed bugs can be difficult, particularly when they are present in small numbers. It is common for occupants of a bed bug infested home to be unaware that the insects are present during the early stages of an infestation. The cryptic and nocturnal behavior of bed bugs aids them in escaping detection. Latency of bite symptoms, failure to develop symptoms, lack of bed bug awareness, and misdiagnosis of symptoms, also contribute to delayed bed bug detection.

Known but unreported bed bug activity is another form of detection failure. Residents living in multi-family housing communities may not report infestations for a variety of reasons, including embarrassment or shame, fear of losing care or in-home services, fear of being ostracized, or negative repercussions from property management, including being held financially responsible for bed bug control efforts (Pinto *et al.*, 2007). Infestations may reach severe levels in residences whose occupants insist on remediating the problem on their own, are tolerant of the bed bug activity, or have cognitive issues such that they are unaware of the infestation.

Failure to detect or report bed bug activity enables infestations to become well-established and promotes the spread of infestations to uninfested locations. This spread complicates control efforts. In countries like the USA, there also is an increased risk of litigation from individuals unknowingly exposed to bed bugs in hotels, in the workplace, movie theatres, or anywhere else where bed bug activity exists (Doggett *et al.*, 2012). For these reasons, the ability to detect low-level bed bug activity is essential in managing (that is, reducing the risk of) the further spread of bed bugs, and minimizes costs associated with their control. Early detection in hotels also protects commercial brands and limits liability exposure. Once detected, the continued monitoring of bed bug activity is important for evaluating the effectiveness of control measures. Monitoring also helps to guide inspections and treatment decisions, as well as helping to determine when an infestation has been eliminated. Various methods of detection are reviewed in this chapter. The different methods are followed by a discussion of bed bug inspection procedures.

27.2 Detection of Bed Bugs

The most common methods for detecting bed bug activity include:

- resident interviews or surveys
- visual inspection
- canine scent detection
- installation of bed bug traps.

Advances in the Biology and Management of Modern Bed Bugs, First Edition.
Edited by Stephen L. Doggett, Dini M. Miller, and Chow-Yang Lee.
© 2018 John Wiley & Sons Ltd. Published 2018 by John Wiley & Sons Ltd.

Each of these methods is discussed, as are their associated advantages, disadvantages, and limitations. Newer technologies that are in development include:

- detection using molecular analysis of DNA (Szalanski *et al.*, 2011)
- solid-phase extraction (Eom *et al.*, 2011)
- sniffer technology
- acoustic detection (Mankin *et al.*, 2010).

These novel methods have not been empirically evaluated under field conditions and therefore are not discussed in this chapter.

Throughout this chapter, unless otherwise stated, the use of the term bed bug refers to the Common bed bug, *Cimex lectularius* L. This is important because comparatively little research has been conducted on the monitoring and detection of the Tropical bed bug, *Cimex hemipterus* (F.). There is an urgent need to investigate detection tools and methods with both species, as virtually all field research to date has been conducted solely on *C. lectularius*.

27.2.1 Resident Interviews or Surveys

Asking simple questions such as, "Are you aware of bed bug activity in your home/apartment?" or "Do you suspect bed bug activity in your home/apartment?" or showing a picture of a bed bug and asking "have you seen these insects?" is an efficient and economical method to identify bed bug activity that has not been reported. In addition to identifying infestations, important information also can be obtained during questioning, such as when the bugs were first detected, how the bugs may have been introduced, where the occupants spend the majority of their time sitting or resting, and corrective actions that have been taken. Answers to these types of questions can be helpful in the inspection and in planning the control strategy.

Surveying residents is only effective if the individual(s) being interviewed are aware of the activity, have some knowledge of what bed bugs are, and are willing to share the information with the person conducting the survey. For example, when interviewed, 50% of residents living in apartments with confirmed bed bug activity indicated they were unaware of bed bug activity at the time the infestations were confirmed by researchers (Wang *et al.*, 2010). In another study, when interviewed, 62% of residents living in apartments with bed bug activity were unaware that their apartments were infested (Cooper *et al.*, 2015a).

Verbal interviews also were found to be somewhat unreliable for determining if infestations had been eliminated after treatments (Wang *et al.*, 2009b; Cooper *et al.*, 2015a). In the study by Wang *et al.* (2009b), half of the residents living in eight apartments treated for bed bugs indicated that they believed their apartments were bed-bug-free, when bed bug activity was still being detected by the researchers. Similarly, Cooper *et al.* (2015a) reported that among 66 residents whose apartments were being treated for existing bed bug activity, 76% believed bed bugs were no longer present. It should be noted that the studies by Wang *et al.* (2009b, 2010) and Cooper *et al.* (2015a) were conducted in communities housing the elderly, a demographic group less likely to develop bite symptoms (Potter *et al.*, 2010), which may partially explain the residents' failure to recognize the bed bug activity in their apartments. Resident interviews may be more effective in family-style housing (housing not specific to elderly and disabled residents) where occupants are more likely to react to bites. Thus, while resident surveys can provide useful information, they may be unreliable and should not be used as the sole method of detection.

27.2.2 Visual Inspection

Visual inspection is the most common method used by pest management professionals (PMPs) for detecting bed bugs (Potter *et al.*, 2015). An advantage of visual inspection is that it provides immediate results and aside from a good flashlight and magnifying lens, no other equipment is required. The presence of live or dead bed bugs, hatched eggs, exuviae, or fecal deposits are all indicators of bed bug activity and can often be easily identified by an experienced person.

However, this method is limited by experience and thoroughness of the inspector and can be a labor-intensive process. More importantly, because of the cryptic, secretive, and nocturnal behavior of bed bugs, it becomes more difficult to determine whether live bed bugs are still present after treatment (Pinto *et al.*, 2007; Vaidyanathan and Feldlaufer, 2013; Cooper *et al.*, 2014, 2015b). The treatment status of infestations also can affect the bed bug distribution. Prior to treatment, bed bugs are more likely to be found on or near host sleeping and resting areas. However, following treatment of an infestation, fewer bed bugs are associated with beds and other host resting areas (Cooper *et al.*, 2015b), making them more difficult to find through visual inspection.

Most inspections focus on the host sleeping and resting areas, where bed bug activity is most easily detected and usually most concentrated prior to treatment. However, depending upon the purpose of the inspection, there is no limit to how much of the structure and its contents may require inspection. Although seemingly obvious, the more time invested and the broader the search, the more likely it is for bed bugs to be detected. For this reason, it is important to understand the goals of the inspection, which in turn will determine the degree of labor investment required. An inspection being conducted simply to detect the presence of bed bugs is different than one intended to determine the extent of an infestation and identify all areas with bed bug activity.

A detection-only inspection for a well-established infestation can often be completed in a few minutes with relative ease, while a low-level infestation may require removal of mattresses, mattress bases (box springs), and bed heads (headboard). If no activity is found, it may be necessary to disassemble bed frames for inspection, and thoroughly inspect upholstered furniture, end tables, and night stands, and then move onto less predictable areas if activity is still not detected. Even very thorough inspections may fail to detect bed bugs present in small numbers (Pinto *et al.*, 2007). An inspection to determine the extent of infestation will usually require a very detailed inspection and more time. In a study by Wang *et al.* (2010), only 52% of 101 apartments with bed bug activity were detected through detailed visual inspections. The inspections took between 42 and 90 min per apartment to complete and included inspection of mattresses, mattress bases, bed frames, bed heads, footboards, all upholstered furniture, curtains, the room perimeter, items stored under the bed and in closets, and wheelchairs (when present). The mean number of bed bugs detected during these inspections was 64.5 ± 15.8 per apartment.

Due to the labor-intensive nature of visual inspections, and the requirement of finding live bugs or eggs, visual inspections are best for settings where bed bug activity is predictably associated with sleeping and resting areas, and where detection of low-level bed bug activity is not required. For a detailed explanation of where and how to inspect for bed bugs, refer to Pinto *et al.* (2007).

27.2.3 Canine Scent Detection

The use of canines for bed bug detection has become increasingly popular in the USA. In a 2015 survey of 236 pest management companies in the USA, 42% reported using bed bug sniffing dogs to detect bed bugs (Potter *et al.*, 2015), compared to 17% in 2010 (Potter *et al.*, 2011). Canine scent detection overcomes the limitations of a visual inspection because the dog is not relying on sight but rather uses its keen olfactory ability to detect the scent of live bed bugs. A properly trained dog can discriminate between the scent of live bed bugs or eggs from other evidence such as bed bug feces, exuviae, carcasses, or hatched eggs, which may not be representative of currently existing bed bug activity (Cooper, 2007; Pinto *et al.*, 2007; Pfiester *et al.*, 2008). Canine inspection can provide immediate results and is often more efficient than visual inspection. It is also especially well-suited for large-scale facilities and inspections in non-traditional settings such as large office buildings, retail stores, and theatres (Cooper, 2007; Pinto *et al.*, 2007; Wang and Cooper, 2011). A major disadvantage of canine scent detection is that it introduces the possibility of false positive results. False positives occur when a dog signals that bed bugs are present when in fact they are not (Cooper, 2007; Pinto *et al.*, 2007; Wang and Cooper, 2011). Additionally, because scent detection is a behavior-based method, a great degree of variability in accuracy exists from one scent detection team to the next (Cooper *et al.*, 2014).

The accuracy of a canine inspection relies upon the individual abilities of the dog and its handler, as well as how well the two work as a team. Pfiester *et al.* (2008) demonstrated the ability of trained dogs to detect bed bugs with a high degree of accuracy under controlled conditions, involving a highly trained professional handler. In this study, dogs were 98% accurate in locating live bed bugs hidden in hotel rooms, with zero false positive alerts. In a separate experiment using a training apparatus under controlled conditions, the dogs were able to discriminate live bed bugs and viable eggs from dead bed bugs, exuviae, and feces with 95% accuracy, while falsely alerting 3% of the time on bed bug feces (Pfiester *et al.*, 2008). While this 2008 study demonstrates the ability of dogs to detect the scent of bed bugs, it was not conducted in a real-world situation where other smells could interfere or distract the dogs from a positive finding, or induce a false alert. In a study by Cooper *et al.* (2014), canine scent detection teams in naturally infested apartments were much less accurate. The mean (min, max) detection and false positive rates, during sixteen inspections conducted by 11 different teams, were 44% (range 10–100%) and 14% (range 0–57%), respectively. Teams with high detection rates tended to have high false positive rates, while those with low detection rates had low false alert rates. Additionally, the probability of a bed bug infestation being detected was not associated with the infestation level. The four teams evaluated on multiple days performed inconsistently from one day to the next (Cooper *et al.*, 2014). The authors suggested that the accuracy of a canine scent detection teams under controlled conditions did not reflect how the team will perform during field inspections. Some of the factors affecting the accuracy of teams in the field are environmental conditions, fatigue of the dog or handler, handler bias, and context shift effects between maintenance training and real-world field inspections (Cooper *et al.*, 2014).

In an effort to improve the consistency and quality of canine scent detection teams, organizations such as the International Association of Canine Scent Inspectors, the World Detector Organization, and the National Entomology Scent Detecting Canine Association, provide certification of dog/handler teams. Additionally, best practices for canine inspection of bed bugs have been developed by the National Pest Management Association (NPMA) in the USA. Interestingly, in the field study by Cooper *et al.* (2014), the accuracy of canine scent (detection and false positive rates) inspections in apartments was no better for teams that had been certified when compared with those without certification. There is a clear need to establish a scientifically sound canine evaluation standard.

A trained bed bug detection dog can typically be purchased for USD 8000–15 000. Several days of handler training and certification may be included in the initial purchase cost. Daily maintenance training of the dog, requires a constant supply of live bed bugs and viable eggs. This supply requires handlers to either maintain a bed bug colony of their own, or to routinely collect bed bugs from field infestations, or purchase bed bugs from an individual or institution that maintains them. Pfiester *et al.* (2008) demonstrated that a pseudoscent prepared from pentane extraction was detected by 100% of the trained dogs tested in an artificial environment, and could be used for detector dog training. However, tests employing pseudoscents have not been undertaken in a real-world situation. The high costs associated with support of a detection team has led many companies to subcontract canine scent detection services rather than providing the service themselves.

In spite of the limitations associated with canine scent detection, it is still a valuable method for large-scale inspections, particularly in environments where the location of bed bugs is less predictable and difficult to inspect visually, such as large office buildings, libraries, retail stores, schools, or theatres (Cooper, 2007; Pinto *et al.*, 2007; Wang and Cooper, 2011). However, due to the concerns about false positive alerts, handlers should conduct visual inspection to verify the presence of bed bugs. If bed bug activity cannot be confirmed visually, additional monitoring to confirm activity should be considered before any treatment is applied.

27.2.4 Bed Bug Monitors

Installation of bed bug monitors is a highly effective method for the detection of bed bugs (Wang *et al.*, 2010; Cooper *et al.*, 2014, 2015a; Wang *et al.*, 2016). Unlike other detection methods, at least two visits are required, one to place devices and another to inspect them. While this may be viewed as a disadvantage, effective monitors detect a greater percentage of infestations than resident interviews (Wang *et al.*, 2010; Cooper *et al.*,

2015a; Wang *et al.*, 2016), visual inspection (Wang *et al.*, 2010, 2016; Boase *et al.*, 2012; Cooper *et al.*, 2015a; Wang *et al.*, 2016), or canine scent detection (Cooper *et al.*, 2014), because monitoring is not dependent upon detection at a single point in time. Monitoring also has been shown to be effective for detecting low-level bed bug activity, which is often missed during visual inspections (Wang *et al.*, 2010; Boase *et al.*, 2012; Cooper *et al.*, 2014, 2015a; Wang *et al.*, 2016).

The effectiveness of a monitor is influenced by how likely a bed bug is to orient to, and then commit to, the device. Orientation and commitment of a bed bug to an object is a subject that is poorly understood, and involves a complex set of behaviors including responses to visual, physical, and chemical cues (Singh *et al.*, 2015a; McNeill *et al.*, 2016). Other behavioral factors include feeding status (Weeks *et al.*, 2011), gender (Weeks *et al.*, 2011; Ulrich *et al.*, 2016), and species (Kim *et al.*, 2017). In a laboratory bioassay, Singh *et al.* (2015a), demonstrated that bed bugs visually orient towards vertical objects, significantly increasing bed bug trap catch in pitfall-style interceptors by adding a vertical rod. In the same study, black traps captured significantly more bed bugs than white traps. As our understanding of bed bug orientation continues to improve, so too will the effectiveness and utility of affordable and readily available bed bug monitoring devices.

In general, monitors can be divided into two types:

1) passive monitors (without a lure)
2) active monitors (with one or more lures to attract bed bugs to the device).

However, most passive devices can be made into an active device by supplementing them with carbon dioxide or a commercially purchased bed bug lure. It is also necessary to make a distinction between devices that do and do not trap bed bugs. Those that do not trap bed bugs are purely a detection tool, while those that trap bed bugs have other functions beyond detection. For example, in addition to detecting bed bugs, pitfall-style traps placed under the legs of beds serve as a barrier that limits bed bug access to the bed (Wang *et al.*, 2009b). Other monitors that trap bed bugs without injury can be used to collect live bed bugs for research purposes. There are many monitors/traps in the marketplace and several examples are shown in Figure 27.1.

While many bed bug monitors are commercially available (Doggett and Russell, 2009; Vaidyanathan and Feldlaufer, 2013), most have never been tested by university researchers for efficacy under field conditions (Doggett *et al.*, 2012; Vaidyanathan and Feldlaufer, 2013). Among those tested, some that were effective, such as CDC3000 (Cimex Science LLC, Portland, OR) and Verifi (FMC Corporation, Philadelphia, PA), are no longer commercially available due to the high purchase and/or maintenance costs. Commercially available passive monitors that have been field tested and found to be effective, include ClimbUp (Wang *et al.*, 2010; Wang *et al.*, 2011; Potter *et al.*, 2013; Cooper *et al.*, 2014, 2015a,b,c) and BlackOut (Potter *et al.*, 2013). In addition, commercially available active monitors that have proven effective include NightWatch (Wang *et al.*, 2011) and SenSci Activ Volcano (M. Merchant, Texas A&M, Dallas, USA, unpublished data).

The need for testing under field conditions was illustrated by Wang and Cooper (2011), who evaluated the "Bed Bug Beacon" under laboratory and field conditions. The monitor produces and releases carbon dioxide at a low rate (2–11 ml/min over 48 h), trapped 80% of bed bugs released in 55.5 × 43.5-cm laboratory arenas. However, this monitor was ineffective under field conditions. It is also important to note that all of the above monitoring devices are pitfall style traps. This style of trap, while effective for *C. lectularius*, may not be as effective for *C. hemipterus*. Differences in leg morphology of the two species enable *C. hemipterus* to more easily to escape pitfall traps than *C. lectularius* (Kim *et al.*, 2017).

A 2015 survey of US PMPs conducted by Potter *et al.* (2015) revealed that 71% of companies surveyed used monitors to aid in the detection of bed bugs. However, no one type of monitor was used by more than 56% of the companies. Survey responses indicated that passive pitfall style traps, sticky traps, active traps (baited with carbon dioxide, heat or chemical lures), and active traps (baited with chemical lures only) were used by 56, 44, 42, and 35% of those surveyed. The survey results suggest that none of the monitoring devices have been widely accepted by PMPs, with nearly a third relying on visual inspection alone. This reluctance to use monitors could be due to uncertainty regarding the efficacy of available monitors or how to effectively incorporate them into their existing bed bug services.

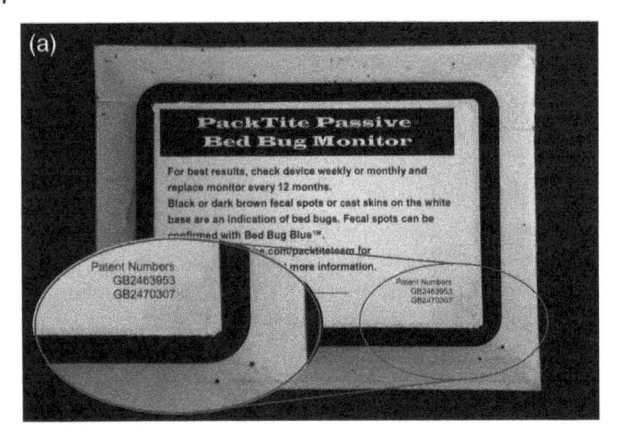

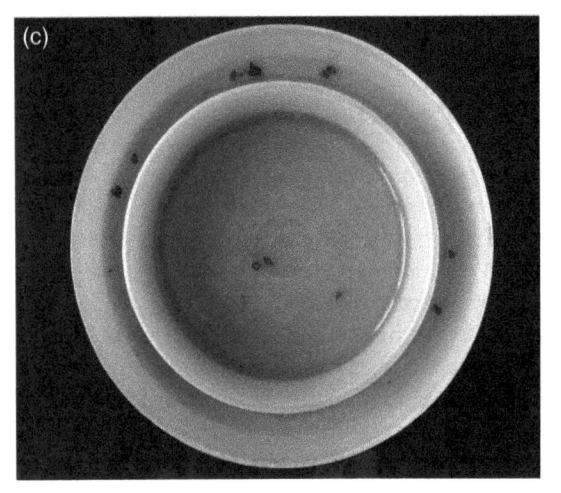

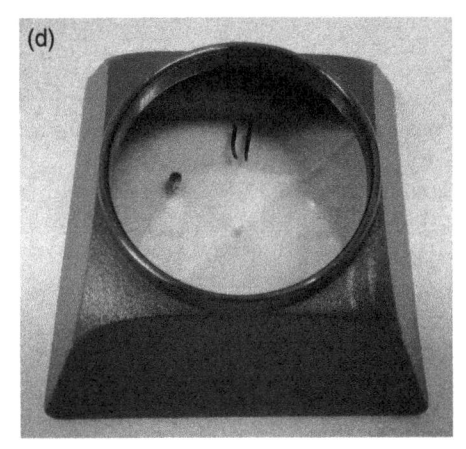

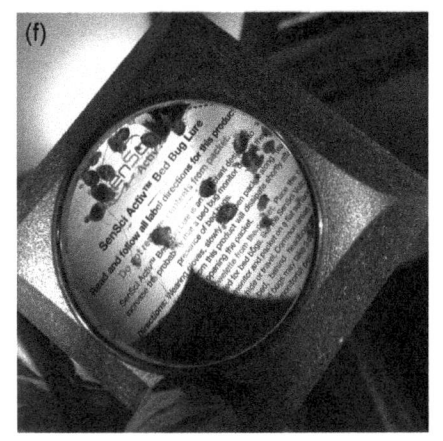

Figure 27.1 Examples of passive and active monitors currently in the marketplace. (a) passive harborage-style monitor (brand name: PackTite™ Passive Bed Bug Monitor, Bed Bugs Limited, London, UK, image courtesy David Cain, Bed Bugs Limited, London); (b) passive pitfall-style monitor designed to be placed under legs of beds (brand name: BlackOut® BedBug Central, Lawrenceville, NJ); (c) passive pitfall-style monitors designed to be placed under legs of beds (brand name: ClimbUp® Insect Interceptor, Susan McKnight, Inc., Memphis, TN); (d) passive monitor designed to be placed near or away from beds (brand name: SenSci Volcano®, BedBug Central, Lawrenceville, NJ); (e) active monitors baited with carbon dioxide, heat, and chemical lure (brand name: NightWatch® Bed Bug Monitor, BioSensory Inc., Putnam, CT); (f) active monitor baited with chemical lure (brand name: SenSci Volcano® with SenSci Activ® Bed Bug Lure, BedBug Central, Lawrenceville, NJ, image courtesy of Susannah Reese, Cornell University, NY).

Passive Monitors

Passive monitors include insect sticky traps, harborage-style traps, and pitfall-style traps. This group contains some of the most and least effective monitors. For example, pitfall-style interceptors have been shown to be the most reliable detection tool for low-level bed bug activity when left in place for 1–2 weeks (Wang et al., 2010; Cooper et al., 2014, 2015a,b; Wang et al., 2016). On the other hand, unbaited insect sticky traps are considered one of the least reliable methods of detection (Pinto et al., 2007; Doggett, 2011; Wang and Cooper, 2011). Hence, they are not even mentioned in review articles that discuss bed bug detection (Doggett et al., 2012; Koganemaru and Miller, 2013; Vaidyanathan and Feldlaufer, 2013). Still, 44% of US (Potter et al., 2015) and 18.5% of Australian (Doggett, 2016) pest control companies surveyed indicated they use sticky traps to detect bed bugs. This is probably due to their being cheap (≤ USD 0.10 each), ease of placement, and ease of maintenance. While bed bugs are sometimes captured on insect sticky traps, it is not recommended that they be relied upon for the detection of bed bugs.

Several harborage-style traps also are commercially available. These traps attempt to exploit the thigmotactic behavior of bed bugs, and tend to be discrete and inexpensive (≤ USD 1 each) but some cost up to USD 21 each. The less expensive units make it affordable to place many out, increasing the probability of detecting bed bugs. Some harborage-style monitors incorporate a sticky adhesive to capture the bed bugs, while others simply offer a suitable environment for bed bugs to seek refuge, but allow the bugs to freely enter and leave the monitor. For this type of monitor, detection is dependent upon seeing the actual bug or fecal deposits left behind by bed bugs that visited the monitor. To date, none of the harborage-style traps have been shown to be effective as an early detection tool in independent scientific trials, with few having been independently tested (Doggett et al., 2012).

Passive pitfall-style monitors, hereafter referred to as "interceptors", work by trapping bed bugs (nymphs and adults of both sexes) in a well from which it is difficult to escape. They have been shown to be highly effective for early detection of bed bugs (Wang et al., 2010, 2011; Cooper et al., 2015a). Some interceptors require a lubricant to prevent bed bugs from escaping and must be relubricated on a regular basis (every two weeks) to maintain their effectiveness. Under laboratory conditions, small (1st instar), large (3rd–5th), and adult *C. lectularius* (male and female) were evaluated for their ability to escape from ClimbUp interceptors that were aged for 14 days under field conditions (Cooper et al., 2015c). None of the large nymphs or adults escaped, but an average of 20% of first instar nymphs escaped from aged/unmaintained monitors after having been trapped for between 10–14 days. Other interceptors have been designed to prevent bed bugs from escaping without regular maintenance, but, with enough dust build up, escape of bed bugs from most pitfall style traps is likely. Typical interceptor cost is USD 3–4 each, making them economically affordable for large-scale use (Cooper et al., 2015a; Wang et al., 2016). While many interceptors exist, few have been independently tested in scientific studies. Of those that have been investigated, interceptors have been shown to be the most effective for detection of low-level infestations when placed under the legs of beds and upholstered furniture for 7–14 days (Wang et al., 2011), according to resident interviews and visual inspections (Wang et al., 2010, Cooper et al., 2015a, Wang et al., 2016).

Interceptors were originally intended to be placed under the legs of beds and upholstered furniture (Wang et al., 2009a), using the sleeping host as a lure. When placed beneath bed legs, the device serves not only as a detection device but also as a barrier, limiting the access of bed bugs to the sleeping host (assuming no bridges exist between the floor, wall, and bed). However, passive pitfall-style traps have also been shown to be effective in detecting bed bug activity when placed along the base of walls in locations away from host sleeping and resting areas (Wang et al., 2010; Potter et al., 2013; Cooper et al., 2014, 2015a,c). Employing interceptors in this manner has been used to study behavioral ecology, by monitoring bed bug movement within and between apartments (Cooper et al., 2015c). Interceptors have been used as a tool for guiding inspections and treatment decisions, and helping to determine when an infestation has been eliminated (Cooper et al., 2015a,b). The utility of interceptors away from beds has resulted in the development of the Volcano, which is intended to be placed next to, rather than under furniture legs, as well as away from sleeping and resting areas. This monitor can be used with or without a lure. Baited Volcano monitors captured a greater number of bed bugs than either unbaited Volcano or ClimbUp devices, but the difference in trap catch over a four-week monitoring period was not significantly different between the baited and two unbaited monitors (M. Merchant, Texas A&M University, Dallas, USA, unpublished data).

Active Monitors

Active monitors using a combination of carbon dioxide, heat, and chemical lure, to attract host-seeking bed bugs were among the first monitors to become commercialized following the bed bug resurgence in the late 1990s (Anderson *et al.*, 2009; Wang *et al.*, 2009a). Because baited traps contain attractants, they can be used in occupied or vacant dwellings (Wang and Cooper, 2011). Monitors baited with carbon dioxide have the advantage of attracting host-seeking bed bugs from a distance of up to 1.5 m, while heat or chemical lures have a much shorter (<5 cm) attractive range (Haynes *et al.*, 2008; Weeks, 2011). Active monitors that use a combination of carbon dioxide, heat, and chemical lure, were effective when tested under field conditions but were expensive and difficult to operate (Wang *et al.*, 2011). More recently, there has been an interest in exploiting the semiochemicals involved in bed bug aggregation and arrestment behavior. Researchers have been investigating ways to synthesize these semiochemicals in the laboratory to use as lures to attract bed bugs into traps (Weeks *et al.*, 2011; Boase *et al.*, 2012; Gries *et al.*, 2015; Vaidyanathan and Feldlaufer, 2013). However, these traps containing aggregation pheromone-type lures have yet to appear on the market.

Among the bed bug attractants, carbon dioxide has been shown to be the most important. Traps baited with carbon dioxide alone captured significantly more bugs than the same trap baited with heat, or a combination of heat and carbon dioxide (Anderson *et al.*, 2009; Wang *et al.*, 2009a; Singh *et al.*, 2012). There is also a positive relationship between carbon dioxide release rate and bed bug trap catch under field conditions. Traps releasing carbon dioxide at a rate of 100 ml/min captured significantly fewer bed bugs than traps releasing 200 or 400 ml/min. Traps releasing carbon dioxide at a rate of 400 ml/min captured significantly fewer bed bugs than those releasing 800 ml/min in occupied apartments (Singh *et al.*, 2013). Wang *et al.* (2010) demonstrated the effectiveness of two active monitors (NightWatch and CDC3000) for the detection of bed bugs in infested apartments. While effective, both monitors were expensive (USD 400–1000), requiring replenishment of the carbon dioxide source (daily to weekly), and electricity to operate. The carbon dioxide release rates from a number of more affordable (USD 20–50) commercial monitors were found to be too low to be competitive with a human host (Singh *et al.*, 2015b). The ability to generate carbon dioxide release rates high enough to be effective, while remaining economically affordable, has been an obstacle that has led to either the loss of effective monitors from the marketplace or the sale of monitors that are either ineffective or have a limited effective range (Singh *et al.*, 2015b).

Homemade traps baited with either dry ice (Wang *et al.*, 2009a, 2010; Singh *et al.*, 2015b) or a large volume (>3 l) of a sugar/yeast/water mixture (Singh *et al.*, 2015b) produce a carbon dioxide release rate similar or higher than that produced by an adult human at rest (250 ml/min). These traps have proven effective for trapping bed bugs in large numbers under both laboratory and field conditions in occupied apartments (Wang *et al.*, 2009a, 2010; Singh *et al.*, 2013, 2015b). However, inherent hazards associated with dry ice (and its variable availability) and the cumbersome nature of the sugar/yeast/water trap, make these traps impractical for commercial use in the present form. However, they could be useful within the research community in the collection of large numbers of bed bugs from infested dwellings. These traps may also have utility in vacant dwellings, where the potential hazards to occupants are not a concern.

Active monitors containing a chemical lure for host-seeking bed bugs also are available. Traps baited with a chemical lure (SenSci Activ Bed Bug Lure), captured significantly more bed bugs than unbaited traps (Singh *et al.*, 2012, 2015). The same lure, when added to a pitfall-style trap, increased the trap catch by 7.2 times compared to unbaited pitfall traps in a field study, demonstrating its effectiveness as a standalone lure (Singh *et al.*, 2015). Commercial traps using this lure, while lacking the long-distance attraction of carbon dioxide, provide an economically affordable active monitor (< USD 6).

The development of lures to attract aggregating bed bugs has gained the attention of scientists and is an area of scientific investigation (Weeks *et al.*, 2011; Gries *et al.*, 2015; Ulrich *et al.*, 2016; see also Chapter 17).

Siljander *et al.* (2008) demonstrated that two known volatile bed bug defensive secretions, (E)-2-hexanal and (E)-2-octenal, are also involved in the aggregation behavior of bed bugs, at lower concentrations than used for defense. In a subsequent study, the bed bug aggregation pheromone was identified as consisting of five volatile components, including (E)-2-hexanal, (E)-2-octenal, and another less volatile histamine that causes arrestment behavior upon contact (Gries *et al.*, 2015). More recently, Ulrich *et al.* (2016) demonstrated that (*E*)-2-hexanal and (*E*)-2-octenal alone attracted both adult males and female bed bugs, but noted that adult male and female bed bugs responded differently to the two aldehydes. At 0.04 µg of a 1:1 mixture, males spent more time close to, or on, aldehyde-treated disks, while females spent more time near but not necessarily at the source of the attractant. The authors suggested the different behavioral responses of adult male and female bed bugs to an aggregation lure could have implications for designs that would maximize the effectiveness of the trap. Additionally, a slow-release formulation would likely be necessary, as the effectiveness of the aldehyde volatiles is short-lived (Ulrich *et al.*, 2016).

While unlikely to attract bed bugs from a long distance (Crawley *et al.*, 2015), traps using an aggregation lure should attract bed bugs of all stages, sexes, and feeding status (Weeks *et al.*, 2011). To date, only one trap using an aggregation-based lure has reached the commercial marketplace. Although not yet tested by university researchers, in a study conducted by an independently contracted laboratory, traps baited with the lure captured both fed and unfed bed bugs and was more effective at detecting bed bugs in infested apartments than visual inspection alone (Boase *et al.*, 2012).

27.3 Field Comparison of Detection Methods

In the peer-reviewed literature, field comparison of canine scent detection and other detection methods is limited to a single study (Cooper *et al.*, 2014),. The same is true for comparison of commercially available active monitors and other detection methods (Wang *et al.*, 2011). On the other hand, there are many comparisons of passive interceptors with resident interviews and visual inspection, with interceptors consistently providing the highest level of detection (Wang *et al.*, 2010, 2011; Cooper *et al.*, 2014, 2015a; Wang *et al.*, 2016).

Canine scent detection was evaluated by Cooper *et al.* (2014), in affordable housing communities. In three separate experiments, passive interceptor traps placed throughout apartments both at and away from beds and furniture for 7–14 days detected more apartments with bed bug activity than inspections by scent detection teams. In three separate experiments, 67 out of 276 apartments were identified as having bed bug infestations. Individual canine scent detection teams inspected between 20 and 106 of the apartments, with varying degrees of accuracy. The mean detection rate among 11 detection teams was 44%, with only two teams detecting more than 50% (75, 88%) of the infested apartments. In addition to those detected, dogs falsely alerted an average of 14% (range 0–57%) of the time. In comparison, 93% of the apartments with bed bug activity were detected by passive pitfall-style interceptors.

Visual inspection, passive interceptors, and two different active monitors were evaluated in apartments by Wang *et al.* (2010). Among 15 lightly infested apartments (≤10 bed bugs detected), the percentage of infested apartments detected by thorough visual inspection, passive pitfall-style interceptors placed under bed legs for 7 days, and active monitors (releasing carbon dioxide, heat and chemical lure) placed next to the head of the bed for 1 day, was 50, 70, and 10–60% respectively. The study demonstrated the effectiveness of passive pitfall-style interceptors for the detection of low-level bed bug activity. Increasing the duration that interceptors are present from 7 days to 14 days can increase detection of low-level infestations to >90% (Cooper *et al.*, 2014).

Passive interceptors were also the most effective method in studies comparing resident interviews, visual inspection, and placement of monitors under bed and furniture legs (Wang *et al.*, 2010). Cooper *et al.* (2015a) identified 55 infested apartments in a 358-unit apartment community housing elderly and disabled residents; only four of the 55 apartments with bed bug activity had been reported to property management. Among the 51 unreported infestations, 71% of residents, when interviewed, said they were unaware of any bed bug activity. Placement of passive interceptors under the legs of beds and sofas for 14 days resulted in detection of 95% of the infested apartments, while visual inspections only identified four apartments missed by monitors. In another study of 2372 apartments in four different housing authorities, Wang *et al.* (2016) compared the detection efficacy of resident interviews, with cursory visual inspections (visible corners of mattresses, mattress bases, and upholstered furniture), and the placement of passive pitfall-style interceptors under beds and upholstered furniture for 10–17 days in apartments where live bugs were visually seen or bed bug evidence was observed. A total of 291 apartments were identified as having bed bug activity. Interestingly, 72% of the infestations were discovered by visual inspection, while 89% were identified by monitoring. Only 47% of the residents interviewed in 193 apartments (291 apartments in total) said they were aware of the bed bug activity.

When selecting a monitor, there are many factors to consider. Some devices cost under USD 1 to purchase, while others can cost several hundred dollars per device. The cost to maintain devices also varies greatly depending upon how often maintenance is required, the amount of physical labor and time required for maintenance and, for active traps, the cost and frequency of replacing lures (Pinto *et al.*, 2007). In some situations, more than one type of device may be necessary to achieve all of the objectives of a detection program.

27.4 Bed Bug Inspections

Inspections play a pivotal role in bed bug management. Inspections are necessary for:

- detection of bed bug activity
- guiding future inspections and treatment decisions
- determining effectiveness of the treatments
- declaring when infestations have been eliminated.

Proactive inspections also play an important role in managing the spread of bed bugs within multi-occupancy dwellings, and from those dwellings into the community, by identifying infestations that are unreported. It is important to recognize that no single method is completely reliable for detecting all bed bug infestations. Each method has advantages, disadvantages, limitations, and situations for which it is best suited. Therefore, the correct inspection tools and methods will vary depending upon the type of environment being inspected, the goals and objectives of the inspection, and the available budget. Understanding the correct combination of methods for a given situation is essential for achieving a high degree of accuracy in an economic manner.

Individuals conducting inspections must be equipped with a good flashlight, magnifying hand lens, inspection mirror, tools for dissembling items, notepad for recording findings, forceps, containers, and labels for collecting specimens. Digital cameras and smart phones also are useful for documentation. The inspector must be trained to recognize physical evidence of bed bugs, including fecal deposits, eggs, and cast skins. It is important that specimens be properly assigned to the correct species and not confused with the other insect pests that are commonly misidentified as bed bugs, such as psocids, dermestid larvae, spider beetles, and immature cockroaches.

Research regarding the inspection and detection of bed bugs is limited to studies conducted in multi-family housing communities. Some of the concepts from these studies can be applied in commercial facilities, such as hotels, motels, hostels, and hospitals.

27.4.1 Detection

Inspections for the detection of bed bugs can be performed reactively in response to a report of suspected activity, or proactively to detect unreported bed bug activity. Proactive inspections are recommended when the risk of infestation is high based on the percentage of apartment units that have been, or are currently infested. For example, in low-income communities, bed bugs often spread rapidly, leading to high infestation rates (Wang *et al.*, 2010; Doggett *et al.*, 2012; Cooper *et al.*, 2015a), which can be economically crippling (Wang *et al.*, 2010; Doggett *et al.*, 2012; Stedfast and Miller, 2014; Cooper *et al.*, 2015a). However, infestation rates can vary widely from one housing community to the next, and even between buildings within a single housing community (Wong et. al. 2013; Wang *et al.* 2016). In low-income housing communities, the number of apartments with bed bug activity is likely to be 3–4 times greater than that reported to property management (Cooper *et al.*, 2015a; Wang *et al.*, 2016). The greatest infestation rates are often found in buildings housing elderly and disabled residents, who are less likely to recognize and report bed bug activity (Wang *et al.*, 2010; Stedfast and Miller, 2014; Cooper *et al.*, 2015a).

Proactive inspections
Building-wide inspections can be conducted initially to identify infestations that have not been reported. Once all of the infestations have been identified and treated, proactive inspections should be conducted periodically (annually or biannually, or more often depending on the frequency of new infestations) to detect newly introduced infestations or recurring infestations in apartments where bed bugs were mistakenly believed to have been eliminated. Following an initial building-wide inspection, Cooper *et al.* (2015a) also recommended proactive inspection of apartments belonging to new residents moving into the community.

Interceptors placed under the legs of beds and upholstered furniture for 14 days have been shown to be more effective than visual inspections and resident interviews combined (Cooper *et al.*, 2015a; Wang *et al.*, 2016). In a study by Cooper *et al.* (2015a), interceptor traps placed under the legs of beds and upholstered furniture for 14 days in 358 apartments took approximately 7 min per apartment to place. These interceptors detected 52 of 55 apartments with bed bug activity that had not been reported to property management. Visual inspection of beds and furniture in apartments with no bed bug activity in interceptor traps took 16 min per apartment to complete, and resulted in the detection of three additional apartments with bed bug activity. The authors concluded that the cost associated with detailed visual inspection was not justified given the low percent of detection compared to interceptors. Wang *et al.* (2016) modified the approach taken by Cooper *et al.* (2015a) in the inspection of 2372 low income apartments. Rather than placing interceptors in all apartments, they conducted brief inspections of the visible areas of mattresses, mattress bases, and upper surfaces of upholstered furniture to quickly identify apartments with activity. Interceptors were only placed in in apartments treated within the past 12 months, and units where signs of bed bug infestation were observed but no live activity detected. Using this alternative approach, Wang *et al.* (2016) placed monitors in 20% of the apartments. This reduced material costs by 80%, and inspections in apartments were completed in less time than with the method used by Cooper *et al.* (2015a). The method used by Cooper *et al.* (2015a) is a more sensitive method for detection of low-level infestations but, due to the high cost, it may not be economically practical for large-scale apartment complexes, particularly in communities where infestation rates are low. Conversely, the method used by Wang *et al.* (2016) will identify most infestations in an economically affordable manner for all types of housing community.

The proactive approach discussed for inspecting multifamily housing also can be applied to other high-risk residential environments, such as group homes and shelters. However, it is impractical in places like hotels, motels, and hostels, where inspection methods must be more discrete. Although infestation rates are typically low in commercial lodging facilities, the constant threat of bed bugs being introduced exists. The costs associated with bed bug activity being identified by paying guests is also potentially high (Doggett *et al.*, 2012). At a minimum, housing and maintenance staff should be trained in how to inspect and recognize bed bug activity. However, low-level bed bug infestations are likely to be restricted to areas that are not being disturbed, such as behind wall-mounted bed heads, under mattress bases, or in bed frames, and thus are unlikely to be detected during daily housekeeping procedures. Additional measures are therefore necessary to detect low-level infestations. Examples include periodic detailed visual inspection of all guest rooms by trained staff or PMPs, installation of bed bug monitors, or canine scent inspections. The cost of visual inspection and use of bed bug monitors may be economically practical in a small hotel, while periodic canine scent inspections may be more economically practical in larger hotels. As with any canine inspection, positive alerts should be confirmed through visual inspection and monitoring, if necessary. A thorough discussion of inspection procedures for lodging facilities is provided by Pinto *et al.* (2007) and Doggett (2011).

Reactive inspections

In spite of the importance of proactive inspections in housing communities, most housing communities wait for infestations to be reported to property management before conducting an inspection (Wong *et al.*, 2013). Reactive inspections performed to confirm reported bed bug activity should not be limited to the apartment (or hotel guest room) where bed bug activity is suspected, but should also extend to the neighboring units, which may have become infested from the unit in question or could be the source of bed bugs in the suspected unit. At a minimum, neighboring-unit inspections should include all units that share a common wall, floor, or ceiling with the suspected unit. However, Cooper *et al.* (2015c) demonstrated that bed bugs also actively migrate from infested apartments to apartments across the hall and suggested these also be included in the scope of neighboring-unit inspections.

Reactive inspections to identify bed bug activity should not be limited to visual inspection due to the poor reliability of this method for detection of low-level infestations. Still, among 26 affordable housing authorities surveyed in Virginia, USA, 65% indicated that they relied upon visual inspection for the detection of bed bugs (Wong *et al.*, 2013). During reactive inspections, sleeping and resting areas also should be monitored when bed bugs cannot be confirmed by visual inspection. The sensitivity of detection can be increased by adding additional monitors away from sleeping and resting areas (Cooper *et al.*, 2014, 2015a,b).

27.4.2 Inspections to Guide and Evaluate Treatment

Following confirmation of bed bug presence, an accurate assessment of the infestation is necessary to provide proper treatment recommendations and requests for client cooperation. Continued assessment of bed bug activity following an initial treatment is necessary to guide treatment decisions. Visual inspections are most useful for assessing an infestation prior to treatment, when bed bugs are most likely to be found in predictable locations. However, following treatment of the infestation, the likelihood of continuing to find surviving bed bugs in these same areas decreases (Cooper *et al.*, 2015b). Placement of effective monitors in the corners of bedrooms and the living room, as well as placement in closets, bathrooms, and the kitchen will increase likelihood of detecting any surviving bed bugs after treatment. (Cooper *et al.*, 2015b). Bed bugs found in interceptors away from host sleeping and resting areas have served as a guide for more detailed inspection and bed bug treatment in unpredictable locations in apartments (Cooper *et al.*, 2015a).

27.4.3 Inspections to Determine Elimination

The premature termination of treatment efforts can lead to a population rebound and promote the spread of bed bugs. Most PMPs rely upon visual detection of bed bugs, which is typically only effective when bed bugs are easily seen in predictable areas such as beds and upholstered furniture. However, following treatment, the majority of bed bugs are no longer found in predictable locations and are often more dispersed (Cooper *et al.*, 2015b). For this reason, detecting bed bugs during at the end of a treatment effort is often more difficult than detecting them at the beginning. In apartments with low-level bed bug infestations that had not yet been treated, interceptors placed at beds and upholstered furniture detected 61% of infestations in apartments compared to less than 40% post-treatment. Interceptors placed away from sleeping and resting areas detected activity in more than 90% of the treated apartments (Cooper *et al.*, 2015b). Cooper *et al.* (2015a) demonstrated the effectiveness of an "elimination protocol" requiring:

- three consecutive 14-day follow-up visits without bed bug activity in interceptors placed throughout apartments
- no observed activity during visual inspection
- no new bugs or bites reported by residents.

An effective elimination protocol is essential for preventing population rebound and managing the spread of bed bugs from treated apartments to adjacent units.

Determining elimination of bed bugs in lodging facilities, such as hotels and motels, presents a number of challenges due to the pressure from management to place treated rooms back into service as quickly as possible. Canine scent detection is one option for inspection of vacant rooms where bed bugs are no longer detected visually. However, even with canine scent detection, in the absence of visual detection, an effective active monitor is necessary to confirm elimination. Active monitors should be free of bed bugs for a few days and may need to be placed at least 10 days apart, to allow for the hatching of eggs, before reasonable certainty exists that a vacant room is bed bug-free. Additional periodic inspection of vacant apartments and guest rooms over several months will help ensure that bed bugs present at previously sub-detectable levels do not still exist (Pinto *et al.*, 2007). In rooms that are vacant, dry ice traps (Wang *et al.*, 2011) or sugar-yeast traps (Singh *et al.*, 2015a) that produce carbon dioxide may be considered given the limited number of commercially available active monitors with proven efficacy.

27.4.4 Inspections in Non-traditional Settings

Bed bugs can be spread virtually anywhere via human-mediated activities. More than 25% of pest management companies surveyed in the USA reported treating for bed bugs in schools, office buildings, doctors' offices, outpatient clinics, hospitals, and various modes of public transportation, while more than 15% treated movie theatres, libraries, and retail stores for bed bug activity (Potter *et al.*, 2015).

Inspections in places where a sleeping human host is not regularly present, such as common areas and offices in apartment buildings, schools, office buildings, and retail stores, are challenging not only due to large size of the areas to be inspected but also because the behavior of bed bugs in non-residential settings is more complicated. One major difference from residential settings is that the host location may constantly change during the day. Bed bug hiding locations and movements are also less predictable. This uncertainty is somewhat reduced in environments that remain dark while the host is present (such as theatres) or where lights are turned off or dimmed (such as workplaces at night).

Visual inspections will be of greater value when hiding places are more predictable, while canine scent detection or installation of monitoring devices may be more practical for large-scale inspections or when the location of bed bugs is unpredictable. Placing pitfall-style monitors was shown to be an effective method for detecting bed bug activity in the office and meeting rooms of a heavily infested, high-rise apartment building (Changlu Wang, unpublished results). Wang placed 27 ClimbUp interceptors along the floor perimeters in four office/meeting rooms of a 15-story apartment building. Bed bugs were detected in all four rooms with a mean number of one bed bug per interceptor over 35 days. Regardless of the predictability of the bed bugs behavior or the size of the area to be inspected, a combination of at least two detection methods will typically be necessary. For example, for a large-scale inspection of an entire office building, retail store, or theatre, a canine scent inspection may be the only economically practical method. However, following a canine inspection, other methods can be utilized on a more localized basis to confirm areas where the dog(s) alerted. These areas can first be inspected visually. If not confirmed through visual inspection, monitoring devices can be installed and checked periodically, in an effort to confirm activity. In employee areas, monitors can be left in place and checked periodically over a number of weeks. Areas accessible to the public may have to be monitored only when the facility is closed, in which case active monitors may be necessary to facilitate trapping bed bugs in the absence of a host.

Conflict of Interest Statement

Richard Cooper has an ownership role in Bedbug Central, the manufacturer of SenSci products, including the BlackOut, Volcano, and Activ Bed Bug Lure. Changlu Wang is listed as a co-inventor on the patent of the ClimbUp interceptor trap and the lure in the SenSci Activ Bed Bug Lure.

References

Anderson, J.F., Ferrandino, F.J., McKnight, S., Nolen, J. and Miller, J. (2009) A carbon dioxide, heat and chemical lure trap for the bedbug, *Cimex lectularius*. *Medical and Veterinary Entomology*, **23** (2), 99–105.

Boase, C., Naylor, R. and Phillips, C. (2012) Laboratory and field evaluation of Suterra's new bed bug monitor. *International Pest Management*, **54** (4), 208–210.

Cooper, R. (2007) Are bed bug dogs up to snuff? *Pest Control*, **1**, 49–51.

Cooper, R., Wang C. and Singh, N. (2014) Accuracy of trained canines for detecting bed bugs (Hemiptera: Cimicidae). *Journal of Economic Entomology*, **107** (6), 2171–2181.

Cooper, R., Wang, C. and Singh, N. (2015a) Evaluation of a model community-wide bed bug management program in affordable housing. *Pest Management Science*, **72** (1), 45–56.

Cooper, R., Wang, C. and Singh, N. (2015b) Effects of various interventions, including mass trapping with passive pitfall traps, on low-level bed bug populations in apartments. *Journal of Economic Entomology*, **109** (2), 762–769.

Cooper, R., Wang, C. and Singh, N. (2015c) Mark-release-recapture reveals extensive movement of bed bugs (*Cimex lectularius* L.) within and between apartments. *PLoS ONE*, **10** (9), e013642.

Crawley, S.E., Potter, M.F. and Haynes, K.F. (2015) Think like a bed bug. *Pest Control Technology*, **43** (12), 96, 99–100, 102, 107–108,141.

Doggett, S.L. (2011) *A Bed Bug Management Policy and Procedural Guide for Accommodation Providers*, Westmead Hospital, Westmead.

Doggett, S.L. (2016) Bed bugs in Australia 2016: are we biting back? Results of the 2016 national bed bug survey. *Professional Pest Manager*, **Aug/Sep**, 28–30.

Doggett, S.L. and Russell, R.C. (2009). Bed bug barriers and traps. *Professional Pest Manager*, **Aug/Sep**, 20–21.

Doggett, S.L., Dwyer, D.E., Peñas, P.F. and Russell, R.C. (2012) Bed bugs: clinical relevance and control options. *Clinical Microbiology Reviews*, **25** (1), 164–185.

Eom, I.Y., Risticevic, S. and Pawliszyn, J. (2011) Simultaneous sampling and analysis of indoor air infested with *Cimex lectularius* L. (Hemiptera: *Cimicidae*) by solid phase microextraction, thin film microextraction and needle trap device. *Analytica Chimica Acta*, **716**, 2–10.

Gries, R., Britton, R., Holmes, M., *et al.* (2015) Bed bug aggregation pheromone finally identified. *Angewandte Chemie*, **54** (4), 1151–1154.

Haynes, K.F., Romero, A., Hassell, R. and Potter, M.F. (2008) The secret life of bed bugs. *Pest World*, **March-April**, 4–8.

Kim, D.Y., Billen, J., Doggett, S. L. and Lee, C.Y. (2017) Differences in climbing ability of *Cimex lectularius* and *Cimex hemipterus* (Hemiptera: Cimicidae). *Journal of Economic Entomology*, **110** (3), 1179–1186.

Koganemaru, R. and Miller, D.M. (2013) The bed bug problem: Past, present, and future control methods. *Pesticide Biochemistry and Physiology*, **106** (3), 177–189.

Mankin, R.W., Hodges, R.D., Nagle, H.T., *et al.* (2010) Acoustic indicators for targeted detection of stored product and urban insect pests by inexpensive infrared, acoustic, and vibrational detection of movement. *Journal of Economic Entomology*, **103** (5), 1636–1646.

McNeill, C.A., Pereira, R.M., Koehler, P.G., McNeill, S.A. and Baldwin, R.W. (2016) Behavioral responses of nymph and adult *Cimex lectularius* (Hemiptera: *Cimicidae*) to colored harborages. *Journal of Medical Entomology*, **53** (4), 760–769.

Pinto, L.J., Cooper, R. and Kraft, S.K. (2007) *Bed Bug Handbook – the Complete Guide to Bed Bugs and Their Control*. Pinto & Associates, Inc., Mechanicsville, MD.

Pfiester, M., Koehler, P. G. and Pereira, M. (2008) Ability of bed bug-detecting canines to locate live bed bugs and viable bed bug eggs. *Journal of Economic Entomology*, **101** (4), 1389–1396.

Potter, M.F., Haynes, K.F., Connelly, K., *et al.* (2010) The sensitivity spectrum: Human reactions to bed bug bites. *Pest Control Technology*, **38** (2), 70–74, 100.

Potter, M.F., Haynes, K. F., Rosenberg, B. and Henrikson, M. (2011) The 2011 bed bugs without borders survey: defining the bed bug resurgence. *PestWorld*, **November/December**, 4–15.

Potter, M.F., Gordon, J. R., Goodman, M.H., and Hardin, T. (2013) Mapping bed bug mobility. *Pest Control Technology*, **41** (6), 72–74, 76, 78, 80.

Potter, M.F., Haynes, K.F. and Fredericks, J. (2015) Bed bugs across America: The 2015 bugs without borders survey. *PestWorld*, **November/December**, 4–14.

Siljander, E., Gries, R., Khaskin, G. and Gries, G. (2008) Identification of the airborne aggregation pheromone of the common bed bug, *Cimex lectularius*. *Journal of Chemical Ecology*, **34** (6), 708–718.

Singh, N., Wang, C. and Cooper, R. (2012) Interactions among carbon dioxide, heat, and chemical lures in attracting the bed bug, *Cimex lectularius* L. (Hemiptera: *Cimicidae*). *Psyche*, **2012**, 1–9.

Singh, N., Wang, C. and Cooper, R. (2013) Effect of trap design, chemical lure, carbon dioxide release rate, and source of carbon dioxide on efficacy of bed bug monitors. *Journal of Economic Entomology*, **106** (4), 1802–1811.

Singh, N., Wang, C. and Cooper, R. (2015a) Role of vision and mechanoreception in bed bug, *Cimex lectularius* L. behavior. *PLoS ONE*, **10** (3), e0118855.

Singh, N., Wang, C. and Cooper, R. (2015b) Effectiveness of a sugar–yeast monitor and a chemical lure for detecting bed bugs. *Journal of Economic Entomology*, **108** (3), 1298–1303.

Stedfast, M.L. and Miller, D.M. (2014) Development and evaluation of a proactive bed bug (Hemiptera: *Cimicidae*) suppression program for low-income multi-unit housing facilities. *Journal of Integrated Pest Management*, **5** (3), E1–E7.

Szalanski, A.L., Tripodi, A.D. and Austin, J.W. (2011) Multiplex polymerase chain reaction diagnostics of bed bug (Hemiptera: *Cimicidae*). *Journal of Medical Entomology*, **48** (4), 937–940.

Ulrich, K.R., Kramer, M. and Feldlaufer, M.F. (2016) Ability of bed bug (Hemiptera: *Cimicidae*) defensive secretions (E)-2-hexenal and (E)-2-octenal to attract adults of the common bed bug *Cimex lectularius*. *Physiological Entomology*, **41** (2), 103–110.

Vaidyanathan, R. and Feldlaufer, M.F. (2013) Bed bug detection: Current technologies and future directions. *American Journal of Tropical Medicine and Hygiene*, **88** (4), 619–625.

Wang, C. and Cooper, R. (2011) Environmentally sound bed bug management solutions. In: *Urban Pest Management: An Environmental Perspective* (ed P. Dhang), CABI International, Oxfordshire, pp. 44–63.

Wang, C., Gibb, T., Bennett, G.W. and McKnight, S. (2009a) Bed bug (Heteroptera: *Cimicidae*) attraction to pitfall traps baited with carbon dioxide, heat, and chemical lure. *Journal Economic Entomology*, **102** (4), 1580–1585.

Wang, C., Gibb, T. and Bennett, G.W. (2009b) Evaluation of two least toxic integrated pest management programs for managing bed bugs (Heteroptera: *Cimicidae*) with discussion of a bed bug intercepting device. *Journal of Medical Entomology*, **46** (3), 566–571.

Wang, C., Saltzmann, K., Chin, E., Bennett, G.W. and Gibb, T. (2010) Characteristics of *Cimex lectularius* (Hemiptera: *Cimicidae*) infestation and dispersal in a high-rise apartment building. *Journal of Economic Entomology*, **103** (1), 172–177.

Wang, C., Tsai, W-T., Cooper, R.A. and White, J. (2011) Effectiveness of bed bug monitors for detecting and trapping bed bugs in apartments. *Journal of Economic Entomology*, **104** (1), 274–278.

Wang, C., Singh, N., Zha, C. and Cooper, R. (2016) Bed bugs: Prevalence in low-income communities, resident's reactions, and implementation of a low-cost inspection protocol *Journal of Medical Entomology*, **53** (3), 639–646.

Weeks, E.N.I., Birkett, M.A., Cameron, M.M., Pickett, J.A. and Logan, J.G. (2011) Semiochemicals of the common bed bug, *Cimex lectularius* L. (Hemiptera *Cimicidae*), and their potential for use in monitoring and control. *Pest Management Science*, **67** (1), 10–20.

Wong, M., Vaidyanathan, N. and Vaidyanathan, R. (2013) Strategies for housing authorities and other lower-income housing providers to control bed bugs. *Journal of Community Housing Development*, **70** (3), 20–28.

28

Non-chemical Control

Stephen A. Kells

28.1 Introduction

The reappearance of bed bugs in mainstream society has resulted in a steep learning curve with regards to controlling infestations. With numerous treatment failures because of insecticide resistance (Moore and Miller, 2006; Romero *et al.*, 2007; see also Chapter 29), people sought ways of managing infestations through non-chemical methods. In North America, additional problems occurred with insecticide product availability, as new insecticides replaced older chemistries, and label instructions on many new products lacked references to bed bugs. There was a lack of understanding about the required thoroughness of insecticide applications necessary to properly control bed bug infestations, which contributed to control failures. Liability concerns regarding human exposure to insecticides were prevalent, especially with owners and managers of pest control companies, and there were concerns that technicians might apply insecticides contrary to label instructions in an effort to provide relief for affected residents. Considering the myriad of personal items that could potentially become infested, repeated control failures resulted in the rapid development of non-chemical control methods to augment or replace insecticide use. In some cases, the prevailing opinion was that non-chemical controls should be the primary control method (Kells, 2006b), or at least there should be some combination of chemical and non-chemical controls (Doggett, 2013; Bennett *et al.*, 2015; Jourdain *et al.*, 2016).

Non-chemical control options offer many choices and advantages for controlling bed bug infestations. In some cases, heating furniture and personal belongings may completely replace insecticide applications. In larger living spaces, vacuuming surfaces and steaming mattresses and other furniture could substitute for insecticide applications where there might be close human contact. Utilizing non-chemical treatments afforded flexibility as to whether insecticides were used as a follow-up, or reserved for inaccessible areas, dependent on the type of non-chemical control methods employed, the size of the infestation, and the availability of insecticides to be used on (in other words, specifically labeled for) infested items. Non-chemical controls could safely and effectively treat items that might be damaged by insecticides, such as books and artwork. In cases of heavy infestation, non-chemical controls offered a more immediate reduction in pest numbers. However, there were some disadvantages discovered when delivering these methods, particularly when homeowners attempted do-it-yourself treatments, or when pest management professionals (PMPs) sought cut-rate alternatives for delivering non-chemical controls (Anonymous, 2011). Even when non-chemical controls were properly implemented, there existed the potential for property damage (for example, from heat warping or melting of belongings). Despite the risks, however, mainstream adoption of non-chemical methods has occurred in most nations, and any successful control of bed bugs will require some form of non-chemical technique.

Advances in the Biology and Management of Modern Bed Bugs, First Edition.
Edited by Stephen L. Doggett, Dini M. Miller, and Chow-Yang Lee.
© 2018 John Wiley & Sons Ltd. Published 2018 by John Wiley & Sons Ltd.

Non-chemical control methods against bed bugs include a diverse collection of techniques:

- establishing barriers between parasite and host
- preventing insects from establishing inaccessible harborages
- physically removing bed bugs
- creating adverse environmental conditions that cause injury or mortality to the insect.

Understanding the complexity associated with the term "non-chemical" is perhaps the first challenge in a discussion about the use of non-chemical controls against infestations. For example, Barzman et al., (2015) considered "natural" biocides (in other words, non-synthetic insecticides), "green chemistries" such as essential oils, and biological agents such as bacteria or fungi to be forms of "non-chemical" control. In other cases, manipulating environmental conditions to synergize insecticide products (such as formalin or dichlorvos) may be considered, at least in part, a non-chemical practice (see, for example, Alderson, 1933; Lehnert et al., 2011). Moreover, there are integrated pest management (IPM) practices, such as education and inspections, that might not directly control bed bugs, but rather may impose pressure on the pest or improve prevention and control practices by changing human behavior (Bennett et al., 2015). Non-chemical control definitions may also be complicated by regulatory intent. For example, diatomaceous earth in Europe is considered non-chemical (in other words, a non-insecticide; EGTOP, 2014), but the same product is registered as an insecticide in North America and Australia. Similarly, nitrogen and carbon dioxide gases are designated as non-chemical asphyxiants in some jurisdictions (Querner and Kjerulff, 2013), but are registerable fumigants in others (Selwitz and Maekawa, 1998).

Technically, any activity that affects bed bug populations, but does not require dispensing an insecticidal liquid, dust, or gas might be considered non-chemical. However, to avoid unnecessary repetition with other chapters in this book, this discussion will focus on five principles directly affecting the insect pest:

- excluding the pest
- physically removing the pest
- creating adverse environmental conditions
- the use of biological agents
- other experimental and novel control methods.

Further, the discussion will focus on specific techniques or technologies that have demonstrated efficacy against bed bug infestations in the field, as well as examples where attempts at non-chemical control did not extend past laboratory studies. Limiting the discussion in this manner also provides an opportunity to assess which methods have been practical and adopted, while also showing the extent to which some remain experimental or unproven.

28.2 Excluding Bed Bugs

Non-chemical techniques have always been a part of controlling bed bugs. However, given the changes in living spaces and lifestyles, the recent resurgence of this pest has initiated a return to and modification of these historical methods. Simplifying the habitat, for example by sealing up bed bug harborages, removing clutter, and using simpler and more durable furnishings, is a historical non-chemical control method that has evolved into a more modern methodology. Such historical techniques include replacing straw and wood sleeping surfaces with iron bed racks (Yuma Territorial Prison and Museum, 2013) or using whitewash to "paint" existing bed bugs in place and seal hiding places (see, for example, Celmina, 1986). In current times, sealing cracks and crevices with an elastomer sealant applied to furniture and around rooms has regularly been recommended (NYC DOHMH, 2012). Additional recommendations to simplify the living space include the use of sealable floor and wall coverings, such as vinyl tiles, instead of carpeting (Doggett, 2013). However, there is little demonstrated efficacy associated with such practices; and the economics have not been studied.

Figure 28.1 Bed bugs and fecal spotting inside a wall void in multi-family housing. (S.A. Kells, University of Minnesota).

There is also a question of a potential shift in bed bug aggregations when a living space has fewer cracks and crevices but still has considerable structural complexity. The presence of large numbers of bed bugs in and around electrical outlets and residing inside wall cavities suggests proper sealing methods may have utility (Figure 28.1). Comparative studies, or IPM studies and economic analysis, are needed to demonstrate the value of sealing cracks and crevices.

Other exclusion techniques have demonstrated more practical efficacy. For example, mattress encasements have been recommended to prevent bed bugs from accessing complex refugia in mattresses sets. Wang *et al.* (2009a, 2013) used encasements in studies where IPM practices were tested. However, the effect of mattress encasements was confounded by other variables in the study and their value as a non-chemical method is typically discussed within the merits of a complete program. If the mattress is harboring bed bugs, encasements have also been recommended for preventing their access to a host, although the encasement must be properly deployed, be biteproof, and remain undamaged so the bed bugs starve (typically requiring several months). Time for mortality by starvation might be extended, as bed bugs readily alter their metabolism to resist the impacts of non-feeding (DeVries *et al.*, 2015), and dehydration (Benoit *et al.*, 2009). Encasements have been further suggested as a barrier against human exposure to pesticide applications on mattresses (Kells, 2006b).

28.3 Physically Removing Bed Bugs

Many historical non-chemical control methods involved physically removing as many bed bugs as possible. For example, during the 1920s, the use of hat pins facilitated bed bug removal and added an element of "competitive entertainment" to the lives of those persons dealing with infestations in North America (Mager, 1997). Manual removal or destruction of bed bugs is still recommended to reduce the biting frequency and the size of the infestation, even though more sophisticated methods are available, such as dislodging them with thin plastic cards and removing their bodies via adhesive tape or vacuuming. Such methods are particularly promoted for people who are challenged in obtaining help or financial resources for dealing with an infestation (Shindelar and Kells, 2011a).

Intercept traps to capture bed bugs before they reach a host have many different forms and formats. The trichomes on bean leaves (*Phaseolus vulgaris* L.) were reported to trap bed bugs, and have been historically used in the Balkans (Richardson, 1943). Attempts at a synthetic product resulted in a similar restraint of bed bugs by hooking body parts, but led to the discovery of a much more complex interaction occurring with barbed trichomes as well (Szyndler *et al.*, 2013). However, this technology has largely been tested on adults and more work is required to ensure it can collect all life stages of bed bugs.

Pitfall intercept traps have also been used to collect bed bugs. Primarily a monitoring device (Wang *et al.*, 2009b), these traps have demonstrated efficacy both in excluding bed bugs attempting to climb onto beds in search of a host and as an insect removal device. Mass placement of traps throughout infested rooms has been shown to be efficacious as part of an IPM program, although Cooper *et al.* (2016) noted an extended period of up to 16 weeks was necessary to show reductions in "low-level" infestations, which they defined as ten bed bugs or fewer. With these methods, efficacy might be negated if bed bugs find aggregation sites that do not have a trap or restraint surface between bed bug and host. The merit of these devices would be seen more as a part of an IPM program rather than as a stand-alone treatment.

Vacuuming bed bugs from surfaces or cracks/crevices requires particular mention as a non-chemical control method. This method has demonstrated utility as a part of IPM programs, for example for German cockroaches (Miller and Meek, 2004), but vacuuming requires a chemical flushing agent to capture inaccessible cockroaches (Wang and Bennett, 2006). The use of a thin plastic card in combination with vacuuming has proved effective as a physical means of flushing and capturing bed bugs from some inaccessible areas (Shindelar and Kells, 2011a). Other benefits of vacuuming include removal of evidence of infestations and simplifying the microhabitat. Vacuum removal of conspecific exuviae-emitting arrestant pheromones (Choe *et al.*, 2016) is appropriate considering small nymphs may harbor in exuviae (Dini Miller, Virginia Tech, Blacksburg, Virginia, unpublished results). However, removal of all arrestment sites will not be complete, as adsorbed feces also contain arrestment pheromones (Olson *et al.*, 2009, 2017). Another potential benefit of vacuuming is the potential for reduction of insect-derived allergens, as demonstrated in control efforts against German cockroaches (Sever *et al.*, 2007), provided the vacuum has a high-efficiency particulate arrestant filter. Development of newer and more portable vacuum cleaners has shown value in experimental sampling of bed bug infestations (Berenger *et al.*, 2015), although more extensive study is required to determine the best vacuum equipment for control purposes.

Complete disposal of infested furnishings is a practice recommended by PMPs, and has been used by hoteliers, landlords, and residents. There is a lack of research as to whether this method is effective in controlling bed bugs, but it does both simplify the area and remove bed bugs from the living space. People paying for professional control services have attempted to reduce costs by removing, instead of treating, items. However, since 2001 there have been numerous incidences reported of bed bugs spreading within buildings and to other sites because of this practice. Spread of infestations has involved infested furniture "seeding" adjacent areas with bed bugs, especially when the items have not been enclosed in plastic prior to removal. Infested furnishings have been scavenged and brought into other living spaces, and people discarding furniture may therefore be at risk of re-introducing bed bugs if they continue to scavenge or purchase infested furniture (Kells, 2006a). Disposing of infested furniture is inadvisable, unless the furniture is severely infested, non-salvageable, and rendered unusable, and to prevent bed bugs from being dispersed.

28.4 Creating Adverse Environmental Conditions against Bed Bugs

A number of methods have been developed for creating adverse environmental conditions for small-scale and large-scale infestation control. In North America, the use of steam applications was one of the earliest techniques developed to supplement or replace insecticides during the bed bug resurgence, starting in the early 2000s. Later, dry heat applications were developed. Other existing techniques adapted for bed bug control included freezing of structural contents or laundering of clothing and other articles. Finally, some novel non-chemical control methods have been tested against bed bugs and other insects, but have yet to be adopted in the mainstream because of practicality, cost, or equipment availability. There are two parts to non-chemical controls that require consideration:

- the basic conditions that will cause lethality to the insect
- the equipment to deliver these conditions in a safe, practical, and efficient manner.

28.4.1 Basic Conditions Causing Lethality

Criteria for extreme environmental conditions depend on the magnitude of temperatures outside the normal survival range for bed bugs. All insects have a range of conditions optimal for growth and development, and for both the Common bed bug, *Cimex lectularius* L., and the Tropical bed bug, *Cimex hemipterus* (F.), this optimal range is approximately 20–30 °C (Mellanby, 1935; Johnson, 1941; How and Lee, 2010). Below or above this optimal range, bed bugs may survive, but growth, development, and fecundity will be affected. For *C. lectularius*, eggs hatch at 15–34 °C (Johnson, 1941). Precise ranges are lacking, because of differences in methods and the research objectives, but the existing collection of research provides a reasonable indication of the temperature range for survival through to optimal development. Above and below the survival range will be a lethal zone, bringing death in days, and proceeding to seconds as the conditions become more severe.

Lethality based on temperatures can be a complicated matter. For example, mortality will depend on how rapidly the temperatures can be transferred to the bed bug, time of exposure, and damage-critical tissues within the insect. Mortality will also depend on the potential for bed bugs to invoke behavioral or physiological methods to resist temperature extremes. Mellanby (1939) was the first to demonstrate the acclimation potential of *C. lectularius* by showing that bed bugs may move at 4.5 °C if pre-exposed to a lower temperature of 15–17 °C. Even under normal feeding conditions, bed bugs are known to produce heat shock proteins (Benoit *et al.*, 2011), indicating their need to be protected from elevated temperatures (starting at approximately 30 °C). However, bed bugs can be killed almost instantly if fed *in vitro* with blood at a temperature of 39 °C (Stephen A. Kells, unpublished results). This finding illustrates how the delivery of even slightly elevated temperatures in the correct circumstances can have a dramatic effect.

28.4.2 Basic Conditions for High-temperature Control

More traditional delivery methods of heating for control of bed bug infestations require higher temperatures than if bed bugs are directly fed blood at 39 °C. Historically, temperature estimates for *in-situ* application have ranged from 43.3 °C for 48 h (Harned and Allen, 1925), and 49–54 °C (up to 71.1 °C; Ross, 1916). In the laboratory, critical temperature estimates have been lower; for example, 40–41 °C for 24 h and 45 °C for 15–60 min (Mellanby, 1935; Johnson, 1941, and several authors therein). Bed bugs will be affected by sub-acute lethal temperatures of 35–40 °C (Rukke *et al.*, 2015), but unless there is a specific habitat limitation restricting temperatures to less than 38–40 °C, control success occurring between 2 and 9 days is of limited practicality. Until further experiments were performed (after 2008), achieving 45 °C seemed to be the industry standard for control with high temperatures. Indeed, measurements of individual bed bugs' critical thermal maximum (CT_{max}) indicated that a temperature of 45.2 °C caused metabolic death (DeVries *et al.*, 2016). However, other estimates demonstrated a higher lethal temperature threshold of 48 °C for controlling adult bed bugs (Benoit et al., 2009; Pereira et al., 2009; Kells and Goblirsch, 2011). One critical consideration for developing these estimates is in the way that bed bugs are exposed to elevated temperatures. Most previous estimates involve immersive exposure, where the insects are placed into existing high temperatures, or the temperature increases rapidly. Kells and Goblirsch (2011) used a temperature increase rate simulating an *in-situ* heat treatment (+3.6 °C/h) and recommended an exposure time of 20 min at 48 °C. Another possibility for the large difference is that clusters of bed bugs may tolerate slightly higher temperatures (DeVries et al., 2016) in addition to resisting dehydration stress (Benoit et al., 2007). Regardless, *C. lectularius* eggs were found to be the most resilient life stage, requiring a temperature of 50 °C for immediate lethality using dry heat (Kells and Goblirsch, 2011). Below 50 °C, the exposure duration becomes important and bed bug eggs require 71.5 min at 48 °C for complete lethality, or 7 h at 45 °C.

Equipment for Delivery of High Temperatures through Steam Applications
Steam has received attention as method of delivering lethal thermal energy to infestations, both in historic and modern times. Large scale steaming of whole rooms has been attempted, with a boiler forcing steam into a building and raising the temperature to 71 °C (Tharaldsen, 1919). The first indication of small-scale bed bug control via steaming was Fewell (1873) who patented a kettle-like steaming device. More recently, steam generators (also known as "vapor steam cleaners") were adapted for control of bed bugs. This happened as early as 2001 in response to control failures encountered by North American PMPs, due to bed bug resistance to pyrethrins and pyrethroids. Vapor steam cleaners are generally used for cleaning on commercial properties, especially commercial kitchens and hospitals. Meek (2003) provided a more recent report of steam as a control method against bed bugs, with applications in hotels resulting in control for between 90 days and more than 12 months.

For delivery of heat onto surfaces via steam, application temperatures immediately after the steam wand has passed should be in the range 71–82 °C (Stephen A. Kells, unpublished results). Within this temperature range, lethal conditions above 50 °C could penetrate 1.1–1.8 cm into fabric (500 thread count to 135 thread count, respectively) over polyester fiber fill. In cracks and crevices, lethal heat would penetrate more than 5.2 cm into the crack (Stephen A. Kells, unpublished results). Puckett et al. (2013) found that more than 84% of bed bugs were controlled when steam was applied at a rate of 10 s per 30.5 cm movement, if temperatures remained between 67 and 74 °C. This confirms a higher temperature range should be considered for ensuring efficacy. A key difference in control success between the Puckett et al. (2013) laboratory study and that of the unpublished data and *in-situ* experiences may be an artifact of the different equipment used. The equipment used by Puckett et al. (2013) was rated for removing wrinkles from clothes, whereas the units most often used by PMPs are actual steam cleaning units. In practice, the use of a higher temperature range substantially reduced control failures for PMPs early in the resurgence (Stephen A. Kells, unpublished results).

Ensuring consistent temperatures in the range of 71–84 °C is critical, because it is human nature to slow the steaming process upon finding an aggregation, or to increase speed towards the end of a control procedure. Slowing the steamer nozzle will increase the temperatures and so add to the risk of surface damage, while rushing the control procedure decreases the delivery temperature resulting in control failure. Periodically verifying surface temperatures with the use of a non-contact infrared thermometer helps maintain a consistent and efficient rate of treatment, reducing damage risks, and ensuring a complete kill (Figure 28.2). Verifying surface temperatures is also critical when using different attachments to disperse the steam across surfaces. Larger attachments (such as a floor brush rather than a smaller triangle brush) could develop the same lethal conditions, but the rate of nozzle travel must be slower, unless the steam flow rate is increased (Shindelar and Kells, 2011c).

In addition to positive field experiences – the reduction in control failures – formalized field research conducted by Wang et al. (2009a) has demonstrated that the use of steam, in combination with a number of other tactics, is successful in controlling bed bugs in infested apartments. Note, however, that the specific benefits of adding steam to the program could not be separated from the overall effort. Even with companies who adopted this means of controlling bed bugs, other chemical and non-chemical means of control were required for success.

Figure 28.2 Use of an infrared thermometer to verify surface temperatures immediately after steam application. (S.A. Kells, University of Minnesota).

Equipment for Delivery of High Temperatures through Dry Heat Applications

Prior to 2005, structural heat treatments were available. In the USA, these were used mainly to control termites (Lewis and Haverty, 1996) and stored-product pests. This method of non-chemical control was not widely employed in other nations, although food plant heat treatments have increased worldwide as the fumigant methyl bromide has been phased out (Pappalardo *et al.*, 2017; R. Hulasare, Temp Air, Inc., Burnsville, MN, unpublished results). Heat systems that deliver temperatures of 54–66 °C typically provide a rapid transfer of heat to the 50 °C threshold into areas where bed bugs aggregate. Higher temperatures (say, 75 °C) may be required if shorter exposures are expected (Loudon, 2017). Dry heat may be created in a number of ways: when air is passed over a heating element (electric), through a steam (or glycol) heat exchanger, or through a direct (or indirect) burner fueled by propane or natural gas. How heat is delivered to the infested space – and the success of the treatment – depends on equipment and power availability, as well as the technical abilities of those conducting the heat treatment. Heat treatments are achieved in three different ways:

- containerized treatments, where items are placed in a container of a fixed dimension
- whole-room heat treatments, where the furnishings are principally left in place in a room, although positioned to enable good heat transfer
- heat treatments using a clothes dryer for smaller items.

Each method has advantages and disadvantages, as well as special considerations for their use that should be summarized.

Containerized heat involves moving infested items from their place within a residence and placing these items into an enclosed area or "container", into which heaters and fans are placed. These containerized heat treatments have used several types of enclosures, of different sizes, including:

- enclosed utility trailers or cube vans
- shipping containers (Figure 28.3)
- portable insulated tents
- homemade insulated boxes (Pereira *et al.*, 2009)
- smaller containers for suitcases and individual items (such as Pack-Tite or ThermalStrike brands)
- "flow-though" heating containers for luggage (Loudon, 2017).

Containerized heat treatments have advantages in that the volume of air to be heated is fixed and consistent for delivery of high temperatures. Therefore, the heating requirements and the resulting control

Figure 28.3 Shipping container fitted with a heater for larger containerized heat treatments. (Courtesy of J. Bruesch, Plunkett's Pest Control, Fridley, MN, USA).

Figure 28.4 Stacking furniture in a portable container for heat treatment. (S.A. Kells, University of Minnesota).

efficacy are more predictable. Also, for items housed in large areas, such as chairs in a large foyer, putting the chairs in the container focuses the treatment on infested items and avoids unnecessary (and therefore more costly) heating of larger spaces (Figure 28.4). Disruption in other locations could also be minimized, especially if the container is well insulated and PMPs can conduct other infestation control measures concurrently (Pereira *et al.*, 2009). Consistency in the treatment volume and equipment requirements means personnel training and obtaining correct power requirements is relatively simple. However, additional labor is required to move and situate items to be heat-treated within the container. With containers that can be set up on site, the labor requirement is less than when moving items out of a residence and into a trailer or truck container. There is a risk of damage while moving and packing furniture into the container and this risk is a consideration for commercial service providers using container heating. Another

Figure 28.5 Whole-room heat treatment with electric furnaces. (Courtesy of Raj Hulasare, Temp-Air, Inc., Burnsville, MN, USA).

important consideration is that anything outside the container, including walls, floors, and other items in the infested room must be addressed using other non-chemical or chemical measures.

The second method of dry heat delivery, termed "whole-room" heat treatments, involves treating the whole living space as the container and keeping all items within the room. Items that may be damaged must be removed (Box 28.1) and there may be some minor repositioning of items to ensure maximal heat transfer (Figure 28.5). Whole-room heating has definite advantages because minimal preparation is required on the part of the resident and much can be simply left in place, including most residents' clothing. As the applicator oversees the moving and arrangement of items in a room, it reduces the chance that untrained personnel may cause a failure in the control procedure. Companies using whole-room equipment claim that complete control can be achieved in one treatment, instead of the two or three treatments that are necessary when using insecticides. Heat treatment failures are rare (<2%) if the application is properly monitored, target sites achieve threshold temperature, and thermal refugia are addressed through other measures (J. Henry, Minneapolis Public Housing Authority, Minneapolis, MN, unpublished results). Whole-room heat treatments usually require a 6–8 h treatment time, though this estimate will depend on the capacity of the heat system, the volume of room to be heated, and the ability to raise the temperature to heat refugia and personal items to the lethal temperature threshold.

Whole-room heat treatments require more training and subsequent monitoring of the process because different room dimensions, building materials and furnishings have a major influence on obtaining lethal temperatures. Simply heating the room air to above 50 °C and holding it for a defined period is not an effective control method and will likely result in failure. Heat sinks, insulating materials, and areas with sub-lethal temperatures immediately adjacent to heated zone need to be monitored and, if necessary other control measures may be required. PMPs must anticipate and monitor thermal consistency, actively seek out cold spots, and manipulate the equipment so that the lethal temperature is achieved in refugia where bed bugs may otherwise survive. Failures from whole-room heat treatments are usually the result of bed bugs finding thermal refugia within the room (J. Henry, Minneapolis Public Housing Authority,

> **Box 28.1 Items that may be damaged by heat treatment.**
>
> Pressurized cans: hairspray, deodorants, insecticidal spray, asthma inhalers, spray paint, oxygen tanks, and cleaning products.
> Ink printer cartridges.
> Chocolate, confectionary, and other food that can melt.
> Medication and medicine.
> Beverages: bottled wine and carbonated beverages.
> Anything made from wax: candles, wax figurines, crayons, and waxy cosmetics.
> Oil paintings.
> Vinyl records, VHS tapes, photo negatives, and film strips.
> Musical instruments; cases will need to be treated.
> House plants.
> All flammable materials: butane lighters, lamp fuel, solvents, fuel for food warmers, ammunition.
> Laminated furniture and laminates (chemically treat if necessary).
> Adapted from Plunkett's, Inc. Fridley, Minnesota, USA.

Minneapolis, MN, USA, unpublished results). Note that bed bugs can be easily re-introduced after heat treatment through dispersal and hitchhiking (See Section 27.4.1). Some of the more unusual re-introductions that have occurred include infested personal items and medical devices: purses, removable casts, prosthetic limbs, wheelchairs, canes, and walkers. These examples would likely defeat any control method, non-chemical or chemical. Heat treatments have no residual efficacy, so other IPM measures are required to ensure ongoing success.

Another disadvantage to whole-room heat treatments is the cost, which can be between 1.5 times and double the cost of an insecticide application. There are substantial capital costs of purchasing the equipment and obtaining the required level of technical expertise necessary for consistent results. Attempts to do without proper equipment and inappropriate use of equipment have led to disastrous results (Anonymous, 2011; Blais, 2013). Technical complexity of buildings contribute to this cost. For example, one issue that has been particularly problematic when using whole-room heat treatments is the potential interaction with fire-suppression equipment, particularly sprinklers. The sprinkler management protocols during heat treatments depend on the property insurance requirements, state laws, and local fire codes. The cost of conducting whole-room heat-treatments often incorporates insurance costs that account for the possible risk of unintentional damage.

The third heat treatment method is through the use of equipment typically available in homes, such as a clothes dryer, a clothes washer, or a hair dryer. Technically a chamber-style heat treatment, a clothes dryer is singled out because it is (often) a readily-available method that affected residents can use for controlling bed bugs themselves, supplementing PMP efforts. The hot laundering of clothes at temperatures above 60 °C, or placing of clothes into a dryer at above 40 °C for at least 30 min, is a mainstay of non-chemical recommendations for bed bug control (Naylor and Boase, 2010). Advice for laundering clothes is the most requested topic on the bedbugs.umn.edu website (Shindelar and Kells, 2017). For do-it-yourself treatments, hair dryers have been suggested for control of bed bugs (NYC DOHMH, 2012). However, observations in laboratory studies have shown that adult bed bugs exhibit a stilting behavior, positioning their body to minimize exposure to the hot air, becoming airborne with the force of the air and landing unscathed 0.5–1 m from their starting point (Stephen A. Kells, unpublished results). Compared to a hair dryer, vacuuming or mechanical removal would be preferable.

Basic Conditions for Delivery of Low-temperature Treatments

Freezing bed bugs is a common control method, and early estimates of lethal conditions against bed bugs included exposures at −17 °C for 2 h and −18 °C for 1 h (Johnson, 1941, and several authors therein).

Benoit *et al.* (2009) indicated that female bed bugs were freeze-intolerant with a super-cooling point (SCP) of −20 °C, and adult bed bugs failed to survive at −16 °C for 1 h. A more comprehensive study on bed bug exposure to cold found that the eggs and unfed first instar nymphs had a SCP averaging −29 °C, and survival could occur after short exposures as low as −25 °C (Olson *et al.*, 2013). Exposure duration to low temperatures is important, and temperature × time × mortality modeling indicates that 3 days' exposure at −18 °C is required to ensure 100% mortality. However, immediate lethality occurs when bed bugs are exposed to −30 °C (Olson *et al.*, 2013). An important point made by Olson *et al.* (2013) is that the majority of bed bugs surviving the freeze episode are able to feed; even the few that survived conditions down to −20 °C. An extended exposure of up to three weeks is required at sub-acute lethal temperatures of between 0 and −10 °C (Rukke *et al.*, 2017), but the practicality of such an approach is questionable unless there is a specific need to avoid temperatures below this range or there are equipment limitations.

Interestingly, it has been found that oscillating between hot and cold temperatures may enhance bed bug mortality. The lethal temperatures stated above were determined in experiments where bed bugs were exposed to a constant temperature (Olson *et al.*, 2013). However, the authors found that warming and cooling through defrost cycles resulted in satisfactory control without attaining a low temperature of −18 °C. Unfortunately, there is presently no practical method of applying controlled temperature cycles *in situ*, and further research would be required define optimal conditions.

Considerations for Delivery of Low Temperatures
Like heat, low-temperature treatments are not a new technology and often have been used to supplement insecticide applications. In some cases, freezing may be the only acceptable treatment method for items that cannot be heat treated or chemically treated, such as firefighting bunker gear that must not be unnecessarily exposed to high temperatures. Naylor and Boase (2010) indicated that 2.5 kg of loosely packed dry laundry can reach −17 °C in approximately 8 h. However, general recommendations are to measure the core temperature of items being frozen to ensure materials reach a lethal temperature × time threshold, ensuring lethality. Due to equipment variability and the insulative capacity of items that might be frozen, the recommendation by Olson *et al.* (2013) of −18 °C for three days is often extended by 24 h for residents dealing with bed bugs (Shindelar and Kells, 2011b).

A question often asked in temperate and colder climates concerns the use of outdoor winter temperatures to freeze items, or even opening a structure to freeze the structure itself. While conceivably whole-structure freezing might be effective, the risk of structural damage, variability of temperatures, solar heating, and several other microclimate issues add to the complexity and increase the risk of control failure. The only time the author has entertained this method as a potentially viable option is a suggestion made regarding structures in Alaska, where there are both consistent low-temperature conditions for extended periods and structures that may accommodate complete freezing with little or no damage. Still, community IPM practices would be critical in preventing re-infestation from neighboring structures that were not undergoing treatment. Several reports of bat bugs, *Cimex adjunctus* Barber, surviving in recreational cabins during Minnesota winters suggest that seasonal temperature changes would be ineffective.

A portable bed bug control technology that uses low temperature is the Cryonite or Rapid Freeze system. There are competing systems using liquid nitrogen, as well. The Cryonite system dispenses liquid carbon dioxide, which rapidly forms fine dry ice crystals. This application method is generally considered non-chemical (although not in all countries) because the carbon dioxide (or nitrogen) applied to surfaces expose bed bugs to temperatures below −78.5 °C, which is well below their normal survival range. The equipment has been largely tested *in situ*, but published data are lacking. However, there has been reason to discourage use of any control method that may displace air in force, thereby ejecting bed bugs and exuvaie bearing small nymphs and eggs from surfaces (Doggett, 2013). Similar to steam, these systems would be a part of a bed bug control program that employs additional methods, including insecticides.

28.5 Biological Agents Tested Against Bed Bugs

Numerous biological agents that are used against bed bugs have been catalogued. Recent work has attempted to test a variety of entomopathogenic fungal strains. Historic documents have catalogued numerous species of rickettsiae, bacteria, fungi, and predators that affect bed bugs (Anonymous, 1964; Strand, 1977). Some of the pathogens cited as effective against bed bugs are also serious human pathogens. Recently, the entomopathogenic fungi *Beauveria bassiana* (Bals.-Criv.) (Barbarin *et al.*, 2012) and *Metarhizium anisopliae* (Metschn.) (Ulrich *et al.*, 2014) have been tested *in vitro* and found to be effective against *C. lectularius*. Further testing would be required to determine the range of activity and humidity at which fungal spores would be active. While *M. anisopliae* was only effective at 98% RH, *B. bassiana* showed activity at 50% RH, a humidity level closer to that in a normal living space. Testing over a wider humidity range would be needed to ensure entomopathogenic fungi were effective at the humidity levels typically found in homes. For example, during Minnesota winters, indoor humidity may drop to 10% RH. An interesting discovery is the potential for the bed bug defensive secretions of (E)-2-hexenal and (E)-2-octenal (which are also pheromones) to inhibit *M. anisopliae* (Ulrich *et al.*, 2015). The complexity of the relationship between entomopathogenic fungi, the environment, and the insect host requires considerable study before becoming a practical method of control. Finally, exposure of humans to high levels of conidia (and any entomopathogen) should be taken into consideration when the organism is applied to living spaces frequented by bed bugs.

In addition to pathogens, many traditional generalist predators have been included in catalogues. Examples inclue the house centipede *Scutigera forceps* (L.), pseudoscorpions, various ant species, and predatory pentatomids and reduviids (Anonymous, 1964). The moth *Pyralis pictalis* L. preys on bed bug eggs, but this insect is also a common household pest (Wattal and Kalra, 1960). In Florida, red imported fire ants (RIFA, *Solenopsis invicta* Buren) have effectively removed bed bugs from a discarded mattress (P. Koehler, University of Florida, Gainesville, FL, unpublished results). In nature, ant predation on swallow bugs, *Oeciacus vicarious* Horvath, has been observed (Brown *et al.*, 2015). An interesting note in this study was that the foraging behavior by ants was limited to outer nests, which is an important consideration with the use of predators – they will optimally forage in their own interests and not necessarily in the interest of the host. Predator use against bed bugs in residences is further limited because predators are also considered pests, and in some cases, such as RIFA, very serious pests.

28.6 Other Non-Chemical Control Methods

A number of other non-chemical techniques have been tested for use against bed bugs and other insects. Although the results show effective control, the equipment required for practical deployment seems limiting. Still, they are worth mentioning because there may be scenarios in which equipment availability, a particular situation, or further development of technology may result in a practical application. Low-oxygen, low-pressure (vacuum) environments are effective for controlling bed bugs with an 8 h treatment at 0.1% oxygen or 12 h at −982 mbar (−29.0 mmHg) (Liu and Haynes, 2016). With other insect pests, particularly those infesting stored products, other technologies have been tested, or are currently being considered for industrial-scale testing. Such recent reports of new control technology include gamma-irradiation (Rempoulakis *et al.*, 2015); radio frequency (27.12 MHZ, 6 kW) heating, and flameless catalytic infrared radiation (Khamis *et al.*, 2011). However, limitations on the practicality of these technologies exist either in the source containment (^{60}Co for gamma radiation, or radio wave emissions), or a lack of penetration (IR radiation). Other non-chemical control methods have been tested and found to be completely ineffective. Solar heating of mattresses wrapped in black plastic was attempted and found to ineffective (Doggett *et al.*, 2006). Ultrasonic repellers continue their track record as being ineffective against blood-feeding insects (Yturralde and Hofstetter, 2012). In the future, such new technologies would have to demonstrate advantages in efficacy, safety, cost, and ease of use comparable to existing control technologies.

28.7 Conclusion

With the resurgence of bed bugs through society, non-chemical control methods have received particular attention, more so than for any other urban household pest. Early failures of insecticides to adequately control infestations and a realization that insecticides could not be applied to all possible locations where bed bugs might reside, led to a rapid search for other techniques of control. The resulting equipment and methods developed for non-chemical control provided practical options for the control of other pests and potential means for managing insecticide resistance. While widespread adoption of effective non-chemical practices still lags in some areas, continued reporting of successes will enable eventual adoption of these techniques as a viable alternative to insecticides.

References

Alderson, A. F. (1933) A practical test of the lethal action of steam and formalin on spore-bearing organisms and bugs. *Journal of the Royal Army Medical Corps*, **60** (5), 374–376.

Anonymous (1964) Bedbugs – Cimicidae. *Bulletin of the World Health Organization*, **30** (Suppl), 91.

Anonymous (2011) *Bedbug Exterminator's Propane Tank Likely Cause of House Fire*, http://www.fox19.com/story/14648449/crews-on-the-scene-of-two-alarm-fire (accessed 17 November 2016).

Barbarin, A.M., Jenkins, N.E., Rajotte, E.G. and Thomas, M.B. (2012) A preliminary evaluation of the potential of *Beauveria bassiana* for bed bug control. *Journal of Invertebrate Pathology*, **111** (1), 82–85.

Barzman, M., Bàrberi, P., Birch, A.N.E., *et al.* (2015) Eight principles of integrated pest management. *Agronomy for Sustainable Development*, **35** (4), 1199–1215.

Bennett, G.W., Gondhalekar, A.D., Wang C., Buczkowski, G. and Gibb, T. (2015) Using research and education to implement practical bed bug control programs in multifamily housing. *Pest Management Science*, **72** (1), 8–14.

Benoit, J.B., Del Grosso, N.A., Yoder, J.A. and Denlinger, D.L. (2007) Resistance to dehydration between bouts of blood feeding in the bed bug, *Cimex lectularius*, is enhanced by water conservation, aggregation, and quiescence. *The American Journal of Tropical Medicine and Hygiene*, **76** (5), 987–993

Benoit, J.B., Lopez-Martinez, G., Teets, N.M., Phillips, S.A. and Denlinger, D.L. (2009) Responses of the bed bug, *Cimex lectularius*, to temperature extremes and dehydration: levels of tolerance, rapid cold hardening and expression of heat shock proteins. *Medical and Veterinary Entomology*, **23** (4), 418–425.

Benoit, J.B., Lopez-Martinez, G., Patrick, K.R., *et al.* (2011) Drinking a hot blood meal elicits a protective heat shock response in mosquitoes. *Proceedings of the National Academy of Sciences*, **108** (19), 8026–8029, and S1.

Berenger, J.M.B., Almeras, L., Leulmi, H. and Parola, P. (2015) A high-performance vacuum cleaner for bed bug sampling: A useful tool for medical entomology. *Journal of Medical Entomology*, **52** (3), 513–515.

Blais, T. (2013) *Edmonton Condo Fire Caused by Calgary Exterminator Prompts Lawsuit From Family Whose Child was Left Permanently Disabled*, http://www.edmontonsun.com/2013/07/11/edmonton-condo-fire-caused-by-calgary-exterminator-prompts-lawsuit-from-family-whose-child-was-left-permanently-disabled (accessed 17 November 2016).

Brown, C.R., Page, C.E., Robison, G.A., O'Brien, V.A. and Booth, W. (2015) Predation by ants controls swallow bug (Hemiptera: Cimicidae: *Oeciacus vicarius*) infestations. *Journal of Vector Ecology*, **40** (1): 152–157.

Celmina, H. (1986) *Women in Soviet Prisons*, Paragon House, New York, NY, http://lpra.vip.lv/celmina/22.html (accessed 17 November 2016).

Choe, D.H., Park H., Vo, C. and Knyshov, A. (2016) Chemically mediated arrestment of the bed bug, *Cimex lectularius*, by volatiles associated with exuviae of conspecifics. *PLoS ONE*, **11** (7), e0159520.

Cooper, R., Wang, C. and Singh, N. (2016) Effects of various interventions, including mass trapping with passive pitfall traps, on low-level bed bug populations in apartments. *Journal of Economic Entomology*, **109** (2), 762–769.

DeVries, Z.C., Kells, S.A. and Appel, A.G. (2015) Effects of starvation and molting on the metabolic rate of the common bed bug, *Cimex lectularius* L. *Physiological and Biochemical Zoology*, **88** (1), 53–65.

DeVries, Z.C., Kells, S.A. and Appel, A.G. (2016) Estimating the critical thermal maximum (CTmax) of bed bugs, *Cimex lectularius*: Comparing thermolimit respirometry with traditional visual methods. *Comparative Biochemistry and Physiology Part A: Molecular & Integrative Physiology*, **197**, 52–57.

Doggett, S.L. (2013) *A Code of Practice for the Control of Bed Bug Infestations in Australia*, 4th edn, http://medent.usyd.edu.au/bedbug/ (accessed 13 September 2016).

Doggett, S.L, Geary, M.J. and Russell, R.C. (2006) Encasing mattresses in black plastic will not provide thermal control of bed bugs, *Cimex* spp. (Hemiptera: Cimicidae). *Journal of Economic Entomology*, **99** (6), 2132–2135.

EGTOP (2014) *Expert Group for Technical Advice on Organic Production Final Report On Plant Protection Products (II). European Commission Directorate-General for Agricultural and Rural Development, Directorate B. Multilateral relations, quality policy, B.4. Organics*, https://ec.europa.eu/agriculture/organic/sites/orgfarming/files/docs/body/egtop-final-report-on-ppp-ii_en.pdf (accessed 3 March 2017).

Fewell, C.L. (1873) Improvement in bed-bug exterminators. U.S. Patent No. **139**,562.

Harned, R.W. and Allen, H.W. (1925) Controlling bedbugs in steam-heated rooms. *Journal of Economic Entomology*, **18** (2), 320–331.

How, Y-F. and Lee, C.Y. (2010) Effects of temperature and humidity on the survival and water loss of *Cimex hemipterus* (Hemiptera: Cimicidae). *Journal of Medical Entomology*, **47** (6), 987–995.

Johnson, C. (1941) The ecology of the bed-bug, *Cimex lectularius* L., in Britain: Report on research, 1935–40. *Journal of Hygiene*, **41** (4), 345–461.

Jourdain, F., Delaunay, P., Bérenger, J.-M., Perrin, Y. and Robert, V. (2016) The common bed bug (*Cimex lectularius*) in metropolitan France. Survey on the attitudes and practices of private- and public-sector professionals. *Parasite*, **23** (38), 1–8.

Khamis, M., Subramanyam, B., Flinn, P.W., Dogan, H. and Gwirtz, J.A. (2011) Susceptibility of *Tribolium castaneum* (Coleoptera: Tenebrionidae) life stages to flameless catalytic infrared radiation. *Journal of Economic Entomology*, **104** (1), 325–330.

Kells, S.A. (2006a) Bed bugs: a systemic pest within society. *American Entomologist*, **52** (2), 107–108.

Kells, S.A. (2006b) Nonchemical control of bed bugs. *American Entomologist*, **52** (2), 109–110.

Kells, S.A. and Goblirsch, M.J. (2011) Temperature and time requirements for controlling bed bugs (*Cimex lectularius*) under commercial heat treatment conditions. *Insects*, **2** (3), 412–422.

Lehnert, M.P., Pereira, R.M., Koehler, P.G., Walker, W. and Lehnert, R.M. (2011) Control of *Cimex lectularius* using heat combined with dichlorvos resin strips. *Medical and Veterinary Entomology*, **25** (4), 460–464.

Lewis, V.R. and Haverty, M.I. (1996) Evaluation of six techniques for control of the Western drywood termite (Isoptera: Kalotermitidae) in structures. *Journal of Economic Entomology*, **89** (4), 922–934.

Liu, Y-B. and Haynes, K.F. (2016) Effects of ultralow oxygen and vacuum treatments on bed bug (Heteroptera: Cimicidae) survival. *Journal of Economic Entomology*, **109** (3), 1310–1316.

Loudon, C. (2017) Rapid killing of bed bugs (*Cimex lectularius* L.) on surfaces using heat: application to luggage. *Pest Management Science*, **73** (1), 64–70.

Mager, M. (1997) Noble family in Fort history. *The Sturgeon Creek Post*, Fort Saskatchewan, AB (3 September), p. 10.

Meek, F. (2003) Focus on public health: Bed bugs bite back. *Pest Control Technology*, **31** (7), 43, 44, 46, 47, 50, 52.

Mellanby, K. (1935) A comparison of the physiology of the two species of bed bug which attack man. *Parasitology*, **27** (1), 111–122.

Mellanby, K. (1939) Low temperature and insect activity. *Proceedings of the Royal Society B: Biological Sciences*, **127** (7), 473–487.

Miller, D.M. and Meek, F. (2004) Cost and efficacy comparison of integrated pest management strategies with monthly spray insecticide applications for German cockroach (Dictyoptera: Blattellidae) control in public housing. *Journal or Economic Entomology*, **97** (2), 559–569.

Moore, D.J. and Miller, D.M. (2006) Laboratory evaluations of insecticide product efficacy for control of *Cimex lectularius*. *Journal of Economic Entomology*, **99** (6), 2080–2086.

Naylor, R.A. and Boase, C.J. (2010) Practical solutions for treating laundry infested with *Cimex lectularius* (Hemiptera: Cimicidae). *Journal of Economic Entomology*, **103** (1), 136–139.

NYC DOHMH (2012) *Preventing and Getting Rid of Bed Bugs Safely: A Guide for Property Owners, Managers and Tenants. New York City Department of Health and Mental Hygiene, New York, NY*, http://www1.nyc.gov/assets/doh/downloads/pdf/vector/bed-bug-guide.pdf (accessed 21 February 2017).

Olson, J.F., Eaton, M., Kells, S.A., Morin, V. and Wang, C. (2013) Cold tolerance of bed bugs and practical recommendations for control. *Journal of Economic Entomology*, **106** (6), 2433–2441.

Olson, J.F., Moon, R.D. and Kells, S.A. (2009) Off-host aggregation behavior and sensory basis of arrestment by *Cimex lectularius* (Heteroptera: Cimicidae). *Journal of Insect Physiology*, **55**, 580–587.

Olson, J.F., Ver Vers, L.M., Moon, R.D., and Kells, S.A. (2017) Two compounds in bed bug feces are sufficient to elicit off-host aggregation by bed bugs, *Cimex lectularius*. *Pest Management Science*, **73**, 198–205.

Pappalardo, G., Chinnici, G. and Pecorino, B. (2017) Assessing the economic feasibility of high heat treatment, using evidence obtained from pasta factories in Sicily (Italy). *Journal of Cleaner Production*, **142** (4), 2435–2445.

Pereira, R.M., Koehler, P.G., Pfiester, M. and Walker, W. (2009) Lethal effects of heat and use of localized heat treatment for control of bed bug infestations. *Journal of Economic Entomology*, **102** (3), 1182–1188.

Puckett, R.T., McDonald, D.L. and Gold, R.E. (2013) Comparison of multiple steam treatment durations for control of bed bugs (*Cimex lectularius* L.). *Pest Management Science*, **69** (9), 1061–1065.

Querner, P. and Kjerulff, A.K., (2013) Non-chemical methods to control pests in museums: An overview. In: *International Congress on Science and Technology for the Conservation of Cultural Heritage, Santiago de Compostela, Spain, Oct 02–05, 2012*, CRC Press, Boca Raton, FL.

Rempoulakis, P., Castro, R., Nemny-Lavy, E. and Nestel, D. (2015) Effects of radiation on the fertility of the Ethiopian fruit fly, *Dacus ciliatus*. *Entomologia Experimentalis et Applicata*, **155** (2), 1570–7458.

Richardson, H.H. (1943) The action of bean leaves against the bedbug. *Journal of Economic Entomology*, **36** (4), 543–545.

Romero, A., Potter, M.F., Potter, D.A. and Haynes, K.F. (2007) Insecticide resistance in the bed bug: a factor in the pest's sudden resurgence? *Journal of Medical Entomology*, **44** (2), 175–178.

Ross, W.A. (1916) Eradication of the bedbug by superheating. *Canadian Entomologist*, **48** (3), 74–76.

Rukke B.A., Aak, A. and Edgar, K.S. (2015) Mortality, temporary sterilization, and maternal effects of sublethal heat in bed bugs. *PLoS ONE*, **10** (5), e0127555.

Rukke, B.A, Hage, M. and Aak, A. (2017) Mortality, fecundity and development among bed bugs (*Cimex lectularius*) exposed to prolonged, intermediate cold stress. *Pest Management Science*, **73** (5), 838–843.

Selwitz, C. and Maekawa, S. (1998) *Inert Gases in the Control of Museum Insect Pests. Research in Conservation. Getty Conservation Institute, Los Angeles, CA*, http://hdl.handle.net/10020/gci_pubs/inert_gases (accessed 4 March, 2014).

Sever, M.L., Arbes, S.J. Jr., Gore, J.C., *et al.* (2007) Cockroach allergen reduction by cockroach control alone in low-income urban homes: A randomized control trial. *Journal of Allergy and Clinical Immunology*, **120** (4), 849–855.

Shindelar, A.K and Kells, S.A. (2011a) *Controlling Bed Bugs by Hand*, https://www.bedbugs.umn.edu/bed-bug-control-in-residences/controlling-bed-bugs-hand (accessed 3 March 2017).

Shindelar, A.K and Kells, S.A. (2011b) *Using Freezing Conditions to Kill Bed Bugs*, https://www.bedbugs.umn.edu/bed-bug-control-in-residences/freezing (accessed 3 March 2017).

Shindelar, A.K and Kells, S.A. (2011c) *Using Steamers To Kill Bed Bugs*, https://www.bedbugs.umn.edu/bed-bug-control-in-residences/steamers (accessed 3 March 2017).

Shindelar, A.K. and Kells, S.A. (2017) The "*Let's Beat the Bug*! campaign" state-wide active public education against bed bugs in Minnesota. *Journal of Environmental Health*, **79** (7), 22–27.

Strand, M.A. (1977) Pathogens of Cimicidae (Bedbugs). *Bulletin of the World Health Organization*, **55** (Suppl 1), 313–315.

Szyndler, M.W., Haynes, K.F., Potter, M.F., Corn, R.M. and Loudon, C. (2013) Entrapment of bed bugs by leaf trichomes inspires microfabrication of biomimetic surfaces. *Journal of the Royal Society Interface*, **10** (83), 2013.0174.

Tharaldsen, T. (1919) Steam as a bedbug eradicator. *Public Health Reports*, **34** (48), 2713–2714.

Ulrich, K.R., Feldlaufer, M.F., Kramer, M. and St. Leger, R.J. (2014) Exposure of bed bugs to *Metarhizium anisopliae* at different humidities. *Journal of Economic Entomology*, **107** (6), 2190–2195.

Ulrich, K.R., Feldlaufer, M.F., Kramer, M. and St. Leger, R.J. (2015) Inhibition of the entomopathogenic fungus *Metarhizium anisopliae* sensu lato *in vitro* by the bed bug defensive secretions (E)-2-hexenal and (E)-2-octenal. *BioControl*, **60** (4), 517–526.

Wang, C. and Bennett, G.W. (2006) Comparative study of integrated pest management and baiting for German cockroach management in public housing. *Journal of Economic Entomology*, **99** (3), 879–885.

Wang, C., Gibb, T. and Bennett, G.W. (2009a) Evaluation of two least toxic integrated pest management programs for managing bed bugs (Heteroptera: Cimicidae) with discussion of a bed bug intercepting device. *Journal of Medical Entomology*, **46** (3), 566–571.

Wang, C., Gibb, T., Bennett, G.W. and McKnight, S. (2009b) Bed bug (Heteroptera: Cimicidae) attraction to pitfall traps baited with carbon dioxide, heat, and chemical lure. *Journal of Economic Entomology*, **102** (4), 1580–1585.

Wang, C., Singh, N., Cooper, R., Liu, C. and Buczkowski, G. (2013) Evaluation of an insecticide dust band treatment method for controlling bed bugs. *Journal of Economic Entomology*, **106** (1), 347–352.

Wattal, B.L. and Kalra, N.L. (1960) *Pyralis pictalis* Curt. (Pyralidae: Lepidoptera) larvae as predators of eggs of bed bug, *Cimex hemipterus* Fab. (Cimicidae: Hemiptera). *Indian Journal of Malariology*, **14**, 77–79.

Yturralde, K.M., and Hofstetter, R.W. (2012) Efficacy of commercially available ultrasonic pest repellent devices to affect behavior of bed bugs (Hemiptera: Cimicidae). *Journal of Economic Entomology*, **105** (6), 2107–2114.

Yuma Territorial Prison and Museum (2013). *Interesting Stories (of the Yuma Territorial Prison)*, www.yumaprison.org/interesting-stories.html (accessed 26 February 2017).

29

Insecticide Resistance

Alvaro Romero

29.1 Introduction

Insecticides have been the principal means of controlling bed bug (*Cimex* spp.) infestations. In the 1800s and early 1900s, sprays for bed bug control were mostly made of arsenic, mercury, and pyrethrum, with the first two being highly toxic to humans (Usinger, 1966; Potter, 2011). These sprays were most effective against early stage infestations, and direct contact of the insecticide solution with the insects was required to cause a lethal effect. Because of the lack of residual activity of these sprays, heavy infestations often required multiple insecticide treatments to eliminate adult bed bugs or nymphs that were missed in previous treatments (Potter, 2011). A more effective way to eliminate bed bug infestations was achieved with a sulfur fumigant and later, at the beginning of 20th century, with hydrocyanic acid (cyanide) gas (Back, 1937; Usinger, 1966; Potter, 2011). The discovery and wide use of the organochlorine DDT in the 1940s (Barnes, 1945), and later the use of other organochlorines and organophosphates, changed the course of the history of bed bug control, at least in many parts of the world. Broad application of these insecticides effectively controlled infestations and caused bed bug populations to decline for decades in the developed world, albeit not Africa (Buxton, 1945; Potter, 2011). However, reports of insecticide resistance to DDT and other compounds in both the Common bed bug, *Cimex lectularius* L., and the Tropical bed bug, *Cimex hemipterus* (F.), emerged after these insecticides became widely used (Johnson and Hill 1948; Busvine, 1958; Mallis and Miller, 1964; see also Table 29.1).

It was around the beginning of the 21st century that a new wave of insecticide resistance reports appeared. Initially, in the UK in 1998, it was suggested that *C. lectularius* had returned and was resistant to commonly used insecticides (Birchard, 1998). Resistance was then identified in *C. hemipterus* in Tanzania (Myamba *et al.*, 2002) and later confirmed in *C. lectularius* in the UK (Boase *et al.*, 2006) and in the USA (Moore and Miller, 2006; Romero *et al.*, 2007). In the USA, extremely high levels of resistance to pyrethroids, namely deltamethrin and lambda-cyhalothrin, two common active ingredients in insecticides, were reported in a population collected in Cincinnati, Ohio (Romero *et al.*, 2007). Resistance ratios (RR) were reported to be above 6000. Currently, pyrethroid resistance in *C. lectularius* (Boase *et al.*, 2006, Romero *et al.*, 2007, Lilly *et al.*, 2009, 2014, 2016a; Watanabe, 2010, Adelman, *et al.*, 2011; Kilpinen *et al.*, 2011; Tawatsin *et al.*, 2011; Durand *et al.*, 2012; Vander Pan *et al.*, 2014; Dang *et al.*, 2015a; Palenchar *et al.*, 2015) and in *C. hemipterus* (Myamba *et al.*, 2002; Karunaratne *et al.*, 2007; Dang *et al.*, 2015b) is widespread and represents a challenge for bed bug management. Furthermore, some populations of *C. lectularius* in the USA that are resistant to pyrethroids have also been recently identified as highly resistant to various neonicotinoids (Romero and Anderson, 2016) and other populations have been demonstrated to have reduced susceptibility to the pyrrole, chlorfenapyr (Ashbrook *et al.*, 2017). Although the extent of neonicotinoid resistance and the reduced susceptibility of chlorfenapyr in the

Advances in the Biology and Management of Modern Bed Bugs, First Edition.
Edited by Stephen L. Doggett, Dini M. Miller, and Chow-Yang Lee.
© 2018 John Wiley & Sons Ltd. Published 2018 by John Wiley & Sons Ltd.

Table 29.1 Reports of resistance to different insecticide classes in bed bugs (*Cimex* spp.), by year, and in different countries.

Country	Reference
Organochlorines	
USA	Johnson and Hill (1948)
Greece	Livadas and Georgopoulos (1953)
Israel	Cwilich *et al.* (1957)
Japan, Korea	Lofgren *et al.* (1958)
French Guiana, Gambia, Hong Kong, Iran, Ivory Coast, Kenya, Singapore, Somalia, Taiwan	Busvine (1958)
Lebanon	Quarterman and Schoof (1958)
Tanganyika	Smith (1958)
Malaysia, Southern Rhodesia (Zimbabwe)	Reid (1960)
Republic of Dahomey (Benin)	Holstein (1960)
Madagascar	Gruchet (1962)
South Africa	Whitehead (1962)
Tunisia	Juminer and Kchouk (1962)
Tanzania	Armstrong *et al.* (1962)
Borneo, Colombia, Hungary, Indonesia, Italy, Poland, Trinidad, Turkey	WHO (1963)
India	Kalra and Krishnamurthy (1965)
Egypt	Enan (1969)
Lybia	Shalaby (1970)
Egypt	Gaaboub (1971)
New Guinea	Bourke (1973)
Venezuela	Tonn *et al.* (1982)
Brazil	Nagem and Williams (1992)
Sri Lanka	Karunaratne *et al.* (2007)
Thailand	Tawatsin *et al.* (2011)
Organophosphates	
India	Sen (1958)
Israel	Barkai (1964)
Sri Lanka	Karunaratne *et al.* (2007)
Denmark	Kilpinen *et al.* (2011)
Thailand	Tawatsin *et al.* (2011)
Carbamates	
UK	Boase *et al.* (2006)
Sri Lanka	Karunaratne *et al.* (2007)
Australia	Lilly *et al.* (2009)
Thailand	Tawatsin *et al.* (2011)
Pyrethroids	
South Africa	Newberry *et al.* (1991)

Table 29.1 (Continued)

	Country	Reference
	Tanzania	Myamba *et al.* (2002)
	UK	Boase *et al.* (2006)
	USA	Romero *et al.* (2007)
	Sri Lanka	Karunaratne *et al.* (2007)
	Japan	Watanabe (2010)
	Denmark	Kilpinen *et al.* (2011)
	Thailand	Tawatsin *et al.* (2011)
	France	Durand *et al.* (2012)
	Germany	Vander Pan *et al.* (2014)
	Australia	Dang *et al.* (2015a)
	Israel	Palenchar *et al.* (2015)
	Malaysia	Dang *et al.* (2015b
Neonicotinoids		
	USA	Romero and Anderson (2016)
Pyrroles		
	USA	Ashbrook *et al.* (2017)

USA is unknown, detection of bed bugs resistant to these groups of insecticides is concerning because there are currently few chemical options available to manage resistant bed bug populations.

There is ample evidence indicating that bed bugs have the ability to develop a complex of adaptive strategies to overcome the effect of insecticides (Yoon *et al.*, 2008; Romero *et al.*, 2009; Zhu *et al.*, 2010; Adelman *et al.*, 2011; Mamidala *et al.*, 2011, 2012; Koganemaru *et al.*, 2013; Zhu *et al.*, 2013; Dang *et al.*, 2015a,b; Lilly *et al.*, 2016b). Although behavioral resistance has been documented in many insects, there is no scientific evidence that this type of resistance occurs in bed bugs. Therefore, this review will focus on the underlying mechanisms in the development of physiological resistance in bed bugs and discusses the need to better understand these mechanisms.

29.2 Insecticides and Insecticide Resistance in Bed Bugs

Resistance is an evolutionary response of organisms to the presence of continued environmental changes, such as exposure to insecticides. Under strong insecticide selection, resistance is a natural and an unavoidable phenomenon because of the preexisting variation of this trait that arises through random mutations (Conway and Comins, 1979; Mallet, 1989). Resistance develops through the selective survival of a few individuals that have inherited mechanisms that can withstand the action of insecticides. If populations with these individuals are continuously exposed to insecticides, susceptible individuals die while resistant ones survive, breed, and pass the resistant traits to their progeny (Staunton *et al.*, 2008). Insect populations generally develop resistance to insecticides faster when these compounds have been used before or share a mode of action with other compounds (Georghiou, 1986). A reduced bioavailability of insecticide residues over time can promote selection on resistant individuals that would have been killed by initial concentrations (Roush and Tabashnik, 1990).

Different classes of insecticides – namely organochlorines, organophosphates, carbamates, pyrethroids, pyrroles, and neonicotinoids – have been used against bed bugs (Romero *et al.*, 2010; Romero, 2011; Davies *et al.*, 2012; Koganemaru and Miller, 2013). Use of insecticides for bed bug control has been generally followed by the emergence of resistant populations. Resistance to various chemical classes has been reported in at least 22 countries (Table 29.1) and this phenomenon is expected to grow due to the high dependency on the few available insecticides for controlling bed bug infestations today, particularly the pyrethroids (Doggett *et al.*, 2004; Potter *et al.*, 2015). Studying and monitoring resistance in field populations is essential for developing strategies that reduce the impact of this phenomenon on bed bug management. In order to achieve this goal, a better understanding of the mechanisms governing insecticide resistance in bed bugs is essential as is the development of methods to overcome the resistance. The mechanisms identified in bed bugs to date include metabolic, reduced penetration, and target-site resistance.

29.3 Metabolic Resistance

Enzymes are important in the metabolism of endogenous substrates as well as the catabolism of xenobiotics such as plant toxins, drugs, and insecticides (Feyereisen, 2006). Increased metabolic detoxification by Cytochrome P450 monooxygenases (P450s) is one mechanism by which insects become resistant to insecticides (Casida, 1970; Scott *et al.*, 1998). Initial insights on the involvement of this detoxifying mechanism in pyrethroid resistance in *C. lectularius* were demonstrated in synergist studies that used the P450 inhibitor piperonyl butoxide (PBO) (Romero *et al.*, 2009). Piperonyl butoxide synergized the pyrethroid, deltamethrin, in two highly pyrethroid-resistant strains collected in Cincinnati, Ohio and Worcester, Massachusetts, but its impact varied. However, the resistance ratio for each strain after PBO treatment was 174 and 39, respectively (Romero *et al.*, 2009). These results suggested that P450s have some involvement in deltamethrin resistance, but other resistance mechanisms were involved as well. Although synergistic studies offer clues about the role of detoxification enzymes in insecticide resistance, synergists may be imperfect inhibitors or may inhibit more than one group of enzymes, and this might be the case for PBO (Liu, 2015; Lilly *et al.*, 2016a). Recent studies demonstrated that the use of specific synergist inhibitors can give a better picture of the groups of enzymes responsible for metabolic pyrethroid resistance in bed bugs. Pre-treatment of pyrethroid resistant bed bugs with a PBO analogue, EN16/5-1, which inhibits only the detoxifying enzyme esterases, greatly improved the effectiveness of deltamethrin (Lilly *et al.*, 2016a). However, for some strains, the addition of EN16/5-1 resulted in little improvement in mortality compared with PBO, demonstrating that some pyrethroid-resistant strains may use more than one metabolic mechanism to detoxify insecticides (Lilly *et al.*, 2016a).

An additional method for detecting the possible involvement of a metabolic enzyme in resistance is the measurement of the catabolic activity of resistance-associated enzymes on specific substrates. When compared with susceptible counterparts, pyrethroid-resistant bed bugs exhibited increased activity of esterases, glutathione S-transferases (GST), and P450s (Karunaratne *et al.*, 2007; Adelman *et al.*, 2011; Romero and Anderson, 2016). The use of advanced molecular techniques has enabled researchers to make significant progress in identifying genes and mechanisms involved in metabolic resistance in bed bugs, and in understanding how these genes function. Overexpression of resistance-associated genes has been detected in pyrethroid-resistant bed bugs (Adelman *et al.*, 2011; Bai *et al.*, 2011; Mamidala *et al.*, 2011; Koganemaru *et al.*, 2013; Benoit *et al.*, 2016). RNA interference (RNAi) – specifically interference of the membrane-bound protein NADPH-dependent P450 reductase (CPR), an activator of the P450 systems – resulted in reduced insecticide resistance in bed bugs and confirmed the involvement of P450-mediated metabolic detoxification as an important mechanism responsible for pyrethroid resistance in bed bugs (Zhu *et al.*, 2012). Results from synergistic studies, from metabolic enzyme activity, and from transcriptomes not only provide clues about the involvement of metabolic enzymes in bed bug resistance and the relative proportion of resistance that is attributable to them, but also offer important information about the potential use of synergist insecticides for the management of resistant bed bug populations.

29.4 Reduced Penetration Resistance

Reduced cuticular penetration of insecticides has been proposed as another mechanism that plays a role in insecticide resistance (Fine *et al.*, 1963). This mode of resistance involves structural and biochemical changes at the insect's cuticular level that reduce the ability of the insecticide to reach its target site (Plapp and Hoyer, 1968; Sawicki and Lord, 1970). Initial studies on gene expression in *C. lectularius* showed higher expression of cuticular protein genes in a pyrethroid-resistant than in a susceptible strain (Bai *et al.*, 2011; Mamidala *et al.*, 2012; Koganemaru *et al.*, 2013). Overexpression of cuticular genes in resistant *C. lectularius* might result in greater deposition of cuticular proteins that would interfere with the penetration of insecticides (Koganemaru *et al.*, 2013). Involvement of cuticular proteins in resistance was additionally inferred by a higher resistance ratio in a pyrethroid-resistant strain when pyrethroid was applied topically rather than sub-cuticularly (Adelman *et al.*, 2011; Koganemaru *et al.*, 2013).

In a recent study in *C. lectularius*, measurement of cuticle thickness at the insect's appendages showed that highly resistant bed bugs possess a thicker cuticle than do less tolerant individuals within the same strain (Lilly *et al.*, 2016b). Furthermore, cuticle thickness was positively correlated with increased time-to-knockdown during forced exposure to a pyrethroid bioassay (Lilly *et al.*, 2016b). In another study in *C. lectularius*, mRNA levels of four P450s (CYP397A1, CP398A1, CYP6DN1, and CYP4-CM1), one esterase (ClCE21331), and two Abc transporters, were overexpressed in the integument of pyrethroid resistant strains when compared to their expression in susceptible strains (Zhu *et al.*, 2013). Knocking down of the four P450s and the two Abc transporters with RNAi increased the susceptibility of a strain to pyrethroids in insecticide bioassays, indicating that different resistance mechanisms can coexist at the insect entry level, preventing or slowing down the penetration of insecticides (Zhu *et al.*, 2013). Future studies on insecticide resistance associated with the cuticle should include comparison of thickness, configuration, and composition of cuticle proteins between resistant and susceptible bed bugs to establish if these differences have a causal link to resistance. This information can set the foundation for improvements of insecticide formulations to enhance insecticide penetration through the cuticle.

29.5 Target-site Resistance

Pyrethroids and DDT exert their toxic effect on arthropods by binding the voltage-gated sodium channels (VGSC) in nerve membranes and altering their function (Dong *et al.*, 2014). Binding of the insecticide to the VGSCs keeps a constant influx of sodium through the membrane, with consequent continued depolarization. This effect on the membrane leads to paralysis and death of the insect (Soderlund, 2008). A frequent resistance mechanism found in insects that are resistant to pyrethroids and DDT consists of substituting the amino acid sequences of the VGSC (Rinkevich *et al.*, 2013). These substitutions in VGSC are known as "knockdown resistance" (*kdr*) and they eliminate or reduce the affinity of the insecticide to their target site, reducing the toxicity of DDT and pyrethroids (Davies *et al.*, 2007). Knockdown resistance is a widely reported phenomenon in many arthropods, causing loss of efficacy of insecticides that has led to pest resurgences and increases in the incidence of many vector-borne diseases (Krogstad, 1996).

Over the last few years, growing evidence from molecular and insecticide studies confirms the involvement of *kdr* mutations in VGSCs as one of the mechanisms responsible for pyrethroid resistance in bed bugs (Yoon *et al.*, 2008; Zhu *et al.*, 2010; Adelman *et al.*, 2011; Dang *et al.*, 2015a,b). The first *kdr* mutations of the VGSC α-subunit gene in bed bugs – the valine to leucine mutation (V419L) and the leucine to isoleucine mutation (L925I) –were detected in a *C. lectularius* sample (the NY-BB population) collected in New York (Yoon *et al.*, 2008). These mutations confer pyrethroid resistance in other arthropods (Dong *et al.*, 2014). Low mortality responses in the NY-BB population during insecticide bioassays provided evidence that these mutations were conferring resistance to pyrethroids in this population (Yoon *et al.*, 2008).

The extent of *kdr* mutations in *C. lectularius* across the USA was subsequently investigated in approximately 100 bed bug populations (Zhu *et al.*, 2010). In this study, four haplotypes were identified according to the presence of each mutation:

- haplotype A indicated no mutations were present at the target site
- haplotype B indicated that only the L925I mutation was present
- haplotype C indicated that the L925I and V419L mutations were present
- haplotype D indicated that only the V419L mutation was present.

Overall, from the 93 bed bug samples screened for *kdr* mutations, 88% possessed at least one of the two *kdr* mutations indicating that target-site insensitivity was widespread across the USA (Zhu *et al.*, 2010). In the same study, 17 bed bug populations with *kdr* mutations were challenged with a discriminating dose of deltamethrin and resistance was indicated, suggesting that the possession of at least one of the mutations was enough to confer resistance to deltamethrin (Zhu *et al.*, 2010). However, two bed bug populations (CIN-1 and CIN-3) collected in Cincinnati Ohio that did not show either of the two mutations (haplotype A) survived the discriminating doses of deltamethrin. This result suggested that *kdr* alone was not responsible for pyrethroid resistance and that other resistance mechanisms were involved (Zhu *et al.*, 2010). Complementary studies with the P450 inhibitor PBO (Romero *et al.*, 2009) and gene silencing of NADPH-dependent P450 reductase (CPR) (Zhu *et al.*, 2012) confirmed that P450s were responsible for pyrethroid resistance in CIN-1.

Molecular studies in bed bug populations from elsewhere not only confirm the presence of *kdr* resistance in *C. lectularius*, but also give insights about when they emerged (Durand *et al.* 2012; Dang *et al.*, 2015a; Palenchar *et al.*, 2015). Analysis of preserved *C. lectularius* that were collected in Korea in 1993, and in Australia between 1994 and 2002, did not contain any of the L925I and V419L *kdr* mutations reported in the USA (Seong *et al.*, 2010). These results contrast with those observed with Australian *C. lectularius* samples collected between 2004 and 2013, which contained at least the L925I mutation (Dang *et al.*, 2015a). The above findings suggest that the *kdr*-mutations V419L and L925I emerged at the beginning of the millennium, during the time when bed bug infestations started to become noticeable in many parts of the world and probably as a result of selection with insecticides (Doggett *et al.*, 2004; Potter, 2006; Romero *et al.*, 2007; Davies *et al.*, 2012; Koganemaru and Miller, 2013). Analysis of the frequency of the resistance gene in *C. lectularius* bed bugs collected at various times within a US military base in South Korea also suggests that some *kdr* mutations have been selected more intensively than others (Seong *et al.*, 2010). Seong and colleagues (2010) analyzed bed bugs collected in 2007 and found a 100% frequency of both the V419L and L925I mutations. However, specimens collected in the same place one year later had a low frequency of the V419L mutation (8%), when compared with the L925I mutation (86%) (Seong *et al.*, 2010). A further segregation seems to have occurred in 2009, when bed bug samples contained only the mutation L925I (Seong *et al.*, 2010). The predominance of L925I was also reported in specimens collected in 2011 in France (Durand *et al.*, 2012) and between 2011 and 2012 in Israel (Palenchar *et al.*, 2015). Taken together, these results support that hypothesis that *kdr* mutations have been subjected to differential selection pressures and some fitness costs might be associated with the presence of certain resistant genes.

In contrast to *C. lectularius*, information on insecticide resistance in *C. hemipterus* is very limited. In recent years, pyrethroid resistance in *C. hemipterus* has been reported in Africa (Myamba *et al.*, 2002), Sri Lanka (Karunaratne *et al.*, 2007), Thailand (Tawatsin *et al.*, 2011; Dang *et al.*, 2015b), Australia, India, Kenya, and Malaysia (Dang *et al.*, 2015b). A recent study identified six point mutations on the VGSC gene in

C. hemipterus samples collected in Australia, Thailand, India, Kenya, and Malaysia (Dang *et al.*, 2015b). Unlike the pyrethroid-resistant strains of *C. lectularius*, *C. hemipterus* did not contain the V419L and L925I *kdr* mutations. Instead, six mutations were identified in *C. hemipterus* (Dang *et al.*, 2015b), two of which – M918I (methionine 918 to isoleucine), and L1014F (leucine 1014 to phenylalanine) – have been previously associated with pyrethroid resistance in other insects (Williamson *et al.*, 1996; Rinkevich *et al.*, 2013).

29.6 Evolution of Resistance and Fitness Costs

Bed bugs have the ability to quickly develop resistance to insecticides when subjected to chemical pressure. In a laboratory study, bed bug populations were shown to have decreased susceptibility to the pyrethroid beta-cyfluthrin after only one generation of selection (Gordon *et al.*, 2014). This rapid evolution of insecticide resistance in bed bugs is a concern, as there are few classes of insecticides available today for bed bug management. In addition, cross resistance between-classes and within-classes of insecticides could jeopardize the effectiveness of new active ingredients (Basit *et al.*, 2011; Mitchell *et al.*, 2012; Romero and Anderson, 2016).

Development of insecticide resistance can also come with significant disadvantages to resistant populations, reducing some life history traits, when compared with susceptible individuals of the same population (Kliot and Ghanim, 2012; Gordon *et al.*, 2015). Fitness costs in resistant insects is the result of the reallocation of resources and energy for the enhanced production of detoxifying enzymes, increased production of cuticular components, and target-site modification, that otherwise would be used for metabolic and developmental processes (Kliot and Ghanim, 2012). A shortened oviposition period, a shortened generation time, and a decreased lifetime reproductive rate have been observed to accompany pyrethroid resistance in *C. lectularius* (Polanco *et al.*, 2011; Gordon *et al.*, 2015). Theoretical estimates in insecticide-free scenarios have shown that reversion of resistant *C. lectularius* to pre-selection levels of susceptibility would occur within 2 to 6.5 generations (Gordon *et al.*, 2015). Such a return to susceptibility opens the possibility of the use of rotation between compounds or the utilization of non-insecticide control tactics to reduce the spread of resistant populations (Gordon *et al.*, 2015). However, a number of biological factors, such as lack of susceptible alleles in the population, lack of gene flow across infestations, cross resistance between compounds, and limited availability of alternative chemical classes, greatly limit the possibility of reversing resistance in bed bugs (Davies *et al.*, 2012; Gordon *et al.*, 2015).

29.7 Conclusions

Ever since the first reports of bed bug resistance to organochlorine insecticides in the 1940s, scientists have attempted to elucidate the mechanisms governing the evolution of insecticide resistance. Bed bug resistance research has progressed significantly in the last decade, from the initial synergistic studies of metabolic detoxification to complex gene expression studies, from examining single mutations in a target protein to characterization of multiple resistance mechanisms. The outcomes of these studies have provided not only a clearer understanding of the variety of mechanisms involved in resistance, but also a basis for the development of strategies for the management of insecticide resistance in bed bugs. The information generated from the last decade of research along with the recent sequencing of the bed bug genome sets a firm foundation for the identification of genetic markers for insecticide resistance. This will guide the development of diagnostic tools for monitoring and predicting the evolution of insecticide resistance in field bed bug populations, which is an important component in developing successful bed bug management programs.

References

Adelman, Z.N., Kilcullen, K.A., Koganemaru, R., *et al.* (2011) Deep sequencing of pyrethroid-resistant bed bugs reveals multiple mechanisms of resistance within a single population. *PLoS ONE*, **6** (10), e26228.

Armstrong, J.A., Bransby-Williams, W.R. and Huddleston, J.A. (1962) Resistance to dieldrin of *Cimex hemipterus* (Fabricius). *Nature*, **193** (4814), 499–501.

Ashbrook, A.R., Scharf, M.E., Bennett, G.W. and Gondhalekar, A.D. (2017) Detection of reduced susceptibility to chlorfenapyr- and bifenthrin-containing products in field populations of the bed bug (Hemiptera: Cimicidae). *Journal of Economic Entomology*, **110** (3), 1195–1202.

Back, E.A. (1937) *Bedbugs Popular Account of Bionomics and Control of Cimex lectularius, L.* Leaflet. United States Department of Agriculture, **146**.

Bai, X., Mamidala, P., Rajarapu, S.P., Jones, S.C. and Mittapalli, O. (2011) Transcriptomics of the bed bug (*Cimex lectularius*). *PLoS ONE*, **6** (1), e16336.

Barkai, A. (1964) A preliminary survey of bed-bug resistance to organo-phosphorus compounds in Israel. *World Health Organization,Vector Control*, **58**, 1–5.

Barnes, S. (1945) The residual toxicity of DDT to bed-bugs (*Cimex lectularius*, L.). *Bulletin of Entomological Research*, **36** (3), 273–282.

Basit, M., Sayyedb, A.H., Saleema, M.A. and Saeeda, S. (2011) Cross-resistance, inheritance and stability of resistance to acetamiprid in cotton whitefly, *Bemisia tabaci* Genn (Hemiptera: Aleyrodidae). *Crop Protection*, **30**, 705–712.

Benoit, J.B., Adelman, Z.N., Reinhardt, K., *et al.* (2016) Unique features of a global human ectoparasite identified through sequencing of the bed bug genome. *Nature Communications*, **7**, 10165.

Birchard, K. (1998) Bed bugs biting in Britain: only rarely used pesticides are effective. *Medical Post*, **34** (38), 55.

Boase, C.J., Small, G. and Naylor, R. (2006) Interim report on insecticide susceptibility status of UK bedbugs. *Professional Pest Controller*, **Summer**, 6–7.

Bourke, T.V. (1973) Some aspects of insecticide application in malaria control programmes other than the effect on the insect vectors. *Papua New Guinea Agricultural Journal*, **24** (1), 33–40.

Busvine, J.R. (1958) Insecticide resistance in bed bugs. *Bulletin of World Health Organization*, **19** (6), 1041–1052.

Buxton, P.A. (1945) The use of the new insecticide DDT in relation to the problems of tropical Medicine. *Transactions of the Royal Society of Tropical Medicine and Hygiene*, **38** (5), 367–400.

Casida, J.E. (1970) Mixed–function oxidase involvement in the biochemistry of insecticide synergists. *Journal of Agricultural and Food Chemistry*, **18**, 753–772.

Conway, G. R. and Comins, H.N. (1979) Resistance to pesticides. 2. Lessons in strategy from mathematical models. *Span*, **22**, 53–55.

Cwilich, R., Mer, G.G. and Meron, A.V. (1957) Bedbugs resistant to Gamma-BHC (Lindane) in Israel. *Nature*, **179** (4560), 636–637.

Dang, K., Toi, C.S., Lilly, D.G., Bu, W. and Doggett, S.L. (2015a) Detection of knockdown resistance mutations in the common bed bug, *Cimex lectularius* (Hemiptera: Cimicidae), in Australia. *Pest Management Science*, **71** (7), 914–922.

Dang, K., Toi, C.S., Lilly, D.G., *et al.* (2015b) Identification of putative *kdr* mutations in the tropical bed bug, *Cimex hemipterus* (Hemiptera: Cimicidae). *Pest Management Science*, **71** (7), 1015–1020.

Davies, T.G.E., Field, L.M., Usherwood, P.N.R. and Williamson, M.S. (2007) DDT, pyrethrins, pyrethroids and insect sodium channels. *IUBMB Life*, **59** (3), 151–162.

Davies, T.G.E., Field, L.M. and Williamson, M.S. (2012) The re-emergence of the bed bug as a nuisance pest: implications of resistance to the pyrethroid insecticides. *Medical and Veterinary Entomology*, **26** (3), 241–254.

Doggett, S.L., Geary, M.J. and Russell, R.C. (2004) The resurgence of bed bugs in Australia: with notes on their ecology and control. *Environmental Health Journal*, **4** (2), 30–38.

Dong, K., Du, Y.Z., Rinkevich, F.D., *et al.* (2014) Molecular biology of insect sodium channels and pyrethroid resistance. *Insect Biochemistry and Molecular Biology*, **50**, 1–17.

Durand, R., Cannet, A., Berdjane, Z., *et al.* (2012) Infestation by pyrethroid resistance bed bugs in the suburb of Paris, France. *Parasite-Journal De La Societe Francaise De Parasitologie*, **19** (4), 381–387.

Enan, O.H. (1969) Susceptibility of bed-bug *Cimex lectularius* to organophosphate and carbamate insecticides in Alexandria, U.A.R. *Journal of the Egyptian Public Health Association*, **44** (6), 607–613.

Feyereisen, R. (2006) Evolution of insect P450. *Biochemical Society Transactions*, **34**, 1252–1255.

Fine, B.C., Godin, P.J. and Thain, E.M. (1963) Penetration of pyrethrin I labelled with carbon-14 into susceptible and pyrethroid-resistant houseflies. *Nature*, **199** (489), 927–928.

Gaaboub, I.A. (1971) Present status of DDT and dieldrin resistance in the bed-bug, *Cimex lectularius*, in Alexandria district, United Arab Republic. *World Health Organization,Vector Control*, **71**, 1–6.

Georghiou, G.P. (1986) The magnitude of the resistance problem. In: *Pesticide Resistance: Strategies and Tactics for Management*, Washington, DC, National Academies Press.

Gordon, J.R., Goodman, M.H., Potter, M.F. and Haynes, K.F. (2014) Population variation in and selection for resistance to pyrethroid-neonicotinoid insecticides in the bed bug. *Scientific Reports*, **4** (3836), 1–7.

Gordon, J.R., Potter, M.F. and Haynes, K.F. (2015) Insecticide resistance in the bed bug comes with a cost. *Scientific Reports*, **5** (10807), 1–7.

Gruchet, H. (1962) Sensibilite de *Cimex hemipterus* Fabr. 1803, au DDT, a la dieldrine et aux melanges DDT + diazinon et dieldrine + diazinon dans la region de Miandrivazo, Madagascar. *Bulletin de la Societe de Pathologie Exotique*, **54** (6), 1358–1365.

Holstein, M.H. (1960) Resistance a la dieldrine chez *Cimex hemipterus* Pab. au Dahomey, Afrique occidentale. *Bulletin de la Societe de Pathologie Exotique*, **52** (3), 664–668.

Johnson, M.S. and Hill, A.J. (1948) Partial resistance of a strain of bed bugs to DDT residual. *Medical News Letter*, **12** (1), 26–28.

Juminer, B. and Kchouk, M. (1962) Resistance des punaises (*C. lectularius*) de Tunis aux insecticides. *Archives Institute Pasteur Tunis*, **4**, 411–416.

Kalra, R.L. and Krishnamurthy, B.S. (1965) Studies on the insecticide susceptibility and control of bed-bugs. *Bulletin of the Indian Society for Malaria and Other Communicable Diseases*, **2** (3), 223–229.

Karunaratne, S.H.P.P., Damayanthi, B.T., Fareena, M.H.J., Imbuldeniya, V. and Hemingway, J. (2007) Insecticide resistance in the tropical bedbug *Cimex hemipterus*. *Pesticide Biochemistry and Physiology*, **88** (1), 102–107.

Kilpinen, O., Kristensen, M. and Vagn Jensen, K.M. (2011) Resistance differences between chlorpyrifos and synthetic pyrethroids in *Cimex lectularius* population from Denmark. *Parasitology Research*, **109** (5), 1461–1464.

Kliot, A. and Ghanim, M. (2012) Fitness costs associated with insecticide resistance. *Pest Management Science*, **68** (11), 1431–1437.

Koganemaru, R. and Miller, D.M. (2013) The bed bug problem: past, present, and future control methods. *Pesticide Biochemistry and Physiology*, **106** (3), 177–189.

Koganemaru, R., Miller, D.M. and Adelman, Z.N. (2013) Robust cuticular penetration resistance in the common bed bug (*Cimex lectularius* L.) correlates with increased steady-state transcript levels of CPR-type cuticle protein genes. *Pesticide Biochemistry and Physiology*, **106** (3), 190–197.

Krogstad, D.J. (1996) Malaria as a reemerging disease. *Epidemiological Reviews*, **18**, 77–89.

Lilly, D.G., Doggett, S.L., Zalucki, M.P., Orton, C.J. and Russell, R.C. (2009) Bed bugs that bite back: confirmation of insecticide resistance in Australia in the common bed bug *Cimex lectularius*. *Professional Pest Manager*, **Aug/Sep**, 22–24, 26.

Lilly, D.G., Zalucki, M.P., Orton, C.J., Russell, R.C. and Doggett, S.L. (2014) Insecticide resistance in *Cimex lectularius* (Hemiptera: Cimidiae) in Australia. In: *Proceedings of the Eighth International Conference on Urban Pests, 2014, July 20–23, Zurich, Switzerland*. OOK-Press, Veszprém.

Lilly, D.G., Dang, K., Webb, C.E. and Doggett, S.L. (2016a) Evidence for metabolic pyrethroid resistance in the common bed bug (Hemiptera: Cimicidae). *Journal of Economic Entomology*, **109** (3), 1364–1368.

Lilly, D.G., Latham, S.L., Webb, C.E. and Doggett, S.L. (2016b) Cuticle thickening in a pyrethroid-resistant strain of the common bed bug, *Cimex lectularius* L. (Hemiptera: Cimicidae). *PLoS ONE*, **11** (4), e0153302.

Liu, N.N. (2015) Insecticide resistance in mosquitoes: impact, mechanisms, and research directions. *Annual Review of Entomology*, **60**, 5337–559.

Livadas, G.A. and Georgopoulos, G. (1953) Development of resistance to DDT by *Anopheles sacharovi* in Greece. *Bulletin of the World Health Organization*, **8**, 497–511.

Lofgren, C.S., Keller, J.C. and Burden, G.S. (1958) Resistance tests with the bed bug and evaluation of insecticides for its control. *Journal of Economic Entomology*, **51** (2), 241–244.

Mallet, J. (1989) The evolution of insecticide resistance: Have the insects won? *Trends in Ecology and Evolution*, **4** (11), 336–340.

Mallis, A. and Miller, A.C. (1964) Prolonged resistance in the house fly and bed bug. *Journal of Economic Entomology*, **57** (4), 608–609.

Mamidala, P., Jones, S.C. and Mittapalli, O. (2011) Metabolic resistance in bed bugs. *Insects*, **2** (1), 36–48.

Mamidala, P., Wijeratne, A.J., Wijeratne, S., et al. (2012) RNA-Seq and molecular docking reveal multi-level pesticide resistance in the bed bug. *BMC Genomics*, **13** (6), 1–16.

Mitchell, S. N., Stevenson, B.J., Müller, P., et al. (2012) Identification and validation of a gene causing cross-resistance between insecticide classes in *Anopheles gambiae* from Ghana. *Proceedings of the National Academy of Sciences of the United States of America*, **109** (16), 6147–6152.

Moore, D.J. and Miller, D.M. (2006) Laboratory evaluations of insecticide product efficacy for control of *Cimex lectularius*. *Journal of Economic Entomology*, **99** (6), 2080–2086.

Myamba, J., Maxwell, C.A., Asidi, A. and Curtis, C. F. (2002) Pyrethroid resistance in tropical bedbugs, *Cimex hemipterus*, associated with use of treated bednets. *Medical and Veterinary Entomology*, **16** (4), 448–451.

Nagem, R.L. and Williams, P. (1992) Teste de susceptibilidade do percevejo, *Cimex lectularius* L. (Hemiptera, Cimicidae) ao DDT em Belo Horizonte, MG (Brasil). *Revista de Saude Publica*, **26** (2), 125–128.

Newberry, K., Mchunu, Z.M. and Cebekhulu, S.Q. (1991) Bedbug reinfestation rates in rural Africa. *Medical and Veterinary Entomology*, **5** (4), 503–505.

Palenchar, D.J., Gellatly, K.J., Yoon, K.S., et al. (2015) Quantitative sequencing for the determination of *kdr*-type resistance allele (V419L, L925I, I936F) frequencies in common bed bug (Hemiptera: Cimicidae) populations collected from Israel. *Journal of Medical Entomology*, **52** (5), 1018–1027.

Plapp, F.W., Jr. and Hoyer, R.F. (1968) Insecticide resistance in the house fly: decreased rate of absorption as the mechanism of action of a gene that acts as an intensifier of resistance. *Journal of Economic Entomology*, **61** (5), 1298–1303.

Polanco, A.M., Brewster, C.C. and Miller, D.M. (2011) Population growth potential of the bed bug, *Cimex lectularius* L.: A life table analysis. *Insects*, **2** (2), 173–185.

Potter, M.F. (2006) The perfect storm: an extension view on bed bugs. *American Entomologist*, **52** (2), 102–104.

Potter, M.F. (2011) The history of bed bug mangement-with lessons from the past. *American Entomologist*, **57** (1), 14–25.

Potter, M.F., Haynes, K.F. and Fredericks, J. (2015) Bed bugs across America. *Pestworld*, **November/December**, 4–14.

Quarterman, K.D. and Schoof, H.F. (1958) The status of insecticide resistance in arthropods of public health importance in 1956. *American Journal of Tropical Medicine and Hygiene*, **7** (1), 74–83.

Reid, J.A. (1960) Resistance to dieldrin and DDT and sensitivity to malathion in the bed-bug *Cimex hemipterus* in Malaya. *Bulletin of the World Health Organization*, **22** (5), 586–587.

Rinkevich, F.D., Du, Y.Z. and Dong, K. (2013) Diversity and convergence of sodium channel mutations involved in resistance to pyrethroids. *Pesticide Biochemistry and Physiology*, **106** (3), 93–100.

Romero, A. (2011) Moving from the old to the new: insecticide research on bed bugs since the resurgence. *Insects*, **2** (2), 210–217.

Romero, A. and Anderson, T.D. (2016) High levels of resistance in the common bed bug, *Cimex lectularius* (Hemiptera: Cimicidae), to neonicotinoid insecticides. *Journal of Medical Entomology*, **53** (3), 727–731.

Romero, A., Potter, M.F., Potter, D.A. and Haynes, K.F. (2007) Insecticide resistance in the bed bug: a factor in the pest's sudden resurgence? *Journal of Medical Entomology*, **44** (2), 175–178.

Romero, A., Potter, M.F. and Haynes, K.F. (2009) Evaluation of piperonyl butoxide as a deltamethrin synergist for pyrethroid-resistant bed bugs. *Journal of Economic Entomology*, **102** (6), 2310–2315.

Romero, A., Potter, M.F. and Haynes, K.F. (2010) Evaluation of chlorfenapyr for control of the bed bug, *Cimex lectularius* L. *Pest Management Science*, **66** (11), 1243–1248.

Roush, R.T. and Tabashnik, B.E. (1990) *Pesticide Resistance in Arthropods*, New York and London, Chapman and Hall.

Sawicki, R.M. and Lord, K.A. (1970) Some properties of a mechanism delaying penetration of insecticides into houseflies. *Pesticide Science*, **1** (5), 213–217.

Scott, J.G., Liu, N. and Wen, Z.M. (1998) Insect cytochromes P450: diversity, insecticide resistance and tolerance to plant toxins. *Comparative Biochemistry and Physiology Part C: Pharmacology, Toxicology and Endocrinology*, **121** (1), 147–155.

Sen, P. (1958) Insecticide resistance in the bedbug, *Cimex hemipterus* (Fabr.), of Calcut. *Bulletin of the Calcutta School of Tropical Medicine*, **6** (4), 163–164.

Seong, K.M., Lee, D.Y., Yoon, K.S., *et al.* (2010) Establishment of quantitative sequencing and filter contact vial bioassay for monitoring pyrethroid resistance in the common bed bug, *Cimex lectularius*. *Journal of Medical Entomology*, **47** (4), 592–599.

Shalaby, A.M. (1970) Insecticide susceptibility of the bedbug *Cimex lectularius* (Hemiptera; Cimicidae), in Libya. *Journal of the Egyptian Public Health Association*, **45** (6), 485–499.

Smith, A. (1958) Dieldrin-resistance in *Cimex hemipterus* Fabricius in the Pare area of north-east Tanganyika. *Bulletin of the World Health Organization*, **19** (6), 1124–1125.

Soderlund, D.M. (2008) Pyrethroids, knockdown resistance and sodium channels. *Pest Management Science*, **64** (6), 610–616.

Staunton, I., J. Gerozisis and P. Hadlington, P. (2008) *Urban Pest Management in Australia*, Sydney, University of New South Wales Press.

Tawatsin, A., Thavara, U., Chompoosri, J., *et al.* (2011) Insecticide resistance in bedbugs in Thailand and laboratory evaluation of insecticides for the control of *Cimex hemipterus* and *Cimex lectularius* (Hemiptera: Cimicidae). *Journal of Medical Entomology*, **48** (5), 1023–1030.

Tonn, R.J., Nelson, M., Espinola, H. and Cardozo, J.V. (1982) Notes on *Cimex hemipterus* and *Rhodnius prolixus* from an area of Venezuela endemic for Chagas disease. *Bulletin of the Society of Vector Ecologists*, 7, 49–50.

Usinger, R.L. (1966) *Monograph of Cimicidae (Hemiptera - Heteroptera)*, Entomological Society of America, College Park.

Vander Pan, A., Kuhn, C., Schmolz, E., *et al.* (2014) Studies on pyrethroid resistance in *Cimex lectularius* (Hemiptera: Cimicidae), in Berlin, Germany. In: *Proceedings of the Eighth International Conference on Urban Pests, 2014, July 20–23, Zurich, Switzerland*. OOK-Press, Veszprém.

Watanabe, M. (2010) Insecticide susceptibility and effect of heat treatment on bedbug, *Cimex lectularius*. *Medical Entomology and Zoology*, **61** (3), 239–244.

Whitehead, G.B. (1962) A study of insecticide resistance in a population of bed bugs, *Cimex lectularius* L., and a method of assessing effectiveness of control measures in houses. *Journal of the Entomological Society of Southern Africa*, **25** (1), 121.

WHO (1963) Insecticide resistance and vector control. Thirteenth report of the WHO expert committee on insecticides. WHO Technical Report Series. No. 265. Geneva: World Health Organization.

Williamson, M.S., Martinez-Torres, D., Hick, C.A. and Devonshire, A.L. (1996) Identification of mutations in the housefly para-type sodium channel gene associated with knockdown resistance (*kdr*) to pyrethroid insecticides. *Molecular and General Genetics MGG*, **252** (1), 51–60.

Yoon, K.S., Kwon, D.H., Strycharz, J.P., *et al.* (2008) Biochemical and molecular analysis of deltamethrin resistance in the common bed bug (Hemiptera: Cimicidae). *Journal of Medical Entomology*, **45** (6), 1092–1101.

Zhu, F., Wigginton, J., Romero, A., *et al.* (2010) Widespread distribution of knockdown resistance mutations in the bed bug, *Cimex lectularius* (Hemiptera: Cimicidae), populations in the United States. *Archives of Insect Biochemistry and Physiology*, **73** (4), 245–257.

Zhu, F., Sams, S., Moural, T., *et al.* (2012) RNA interference of NADPH-cytochrome P450 reductase results in reduced insecticide resistance in the bed bug, *Cimex lectularius. PLoS ONE*, **7** (2), e31037.

Zhu, F., Gujar, H., Gordon, J.R., *et al.* (2013) Bed bugs evolved unique adaptive strategy to resist pyrethroid insecticides. *Scientific Reports*, **3** (1456), 1–8.

30

Chemical Control

Chow-Yang Lee, Dini M. Miller and Stephen L. Doggett

30.1 Introduction

In a survey of pest management professionals (PMPs), bed bugs were ranked as the most difficult of all urban pests to control (Potter *et al.*, 2015), largely due to issues of insecticide resistance. Successful bed bug management relies on integrated pest management, encompassing non-chemical means of control, as well as the judicious use of insecticides. Despite the resistance issues, chemicals are often necessary to completely eradicate an infestation. Using the right insecticide in the correct formulation is crucial for successful bed bug management. Throughout history, there have been a number of insecticide classes used against bed bugs. However, some of the most effective insecticides are no longer permitted for use due to environmental and human safety concerns. To date, the majority of insecticide efficacy evaluations have been undertaken on the Common bed bug, *Cimex lectularius* L., with only a limited number of trials on the Tropical bed bug, *Cimex hemipterus* (F.). The published evaluations on *C. lectularius* have mostly been undertaken in the USA, while the majority of those on *C. hemipterus* were completed in Africa, and more recently, in Thailand and Malaysia.

This chapter reviews the literature on chemical control relevant to the modern bed bug resurgence, namely from 1990 to early 2017, focusing on insecticidal classes that are in common use today. For the earlier literature, the reviews of Busvine (1958, 1959), Usinger (1966), and Potter (2011), should be consulted.

30.2 Insecticide Classes used Against Bed Bugs

There are at least 12 classes of insecticides (namely pyrethroids, organophosphates, carbamates, chlorinated hydrocarbons, neonicotinoids, spinosyn, oxidiazolone, phenyl pyrazole, pyrroles, insect growth regulators, botanicals, inorganic compounds, and others) that have been evaluated for the control of *C. lectularius* and *C. hemipterus* since 1990 (Table 30.1). In this chapter, only the major groups employed for the control of bed bugs will be discussed.

30.2.1 Pyrethroids

Pyrethroids are probably amongst the most common compounds used in insecticide products for the control of bed bugs today, despite the high level of resistance to this group (Dang *et al.*, 2017a). To date, 15 pyrethroids have been evaluated (Table 30.1). Amongst these compounds, deltamethrin and permethrin are probably the most common pyrethroids employed, and they occur in various formulations (liquid sprays, pressurized aerosols, dusts, mattress liners, and so on) (Table 30.2).

Advances in the Biology and Management of Modern Bed Bugs, First Edition.
Edited by Stephen L. Doggett, Dini M. Miller, and Chow-Yang Lee.
© 2018 John Wiley & Sons Ltd. Published 2018 by John Wiley & Sons Ltd.

Table 30.1 Insecticide classes and their compounds evaluated for the control of *C. lectularius* and *C. hemipterus* during 1990–2016.

Class	Insecticide(s)	Formulation[1]	Type of evaluation[2]	Species[3]	Bed bug strain(s)[4]	Concentration (mg/m²)[5]	Findings (mortality)	Reference(s)
Chlorinated hydrocarbon	DDT	TG	S-FP	Cl, Ch	field	4%	LT_{50} = 9.1–14.8 d, mortality 0–20%	Tawatsin et al. (2011)
	DDT	TG	S-T	Ch	field	500	No mortality recorded after 72 h	How and Lee (2011)
	dieldrin	TG	S-FP	Cl, Ch	field	0.8, 4%	LT_{50} = 0.3–0.9 d, mortality 0–15%	Tawatsin et al. (2011)
Pyrethroid	alphacypermethrin	SC	S-W	Cl	Sheffield	0.03%	100% mortality at 6 h	Turner and Brigham (2008)
	alphacypermethrin	SC	S, DS	Cl	field		LC_{50} = 353.6 mg/m² at 72 h	Suwannayod et al. (2010)
	betacyfluthrin	LS	TP, S	Cl	N/S	label rate	TP: 65% at 24 h, S: 45% at 10 d	Lilly et al. (2009a, b)
	bifenthrin	LS	S	Cl	Harlan	0.02%	LT_{50} = 0.89 h, 100% in 120 min	Moore and Miller (2006)
	bifenthrin	TG	S	Cl	field	0.5–500 ppm	LC_{50} = 2.3–21.7 ppm	Steelman et al. (2008)
	bifenthrin	WP	S	Cl	Sheffield	0.05%	100% mortality after 6 h	Turner and Brigham (2008)
	bifenthrin	WP	S, DS	Cl	field		LC_{50} = 1767.8 mg/m² at 72 h	Suwannayod et al. (2010)
	bifenthrin	SC	S-T	Ch	field	50	LT_{50} = 1.4–4.3 h	How and Lee (2011)
	bifenthrin	EC	DS	Cl, Ch	field	1000 mg/l	LT_{50} = 7.5 d to no mortality after 7 d	Tawatsin et al. (2011)
	bifenthrin	LS	S-FP S-GV	Cl	Harlan, field	0.014 µg/cm²	40–100% mortality after 3 d	Ashbrook et al. (2017)
	cypermethrin	EC	DS	Cl, Ch	field	1000 mg/l	LT_{50} = 6.4 d to no mortality after 7 d	Tawatsin et al. (2011)
	cyfluthrin	TG	S-FP	Cl, Ch	field	0.15%	LT_{50} = 11.5–15.1 d	Tawatsin et al. (2011)
	cyfluthrin	dust	S	Cl	field	200 mg/cm²	100% in 24 h for all 4 field strains	Romero et al. (2009a)
	cyfluthrin	dust	L, F	Cl	Lab, field		84–99% (lab), 95% field reduction	Wang et al. (2013b)
	cyphenothrin	TG	TP, S, DS	Cl	JESC	0.001–1.0 µg	TP: 100% (0.001 µg), S: 100% (mg/m²)	Okamoto et al. (2010)
	deltamethrin	WP, MC	F	Cl, Ch	field	15–25	Elimination recorded in 33–80% huts	Newberry (1991)
	deltamethrin	SC	S	Cl	Harlan, Field	0.06%	$LT_{50\ Harlan}$ = 1.0 h, $LT_{50\ field}$ = 343.5 h	Moore and Miller (2006)

Pyrethroid	deltamethrin	SC	S	Cl	Monheim	15	LT_{50} = 57–132 min	Barile et al. (2008)
	deltamethrin	SC	F	Cl	field	22.7	50–75% reduction after 3 d	Barile et al. (2008)
	deltamethrin, pyrethrin	EC	F	Cl	field	22.7/40.6	>95% reduction after 3 d	Barile et al. (2008)
	deltamethrin	WG	S	Cl	Sheffield	0.02%	100% mortality after 4 h	Turner and Brigham (2008)
	deltamethrin	LS	TP, S	Cl	N/S	Label rate	TP: 60% at 24 h, S: 38% at 10 d	Lilly et al. (2009a, b)
	deltamethrin	dust	S	Cl	field	200 mg/cm^2	5–100% among 4 strains tested	Romero et al. (2009a)
	deltamethrin	LS	BR	Cl	field	0.06%	Avoided resting on treated filter paper	Romero et al. (2009a)
	deltamethrin	LS	TP, S	Cl	field		LC_{50} = 0.00051 to more than 1.32 mg/cm^2	Romero et al. (2009a)
	deltamethrin	TG	S-FP	Cl, Ch	field	0.05%	LT_{50} = 10.7–18.0 d	Tawatsin et al. (2011)
	deltamethrin	N/A	S	Cl	field	0.03%	84–100% mortality	Tahir and Malik (2014)
	deltamethrin	PA	S	Cl	Earl, Jersey	0.01 mg/cm^2	92–100% mortality in 10 d	Choe and Campbell (2014)
	deltamethrin	PA	TP	Cl	Earl, Jersey	1, 2, 5 ng	50–90% mortality in 10 d	Choe and Campbell (2014)
	deltamethrin	TG	TP	Cl	i2L		LD_{50} = 0.221–0.572 ng	DeVries et al. (2015)
	esfenvalerate	SC	DS	Cl, Ch	field	500 mg/l	LT_{50} = 7.7–54.9 d	Tawatsin et al. (2011)
	etofenprox	TG	S-FP	Cl, Ch	field	0.5%	LT_{50} = 6.8–20.0 d	Tawatsin et al. (2011)
	etofenprox	EW	DS	Cl, Ch	field	4000 mg/l	LT_{50} = 8.0–37.7 d	Tawatsin et al. (2011)
	imiprothrin	TG	DS	Cl	JESC	0.008–0.2%	26.7–90% mortality in 24 h	Okamoto et al. (2010)
	lambdacyhalothrin	WP	F	Cl	field	0.038 g/m^2	0–100% mortality	Le Sueur et al. (1993)
	lambdacyhalothrin	CS	S	Cl	Harlan	0.03%	LT_{50} = 0.34 h, 100% in 90 min	Moore and Miller (2006)
	lambdacyhalothrin	TG	S	Cl	field	0.05–500 ppm	LC_{50} = 0.7–5.0 ppm	Steelman et al. (2008)
	lambdacyhalothrin	ME	S	Cl	Sheffield	0.05%	100% mortality after 45 min	Turner and Brigham (2008)
	lambdacyhalothrin	TG	S-FP	Cl, Ch	field	0.05%	LT_{50} = 8.3–15.2 d	Tawatsin et al. (2011)

(*Continued*)

Table 30.1 (Continued)

Class	Insecticide(s)	Formulation[1]	Type of evaluation[2]	Species[3]	Bed bug strain(s)[4]	Concentration (mg/m²)[5]	Findings (mortality)	Reference(s)
Pyrethroid	lambdacyhalothrin	MC	S-T	Ch	field	10	$LT_{50} = 0.5$–1.3 h	How and Lee (2011)
	lambdacyhalothrin	CS	DS, S	Cl	field		DS: 78% (insect), 67.4% (egg), S: 56.7%	Singh et al. (2014)
	lambdacyhalothrin	CS	DS	Cl	field	4.07 mg/cm²	45–68% mortality at 14 d	Wang et al. (2016a)
	permethrin	BN	F	Ch	field	500 mg/m²	Total eradication in all houses after 14 d	Temu et al. (1999)
	permethrin	LS	S	Cl	Harlan	0.05%	$LT_{50} = 1.46$ h	Moore and Miller (2006)
	permethrin	BN	S	Cl	N/S	2% w/w	25% (30 min exposure), 100% (60 min)	Sharma et al. (2006)
	permethrin	TG	S	Cl	field	5–500 ppm	$LC_{50} = 8.2$–37.1 ppm	Steelman et al. (2008)
	permethrin	EC	TP, S	Cl	N/S	label rate	TP: 10% at 24 h, S: 20% at 10 d	Lilly et al. (2009a, b)
	permethrin	TG	TP, S, DS	Cl	JESC	0.001–1.0 µg	TP: 100% (0.01 µg), S: 100% (125 mg/m²)	Okamoto et al. (2010)
	permethrin	EC	S, TP	Cl	lab, field	1%	TP: 100% in 11–99 min, S: 246–1500 min	Watanabe (2010)
	permethrin	TG	S-FP	Cl, Ch	field	0.75%	$LT_{50} = 10.0$–15.3 d	Tawatsin et al. (2011)
	permethrin	ML	S, RP, FD	Cl	field	1.64%	22–100% in 10 d, not repellent, FD: 4–83%	Jones et al. (2013)
	d-phenothrin	OW	S	Cl	Sheffield	0.15%	90% mortality after 48 h	Turner and Brigham (2008)
	d-phenothrin	TG	TP, S, DS	Cl	JESC	0.001–1.0 µg	TP: 100% (0.01 µg), S: 100% (125 mg/m²)	Okamoto et al. (2010)
	phenothrin	EC	S, TP	Cl	lab, field	1%	TP: 100% in 22–182 min, S: 312–13140 min	Watanabe (2010)
	phenothrin	PA	S	Cl	field	1.67 g	12–41% mortality at 14 d	Wang et al. (2016a)
	d-tetramethrin	TG	DS	Cl	JESC	0.04–1.0%	23.3–86.7% mortality in 24 h	Okamoto et al. (2010)
Organophosphate	diazinon	N/A	F	N/A	field	N/S	Effective in reducing populations	Ijumba et al. (2005)
	diazinon	TG	S	Cl	field	5–1000 ppm	$LC_{50} = 9.8$–561.5 ppm	Steelman et al. (2008)
	diazinon	LS	TP, S	Cl	N/S	label rate	TP: 100% in 6 h, S: 100% in 24 h	Lilly et al. (2009a, b)
	diazinon	MC	DS	Ch	field	4600 mg/l	$LT_{50} = 7.7$–34.2 d	Tawatsin et al. (2011)
	dichlorvos	TG	S	Cl	field	5–500 ppm	$LC_{50} = 170.9$–750.1 ppm	Steelman et al. (2008)

	dichlorvos	LS	F	Ch	field	0.5%	100% reduction in 15 d	Dhadwal (2009)
	dichlorvos	RS	L	Cl	field	N/S	100% (insects and eggs) in 14 d	Potter et al. (2010a)
	dichlorvos	EC	S, TP	Cl	lab, field	1%	TP: 100% in 3.8–7.5 min, S: 40–126 min	Watanabe (2010)
	dichlorvos	RS	SF	Cl	Harlan	18.6%	100% (eggs and insects) in 3–7 d	Lehnert et al. (2011)
	dichlorvos	RS	SF	Cl	N/S	18.6%	100% (eggs and insects) in 3–7 d	Pereira and Koehler (2011)
Organophosphate	fenitrothion	LS	F	Cl, Ch	field	1.06/1.47 g/m^2	100% in 30 d	Newberry (1991)
	fenitrothion	TG	TP, S, DS	Cl	JESC	0.001–1.0 µg	TP: 100% (0.1 µg), S: 100% (25 mg/m^2)	Okamoto et al. (2010)
	fenitrothion	EC	TP, S	Cl	lab, field	1%	TP: 100% in 11.1–24.8 min, S: 117–420 min	Watanabe (2010)
	fenitrothion	MC	S-T	Ch	field	50	LT$_{50}$ = 5.4–18.2 h	How and Lee (2011)
	fenitrothion	TG	S-FP	Cl, Ch	field	1%	LT$_{50}$ = 2.9–17.1 d	Tawatsin et al. (2011)
	malathion	LS	F	Ch	field	2%	100% reduction in 15 d	Dhadwal (2009)
	malathion	TG	S-FP	Cl, Ch	field	5%	LT$_{50}$ = 4.8–11.7 d	Tawatsin et al. (2011)
	pirimiphos-methyl	LS	TP, S	Cl	N/S	label rate	TP: 100% in 6 h, S: 100% in 5 h	Lilly et al. (2009a, b)
	pirimiphos-methyl	EC	S, DS	Cl	field		LC$_{50}$ = 14.9 mg/m^2 at 72 h	Suwannayod et al. (2010)
	propetamphos	CS	S, DS	Cl	field		LC$_{50}$ = 6.67 mg/m^2 at 72	Suwannayod et al. (2010)
	tetrachlorvinphos	TG	S	Cl	field	5–500 ppm	N/S	Steelman et al. (2008)
Carbamate	bendiocarb	LS	F	Ch	field	400	Elimination recorded in 9% of huts	Newberry (1991)
	bendiocarb	WP	S	Cl	Monheim	96/120	LT$_{50}$ = 29–59 min	Barile et al. (2008)
	bendiocarb	WP	S	Cl	Sheffield	0.3%	100% mortality after 2 h	Turner and Brigham (2008)
	bendiocarb	LS	TP, S	Cl	N/S	label rate	TP: 65% at 24 h, S: 50% at 10 d	Lilly et al. (2009a, b)
	bendiocarb	TG	S-FP	Cl, Ch	field	0.1%	LT$_{50}$ = 11.0–13.5 d	Tawatsin et al. (2011)
	fenobucarb	EC	DS	Cl, Ch	field	5000 mg/l	LT$_{50}$ = 6.0–42.8 d	Tawatsin et al. (2011)

(Continued)

Table 30.1 (Continued)

Class	Insecticide(s)	Formulation[1]	Type of evaluation[2]	Species[3]	Bed bug strain(s)[4]	Concentration (mg/m³)[5]	Findings (mortality)	Reference(s)
Carbamate	carbaryl	TG	S	Cl	field	5–500 ppm	LC_{50} = 3.4–27.7 ppm	Steelman et al. (2008)
	propoxur	LS	F	Ch	field	2%	100% reduction in 15 d	Dhadwal (2009)
	propoxur	TG	TP, S, DS	Cl	JESC	1.0 µg, 125 mg/m²	TP: 3.3%, S: 4.2%	Okamoto et al. (2010)
	propoxur	OL	TP, S	Cl	lab, field	1%	TP: 100% in 26.5–52 min, S: 6–16.5 min	Watanabe (2010)
	propoxur	TG	S-FP	Cl, Ch	field	0.1%	LT_{50} = 10.4–12.8 d	Tawatsin et al. (2011)
	propoxur	EC	DS	Cl, Ch	field		LT_{50} = 7.1 to no mortality after 7 d	Tawatsin et al. (2011)
Neonicotinoid	imidacloprid	TG	S	Cl	field	0.5–500 ppm	LC_{50} = 0.15–6.17 ppm	Steelman et al. (2008)
	imidacloprid	SC	S	Ch	field	5	LT_{50} = 1.4–22.6 h	How and Lee (2011)
	imidacloprid	SC	DS	Cl, Ch	field	500 mg/l	LT_{50} = 0.03–0.9 d	Tawatsin et al. (2011)
	dinotefuran	PA	S	Cl	field	1.22 g	37–97% mortality at 14 d	Wang et al. (2016a)
Pyrrole	chlorfenapyr	LS	S	Cl	Harlan	0.5%	LT_{50} = 243.7 h	Moore and Miller (2006)
	chlorfenapyr	TG	S	Cl	field	5–1000 ppm	LC_{50} = 104.6–617.4 ppm	Steelman et al. (2008)
	chlorfenapyr	SC	S	Cl	Sheffield	0.49%	<40% mortality after 48 h	Turner and Brigham (2008)
	chlorfenapyr	LS	F	Cl	field	0.5% (0.95–2.84 l)	78% reduction in 4 wk	Potter et al. (2008)
	chlorfenapyr	LS	BR	Cl	field	0.6%	Did not avoid treated surfaces	Romero et al. (2009b)
	chlorfenapyr	LS	S, DS	Cl	field	5 g/l	S: LT_{50} = 3.5–4.9 d, DS: LT_{50} = 5.6 d	Romero et al. (2009b)
	chlorfenapyr	PA	S, DS	Cl	field	5 g/l	S: LT_{50} = 2.5 d, DS: LT_{50} = 1.5 d	Romero et al. (2009b)
	chlorfenapyr	LS	DS	Cl, Ch	field	624 mg/l	LT_{50} = 1.5–8.7 d	Tawatsin et al. (2011)
	chlorfenapyr	PA	S	Cl	Earl, Jersey	0.6 mg/cm²	64–88% mortality in 10 d	Choe and Campbell (2014)
	chlorfenapyr	PA	TP	Cl	Earl, Jersey	100, 200, 500 ng	25–100% mortality in 10 d	Choe and Campbell (2014)
	chlorfenapyr	PA	S	Cl	field	0.14 g	75–100% mortality in 14 d	Wang et al. (2016a)
	chlorfenapyr	SC	S-FP, S-GV	Cl	Harlan, field	GV: 78.7 µg/cm²	56–100% mortality in 7 d	Ashbrook et al. (2017)

Insect growth regulator (IGR)	S-methoprene	LS	S	Cl	Camb., field	8, 16, 30, 50 mg/m², 30 and 50 mg/m²	Inhibited development	Naylor et al. (2008)
	S-methoprene	LS	S, DS	Cl	field	3.6/7.4	No effects on development and fecundity	Goodman et al. (2013)
	S-methoprene, pyrethrin	LS	S	Cl	susc., field	30/21.4	30 mg/m² inhibited development	Bajomi et al. (2011, 2012)
	hydroprene	LS	S, DS	Cl	field	65.1/89.2/317.8	89.2 and 317.8 inhibited development and fecundity	Goodman et al. (2013)
Spinosyn	spinosad	TG	S	Cl	field	5–500 ppm	$LC_{50} = 69.3–650.9$ ppm	Steelman et al. (2008)
Oxidiazolone	metoxadiazone	TG	TP, S, DS	Cl	JESC	1.0 µg, 125 mg/m²	TP: 3.3%, S: 4.2% mortality	Okamoto et al. (2010)
Phenyl pyrazole	fipronil	TG	S	Cl	field	0.5–50 ppm	$LC_{50} = 7.8–34.8$ ppm	Steelman et al. (2008)
	fipronil	EC	S	Ch	field	10	$LT_{50} = 6.1–136.8$ h	How and Lee (2011)
	fipronil	EC	DS	Cl, Ch	field	250 mg/l	$LT_{50} = 0.8–12.7$ d	Tawatsin et al. (2011)
Botanical	cedar oil	LS	DS	Cl	Harlan, field	10%	100% mortality in ≤1 h	Hinson et al. (2014)
	cedar oil	LS	DS	Cl	N/S	N/S	80–100% mortality in 1 d	Donahue et al. (2015)
	cedar oil	FM	PD	Cl	Harlan	0.3 ml	No mortality at 5 d	Feldlaufer and Ulrich (2015)
	cinnamon oil	FM	PD, TB	Cl	Harlan	0.3, 30 ml	PD: 100% at 5 d, TB: 0.4% at 5 d	Feldlaufer and Ulrich (2015)

(*Continued*)

Table 30.1 (Continued)

Class	Insecticide(s)	Formulation[1]	Type of evaluation[2]	Species[3]	Bed bug strain(s)[4]	Concentration (mg/m^2)[5]	Findings (mortality)	Reference(s)
Botanical	clove, rosemary, thyme, peppermint	LS	DS	Cl	N/S	N/S	80–100% mortality in 1 d	Donahue et al. (2015)
	clove oil	FM	PD, TB	Cl	Harlan	0.3, 30 ml	PD: 98.7% at 5 d, TB: 0% at 5 d	Feldlaufer and Ulrich (2015)
	clove, peppermint oils	LS	DS	Cl	Harlan, field	0.03/1.0%	100% mortality in 1–2 wk	Hinson et al. (2014)
	geranium oil	FM	PD, TB	Cl	Harlan	0.3, 30 ml	PD: 100% at 5 d, TB: 0.4% at 5 d	Feldlaufer and Ulrich (2015)
	lemon grass oil	FM	PD, TB	Cl	Harlan	0.3, 30 ml	PD: 97.8% at 5 d, TB: 7.9% at 5 d	Feldlaufer and Ulrich (2015)
	d-limonene	LS	DS	Cl	N/S	N/S	80–100% mortality in 1 d	Donahue et al. (2015)
	neem	dust	HT	Cl	field	20 mg	40–70% mortality at 48–144 h	Akhtar and Isman (2013)
	neem	LS	DS	Cl	Harlan, field	70%	<70% mortality in 2 wk	Hinson et al. (2014)
	neem oil	LS	DS	Cl	N/S	N/S	80–100% mortality in 1 d	Donahue et al. (2015)
	neem	FM	PD, TB	Cl	Harlan	0.3, 30 ml	PD: 2.4% at 5 d, TB: 0.8% at 5 d	Feldlaufer and Ulrich (2015)
	peppermint oil	LS	DS	Cl	N/S	N/S	80–100% mortality in 1 d	Donahue et al. (2015)
	peppermint oil	FM	PD, TB	Cl	Harlan	0.3, 30 ml	PD: 100% at 5 d, TB: 18.8% at 5 d	Feldlaufer and Ulrich (2015)
	pyrethrin	LS	TP, S	Cl	N/S	label rate	TP: 0% at 24 h, S: 35% at 10 d	Lilly et al. (2009a, b)
	rosemary oil	FM	PD, TB	Cl	Harlan	0.3, 30 ml	PD: 100% at 5 d, TB: 99.5% at 5 d	Feldlaufer and Ulrich (2015)
	rotenone	dust	HT	Cl	field	20 mg	30–58% mortality in 48–144 h	Akhtar and Isman (2013)
	ryania	dust	HT	Cl	field	20 mg	20–45% mortality in 48–144 h	Akhtar and Isman (2013)
	thyme oil	FM	PD, TB	Cl	Harlan	0.3, 30 ml	PD: 100% at 5 d, TB: 30.7% at 5 d	Feldlaufer and Ulrich (2015)
Inorganic	diatomaceous earth	dust	S	Cl	field	200 mg/cm^2	>90% in 4 d against 4 field strains	Romero et al. (2009a)
	diatomaceous earth	dust	HT	Cl	field	20 mg	85–100% in 48–96 h	Akhtar and Isman (2013)
	diatomaceous earth	dust	F	Cl	field	55–85 g/ apartment	78% in 16 d	Potter et al. (2014a, b)
	limestone	dust	S	Cl	field	200 mg/cm^2	50% in 13 d against 4 field strains	Romero et al. (2009a)

Inorganic	silica gel	dust	S	Cl	N/S	N/S	~100% mortality in 1 d	Donahue et al. (2015)
	silica gel	dust	F	Cl	field	0.13/0.4/ 1.34 mg/cm²	85–100% in 1 d	Potter et al. (2014b)
	silica, pyrethrin, PBO	dust	S	Cl	field	200 mg/cm²	100% in 3 d against 4 field strains	Romero et al. (2009a)
	silica, pyrethrin, PBO	dust	S	Cl	Earl, Jersey	8 mg dust	98–100% in 10 d	Choe and Campbell (2014)
	chlorine dioxide	FM	L	Cl	field	362, 724, 1086 ppm	100% mortality on insects in 0–6 h	Gibbs et al. (2012)
	carbon dioxide	FM	L	Cl	lab, field	19–100%	100% mortality in 24–48 h	Wang et al. (2012)
	carbon dioxide	FM	FC	Cl	N/S	75/90% CO_2	100% in 12 h with 90+% CO_2	Martin and Henderson (2013)
	carbon dioxide	FM	L	Ch	N/S	41–46%	Egg (22.5%), insects (90–100%)	Nanoudon and Chanbang (2014)
Others	DEET	RP	L	Cl	field	2.5, 5.0%	18% (24 h, 2.5%), 100% (2 h, 5%)	Wang et al. (2012)
	disodium octaborate tetrahydrate	LS	DS	Cl	Harlan, field	8.5%	<70% mortality in 2 wk	Hinson et al. (2014)
	picaridin	RP	L	Cl	field	7%	20% (24 h, 7%)	Wang et al. (2012)
	sulfuryl fluoride	FM	F	Cl	field	206.7, 2123.8 cm²	100% mortality in 14 d	Miller and Fisher (2008)
	sulfuryl fluoride	FM	L	Cl	field	285 g-h/m³	100% mortality of eggs and insects	Phillips et al. (2014)
	ozone	FM	L	Cl	N/S		CT_{99} = 294212–2323093 ppm-min	Feston (2015)
Mixture	beta-cyfluthrin, imidacloprid	SC	S	Cl	N/S	21/84 mg/m²	100% mortality in 24 h	Bendeck et al. (2011)
	beta-cyfluthrin, imidacloprid	SC	S, F	Cl	field	0.075%	S: 85–95%, F: 99% reduction in 12 wk	Potter et al. (2012)
	beta-cyfluthrin, imidacloprid	SC	S	Cl	Earl, Jersey	0.075%	60–76% mortality in 10 d	Choe and Campbell (2014)
	beta-cyfluthrin, imidacloprid	SC	DS, S	Cl	field	N/S	DS: 100% (insect), 58% (egg), S: 63.3%	Singh et al. (2014)

(Continued)

Table 30.1 (Continued)

Class	Insecticide(s)	Formulation[1]	Type of evaluation[2]	Species[3]	Bed bug strain(s)[4]	Concentration (mg/m²)[5]	Findings (mortality)	Reference(s)
	beta-cyfluthrin, imidacloprid	SC	F	Cl	field	0.075%	92.9% in 12 wk	Wang et al. (2014)
	beta-cyfluthrin, imidacloprid	SC	F	Cl	field	0.025/0.05%	87% in 14 d	Wang et al. (2015)
	beta-cyfluthrin, imidacloprid	SC	S	Ch	N/S	0.0063, 0.063%	83.3–87.5% (nymphs), 58.3–70.8% (adults)	Majid and Zahran (2015)
	beta-cyfluthrin, imidacloprid	SC	S	Cl	field	4.07 mg/cm²	48–70% mortality at 14 d	Wang et al. (2016a)
	beta-cyfluthrin, imidacloprid	SC	DS	Cl	Harlan, Jersey	0.075%	Reduced egg hatch (87–100%)	Hinson et al. (2016)
	bifenthrin, acetamiprid	LS	S, DS	Cl	field	0.11%	S: 84–100%, DS: 100% in 4 h	Potter et al. (2012)
	bifenthrin, acetamiprid	LS	F	Cl	field	0.06/0.05%	98% reduction in 14 d	Wang et al. (2015)
	bifenthrin, acetamiprid	WP	S	Cl	field	4.07 mg/cm²	90–98% mortality at 14 d	Wang et al. (2016a)
Mixture	lambda-cyhalothrin, thiamethoxam	EC	F	Cl	field	0.03/0.10%	89% reduction in 8 wk	Wang et al. (2015)
	lambda-cyhalothrin, thiamethoxam	EC	S	Cl	field	4.07 mg/cm²	48–95% mortality at 14 d	Wang et al. (2016a)
	kerosene, pyrethrum, malathion	LS	F	Ch	field	0.1/1%	100% reduction in 15 d	Dhadwal (2009)
	pyrethrin, esfenvelerate	TRF	SF	Cl	Harlan, field		2.1–28% (field strains)	Jones and Bryant (2012a, b)
	tetramethrin, cypermethrin	TRF	SF	Cl	Harlan, field		0–66% (field strains)	Jones and Bryant (2012a, b)
	deltamethrin, tetramethrin	LS	TP, S	Cl	N/S	label rate	TP: 90% in 7 d, S: 35% in 10 d	Lilly et al. (2009a, b)
	phenothrin, imidacloprid	PA	S	Cl	field	2.61 g	33–83% mortality at 14 d	Wang et al. (2016a)
	geraniol, cedar oil, sodium lauryl sulfate	LS	F	Cl	field	N/S	92.5% reduction in 12 wk	Wang et al. (2014)

[1] BN, bed net; CS, microencapsulated; EC, emulsifiable concentrate; EW/OW, emulsion; oil in water; FM, fumigant; LS, liquid spray; ME, microemulsion; MC, microencapsulated; ML, mattress liner; N/A, not available; OL, oil liquid; PA, pressurized aerosol; RP, repellent; RS, resin strip; SC, suspension concentrate; TG, technical grade; TRS, Total release fogger; WG, water-dispersible granule; WP, wettable powder.
[2] BR, behavioral response; DS, direct spray; FC, fumigation chamber; FD, feeding; FP, filter paper; F, field evaluation; GV, glass vial; HT, horizontal transfer; L, lab evaluation; PD, petri dish; RP, repellency; S, surface contact evaluation in the lab; SF, lab evaluation at larger scale; TP, topical application; T, tile; TB, trash bag; W, wood.
[3] Cl, C. lectularius; Ch, C. hemipterus.

Table 30.2 Insecticide formulations evaluated for the control of *C. lectularius* and *C. hemipterus* during 1990–2016.

Formulation	Active ingredient(s)	References
Liquid spray	alphacypermethrin	Turner and Brigham (2008), Suwannayod *et al.* (2010)
	bendiocarb	Barile *et al.* (2008), Lilly *et al.* (2009a, b), Newberry (1991), Turner and Brigham (2008)
	betacyfluthrin	Lilly *et al.* (2009a, b), Moore and Miller (2009)
	betacyfluthrin/ imidacloprid	Bendeck *et al.* (2011), Potter *et al.* (2012), Goddard (2013), Choe and Campbell (2014), Singh *et al.* (2014), Wang *et al.* (2014, 2015, 2016), Campbell and Miller (2015), Majid and Zahran (2015), Hinson *et al.* (2016)
	bifenthrin	Moore and Miller (2006), Turner and Brigham (2008), Suwannayod *et al.* (2010), How and Lee (2011), Tawatsin *et al.* (2011), Ashbrook *et al.* (2017)
	bifenthrin/ acetamiprid	Potter *et al.* (2012), Campbell and Miller (2015), Wang *et al.* (2015, 2016)
	cedar oil	Hinson *et al.* (2014), Donahue *et al.* (2015)
	chlorfenapyr	Moore and Miller (2006, 2009), Turner and Brigham (2008), Potter *et al.* (2008), Romero *et al.* (2009b, 2010), Doggett *et al.* (2011), Tawatsin *et al.* (2011), Goddard (2013), Ashbrook *et al.* (2017)
	clove/ peppermint oils	Hinson *et al.* (2014), Singh *et al.* (2014)
	cypermethrin	Turner and Brigham (2008), Tawatsin *et al.* (2011)
	cyphenothrin	Okamoto *et al.* (2010)
	deltamethrin	Lilly *et al.* (2009a, b), Goddard (2013), Newberry (1991), Moore and Miller (2006, 2009), Turner and Brigham (2008), Romero *et al.* (2009c, b)
	deltamethrin/ pyrethrin	Barile *et al.* (2008)
	deltamethrin/ tetramethrin	Lilly *et al.* (2009a, b)
	diazinon	Lilly *et al.* (2009a, b), Tawatsin *et al.* (2011)
	dichlorvos	Dhadwal (2009)
	disodium octaborate tetrahydrate	Hinson *et al.* (2014)
	esfenvalerate	Tawatsin *et al.* (2011)
	etofenprox	Tawatsin *et al.* (2011)
	eugenol	Donahue *et al.* (2015)
	fenitrothion	Newberry (1991), Okamoto *et al.* (2010), How and Lee (2011)
	fenobucarb	Tawatsin *et al.* (2011)
	fipronil	How and Lee (2011), Tawatsin *et al.* (2011)
	geraniol/ cedar oil/ sodium lauryl sulfate	Wang *et al.* (2014)
	hydroprene	Goodman *et al.* (2013)
	imidacloprid	How and Lee (2011), Tawatsin *et al.* (2011)
	imiprothrin	Okamoto *et al.* (2010)
	isopropyl alcohol (Steri-Fab)	Moore and Miller (2009)
	kerosene/ pyrethrum/ malathion	Dhadwal (2009)
	lambdacyhalothrin	Le Sueur *et al.* (1993), Turner and Brigham (2008), Moore and Miller (2006), How and Lee (2011), Anderson and Cowles (2012), Singh *et al.* (2014), Wang *et al.* (2016a)

(Continued)

Table 30.2 (Continued)

Formulation	Active ingredient(s)	References
Liquid spray	lambdacyhalothrin/ thiamethoxam	Wang et al. (2015, 2016)
	d-limonene	Donahue et al. (2015)
	malathion	Dhadwal (2009)
	S-methoprene	Naylor et al. (2008), Goodman et al. (2013)
	S-methoprene/ pyrethrin	Bajomi et al. (2011, 2012)
	metoxadiazone	Okamoto et al. (2010)
	mint, clove, citronella, rosemary	Singh et al. (2014)
	neem oil	Donahue et al. (2015), Feldlaufer and Ulrich (2015), Hinson et al. (2014)
	peppermint oil	Donahue et al. (2015)
	Peppermint, sodium lauryl sulfate	Singh et al. (2014)
	permethrin	Moore and Miller (2006), Lilly et al. (2009a, b), Okamoto et al. (2010)
	2-phenetyl propionate, geraniol, cedar, eugenol, citronella oil	Singh et al. (2014)
	d-phenothrin	Turner and Brigham (2008), Okamoto et al. (2010)
	pirimiphos-methyl	Lilly et al. (2009a, b), Suwannayod et al. (2010)
	propetamphos	Suwannayod et al. (2010)
Liquid spray	propoxur	Dhadwal (2009), Okamoto et al. (2010), Tawatsin et al. (2011), Goddard (2013)
	pyrethrin	Lilly et al. (2009a, b)
	rosemary oil	Donahue et al. (2015)
	sodium lauryl sulfate, sodium chloride, citric acid, cinnamon, lemon grass, clove	Singh et al. (2014)
	d-tetramethrin	Okamoto et al. (2010)
Pressurized aerosol	chlorfenapyr	Romero et al. (2010), Goddard (2013), Choe and Campbell (2014), Wang et al. (2016a)
	cypermethrin, prallethrin, imiprothrin	Tawatsin et al. (2011)
	cypermethrin, imiprothrin	Tawatsin et al. (2011)
	imiprothrin, permethrin, esbiothrin	Tawatsin et al. (2011)
	deltamethrin	Anderson and Cowles (2012), Choe and Campbell (2014)
	dinotefuran	Goddard (2013), Wang et al. (2016a)
	esbiothrin	Tawatsin et al. (2011)
	etofenprox	Tawatsin et al. (2011)
	hydroprene	Moore and Miller (2009)
	imiprothrin, cyfluthrin, d-allethrin, permethrin	Tawatsin et al. (2011)
	permethrin, d-allethrin, S-bioallethrin, d-tetramethrin	Tawatsin et al. (2011)
	phenothrin	Wang et al. (2016a)
	phenothrin, imidacloprid	Wang et al. (2016a)
	d-tetramethrin, prallethrin	Tawatsin et al. (2011)
	tetramethrin, cypermethrin	Tawatsin et al. (2011)

Table 30.2 (Continued)

Formulation	Active ingredient(s)	References
Total-release fogger	cypermethrin	Jones and Bryant (2012a, b)
	pyrethrin, esfenvelerate	Jones and Bryant (2012a, b)
	tetramethrin, cypermethrin	Jones and Bryant (2012a, b)
Resin strip	dichlorvos	Potter *et al.* (2010a), Pereira and Koehler (2011)
Bed net	permethrin	Temu *et al.* (1999)
	permethrin (Olyset net)	Sharma *et al.* (2006)
Fumigant	acetaphenone	Feldlaufer and Ulrich (2015)
	carbon dioxide	Martin and Henderson (2013), Nanoudon and Chanbang (2014)
	cedar wood	Feldlaufer and Ulrich (2015)
	chlorine dioxide	Gibbs *et al.* (2012)
	cinnamon	Feldlaufer and Ulrich (2015)
	citronella	Feldlaufer and Ulrich (2015)
	clove	Feldlaufer and Ulrich (2015)
	geranium	Feldlaufer and Ulrich (2015)
	lemon grass	Feldlaufer and Ulrich (2015)
	neem	Feldlaufer and Ulrich (2015)
	peppermint	Feldlaufer and Ulrich (2015)
	rosemary	Feldlaufer and Ulrich (2015)
	sulfuryl floride	Miller and Fisher (2008)
	thyme	Feldlaufer and Ulrich (2015)
Dust	cyfluthrin	Romero *et al.* (2009a), Wang *et al.* (2013b)
	deltamethrin	Romero *et al.* (2009a), Anderson and Cowles (2012)
	diatomaceous earth	Potter *et al.* (2009a, 2013, 2014a), Akhtar and Isman (2013, 2016)
	limestone	Moore and Miller (2009), Romero *et al.* (2009a)
	neem	Akhtar and Isman (2013)
	pyrethrin	Anderson and Cowles (2012)
	rotenone	Akhtar and Isman (2013)
	ryania	Akhtar and Isman (2013)
	silica gel	Anderson and Cowles (2012), Potter *et al.* (2014b), Donahue *et al.* (2015)
	silica, pyrethrin, PBO	Romero *et al.* (2009a), Choe and Campbell (2014)
Repellent	DEET	Wang *et al.* (2013a)
	Gamma-methyl tridecalactone	Wang *et al.* (2013a)
	isolongifolenone	Wang *et al.* (2013a)
	picaridin	Wang *et al.* (2013a)
	permethrin	Wang *et al.* (2013a)
	propyl dihydrojasmonate	Wang *et al.* (2013a)
Mattress liner	permethrin	Jones *et al.* (2013, 2015), Hampshire (2015)

Pyrethroids disrupt the normal nerve function of the axon, targeting the sodium channel by affecting the rate of closing of the sodium gate, leading to increasing flow of sodium ions into the cell. The induced repetitive firing of nerves causes excitatory paralysis of insects followed by death (Yu, 2008). The addition of a synergist, such as piperonyl butoxide or MGK-264, may increase the toxicity of pyrethroids by turning off metabolic resistance mechanisms, but the addition of these synergists does not always counteract such mechanisms (Romero et al., 2009c; How and Lee, 2011). Pyrethroids also are now incorporated into mixtures with insecticide compounds of other classes such as neonicotinoids, for example, beta-cyfluthrin with imidacloprid, bifenthrin with acetamiprid, and lambda-cyhalothrin with thiamethoxam. These combinations show a greater level of efficacy when applied for the control of pyrethroid-resistant bed bugs (Gordon et al., 2014; Wang et al., 2016a), as well as their eggs (Hinson et al., 2016).

Despite numerous reports of resistance to pyrethroids in both *C. lectularius* and *C. hemipterus*, pyrethroid-based products continued to be widely employed, and are often misused by consumers who cannot afford the services provided by PMPs (EPA, 2010; see also Chapter 14). Wang et al. (2016b) reported that 72% of 245 residents who applied insecticides used pyrethroids. For detailed discussion on insecticide resistance, see Chapter 29.

30.2.2 Organophosphates and Carbamates

Organophosphates (OPs) and carbamates (CARBs) were widely used for the control of bed bugs in the past, but have now been withdrawn in many countries. They remain available in most Asian countries and are often regarded as a good tool to overcome pyrethroid-resistant bed bugs (see Chapter 6). OPs and CARBs affect the normal function of acetylcholine neurotransmitters post-synapse by inhibiting the function of acetylcholinesterase.

Seven organophosphate compounds had been evaluated, namely diazinon, dichlorvos, fenitrothion, malathion, propetamphos, pirimiphos-methyl, and tetrachlorvinphos (Table 30.1). Generally, organophosphate compounds are applied either as liquid sprays or in the form of impregnated strips (in this case, dichlorvos). In recent years, OPs have no longer been used in Europe and the USA, except as dichlorvos-impregnated strips (Potter et al., 2010b). In contrast to Asia, where OPs are more accepted, Doggett et al. (2012) noted that OPs in the form of liquid sprays are not widely used by PMPs in Australia because of staining and odor issues, although a small number are still registered for use. Despite this, impregnated strips containing the volatile OP, dichlorvos, can be effective for small-scale fumigation of items in garbage bags (Potter et al., 2010a; Pereira and Koehler, 2011). As for the CARBs, between 1991 and 2011, several evaluations of residual sprays of bendiocarb, fenobucarb, and propoxur have been reported.

Between the OPs and CARBs, resistance to the OPs in bed bugs has only rarely been documented. Nevertheless, resistance has been reported for both *C. lectularius* in Thailand (Tawatsin et al., 2011) and Denmark (Kilpinen et al., 2011), and in *C. hemipterus* in Thailand (Tawatsin et al., 2011). Resistance to the CARBs in *C. lectularius* was reported by Boase et al. (2006) and Lilly et al. (2009c), which was likely due to a metabolic-based resistance mechanism (although cuticular resistance cannot be discounted). Field-collected populations of *C. hemipterus* from Thailand also were found to be resistant to propoxur (Tawatsin et al., 2011).

30.2.3 Neonicotinoids

Neonicotinoids, such as:

- imidacloprid in the form of liquid sprays and technical grade solutions (Steelman et al., 2008; How and Lee, 2011; Tawatsin et al., 2011; Romero and Anderson, 2016)
- dinotefuran in the form of
  - pressurized aerosols (Wang et al., 2016a)
  - dusts (Singh et al., 2013)

have been evaluated against one or both species of bed bugs. Neonicotinoids affect the insects by mimicking acetylcholine, and bind to the acetylcholine receptors at post-synapse nerve cells resulting in nerve hyperstimulation and death (Yu, 2008). The performance of imidacloprid alone against *C. hemipterus* was less effective than other insecticide classes, such as the pyrethroids and phenyl pyrazole (How and Lee, 2011). At present, there are at least three formulated products containing a pyrethroid–neonicotinoid mixture (Table 30.1). Recently, Romero and Anderson (2016) reported a high level of neonicotinoid resistance in two strains of *C. lectularius*. If such resistance is widespread, this will have major implications on how pyrethroid–neonicotinoid mixtures are used against already widespread pyrethroid-resistant populations.

30.2.4 Halogenated Pyrroles

Only one halogenated pyrrole, chlorfenapyr, has been evaluated for efficacy and is being used for bed bug management. This is a slow-acting compound that must first be converted into an active form via oxidative removal of the N-ethoxymethyl group, before the metabolite acts on the insect by impairing the ability of mitochondria to produce ATP (Yu, 2008). Chlorfenapyr is available either as a liquid spray formulation (Moore and Miller, 2006; Potter *et al.*, 2008; Turner and Brigham, 2008; Moore and Miller, 2009; Romero *et al.*, 2010; Tawatsin *et al.*, 2011; Goddard, 2014) or as a pressurized aerosol (Romero *et al.*, 2010; Choe and Campbell, 2014; Wang *et al.*, 2016a). Evaluations have shown that the effectiveness of chlorfenapyr against bed bugs is mixed, ranging from no or low performance (Moore and Miller, 2006; Doggett *et al.*, 2011, 2012; Ashbrook *et al.*, 2017), to moderate performance (Potter *et al.*, 2008), to highly effective (Romero *et al.*, 2010; Tawatsin *et al.*, 2011; Wang *et al.*, 2016a). It is important to note that differences in test protocols, insecticide formulations, and bed bug strains may have attributed to this variation, as discussed below.

30.2.5 Insect Growth Regulators

Insect growth regulators (IGRs) disrupt the growth processes of the insects, either by:

- preventing immature insects from becoming adults, or affecting the reproduction of the adults; these are known as "juvenile hormone analogues" (JHAs)
- inhibiting the function of chitin synthetase, which is involved in the synthesis of chitin (an important component of insect exoskeleton), and hence disrupt the normal molting process in insects and cause mortality; these are known as 'chitin synthesis inhibitors' (CSIs).

To date, only a few IGRs have been evaluated, including the JHAs, S-methoprene (Naylor *et al.*, 2008; Goodman *et al.*, 2013), hydroprene (Goodman *et al.*, 2013), and the CSI, triflumuron (Doggett and Russell, 2007). Goodman *et al.* (2013) found that hydroprene was only effective against *C. lectularius* when applied at ten times the label rate. Doggett *et al.* (2012) highlighted the ethical issues associated with the use of IGRs against bed bugs. As the immature insects are required to feed on their host for the insecticide to work, this means that the success of the insecticide fundamentally relies on people being bitten.

30.2.6 Inorganic and Mineral Compounds

Inorganic and mineral compounds such as:

- diatomaceous earth (Todd, 2006; Doggett and Russell, 2007; Romero *et al.*, 2009a; Akhtar and Isman, 2013; Singh *et al.*, 2016a)
- silica gel (Choe and Campbell, 2014; Potter *et al.*, 2014b; Donahue *et al.*, 2015; Singh *et al.* 2016a)
- limestone dusts (Romero *et al.*, 2009a)

are thought to act by absorbing the waxy layer in the insect exoskeleton (Yu, 2008). This layer normally prevents the insect from losing water, so its loss results in increased permeability of the exoskeleton, thereby leading to death via dehydration. The compounds also remove moisture from bed bug harborages, making those spaces less suitable for aggregation (Pinto *et al.*, 2007). For this reason, this group of materials are generally known as "desiccant dusts", although aerosolized formulations are available. These materials are relatively slow-acting and may take between 24 h and several weeks (depending on concentration and particular product used) to provide 100% mortality of the different stages of bed bugs. More recently, Lilly *et al.* (2016a) reported that the mechanisms conferring resistance to pyrethroids, such as cuticular thickening, may have potential secondary impacts on non-synthetic insecticides, including desiccant dusts, by slowing down the rate of kill. Silica gel tends to be faster acting than diatomaceous earth, with the former typically producing a complete kill usually within 24 h (Singh *et al.*, 2016a). Even fully engorged bed bugs may die in a few hours when exposed to silica gels, which raises questions about the mode of action (Potter *et al.* 2014b).

30.2.7 Botanical Insecticides

Botanical insecticide formulations are increasingly popular, especially those containing the essential oils of neem, cedar, clove, peppermint, geranium, lemon grass, cinnamon, rosemary, and thyme; others also exist. In the USA, many formulations for bed bug management are exempt from US EPA registration. Evaluations have shown that most of these products are ineffective as direct spray formulations (Singh *et al.*, 2014) or as fumigants (Feldlaufer and Ulrich, 2015). Singh *et al.* (2014) found two products to be effective as direct sprays, namely EcoRaider and Bed Bug Patrol. The mode of action of these insecticides is generally unknown (Doggett *et al.*, 2012).

30.2.8 Poisonous Gases

Poisonous gases, such as sulfuryl fluoride (Miller and Fisher, 2008; Phillips *et al.*, 2014) and methyl bromide, and inert gases, such as carbon dioxide and ozone (Wang *et al.*, 2012; Martin and Henderson, 2013; Nanoudon and Chanbang, 2014; Feston, 2015) have been used as fumigants for the control of bed bugs. Poisonous gases have the great advantage of potentially being able to penetrate deep into all the hidden areas and harborages of the insect, and can kill eggs and all bed bug stages. Fumigants can effectively kill all bed bugs in infested materials and structures within 18–24 h, depending on the fumigant concentration and respiration rate of the targeted pest. Sulfuryl fluoride is generally the acceptable choice of fumigant in most countries at present. Methyl bromide (bromomethane), an ozone-depleting compound, is still in use as fumigant in several countries for bed bug control in commercial aircraft. Its use was phased out in 2017, except for critical and quarantine uses only. Unlike poisonous gases, the penetration ability of most inert gases is poor and would require a longer period to reduce the bed bug population. Carbon dioxide is an exception, and has been shown to be able to penetrate dense material easily (Wang *et al.*, 2012).

30.3 Insecticide Formulations for Bed Bug Management

At least nine insecticide formulations have been evaluated for the purpose of bed bug management (Table 30.2).

30.3.1 Liquid Sprays

Liquid sprays have been the most common formulation used, mainly for the application of a toxic residue on a surface, aiming to kill bed bugs for days to weeks when the insect walks over a treated surface. They are also used as sprays for direct application on visible insects. Residual insecticide application has been the main method of treatment to control bed bugs since the use of DDT began some seven decades ago, and it continues to be the most common method, albeit with variable efficacy. Formulations such as emulsifiable concentrates, suspension concentrates, microemulsions, microencapsulates, and wettable powders, are normally diluted in water (except a few which require dilution in an organic solvent), and are applied using a compressed air sprayer. Typically, a nozzle producing a fan-spray will be used for flat surfaces such as beds, sofas, baseboards, carpet edges and so on, while a pin-spray nozzle is used for bed bug harborages such as cracks and crevices, and other narrow openings. To date, most of the major groups of insecticides (excluding the mineral based ones) have been formulated as liquid sprays for the control of bed bugs.

30.3.2 Pressurized Aerosols

Pressurized aerosols are self-contained systems that emit a space-spray in a droplet form, which consists of a residual insecticide (which may also include a synergist such as piperonyl butoxide), an oil solvent, and a hydrocarbon-based propellant. They are useful for application into cracks and crevices, and other tight openings where bed bugs harbor. Normally an extension nozzle is included with the aerosol, which allows for targeted dispensing into narrow spaces and voids. Interestingly, pyrethroid-based aerosol formulations applied directly to bed bugs are normally more effective than their liquid-spray counterparts (even with the same active ingredient). Often the aerosol is capable of a complete kill of test insects within 120 min (Doggett *et al.*, 2012), irrespective of resistance level, whereas the liquid spray often results in a minimal kill. The reason for the discrepancy in efficacy between the two formulations is presently unknown.

30.3.3 Dusts

Dusts are dry mixtures of insecticide(s) and a diluent (such as talc or clay particles), although desiccant dusts may have no diluents. Dusts can be puffed into bed bug hiding places to provide a long-lasting effect. Dusts are useful in locations where sprays are not permissible. However, the dust may leave an unsightly white deposit, limiting their use to hidden areas. When formulated with insecticides for the control of bed bugs, dusts primarily contain pyrethroids such as deltamethrin, permethrin, cyfluthrin, or botanical insecticides. Silicon dioxide (also known as silica) is the most efficacious product that can be employed as a residual for the control of resistant bed bugs (Potter *et al.*, 2014b; Singh *et al.*, 2016a) and can be applied as a prophylactic to minimize potential bed bug infestations.

30.3.4 Fumigants

Fumigants, such as sulfuryl fluoride, carbon dioxide and methyl bromide, are used to kill all stages of bed bugs within a contained area. Control is fast with just a single treatment, because fumigants have excellent penetrative ability. However, fumigation provides no residual effect for controlling bed bugs that may be introduced after the treatment process is completed. As it is of high risk to humans, fumigation can only be carried out by specially trained personnel (often only those with a specific license) and usually off site away from the infestation. This poses logistical issues, as items must be moved for treatment. Fumigation is an expensive and tedious exercise, and involves building specially designed trailers, and complex coordination, especially when treating entire apartment buildings, when it must be ensured that all the residents and pets are away during the treatment. When treating a whole building, extreme caution has to be used to ensure it is tightly sealed. Off-site fumigation, which poses lower risks, involves transferring items, such as infested furniture, into a mobile fumigation chamber. Fumigation may also involve the use of volatile insecticides (such as dichlorvos) and some natural products. Feldlaufer and Ulrich (2015) evaluated the potential of several essential oils and a commercial product containing cold-pressed neem oil as fumigants against *C. lectularius*. They tested bed bugs in Petri dishes and in garbage bags with various fumigants, and found that only rosemary oil and a neem-based product caused greater than 99% mortality of bed bugs. Phosphine gas also has been used against bed bugs by untrained individuals and sadly has resulted in a number of deaths (see Chapter 14).

30.3.5 Permethrin-impregnated Fabrics

Permethrin-impregnated fabrics have been manufactured with the aim of killing bed bugs. They can be in two forms: mattress ticking (in other words, the fabric of the mattress itself) or mattress covers. Both are impregnated with permethrin. Most manufacturers claim that the product will kill all insects within 24–48 h, while some manufacturers claim that they repel bed bugs. In a laboratory study, Jones et al. (2013) reported that continuous exposure to a mattress liner containing 1.64% permethrin resulted in 87–100% mortality in 24 h for populations of moderately pyrethroid-resistant and susceptible strains of *C. lectularius*. In contrast, only 22% mortality was achieved when it was tested against a highly pyrethroid-resistant population. Another study, testing the same mattress cover used in Jones et al. (2013), evaluated a moderately resistant strain of *C. lectularius* and found that even continual exposure over 16 days failed to completely kill all bed bugs, and test mortality was ostensibly similar to control mortality (Doggett et al., 2011). Jones et al. (2015) revealed that short exposure (10 min) to the mattress liner decreased the feeding and fecundity of both pyrethroid-resistant and susceptible strains of *C. lectularius*, although the highly resistant strain in the former study was not evaluated in these experiments. Judging from the number of field populations around the world of both *C. lectularius* and *C. hemipterus* that are highly resistant to pyrethroids, it is likely that there will be limited benefit of using such products to manage bed bug infestations in the field. Furthermore, it is normally recommended than any impregnated material is covered with a non-insecticide impregnated fabric to avoid direct contact with the human skin (and possible harmful effects). This will further obviate any efficacy of the treated fabrics in potentially controlling bed bugs. The continual use of pyrethroids in such a manner is also likely to further select bed bug populations for resistance, which could lead to cross resistance to many insecticide classes (Doggett et al., 2011, 2012; Lilly et al., 2016b,c).

Another use of permethrin-impregnated fabric is in mosquito bed nets. For the most part, bed nets impregnated with pyrethroids are unlikely to affect resistant strains of the bed bug.

30.3.6 Insect Repellents

Insect repellents are formulations that aim to prevent the bites of blood-sucking insects such as mosquitoes when applied onto the skin or clothing. They are normally available as liquids (pressurized and pump packs), lotions, creams, foams, wipe-on towelettes, or as solid waxes (Anonymous, 1997). The potential of insect repellents as a personal protection tool to prevent bed bugs bites is worth exploring. Wang et al. (2013a) evaluated a number of novel and conventional repellent active ingredients, and found that DEET (*N, N*-diethyl-*m*-toluamide), isolongifolanone, propyl di-hydrojasmonate, and gamma-methyl tridecalactone, were highly repellent to *C. lectularius* for at least 9 h. They found a longer lasting repellency effect only at a concentration of 25% DEET. At this concentration, DEET may dissolve and damage plastic and synthetic fabric materials.

30.3.7 Total Release Foggers

Total release foggers (bug bombs) were found to be ineffective against *C. lectularius* (Jones and Bryant, 2012b) because the aerosolized particles were unable to penetrate into bed bug harborages. Other insecticide formulations that have or may be used for the control of bed bugs include impregnated strips (with dichlorvos, as mentioned above).

30.4 Factors Affecting Insecticide Efficacy

In published insecticide efficacy studies evaluating bed bug control, it is often difficult to compare individual reported results. Many factors are known to influence the test outcomes of insecticide evaluations.

30.4.1 Test Method

A number of laboratory testing methods have been used for bed bug bioassays (Table 30.1), including:

- topical applications (Okamoto *et al.*, 2010; Choe and Campbell, 2014; Lilly *et al.*, 2016a)
- surface contact (Moore and Miller, 2006; How and Lee, 2011; Singh *et al.*, 2014)
- direct spray methods (Romero *et al.*, 2010; Tawatsin *et al.*, 2011; Hinson *et al.*, 2014).

Topical applications have been used to establish the lethal dose (such as the LD_{50}), while the latter two methods, normally at pre-determined rates, have been used to determine the lethal time (LT_{50}). Insecticide tests have been conducted using both technical grade and formulated products. Technical grade insecticide is normally diluted in an organic solvent such as acetone or ethanol, while formulated products are diluted in water as per label directions, or in a ready-to-use form. Technical grade insecticide provides a better comparison of the efficacy of active ingredients between strains, as well as over time, because formulated products vary in their compositions, which can substantially influence overall efficacy. The compositions are also usually undisclosed and may change without notice.

30.4.2 Test Substrate

When evaluating the efficacy of a product, different types of substrate should be tested, as they may significantly affect the outcome (Wang *et al.*, 2016a). Most researchers use insecticide applied on cellulose-based filter paper (Okamoto *et al.*, 2010; Romero *et al.*, 2010; Tawatsin *et al.*, 2011). Others coat the test compound on glass surfaces and tiles (Turner and Brigham, 2008; How and Lee, 2011), fabric (Fletcher and Axtell, 1993), hardboard panels (Moore and Miller, 2006), or vials (Steelman *et al.*, 2008). Dang *et al.* (2014) used a commercially available pyrethroid-impregnated mosquito mat as a convenient tool to evaluate pyrethroid susceptibility in *C. lectularius*. Different substrates (especially between porous and non-porous surfaces) may result in vastly different test outcomes (Ashbrook *et al.*, 2017; Dang *et al.*, 2017b). For example, Alpine and Bedlam Plus aerosols applied to unpainted wood, had significantly lower residual efficacies than when applied to fabric, painted wood, and vinyl substrates (Wang *et al.*, 2016a). In addition, when acetone-diluted DDT was applied to filter paper during insecticide resistance screening, the irregular size and distribution of DDT crystals after evaporation of acetone resulted in erratic results (Busvine and Barnes, 1947). Lyon (1965) reported that insecticide diluted in a volatile solvent such as acetone, and applied to an absorbent substrate, can result in a crystal bloom on the surface after evaporation, with reduced efficacy resulting. The insecticide molecule also may crystallize between the filter paper fibers, leading to reduced pick-up by the insects (Lee *et al.*, 1997). Furthermore, in efficacy trials for product registration, it is usually necessary to apply products according to label instructions. Thus if the instruction is to "apply to point of runoff" then typically larger volumes will be applied to porous surfaces, which may lead to enhanced efficacy compared with a non-porous surface.

30.4.3 Test Arena

Laboratory tests evaluating insecticide efficacy bed bugs have been conducted in very small areas such as vials (Steelman *et al.*, 2008) and cell culture plates (Romero *et al.*, 2009c). Also used have been Petri dishes (Choe and Campbell, 2014; Feldlaufer and Ulrich, 2015) and 15 × 15 cm ceramic tiles (How and Lee, 2011). Evaluation has also been undertaken in plastic garbage bags when evaluating volatile compounds (Wang *et al.*, 2012; Feldlaufer and Ulrich, 2015). It is presently unknown if arena size affects efficacy.

30.4.4 Experimental Details

Insecticide evaluations have been performed using different numbers of insects. Some researchers use different stages and sexes of bed bugs in their evaluations, while others only use either adult males or late instars.

Others feed their bed bugs some days before the experiment, or evaluate them when starved. Most studies have not offered bed bugs blood meals after treatment, which could produce higher efficacy than if they were fed (Singh *et al.*, 2016b). Some insecticide evaluations expose the test insects on a treated substrate for a designated time period (say, 2 h) and subsequently move the insects to a clean container (without insecticide) to assess the mortality rate at 24 or 48 h. Other studies continuously expose the test insects on the substrate until all insects are dead, or until control mortality becomes unacceptable. The assessment period to evaluate the mortality could be at 24 or 48 h for fast-acting insecticides such as pyrethroids and organophosphates, and 72 h for botanical compounds. Some actives, such as chlorfenapyr, which is very slow-acting and involves bioactivation of the parental compound, may require mortality to be recorded up to 14 days post-treatment (Singh *et al.*, 2016b). Some desiccant dusts may even take longer to produce a complete kill. In addition, it is also a common practice to count morbid and moribund (M/M) insects after the evaluation as dead insects. The problem with this approach was highlighted in a study in which *C. lectularius* were treated with deltamethrin. Feldlaufer *et al.* (2013) demonstrated that some 53% of the insects that were classified as M/M after 24 h exposure were able to recover and feed after they were transferred into a clean container over a two-week period. The authors also demonstrated that there were no significant differences in the mortality rates of adult males and females, hence adult males were recommended for insecticide evaluation to preserve the adult females for colony maintenance purposes. Another option would be to include nymphs in the evaluation, as they represent 80% of the field population, and they may respond differently to the insecticides (Changlu Wang, Rutgers University, unpublished results)

30.4.5 Strain and Bed Bug Species

Almost all evaluations of insecticides have been carried out using *C. lectularius* (Moore and Miller, 2006; Lilly *et al.*, 2009a,b; Romero *et al.*, 2010; Jones and Bryant, 2012b; Wang *et al.*, 2013b, 2016; Choe and Campbell, 2014; Feldlaufer and Ulrich, 2015;) and only a handful using *C. hemipterus* (How and Lee, 2011; Tawatsin *et al.*, 2011). At present, there are several susceptible strains of *C. lectularius* such as the Harlan (Moore and Miller, 2006; Feldlaufer *et al.*, 2013; Feldlaufer and Ulrich, 2015), Monheim (Dang *et al.*, 2014, 2015a), and Earl (Romero *et al.*, 2010; Choe and Campbell, 2014), but no susceptible strain of *C. hemipterus*. How and Lee (2011) collected a field strain of *C. hemipterus* from Kuala Lumpur and cultured the strain in the laboratory for more than eight years, but its pyrethroid resistance level remained high (Dang *et al.*, 2015b). The same researchers also found a Queensland strain of *C. hemipterus* that has maintained high levels of pyrethroid resistance despite being reared in the laboratory for over ten years. Hence all "field" strains should be carefully tested for resistance even though they may have been reared in the laboratory for a number of years.

Bed bugs strains in colonies can be highly heterogeneous; the resistance mechanisms vary between strains, and even within strains, and there can be considerably variability in the number of individuals with a particular resistance mechanism. Over time, if not challenged with insecticides, strains may lose their resistance. Therefore, in published efficacy trials, a value for resistance should always be included, ideally calculated using technical grade insecticide via a probit analysis in a dose–response curve.

Susceptible, Resistant, or Both Strains

For evaluation of insecticide products, especially for the purpose of product registration, both susceptible and resistant strains should be tested. The susceptible strains should be included to ensure that the insecticide is active, in case it fails to control the resistant strain/s. In the past, studies have often employed only a susceptible strain, or field strains that have been maintained in the laboratory for some years (as noted, resistant mechanisms can be lost over time). Most registration authorities do not require that resistant strains are used in efficacy evaluations.

To date, there is no standard protocol for the evaluation of insecticide products for the control of insecticide-resistant bed bugs. Most laboratories have their own protocols, and discrepancies in test results between laboratories are very common. This is probably due in part to the different methodologies employed, but can also be explained by the fact that individual strains of bed bugs possess different resistance mechanisms, which confer variable levels of resistance. Attempts by the US EPA to develop efficacy guidelines in 2013 resulted in an algorithm that was not logistically feasible nor commercially viable. Following public comment, the guidelines were updated in 2017, but the requirements on the bed bug strains to be used still presents a serious challenge and may not be practicable (Doggett, 2017). There is an urgent need to develop standard protocols, or guidelines that are practical, workable, and effective (Doggett, 2015). These require input from experts from academia and industry, as well as from PMPs.

References

Akhtar, Y. and Isman, M.B. (2013) Horizontal transfer of diatomaceous earth and botanical insecticides in the common bed bug, *Cimex lectularius* L.; Hemiptera: Cimicidae. *PloS ONE*, **8** (9), e75626.

Akhtar, Y. and Isman, M.B. (2016) Efficacy of diatomaceous earth and a DE-aerosol formulation against the common bed bug, *Cimex lectularius* Linnaeus in the laboratory. *Journal of Pest Science*, **89** (4), 1013–1021.

Anderson, J.F. and Cowles, R.S. (2012) Susceptibility of *Cimex lectularius* (Hemiptera: Cimicidae) to pyrethroid insecticides and to insecticidal dusts with or without pyrethroid insecticides. *Journal of Economic Entomology*, **105** (5), 1789–1795.

Anonymous (1997) *Chemical Methods for the Control of Vectors and Pests of Public Health Importance*, World Health Organization, Geneva.

Ashbrook, A.R., Scharf, M.E., Bennett, G.W. and Gondhalekar, A.D. (2017). Detection of reduced susceptibility to chlorfenapyr- and bifenthrin-containing products in field populations of the bed bug (Hemiptera: Cimicidae). *Journal of Economic Entomology*, **110** (3), 1195–1202.

Bajomi, D., Szilagyi, J., Naylor, R. and Takacs, L. (2011) S-methoprene formulations: laboratory tests for efficacy against bed bugs. In: *Proceedings of the Eight International Conference on Urban Pests, August 7–10, 2011, Ouro Preto, Brazil*. Instituto Biológico, São Paulo.

Bajomi, D., Papp, G. and SzilagyI, J. (2012) Control of resistant bed bugs with new s-methoprene-based insecticide formulations. *International Pest Control*, **54** (4), 190–194.

Barile, J., Nauen, R., Nentwig, G., Pospischil, R. and Reid, B. (2008) Laboratory and field evaluation of deltamethrin and bendiocarb to control *Cimex lectularius* (Heteroptera: Cimicidae). In: *Proceedings of the Sixth International Conference on Urban Pests, July 13–16, 2008, Budapest, Hungary*. OOK-Press, Hungary, Veszprém.

Bendeck, O.R.P., Bernardini, J.F., Belluco, F., *et al.* (2011) Effectiveness of imidacloprid + betacyfluthrin to control *Cimex lectularius* (Hemiptera, Cimicidae). In: *Proceedings of the Eight International Conference on Urban Pests, August 7–10, 2011, Ouro Preto, Brazil*. Instituto Biológico, São Paulo.

Boase, C.J., Small, G. and Naylor, R. (2006) Interim report on insecticide susceptibility status of UK bedbugs. *Professional Pest Controller*, **Summer**, 6–7.

Busvine, J.R. (1958) Insecticide-resistance in bed-bugs. *Bulletin of the World Health Organization*, **19** (6), 1041–1052.

Busvine, J.R. (1959) Insecticide resistance in bed bugs. *Annals of Applied Biology*, **47** (3), 618–620.

Busvine, J.R. and Barnes, S. (1947) Observations on mortality among insects exposed to dry insecticidal films. *Bulletin of Entomological Research*, **38** (1), 81–90.

Campbell, B.E. and Miller, D.M. (2015) Insecticide resistance in eggs and first instars of the bed bug, *Cimex lectularius* (Hemiptera: Cimicidae). *Insects*, **6** (1), 122–132.

Choe, D. and Campbell, K. (2014) Effect of feeding status on mortality response of adult bed bugs (Hemiptera: Cimicidae) to some insecticide products. *Journal of Economic Entomology*, **107** (3), 1206–1215.

Dang, K., Lilly, D.G., Bu, W. and Doggett, S.L. (2014) Simple, rapid and cost-effective technique for the detection of pyrethroid resistance in bed bugs, *Cimex* spp. (Hemiptera: Cimicidae). *Austral Entomology*, **54** (2), 191–196.

Dang, K., Toi, C.S., Lilly, D.G., Bu, W. and Doggett, S.L. (2015a) Detection of knockdown resistance mutations in the common bed bug, *Cimex lectularius* (Hemiptera: Cimicidae), in Australia. *Pest Management Science*, **71** (7), 914–922.

Dang, K., Toi, C. S., Lilly, D. G., *et al.*, (2015b) Identification of putative *kdr* mutations in the tropical bed bug, *Cimex hemipterus* (Hemiptera: Cimicidae). *Pest Management Science*, **71** (7), 1015–1020.

Dang, K., Doggett, S.L., Singham, G.V. and Lee C.Y. (2017a) Insecticide resistance and resistance mechanisms in bed bugs, *Cimex* spp. (Hemiptera: Cimicidae). *Parasites & Vectors*, **10** (1), 318.

Dang, K., Singham, G.V., Doggett, S.L., Lilly, D.G., and Lee, C.Y. (2017b) Effects of different surfaces and insecticide carriers on insecticide bioassays against bed bugs, *Cimex* spp. (Hemiptera: Cimicidae). *Journal of Economic Entomology*, **110** (2), 558–566.

DeVries, Z.C., Reid, W.R., Kells, S.A. and Appel, A.G. (2015) Effects of starvation on deltamethrin tolerance in bed bugs, *Cimex lectularius* L. (Hemiptera: Cimicidae). *Insects*, **6** (1), 102–111.

Dhadwal, B.S. (2009) Field trial of various insecticides for the control of bedbugs. *Journal of Communicable Diseases*, **41** (1), 57–60.

Doggett, S.L. (2015) Is poor bed bug control due to inadequate efficacy testing? *Pest*, **Aug/Sep**, 26–27.

Doggett, S.L. (2017) Bed bug insecticides, do they do what it says on the tin? *Pest*, **Aug/Sep**, 29–31.

Doggett, S.L. and Russell, R.C. (2007) Bed Bugs – latest trends and development. In: *Proceedings of the Australian Environmental Pest Managers Association National Conference, June 4–6, 2007, Coffs Harbour, Australia.* Australian Environmental Pest Managers Association.

Doggett, S.L., Orton, C.J., Lilly, D.G. and Russell, R.C. (2011) Bed bugs – A growing concern worldwide. Australian and international trends update and causes for concern. In: *Proceedings of the Australian Environmental Pest Managers Association NSW Conference, June 2, 2011, Homebush Bay, Australia.* AEPMA.

Doggett, S.L., Dwyer, D.E., Peñas, P.F. and Russell, R.C. (2012) Bed bugs: clinical relevance and control options. *Clinical Microbiology Reviews*, **25** (1), 164–192.

Donahue, W.A., Showler, A.T., Donahue, M W., *et al.* (2015) Knockdown and lethal effects of eight commercial nonconventional and two pyrethroid insecticides against moderately permethrin-resistant adult bed bugs, *Cimex lectularius* (L.)(Hemiptera: Cimicidae). *Biopesticides International*, **11** (2), 108–117.

EPA (2010) Joint statement on bed bug control in the United States from the U.S. Centers for Disease Control and Prevention (CDC) and the U.S. Environmental Protection Agency (EPA). US Department of Health and Human Services, Atlanta.

Feldlaufer, M.F. and Ulrich, K.R. (2015) Essential oils as fumigants for bed bugs (Hemiptera: Cimicidae). *Journal of Entomological Science*, **50** (2), 129–137.

Feldlaufer, M.F., Ulrich, K.R. and Kramer, M. (2013) No sex-related differences in mortality in bed bugs (Hemiptera: Cimicidae) exposed to deltamethrin, and surviving bed bugs can recover. *Journal of Economic Entomology*, **106** (2), 988–994.

Feston, J. (2015) The effects of ozone gas on the common bed bug (*Cimex lectularius* L.). Masters thesis, Purdue University.

Fletcher, M.G. and Axtell, R.C. (1993) Susceptibility of the bedbug, *Cimex lectularius*, to selected insecticides and various treated surfaces. *Medical and Veterinary Entomology*, **7** (1), 69–72.

Gibbs, S.G., Lowe, J.J., Smith, P.W. and Hewlett, A.L. (2012) Gaseous chlorine dioxide as an alternative for bedbug control. *Infection Control and Hospital Epidemiology*, **33** (5), 495–499.

Goddard, J. (2013) Laboratory assays of various insecticides against bed bugs (Hemiptera: Cimicidae) and their eggs. *Journal of Entomological Science*, **48** (1), 65–69.

Goddard, J. (2014) Long-term efficacy of various natural or "green" insecticides against bed bugs: a double-blind study. *Insects*, **5** (4), 942–951.

Goodman, M.H., Potter, M.F. and Haynes, K.F. (2013) Effects of juvenile hormone analog formulations on development and reproduction in the bed bug *Cimex lectularius* (Hemiptera: Cimicidae). *Pest Management Science*, **69** (2), 240–244.

Gordon, J.R., Goodman, M.H., Potter, M.F. and Haynes, K.F. (2014) Population variation in and selection for resistance to pyrethroid-neonicotinoid insecticides in the bed bug. *Scientific Reports*, **4** (3836), 1–7.

Hampshire, K. (2015) Mattress matters. *Pest Control Technology*, **43** (6), 88, 90.

Hinson, K.R., Benson, E.P., Zungoli, P.A., Bridges, W.C. and Ellis, B.R. (2014) Assessment of natural-based products for bed bug (Hemiptera: Cimicidae) control. In: *Proceedings of the Eighth International Conference on Urban Pests, 2014, July 20–23, Zurich, Switzerland.* OOK-Press, Veszprém.

Hinson, K.R., Benson, E.P., Zungoli, P.A., Bridges, W.C. and Ellis, B.R. (2016) Egg hatch rate and nymphal survival of the bed bug (Hemiptera: Cimicidae) after exposure to insecticide sprays. *Journal of Economic Entomology*, **109** (6), 2495–2499.

How, Y.F. and Lee, C.Y. (2011) Surface contact toxicity and synergism of several insecticides against different stages of the tropical bed bug, *Cimex hemipterus* (Hemiptera: Cimicidae). *Pest Management Science*, **67** (6), 734–740.

Ijumba, J.N., Lyatuu, E., Lawrence, B. and Masenga, C. (2005) Bio-efficacy of diazinon (0,0 diethyl 0-(2-isopropyl-6-methyl-4-pyrimidnyl) phosphorothioate) against *Cimex* and *Pediculus* species at a social welfare camp, Magugu, Babati District, northern Tanzania. *Tanzania Health Research Bulletin*, **7** (1), 16–19.

Jones, S.C. and Bryant, J L. (2012a) Bed bug "bug bombs". *Pest Control Technology*, **40** (10), 106, 112.

Jones, S.C. and Bryant, J.L. (2012b) Ineffectiveness of over-the-counter total-release foggers against the bed bug (Heteroptera: Cimicidae). *Journal of Economic Entomology*, **105** (3), 957–963.

Jones, S.C., Bryant, J.L. and Harrison, S.A. (2013) Behavioral Responses of the Bed Bug to Permethrin-Impregnated *Active*Guard$^{(TM)}$ Fabric. *Insects*, **4** (2), 230–240.

Jones, S.C., Bryant, J.L. and Sivakoff, F.S. (2015) Sublethal effects of *Active*Guard exposure on feeding behavior and fecundity of the bed bug (Hemiptera: Cimicidae). *Journal of Medical Entomology*, **52** (3), 413–418.

Kilpinen, O., Kristensen, M. and Vagn Jensen, K.M. (2011) Resistance differences between chlorpyrifos and synthetic pyrethroids in *Cimex lectularius* population from Denmark. *Parasitology Research*, **109** (5), 1461–1464.

Le Sueur, D., Sharp, B.L., Fraser, C. and Ngxongo, S.M. (1993) Assessment of the residual efficacy of lambda-cyhalothrin. 1. A laboratory study using *Anopheles arabiensis* and *Cimex lectularius* (Hemiptera: Cimicidae) on treated daub wall substrates from Natal, South Africa. *Journal of the American Mosquito Control Association*, **9** (4), 408–413.

Lee C.Y., Loke, KM., Yap, H.H. and Chong, ASC. (1997) Baseline malathion and permethrin susceptibility in field collected *Culex quinquefasciatus* (Say) mosquitoes from Penang, Malaysia. *Tropical Biomedicine*, **14** (1&2), 87–91.

Lehnert, M.P., Pereira, R.M., Koehler, P.G., Walker, W. and Lehnert, M.S. (2011) Control of *Cimex lectularius* using heat combined with dichlorvos resin strips. *Medical and Veterinary Entomology*, **25** (4), 460–464.

Lilly, D.G., Doggett, S.L., Orton, C.J. and Russell, R.C. (2009a) Bed bug product efficacy under the spotlight – Part 1. *Professional Pest Manager*, **13** (February/March), 14–16.

Lilly, D.G., Doggett, S.L., Orton, C. J. and Russell, R. C. (2009b) Bed bug product efficacy under the spotlight – Part 2. *Professional Pest Manager*, **13** (April/May), 14–15, 18.

Lilly, D.G., Doggett, S.L., Zalucki, M.P., Orton, C.J. and Russell, R.C. (2009c) Bed bugs that bite back: confirmation of insecticide resistance in Australia in the common bed bug *Cimex lectularius. Professional Pest Manager*, **13** (August/September), 22–24, 26.

Lilly, D.G., Webb, C.E. and Doggett, S.L. (2016a). Evidence of tolerance to silica-based desiccant dusts in a pyrethroid-resistant strain of *Cimex lectularius* (Hemiptera: Cimicidae). *Insects*, **7**, 74.

Lilly, D.G., Dang, K., Webb, C.E. and Doggett, S.L. (2016b). Evidence for metabolic pyrethroid resistance in the common bed bug (Hemiptera: Cimicidae). *Journal of Economic Entomology*, **109** (3), 1364–1368.

Lilly, D.G., Latham, S.L., Webb, C.E. and Doggett, S.L. (2016c) Cuticle thickening in a pyrethroid-resistant strain of the common bed bug, *Cimex lectularius* L. (Hemiptera: Cimicidae). *PloS ONE*, **11** (4), e0153302.

Lyon, R.L. (1965) Structure and toxicity of insecticide deposits for control of bark beetles. *USDA Technical Bulletins*, **1343**, 1–54.

Majid, A.H.A. and Zahran, Z. (2015) Laboratory bioassay on efficacy of dual mode of action insecticides (beta-cyfluthrin and imidacloprid) towards tropical bed bugs, *Cimex hemipterus* (Hemiptera: Cimicidae). *Journal of Entomology and Zoology Studies*, **3** (5), 217–220.

Martin, J. and Henderson, G. (2013) CO_2 is it a viable option? *Pest Control Technology*, **41** (12), 86–92.

Miller, D.M. and Fisher, M.L. (2008) Bed bug (Hemiptera: Cimicidae) response to fumigation using sulfuryl fluoride. In: *Proceedings of the Sixth International Conference on Urban Pests, July 13–16, 2008, Budapest, Hungary*. OOK-Press, Hungary, Veszprém.

Moore, D.J. and Miller, D.M. (2006) Laboratory evaluations of insecticide product efficacy for control of *Cimex lectularius*. *Journal of Economic Entomology*, **99** (6), 2080–2086.

Moore, D.J. and Miller, D.M. (2009) Field evaluations of insecticide treatment regimens for control of the common bed bug, *Cimex lectularius* (L.). *Pest Management Science*, **65** (3), 332–338.

Nanoudon, S. and Chanbang, Y. (2014) Use of solid carbon dioxide for the controlling of *Cimex hemipterus* (Fabricius) under laboratory conditions. In: *Proceedings of the Joint International Tropical Medicine Meeting, December 2–4, 2014, Bangkok, Thailand*. Faculty of Tropical Medicine, Mahidol University.

Naylor, R., Bajomi, D. and Boase, C. (2008) Efficacy of (s)-methoprene against *Cimex lectularius* (Hemiptera: Cimicidae). In: *Proceedings of the Sixth International Conference on Urban Pests, July 13–16, 2008, Budapest, Hungary*. OOK-Press, Hungary, Veszprém.

Newberry, K. (1991) Field trials of bendiocarb, deltamethrin and fenitrothion to control DDT-resistant bedbugs in KwaZulu, South Africa. *International Pest Control*, **33** (3), 64–68.

Okamoto, H., Sembo, S., Ishiwatari, T. and Miyaguchi, J. (2010) Insecticidal activity of 8 household and hygiene insecticides against bedbug (*Cimex lectularius*). *Medical Entomology and Zoology*, **61** (3), 245–250.

Pereira, R.M. and Koehler, P.G. (2011) Use of heat, volatile insecticide, and monitoring tools to control bed bugs (Heteroptera: Cimicidae). In: *Proceedings of the Eight International Conference on Urban Pests, August 7–10, 2011, Ouro Preto, Brazil*. Instituto Biológico, São Paulo.

Phillips, T.W., Aikins, M.J., Thoms, E., Demark, J. and Wang, C. (2014) Fumigation of bed bugs (Hemiptera: Cimicidae): effective application rates for sulfuryl fluoride. *Journal of Economic Entomology*, **107** (4), 1582–1589.

Pinto, L.J., Cooper, R. and Kraft, S.K. (2007) *Bed Bug Handbook. The Complete Guide to Bed Bugs and their Control*, Pinto & Associates, Mechanicsville, MD.

Potter, M.F. (2011) The history of bed bug mangement-with lessons from the past. *American Entomologist*, **57** (1), 14–25.

Potter, M.F., Haynes, K.F., Romero, A., Hardebeck, E. and Wickemeyer, W. (2008) Is there a new answer? *Pest Control Technology*, **36** (6), 118–124.

Potter, M.F., Haynes, K.F., Goodman, M., Stamper, S. and Sams, S. (2010a) Blast from the past. *Pest Management Professional*, **78** (3), 46–52.

Potter, M.F., Rosenberg, B. and Henriksen, M. (2010b) Bugs without borders. Defining the global bed bug resurgence. *Pestworld*, **Sep/Oct**, 1–12.

Potter, M.F., Haynes, K.F., Gordon, J.R., Hardebeck, E. and Wickemeyer, W. (2012) Dual-action bed bug killers. *Pest Control Technology*, **40** (3), 62, 76.

Potter, M.F., Haynes, K.F., Christensen, C., *et al.* (2013) Where do bed bugs stand: when the dust settles? *Pest Control Technology*, **41** (12), 72–80.

Potter, M.F., Haynes, K.F., Christensen, C., *et al.* (2014a) Where do bed bugs stand when the dust settles? *Pest*, **34**, 19–22.

Potter, M.F., Haynes, K.F., Gordon, J.R., *et al.* (2014b) Silica gel: a better bed bug desiccant. *Pest Control Technology*, **42** (8), 76, 84.

Potter, M.F., Haynes, K.F. and Fredericks, J. (2015) Bed bugs across America. *Pestworld*, **November/December**, 4–14.

Romero, A. and Anderson, T.D. (2016) High levels of resistance in the common bed bug, *Cimex lectularius* (Hemiptera: Cimicidae), to neonicotinoid insecticides. *Journal of Medical Entomology*, **53** (3), 727–731.

Romero, A., Potter, M.F. and Haynes, K.F. (2009a) Are dusts the bed bug bullet? *Pest Management Professional*, **77** (5), 22–23, 26, 28, 30.

Romero, A., Potter, M.F. and Haynes, K.F. (2009b) Behavioral responses of the bed bug to insecticide residues. *Journal of Medical Entomology*, **46** (1), 51–57.

Romero, A., Potter, M.F. and Haynes, K.F. (2009c) Evaluation of piperonyl butoxide as a deltamethrin synergist for pyrethroid-resistant bed bugs. *Journal of Economic Entomology*, **102** (6), 2310–2315.

Romero, A., Potter, M.F. and Haynes, K.F. (2010) Evaluation of chlorfenapyr for control of the bed bug, *Cimex lectularius* L. *Pest Management Science*, **66** (11), 1243–1248.

Sharma, S.K., Upadhyay, A.K., Haque, *et al.* (2006) Wash resistance and bioefficacy of Olyset net – a long-lasting insecticide-treated mosquito net against malaria vectors and nontarget household pests. *Journal of Medical Entomology*, **43** (5), 884–888.

Singh, N., Wang, C. L. and Cooper, R. (2013) Effectiveness of a reduced-risk insecticide based bed bug management program in low-income housing. *Insects*, **4** (4), 731–742.

Singh, N., Wang, C. and Cooper, R. (2014) Potential of essential oil-based pesticides and detergents for bed bug control. *Journal of Economic Entomology*, **107** (6), 2163–2170.

Singh, N., Wang, C., Wang, D., Cooper, R. and Zha, C. (2016a) Comparative efficacy of selected dust insecticides for controlling *Cimex lectularius* (Hemiptera: Cimicidae). *Journal of Economic Entomology*, **109** (4), 1819–1826.

Singh, N., Wang, C. and Cooper, R. (2016b) Posttreatment feeding affects mortality of bed bugs (Hemiptera: Cimicidae) exposed to insecticides. *Journal of Economic Entomology*, **109** (1), 273–283.

Steelman, C.D., Szalanski, A.L., Trout, R., *et al.* (2008) Susceptibility of the bed bug *Cimex lectularius* L. (Heteroptera: Cimicidae) collected in poultry production facilities to selected insecticides. *Journal of Agricultural and Urban Entomology*, **25** (1), 41–51.

Suwannayod, S., Chanbang, Y. and Buranapanichpan, S. (2010) The life cycle and effectiveness of insecticides against the bed bugs of Thailand. *Southeast Asian Journal of Tropical Medicine and Public Health*, **41** (3), 548–554.

Tahir, H.M. and Malik, H.T. (2014) Susceptibility of *Cimex lectularius* L. (Heteroptera: Cimicidae) to deltamethrin. *Pakistan Journal of Zoology*, **46** (1), 288–290.

Tawatsin, A., Thavara, U., Chompoosri, J., *et al.* (2011) Insecticide resistance in bedbugs in Thailand and laboratory evaluation of insecticides for the control of *Cimex hemipterus* and *Cimex lectularius* (Hemiptera: Cimicidae). *Journal of Medical Entomology*, **48** (5), 1023–1030.

Temu, E.A., Minjas, J.N., Shiff, C.J. and Majala, A. (1999) Bedbug control by permethrin-impregnated bednets in Tanzania. *Medical and Veterinary Entomology*, **13** (4), 457–459.

Todd, R.G. (2006) Efficacy of bed bug control products in lab bioassays: do they make it past the starting gate? *American Entomologist*, **52** (2), 113–116.

Turner, K.L. and Brigham, A.J. (2008) Efficacy of seven commercial pest control products against *Cimex lectularius* (Hemiptera: Cimicidae). In: *Proceedings of the Sixth International Conference on Urban Pests, July 13–16, 2008, Budapest, Hungary*. OOK-Press, Hungary, Veszprém.

Usinger, R.L. (1966) *Monograph of Cimicidae (Hemiptera – Heteroptera)*, Entomological Society of America, College Park.

Wang, C., Wang, C., Lu, L. and Xu, M. (2012) Carbon dioxide fumigation for controlling bed bugs. *Journal of Medical Entomology*, **49** (5), 1076–1083.

Wang, C., Lu, L.H., Zhang, A.J. and Liu, C.F. (2013a) Repellency of selected chemicals against the bed bug (Hemiptera: Cimicidae). *Journal of Economic Entomology*, **106** (6), 2522–2529.

Wang, C., Singh, N., Cooper, R.A., Liu, C.G. and Buczkowski, G. (2013b) Evaluation of an insecticide dust band treatment method for controlling bed bugs. *Journal of Economic Entomology*, **106** (1), 347–352.

Wang, C., Singh, N. and Cooper, R. (2014) Efficacy of an essential oil-based pesticide for controlling bed bug (*Cimex lectularius*) infestations in apartment buildings. *Insects*, **5** (4), 849–859.

Wang, C., Singh, N. and Cooper, R. (2015) Field study of the comparative efficacy of three pyrethroid/neonicotinoid mixture products for the control of the common bed bug, *Cimex lectularius*. *Insects*, **6** (1), 197–205.

Wang, C., Singh, N., Zha, C. and Cooper, R. (2016a) Efficacy of selected insecticide sprays and aerosols against the common bed bug, *Cimex lectularius* (Hemiptera: Cimicidae). *Insects*, **7** (1), 5.

Wang, C., Singh, N., Zha, C. and Cooper, R. (2016b) Bed bugs: Prevalence in low-income communities, resident's reactions, and implementation of a low-cost inspection protocol. *Journal of Medical Entomology*, **53** (3), 639–646.

Watanabe, M. (2010). Insecticide susceptibility and effect of heat treatment on bedbug, *Cimex lectularius*. *Medical Entomology and Zoology*, **61** (3), 239–244.

Yu, S.J. (2008) *The Toxicology and Biochemistry of Insecticides*, CRC Press, New York.

31

Limitations of Bed Bug Management Technologies

Stephen L. Doggett and Mark F. Feldlaufer

31.1 Introduction

The global resurgence of bed bugs has been well-documented in the literature (see Doggett *et al.*, 2012; Benoit and Attardo, 2013) and has been extremely costly to society. In the USA alone, revenue derived by pest management companies related to bed bug infestations rose to USD 611.2 million in 2016 (Anonymous, 2017). It stands to reason that the two key aspects in mitigating the impact of the Common bed bug, *Cimex lectularius* L., and the Tropical bed bug, *Cimex hemipterus* (F.), revolve around:

- *Detection:* Is a bed bug infestation actually present?
- *Control:* What is the best way to eliminate a bed bug infestation?

In spite of the advances in detection and the control of bed bugs, certain limitations still exist in bed bug management technologies. This chapter attempts to address many of these limitations, and to critically review the potential of current and proposed bed bug management technologies. The reader is also referred to specific chapters in this book that deal with detection and control (see Part 5 of this book).

31.2 Bed Bug Detection

31.2.1 Traps

As bed bug infestations cannot be diagnosed from bites with any level of certainty, detection of bed bugs or their evidence (spotting due to defecation or shed skins) are of the utmost importance (Pinto *et al.*, 2007). Early detection means that the infestation is smaller and usually more concentrated, with a higher guarantee of a successful control outcome. Early detection is especially important in the hospitality sector; for brand protection and to minimize the risk of potential bed-bug-related litigation. Visual inspection by trained pest management professionals (PMPs) is one method used to detect a bed bug infestation. Numerous monitors ("active" monitors that utilize a lure and "passive" monitors that provide a simple harborage or trap bed bugs in a pit) have been developed to detect bed bug activity. The use of chemical lures and monitors for bed bugs has been addressed in a several recent reviews (Wang and Cooper, 2011; Weeks *et al.*, 2011; Vaidyanathan and Feldlaufer, 2013) as well as in Chapter 27 of this book. Artificial pheromones based on those produced by either the host or by bed bugs have been incorporated into monitors. Such "active" monitors arguably suffer from the fact that they are competing with the natural signals released by the host, other bed bugs, or a mixture of pheromones present in bed bug harborages. However, Wang and colleagues (2010) did find that some active monitors could actually attract bed bugs when present in low-level populations.

Advances in the Biology and Management of Modern Bed Bugs, First Edition.
Edited by Stephen L. Doggett, Dini M. Miller, and Chow-Yang Lee.
© 2018 John Wiley & Sons Ltd. Published 2018 by John Wiley & Sons Ltd.

There is a report that using bed bug-produced compounds (notably defensive secretions) in conjunction with an insecticide such as diatomaceous earth, will enhance the efficacy of control (Benoit et al., 2009). In an integrated program, perhaps this may further complicate the use of an active monitor that may incorporate the same semiochemicals into the detection device. Ultimately, it would be expected that if one bed bug trap was found to be superior with regard to attractiveness to bed bugs, PMPs would use it to the exclusion of others. However, this is simply not yet the case. Superior marketing and a lack of bed bug understanding and education ensures that even inefficient devices such as sticky traps are still widely employed (Potter et al., 2015).

Passive monitors (those without pheromone lures), such as pitfall traps and simple harborages (usually made of cardboard), have also been used to monitor for potential bed bug infestations and to assess the efficacy of eradication programs. Pitfall traps have limitations, in that a buildup of dust and debris inside the trap can lead to their failing (Doggett, 2013). Thus regular maintenance in the form of cleaning and the addition of talc is required to ensure bed bugs are unable to climb up the interior of the trap.

Some passive monitors on the market are simple pieces of cardboard that only provide a limited harborage area (Doggett, 2012). A typical infestation contains numerous locations where bed bugs can hide, so the utility of such devices in bed bug detection must be questioned. These types of trap are further compromised by the behavior of bed bugs, which tend to return to sites where aggregation pheromones have been deposited (Mendki et al., 2014) and most simple traps have no attractant incorporated. Many traps also rely on sticky gels for trapping the insect, yet it is known that such surfaces tend to repel bed bugs (Doggett, 2012).

Bed bug traps will need to take into account biological differences between the two bed bug species and potentially need to be marketed as species-specific, if key differences are found. For example, it is not presently known if semiochemicals attractive to *C. lectularius* may also attract *C. hemipterus*. Currently, the existing lures on the market have been developed for the former species. Furthermore, *C. hemipterus* is capable of climbing smooth surfaces, whereas *C. lectularius* is not. Investigations have found that *C. hemipterus* has a greater number of tenent hairs on the tibial pad, enabling them to cling better onto smooth substrates (Kim et al., 2017). Pitfall traps may therefore have limited value for the trapping and monitoring of *C. hemipterus*, unless talc or some other lubricant is constantly present to reduce frictional forces on the tarsi.

Beyond these more biologically-based constraints that can limit the effectiveness of a bed bug trap, there are also commercial realities that need to be considered. A trap which is not discreet is unlikely to be used by the hospitality sector, as it almost advertises that bed bugs may be present in the facility. A trap that requires regular maintenance or prolonged and detailed inspection could increase housekeeping costs to unsustainable levels. Finally, a trap that is too expensive is less likely to be used, which is probably why several of the more costly varieties are no longer on the market (Doggett, 2012).

31.2.2 Canines

Because visual inspections are time-consuming and small infestations easily missed, dogs trained to detect bed bug infestations can be employed (Pfiester et al., 2008). The use of trained canines, however, is not without issues, including handler interactions that can affect a dog's ability to accurately detect bed bugs (Lit et al., 2011; Cooper et al., 2014).

The chemical basis of canine detection of bed bugs was addressed previously (Vaidyanathan and Feldlaufer, 2013), and the major defensive secretions of *C. lectularius* and *C. hemipterus*, namely the alkenes E-2-hexenal and E-2-octenal, were found to cause an alerting response by a dog. Interestingly, the dog also alerted on live nymphs and adults of the neotropical stink bug *Euschistus heros* (F.) (Mark Feldlaufer, unpublished results). While this species is not found in the USA, it contains both E-2-hexenal and E-2-octenal (Moraes et al., 2008). The overall performance of bed bug detection canines to detect bed bugs in the field is covered in Chapter 27.

31.2.3 Novel Detection Methods

Newer technologies available for the detection of bed bugs have also been addressed in a recent review (Vaidyanathan and Feldlaufer, 2013). Most of these technologies, such as sampling for bed bug DNA and then analyzing the sample using gas-liquid chromatography or via molecular techniques, require the use of relatively sophisticated laboratory equipment. As a result, such systems are not used for detecting bed bug infestations on a routine basis.

Electronic noses, often referred to as "e-noses", have been used in the food and fragrance industry along with "e-tongues" (see Wilson and Baietto, 2009; Baldwin *et al.*, 2011, and references therein). These devices have also been used to detect insect crop pests (Lan *et al.*, 2008). Electronic detectors need to be "tuned" to recognize particular chemicals or mixtures of chemicals, and then require an operator to sample the air for a potential bed bug infestation. Electronic detectors for bed bug detection have come onto the market, but for the device to respond the current units need to be placed so close to the insect (BBD, 2011) that there appears to be little practicality in their use over visual inspections.

Unfortunately in terms of bed bug detection technologies, for the most part, independent comparisons of efficacy and cost effectiveness have yet to be undertaken using a rigorous scientific approach.

31.3 Bed Bug Control

31.3.1 Housing Types

While all housing types can harbor bed bugs, multiple occupancy buildings are considered to have the greatest potential for infestation. This is due to frequent turnover of units, high density occupancy, and adjoining walls, which can facilitate the active movement of bed bugs through electrical conduits, piping, cracks and crevices, and other less obvious passageways (Wang *et al.*, 2010). Infestations in common areas of any form of accommodation can lead to a rapid and widespread dispersal of bed bugs. Problems are often exacerbated for low-income residents, who may not have the financial resources to effectively deal with infestations. Likewise, tenants sometimes have to rely upon a property manager to decide when treatment is necessary. The manager is often very likely to choose the least expensive control option, often with marginal success. The lack of timely treatments may foster the growth and spread of infestations within facilities.

31.3.2 Chemical Control Methods

The application of synthetic chemical insecticides or pesticides is the most common method employed to treat bed bug infestations. The insecticidal action of DDT is well-known, and this pesticide saw use for bed bug control until it was largely banned internationally for environmental, health, and political reasons. While DDT and other organochlorines (OCs) were certainly effective against bed bugs when first introduced, populations of bed bugs resistant to the OCs soon arose after use of this class of compounds became widespread (Busvine, 1958; Davies *et al.*, 2012). A study evaluating the efficacy of DDT for bed bug control in poultry houses in Arkansas (USA) revealed that high levels of bed bug resistance to DDT still exist (Steelman *et al.*, 2008). Pyrethroids are another class of pesticides that have been successfully used to treat bed bug infestations, although widespread pyrethroid-resistance in both *C. lectularius* and *C. hemipterus* has been documented in field-collected populations (Romero *et al.*, 2007; Tawatsin *et al.*, 2011; Davies *et al.*, 2012; Zhu *et al.*, 2010, 2012; see also Chapter 29 and the references therein). Because DDT and pyrethroid insecticides have similar modes of action (Davies *et al.*, 2007), it has been proposed that the previous frequent use of DDT to control bed bugs may have actually predisposed current populations of bed bugs to pyrethroid resistance (Yoon *et al.*, 2008). Therefore, given the widespread resistance of bed bugs to the pyrethroids, it is unlikely that DDT would be effective as a control measure.

One of the more controversial applications of pyrethroids is the use of permethrin-impregnated encasements and mattress ticking (the covering fabric used on mattresses). With the widespread and high levels of resistance reported from around the world, the use of an older generation pyrethroid for controlling modern bed bug strains would appear counterintuitive. One research group (funded by the product manufacturer), showed that most bed bugs in three of four *C. lectularius* strains continually exposed to a permethrin-treated fabric died within one day. However, the highly resistant "Marcia'" strain only had 22% mortality after ten days, despite continuous exposure (Jones et al., 2013). The same research group then demonstrated that short-term exposure to the identical treated fabric resulted in decreased fecundity in five strains of *C. lectularius* (Jones et al., 2015). The latter study, however, did not include the Marcia strain, so information about the impact on fecundity of this pyrethroid-resistant strain was unavailable. To date, the only independent study evaluating permethrin-impregnated fabrics demonstrated that such products have little efficacy for controlling a modern resistant strain of *C. lectularius* (Doggett et al., 2011).

The problem in using pyrethroids for residual control of potentially pyrethroid-resistant bed bug strains is that the exposure selects for an even more resistant population. Investigations into modern and museum strains of *C. lectularius* in Australia found that strains early in the resurgence possessed a knockdown mutation (I936F) that conferred only mild pyrethroid resistance (Dang et al., 2014, 2015). This mutation largely disappeared, only to be replaced by mutations that conferred a higher level of resistance. Bed bug populations then possessed multiple resistance mechanisms (Lilly et al., 2016a,b). Even within the same resistant strain there are always varying levels of resistance amongst individuals. In addition, some individuals lose resistance mechanisms over time (Dang et al., 2015), while others are simply more resistant. In exposing one strain of *C. lectularius* to a label-rate dose of lambda-cyhalothrin, the survivors after 24 h of continuous exposure were found to have significantly thicker cuticles than those that succumbed (Lilly et al., 2016b). Thus continuous exposure of bed bugs to sublethal doses of pyrethroids is likely to lead to enhanced resistance and thus greater challenges in controlling field infestations.

As pyrethroid resistance has been shown to be extremely widespread, and susceptible bed bug populations are rarely if ever found, the suggestion that this class of insecticide can be employed effectively against modern bed bugs under any circumstance would seem odd. However, the formulation of a product greatly influences efficacy. For example, any pyrethroid-based aerosol sprayed directly onto resistant bed bugs can produce a complete kill, yet the same active usually fails to do so when applied in a liquid form (Doggett et al., 2012). It is presently unknown why this is, as the excipient (the aerosol without the active) provides no control at all (Stephen Doggett, unpublished results), although it has to be assumed that the propellant plays some synergistic role in product efficacy. Therefore, despite the high level of resistance to the pyrethroids, new formulations may be developed to overcome this resistance.

"Dual-action insecticides" are now widely being used by PMPs for bed bug control (Potter et al., 2013a). Typically, these products contain two active ingredients combining insecticides from different chemical classes, while some may also be formulated with a synergist such as piperonyl butoxide. The most common dual-action products employed against bed bugs are a mixture of a pyrethroid and a neonicotinoid. However, there was a recent report of bed bug resistance to the neonicotinoids (Romero and Anderson, 2016) and, in light of pyrethroid resistance, this raises questions on the future use of dual-action insecticides with actives from these chemical classes.

The use of desiccant dusts for bed bug management is now routinely employed (Doggett et al., 2008; Romero et al., 2009; Akhtar and Isman, 2013, 2016; Potter et al., 2013b, 2014; Stedfast and Miller, 2014). Some researchers have highlighted the limitations of one desiccant dust, namely diatomaceous earth (DE), which can be very slow acting (Potter et al., 2013b, 2014; Singh et al., 2016), but silicon dioxide has been demonstrated to be highly efficacious and capable of killing all bed bugs within 24 h or less (Potter et al. 2014; Singh et al. 2016). Investigations have been undertaken to enhance the efficacy of DE dust and the addition of alarm pheromones were found to do so by causing an excitatory response in the insect, thereby increasing contact exposure with the insecticide (Benoit et al., 2009).

Another more recent study examined the addition of carbon dioxide, via dry ice, to rooms treated with desiccant dust. The carbon dioxide acted as a bed bug stimulant, leading to increased contact with the dust and enhanced efficacy (Roligheten, 2014; Aak *et al.*, 2017). However, in spite of promising results, the lack of readily available dry ice (and the logistical issues of adding it daily) or the occupational hazards associated with gas cylinders may limit the widespread use of carbon dioxide for routine bed bug control.

Insect growth regulators (IGRs) are registered in many countries for bed bug management, but their use is not without controversy. Since IGRs affect the molting process, the bed bug would still require a blood meal. Therefore, a client would still have to be bitten in order for a product to work (Doggett, 2012). In laboratory trials, IGRs have also shown limited efficacy in reducing bed bug egg production (Goodman *et al.*, 2012).

The use of different delivery systems for insecticides to control bed bugs has also been explored. Several over-the-counter total release aerosols ("foggers" or "bombs") were found to be ineffective against bed bugs (Jones and Bryant, 2012). Poor performance was attributed to either pyrethroid-resistant bed bugs (a pyrethroid insecticide was the active ingredient in the foggers) or the inability of the fogging material to directly reach susceptible bed bugs that were residing in an (artificial) harborage. This latter point raises the issue of direct application. A statement that a product kills 100% of bed bugs on direct contact may be accurate. However, given the cryptic nature of bed bugs, it may be unlikely that a product actually contacts a bed bug directly under field conditions, thereby weakening the claim (see Doggett *et al.*, 2012 and references therein). Furthermore, such products usually provide little residual control.

For bed bug-infested items not amenable to liquid spray, fumigation is an option. While sulfuryl fluoride is lethal to bed bugs (Miller and Fisher, 2008; Phillips *et al.*, 2014) it is also quite lethal to humans and in most countries the use of sulfuryl fluoride to control bed bugs (and other vermin) is restricted to fumigation-licensed PMPs. Ozone has been used as a fumigant for bed bug control by non-professionals in Australia since the early 2000s (Doggett, 2009). The machines that produce ozone are inexpensive and widely available. Ozone gas has only recently become the center of scientific investigations (Feston, 2015) and its potential field application and safe use have yet to be accurately assessed in bed bug management. On a smaller scale, Wise (2013) demonstrated that dichlorvos-impregnated strips can be effective at controlling *C. lectularius* in small spaces with good air flow.

The fumigation efficacy of several essential oils for the control of *C. lectularius* was compared to Cirkil (Terramera Inc., Ferndale, WA, USA), which is a neem-based fumigant available in the USA for bed bug control (Feldlaufer and Ulrich, 2015). Only rosemary oil approached the mortality achieved using Cirkil (99.5% vs 100%, respectively), when tested under conditions that mimicked field use. Interestingly, acetophenone was also quite active in this fumigation study, producing 98% bed bug mortality. Although not listed as an ingredient in Cirkil, acetophenone was included in the study because it had an odor similar to Cirkil, which is widely regarded as being unpleasant. While "odor" is certainly subjective, it could possibly influence an individual's choice of an over-the-counter product.

31.3.3 Other Chemical Control Products

Chemicals consumed by humans, such as ibuprofen for pain relief or caffeine in coffee, teas, and sodas, have also been investigated for their effect on bed bug development (Narain and Kamble, 2015). These *in-vitro* studies using adult *C. lectularius* showed that both ibuprofen and caffeine could negatively impact bed bug fecundity, although the authors of the study point out that it is unlikely that ibuprofen would be incorporated into a bed bug management strategy, based on the human health concerns of taking this medication on an ongoing basis and at a dose that could affect bed bugs. The use of ivermectin, a bacteria-derived macrocyclic lactone that is used as an anti-parasitic medication for the control of worms (WHO, 2015), has also been investigated for its ability to adversely affect *C. lectularius*. In one study, *C. lectularius* either died, became paralyzed, or suffered delays in molting when fed on ivermectin-treated mice (Ostlind *et al.*, 2001). *In-vitro* and *in-vivo* studies demonstrated that pharmacological doses of ivermectin caused morbidity and mortality in *C. lectularius* (Sheele *et al.*, 2013). *In-vitro* studies were extended to include moxidectin, another related anti-helminthic drug (Sheele and Ridge, 2016). Arguably, however, a systemic approach to bed bug management raises ethical issues because effective means of control are available other than the oral administration of drugs to humans. Presumably, such medications would only be employed in situations of chronic infestation where control efforts had not been successful. The compliance of residents in taking such drugs would need to be assessed prior to implementing a building-wide program and such evaluations have yet to be undertaken.

31.3.4 Non-chemical Bed Bug Control Methods

While chemical insecticides continue to be used for bed bug control, the potential for insecticide resistance to additional classes of chemicals beyond those already identified remains (Gordon *et al.*, 2014). Therefore, the use of non-chemical control methods for bed bug infestations is imperative, both to maximize the chance of treatment success, but also to delay resistance developing to other insecticides. The reader is referred to Chapter 28 for an in-depth coverage of these methods, and to previous publications that have focused on the non-chemical control of bed bugs (Wang and Cooper, 2011; Doggett *et al.*, 2012).

31.3.5 Heat

One alternative to chemical insecticide use is whole-room heat treatments. This technique is not new, as room treatments using steam for bed bug control was reported prior to the development of modern (non-steam) heating devices (Harned and Allen, 1925). Modern heat-producing machines have been used to eliminate bed bug infestations (Pereira *et al.*, 2009; Kells and Goblirsch, 2011). Sub-lethal temperatures, while not killing bed bugs immediately, also may be more deleterious than originally thought (Rukke *et al.*, 2015). It is known that prolonged heat leads to a reduction in symbiont levels, which in turn negatively effects the fecundity of *C. lectularius* (Chang, 1974; Heaton, 2013).

Using heat to kill bed bugs offers the advantages of immediate efficacy and the ability to treat household items not amenable to a liquid (chemical) treatment. However, the relatively high cost of heat treatment makes it less feasible for many residents and for multi-unit or large commercially-managed properties with limited pest control budgets. When the heating system used is inadequate for the size of the space or the heat treatments are not sustained for sufficient lengths of time, there will be bed bug survivors. In addition, if there are large competing heat sinks, voluminous clutter, or heated air is not uniformly distributed, bed bugs may seek localized pockets of cooler air, or move from the heated area into adjacent rooms or units, enabling them to survive and disperse. Circulation of heat, usually accomplished by fans, is thought to be crucial to an effective treatment for bed bugs (Pereira *et al.*, 2009). While the use of heating units that rely on conduction are available, their use has generally been restricted to treating small personal items (Wang and Cooper, 2016). Other types of heat treatment that rely upon sunlight and ambient (hot) temperatures, have been shown to be largely ineffective (Doggett *et al.*, 2006).

31.3.6 Other Non-chemical Controls

The use of steam as a form of heat treatment has been used to effectively control bed bugs on furniture and other personal items (see Kells, 2006; Wang *et al.*, 2009; Shindelar and Kells, 2011). However, steaming seems to be better incorporated into an integrated approach for controlling bed bugs rather than used as a stand-alone procedure. The same can be said of laundering regimens to eliminate bed bugs from clothing and linens (Naylor and Boase, 2010). Likewise, the use of low temperatures (in other words, freezing) to kill bed bugs has tended to be of limited use, and only in specific instances (see Wang and Cooper, 2011; Olson *et al.*, 2013, and references therein). Similarly, devices that use gas under high pressure to freeze bed bugs on contact also appear to have limited use, because the air displaced by the gas stream can result in the non-lethal scattering of insects (Doggett *et al.*, 2012). In contrast, vacuuming is less likely to result in insect dispersal and can rapidly reduce an insect biomass. Vacuum equipment is relatively inexpensive, and the process requires minimal operator training (Doggett, 2013; Pinto *et al.*, 2007).

An interesting approach to monitoring and controlling bed bugs is based upon the ability of the leaf trichomes of bean leaves (*Phaselous vulgaris* L.) to entrap bed bugs (Szyndler *et al.*, 2013). Bean leaves have long been used for bed bug control in the Balkan countries, from well before the advent of modern insecticides (Bogdandy, 1927; Richardson, 1943). A recent translation of this method has led to a patent application using microfibers that mimic the action of the leaf trichomes (Regents of the University of California, 2013). Whether a commercially-viable bed bug control product will result from this research is speculative at this time.

Ultraviolet light (UV) has been investigated for the control of *C. lectularius* eggs and first instar nymphs (Iten, 2013). It was found that UV could kill eggs and also impair host searching behavior. However, when other methods are available that can kill all stages or remove them (such as steam and vacuuming), the utility of UV light appears limited.

Exposing bed bugs to ultralow oxygen levels is another form of effective non-chemical bed bug management. Ultralow oxygen levels can be achieved using vacuums to remove air within a container (Liu and Haynes, 2016) or via the addition of oxygen-depleting molecules to a confined area (Brandenburg, 2015). The latter system has been commercially available for some years and can be used to control bed bugs on mattresses, but some days are required to achieve a complete kill. Carbon dioxide via dry ice or compressed gas can also be added to containers that are subsequently sealed to kill bed bugs within (Wang *et al.* 2012). Clearly such systems are aimed at small-scale treatments due to the logistics involved in bagging and sealing infested items, in particular the treatment of sensitive objects such as electronic devices, which heating or freezing may damage.

Ultrasonic pest repellers are claimed by the manufacturer to deter a variety of household pests, including vertebrates such as rats and mice and invertebrates such as bed bugs. Many of these devices can be found for sale on the internet. In the only study published to date, four such devices were evaluated for efficacy and none were found to repel bed bugs (Yturralde and Hofstetter, 2012).

Finally, the marketplace has been flooded with numerous detection and control devices of unproven efficacy. Often the products are accompanied by professionally-produced web sites and customer testimonials. Yet few have been subjected to independent research. It is not uncommon for many of these products to promote integrated pest management (IPM). As such, the relative efficacy of each individual IPM component is difficult to independently assess, thereby masking a poorly performing product.

Currently, every bed bug management product on the market has inherent limitations. If there was a magical silver bullet for bed bug control, the resurgence would have already ended, yet it has not. Thus it is important for the PMP to know not only what a product can do, but what it cannot do as well.

References

Aak, A., Roligheten, E., Rukke, B.A. and Birkemoe, T. (2017) Desiccant dust and the use of CO_2 gas as a mobility stimulant for bed bugs: a potential control solution? *Journal of Pest Science*, **90** (1), 249–259.

Akhtar, Y. and Isman, M.B. (2013) Horizontal transfer of diatomaceous earth and botanical insecticides in the common bed bug, *Cimex lectularius* L.; Hemiptera: Cimicidae. *PloS ONE*, **8** (9), e75626.

Akhtar, Y. and Isman, M.B. (2016) Efficacy of diatomaceous earth and a DE-aerosol formulation against the common bed bug, *Cimex lectularius* Linnaeus in the laboratory. *Journal of Pest Science*, **89** (4), 1013–1021.

Anonymous (2017) *Specialty Consultants: U.S. Structural Pest Control Market Surpasses $8 Billion*, http://www.pctonline.com/article/specialty-consultants-research-2017-market-report/ (accessed 27 April 2017).

Baldwin, E.A., Bai, J., Plotto, A. and Dea, S. (2011) Electronic noses and tongues: Applications for the food and pharmaceutical industries. *Sensors*, **11** (5), 4744–4766.

BBD [Bed Bug Detector] (2011) *The BBD-100 Personal Bed Bug Detector Finds Bed Bugs Now*, https://www.youtube.com/watch?v=Yl3eklUwOx0 (accessed 9 August 2016).

Benoit, J.B. and Attardo, G.M. (2013) Mechanisms that contribute to the establishment and persistence of bed bug infestations. *Terrestrial Arthropod Reviews*, **6** (3), 227–246.

Benoit, J.B., Phillips, S.A., Croxall, T.J., et al. (2009) Addition of alarm pheromone components improves the effectiveness of desiccant dusts against *Cimex lectularius*. *Journal of Medical Entomology*, **46** (3), 572–579.

Bogdandy, S. (1927) Ausrottung von bettwaiizen mit bohnenblaitern [Exterminating bed-bugs with bean leaves]. *Naturwissenschaften*, **15** (22), 474.

Brandenburg (2015) *BB Alert 02 PM*, www.b-one.com/product/details/bbalert-o2pm (accessed 9 August 2016).

Busvine, J.R. (1958) Insecticide-resistance in bed-bugs. *Bulletin of the World Health Organization*, **19** (6), 1041–1052.

Chang, K.P. (1974) Effects of elevated temperature on mycetome and symbiotes of the bed bug, *Cimex lectularius* (Heteroptera). *Journal of Invertebrate Pathology*, **23** (3), 333–340.

Cooper, R., Wang C. and Singh, N. (2014) Accuracy of trained canines for detecting bed bugs (Hemiptera: Cimicidae). *Journal of Economic Entomology*, **107** (6), 2171–2181.

Dang, K., Toi, C.S., Lilly, D.G., et al. (2014) Detection of knockdown resistance (kdr) in *Cimex lectularius* and *Cimex hemipterus* (Hemiptera: Cimicidae). In: *Proceedings of the Eighth International Conference on Urban Pests, 2014, July 20–23, Zurich, Switzerland*. OOK-Press, Veszprém.

Dang, K., Toi, C.S., Lilly, D.G., Bu, W. and Doggett, S.L. (2015) Detection of knockdown resistance mutations in the common bed bug, *Cimex lectularius* (Hemiptera: Cimicidae), in Australia. *Pest Management Science*, **71** (7), 914–922.

Davies, T.G.E., Field, L.M., Usherwood, P.N.R. and Williamson, M.S. (2007) DDT, pyrethrins, pyrethroids and insect sodium channels. *IUBMB Life*, **59** (3), 151–162.

Davies, T.G.E, Field, L.M. and Williamson, M.S. (2012) The reemergence of the bed bug as a nuisance pest: implications of resistance to pyrethroid insecticides. *Medical and Veterinary Entomology*, **26** (3), 241–254.

Doggett, S.L. (2009) The financial impacts of bed bugs. In: *Proceedings of the AEPMA Bed Bug Workshop, May 7, 2009, Westmead Hospital, Australia*. Westmead Hospital.

Doggett, S.L. (2012) Caveat emptor! Bed bug product buyers beware. *Pest*, **Jul/Aug**, 16–17.

Doggett, S.L. (2013) *A Code of Practice for the Control of Bed Bug Infestations in Australia*, 4th edn, Department of Medical Entomology and The Australian Environmental Pest Managers Association, Sydney.

Doggett, S.L., Geary, M.J. and Russell, R.C. (2006) Encasing mattresses in black plastic will not provide thermal control of bed bugs, *Cimex* spp. (Hemiptera: Cimicidae). *Journal of Economic Entomology*, **99** (6), 2132–2135.

Doggett, S.L., Geary, M.J., Lilly, D. and Russell, R.C. (2008) *The Efficacy of Diatomaceous Earth Against the Common Bed Bug, Cimex lectularius*, Department of Medical Entomology, Sydney.

Doggett, S.L., Orton, C.J., Lilly, D.G. and Russell, R.C. (2011) Bed bugs – a growing concern worldwide. Australian and international trends update and causes for concern. In: *Proceedings of the Australian Environmental Pest Managers Association NSW Conference, June 2, 2011, Homebush Bay, Australia*.

Doggett, S.L., Dwyer, D.E., Peñas, P.F. and Russell, R.C. (2012) Bed bugs: Clinical Relevance and control options. *Clinical Microbiology Reviews*, **25** (1), 164–192.

Feldlaufer, M.F. and Ulrich, K.R. (2015) Essential oils as fumigants for bed bugs (Hemiptera: Cimicidae), *Journal of Entomological Science*, **50** (2), 129–137.

Feston, J. (2015) The effects of ozone gas on the common bed bug (*Cimex lectularius* L.). Master thesis, Purdue University.

Goodman, M.H., Potter, M.F. and Haynes, K.F. (2012) Shedding light on IGRs and bed bugs. *Pest Control Technology*, **40** (8), 38, 46.

Gordon, J.R., Goodman, M.H., Potter, M.F. and Haynes, K.F. (2014) Population variation in and selection for resistance to pyrethroid-neonicotinoid insecticides in the bed bug. *Scientific Reports*, **4** (3836), 1–7.

Harned, R.W. and Allen, H.W. (1925) Controlling bedbugs in steam-heated rooms. *Journal of Economic Entomology*, **18** (2), 320–331.

Heaton, L.L. (2013) *Wolbachia* in bedbugs *Cimex lectularius*. PhD thesis, University of Sheffield.

Iten, J. (2013) The impact of ultraviolet light on survival and behavior of the human bed bug, *Cimex lectularius* Linnaeus. Graduate thesis, The Ohio State University.

Jones, S.C. and Bryant, J.L. (2012) Ineffectiveness of over-the-counter total-release foggers against the bed bug (Heteroptera: Cimicidae). *Journal of Economic Entomology*, **105** (3), 957–963.

Jones, S.C., Bryant, J.L. and Harrison, S.A. (2013) Behavioral responses of the bed bug to permethrin-impregnated *Active*Guard(TM) fabric. *Insects*, **4** (2), 230–240.

Jones, S.C., Bryant, J.L. and Sivakoff, F.S. (2015) Sublethal effects of *Active*Guard exposure on feeding behavior and fecundity of the bed bug (Hemiptera: Cimicidae). *Journal of Medical Entomology*, **52** (3), 413–418.

Kells, S.A. (2006) Nonchemical control of bed bugs. *American Entomologist*, **52** (2), 109–110.

Kells, S.A. and Goblirsch, M.J. (2011) Temperature and time requirements for controlling bed bugs (*Cimex lectularius*) under commercial heat treatment conditions. *Insects*, **2** (3), 412–422.

Kim, D.Y., Billen, J., Doggett, S.L. and Lee, C.Y. (2017) Differences in climbing ability of *Cimex lectularius* and *Cimex hemipterus* (Hemiptera: Cimicidae). *Journal of Economic Entomology*, **110** (3), 1179–1186.

Lan Y.B., Zheng, X.Z., Westbrook, J.K., *et al.* (2008) Identification of stink bugs using an electronic nose. *Journal of Bionic Engineering*, **Suppl.**, 172–180.

Lilly, D.G., Dang, K., Webb, C.E. and Doggett, S.L. (2016a) Evidence for metabolic pyrethroid resistance in the common bed bug (Hemiptera: Cimicidae). *Journal of Economic Entomology*, **109** (3), 1364–1368.

Lilly, D.G., Latham, S.L., Webb, C.E. and Doggett, S.L. (2016b) Cuticle thickening in a pyrethroid-resistant strain of the common bed bug, *Cimex lectularius* L. (Hemiptera: Cimicidae). *PloS ONE*, **11** (4), e0153302.

Lit, L., Schweitzer, J.B. and Oberbauer, A.M. (2011) Handler beliefs affect scent detection dog outcomes. *Animal Cognition*, **14**, 387–394.

Liu, Y.B. and Haynes, K.F. (2016) Effects of ultralow oxygen and vacuum treatments on bed bug (Heteroptera: Cimicidae) survival. *Journal of Economic Entomology*, **109** (3), 1310–1316.

Mendki, M.J., Ganesan, K., Parashar, B.D., Sukumaran, D. and Prakash, S. (2014) Aggregation responses of *Cimex hemipterus* F. [sic] to semiochemicals identified from their excreta. *Journal of Vector Borne Diseases*, **51** (3), 224–229.

Miller, D.M. and Fisher, M.L. (2008) Bed bug (Hemiptera: Cimicidae) response to fumigation using sulfuryl fluoride. In: *Proceedings of the Sixth International Conference on Urban Pests, July 13–16, 2008, Budapest, Hungary*. OOK-Press, Hungary, Veszprém.

Moraes, M.C.B., Pareja, M., Laumann, R.A. and Borges, M. (2008) The chemical volatiles (semiochemicals) produced by neotropical stink bugs (Hemiptera: Pentatomidae). *Neotropical Entomology*, **37**, 489–505.

Narain, R.B. and Kamble, S.T. (2015) Effects of ibuprofen and caffeine concentrations on the common bed bug (*Cimex lectularius* L.) feeding and fecundity. *Entomology, Ornithology & Herpetology*, **4** (2), http://dx.doi.org/10.4172/2161-0983.1000152

Naylor, R.A. and Boase, C.J. (2010) Practical solutions for treating laundry infested with *Cimex lectularius* (Hemiptera: Cimicidae). *Journal of Economic Entomology*, **103** (1), 136–139.

Olson, J.F., Eaton, M., Kells, S.A., Morin, V. and Wang, C. (2013) Cold tolerance of bed bugs and practical recommendations for control. *Journal of Economic Entomology*, **106** (6), 2433–2441.

Ostlind, D.A., Cifelli, S., Conroy, J.A., *et al.* (2001) A novel *Cimex lectularius*–rodent assay for the detection of systemic ectoparasiticide activity. *Southwestern Entomologist*, **26** (3), 181–186.

Pereira, R.M., Koehler, P.G., Pfiester, M. and Walker, W. (2009) Lethal effects of heat and use of localized heat treatment for control of bed bug infestations. *Journal of Economic Entomology*, **102** (3), 1182–1188.

Pfiester, M., Koehler, P.G. and Pereira, R.M. (2008) Ability of bed bug-detecting canines to locate live bed bugs and viable bed bug eggs. *Journal of Economic Entomology*, **101** (4), 1389–1396.

Phillips, T.W., Aikens, M.J., Thoms, E., Demark, J. and Wang, C. (2014) Fumigation of bed bugs (Hemiptera: Cimicidae): Effective application rates for sulfuryl fluoride. *Journal of Economic Entomology*, **107** (4), 1582–1589.

Pinto, L.J., Cooper, R. and Kraft S.K. (2007) *Bed Bug Handbook: The Complete Guide to Bed Bugs and Their Control*, Pinto and Associates, Mechanicsville, Maryland.

Potter, M. F., Fredericks, J. and Henriksen, M. (2013a) *The 2013 Bugs Without Borders Survey: Bed Bug Infestations Increasing Across America*, http://npmapestworld.org/releases/bedbugsurvey.cfm (accessed 2 May 2016).

Potter, M.F., Haynes, K.F., Christensen, C., et al. (2013b) Where do bed bugs stand: when the dust settles? *Pest Control Technology*, **41** (12), 72–80.

Potter, M.F., Haynes, K.F., Gordon, J.R., et al. (2014) Silica gel: a better bed bug desiccant. *Pest Control Technology*, **42** (8), 76, 84.

Potter, M.F., Haynes, K.F. and Fredericks, J. (2015) Bed bugs across America. *Pestworld*, Nov/Dec, 4–14.

Richardson, H.H. (1943) The action of bean leaves against the bedbug. *Journal of Economic Entomology*, **36** (4), 543–545.

Roligheten, E. (2014) Management of bed bugs using desiccant dust and CO_2 activation, a laboratory study and a field trial. MSc thesis, Norges miljø- og Biovitenskapelige Universitet.

Romero, A. and Anderson, T.D. (2016) High levels of resistance in the common bed bug *Cimex lectularius* (Hemiptera: Cimicidae), to neonicotinoid insecticides. *Journal of Medical Entomology*, **53** (3), 727–731.

Romero, A., Potter, M.F., Potter, D.A. and Haynes, KF. (2007) Insecticide resistance in the bed bug: A factor in the pest's sudden resurgence? *Journal of Medical Entomology*, **44** (2), 175–178.

Romero, A., Potter, M.F. and Haynes, K.F. (2009) Are dusts the bed bug bullet? *Pest Management Professional*, **77** (5), 22–23, 26, 28, 30.

Rukke, B.A., Aak, A. and Edgar, K.S. (2015) Mortality, temporary sterilization, and maternal effects of sublethal heat in bed bugs. *PLoS ONE*, **10** (5), e0127555.

Sheele, J.M. and Ridge, G.E. (2016) Toxicity and potential utility of ivermectin and moxidectin as xenointoxicants against the common bed bug, *Cimex lectularius* L. *Parasitology Research*, **115** (8), 3071–3081.

Sheele, J.M., Anderson, J.F., Tran, T.D., et al. (2013) Ivermectin causes *Cimex lectularius* (bedbug) morbidity and mortality. *The Journal of Emergency Medicine*, **45**, 433–440.

Shindelar, A. and Kells, S.A. (2011) *Let's beat the bug! Using steamers to kill bed bugs*, www.bedbugs.umn.edu/files/2012/07/Y2014M05D27-Using-Steamers-to-Kill-Bed-Bugs.pdf (accessed 9 May 2016).

Singh, N., Wang, C., Wang, D. Cooper, R. and Zha, C. (2016) Comparative efficacy of selected dust insecticides for controlling *Cimex lectularius* (Hemiptera: Cimicidae). *Journal of Economic Entomology*, **109** (4), 1819–1826.

Stedfast, M.L. and Miller, D.M. (2014) Development and evaluation of a proactive bed bug (Hemiptera: Cimicidae) suppression program for low-income multi-unit housing facilities. *Journal of Integrated Pest Management*, **5** (3), E1–E7.

Steelman, C.D., Szalanski, A.L., Trout, R., et al. (2008) Susceptibility of the bed bug *Cimex lectularius* L. (Heteroptera: Cimicidie) collected in poultry production facilities to selected insecticides. *Journal of Agriculture and Urban Entomology*, **25** (1), 41–51.

Szyndler, M.W., Haynes, K.F., Potter, M.F., Corn, R.M. and Loudon, C. (2013) Entrapment of bed bugs by leaf trichomes inspires microfabrication of biomimetic surfaces. *Journal of the Royal Society Interface*, **10** (83), 1–9.

Tawatsin, A., Thavara, U., Chompoosri, J., et al. (2011) Insecticide resistance in bedbugs in Thailand and Laboratory evaluation of insecticides for the control of *Cimex hemipterus* and *Cimex lectularius* (Hemiptera: Cimicidae). *Journal of Medical Entomology*, **48** (5), 1023–1030.

Regents of the University of California (2013) Microfabricated surfaces for the physical control of insects. *Patent application*, US 20150013213A1.

Vaidyanathan, R. and Feldlaufer, M.F. (2013) Bed bug detection: Current technologies and future directions. *American Journal of Tropical Medicine and Hygiene*, **88** (4), 619–625.

Wang, C. and Cooper, R. (2011) Environmentally sound bed bug management solutions, in *Urban Pest Management: An Environmental Perspective*, (ed. P. Dhang), CABI, Cambridge, pp. 44–63.

Wang, C. and Cooper, R. (2016) Research update: portable heat chambers. *Pest Control Technology*, **44** (3), 56, 58–61.

Wang, C., Gibb, T. and Bennett, G.W. (2009) Evaluation of two least toxic integrated pest management programs for bed bugs (Heteroptera: Cimicidae) with discussion of a bed bug intercepting device. *Journal of Medical Entomology*, **46** (3), 566–571.

Wang, C., Saltzmann, K., Chin, E., Bennett, G.W. and Gibb, T. (2010) Characteristics of *Cimex lectularius* (Hemiptera: Cimicidae), infestation and dispersal in a high-rise apartment building. *Journal of Economic Entomology*, **103** (1), 172–177.

Weeks, E.N.I., Birkett, M.A., Cameron, M.M., *et al.* (2011) Semiochemicals of the common bed bug, *Cimex lectularius* L. (Hemiptera: Cimicidae), and their potential for use in monitoring and control. *Pest Management Science*, **67** (1), 10–20.

Wang, C., Lu, L. and Xu, M. (2012) Carbon dioxide fumigation for controlling bed bugs. *Journal of Medical Entomology*, **49** (5), 1076–1083.

WHO (2015) *WHO Model List of Essential Medicines*. World Health Organization, http://who.int/medicines/publications/essentialmedicines/en/ (accessed 26 May 2016).

Wilson, A.D. and Baietto, M. (2009) Applications and advances in electronic-nose technologies. *Sensors*, **9** (7), 5099–5148.

Wise, L. (2013) Temperature effects on vapor toxicity of volatile compounds to bed bugs, *Cimex lectularius* L. MSc thesis, University of Florida.

Yoon, K.S., Kwon, D.H., Strycharz, J.P., *et al.* (2008) Biochemical and molecular analysis of deltamethrin resistance in the common bed bug (Hemiptera: Cimicidae). *Journal of Medical Entomology*, **45** (6), 1092–1101.

Yturralde, K.M. and Hofstetter, R.W. (2012) Efficacy of commercially available ultrasonic pest repellent devices to affect behavior of bed bugs (Hemiptera: Cimicidae). *Journal of Economic Entomology*, **105** (6), 2107–2114.

Zhu, F., Wigginton, J., Romero, A., *et al.* (2010) Widespread distribution of knockdown resistance mutations in the bed bug, *Cimex lectularius* (Hemiptera: Cimicidae), populations in the United States. *Archives of Insect Biochemistry and Physiology*, **73** (4), 245–257.

Zhu, F., Sams, S., Mournal, T., *et al.* (2012) RNA interference of NADPH-cytochrome P450 reductase results in reduced insecticide resistance in the bed bug, *Cimex lectularius. PLoS ONE*, **7** (2), e31037.

32

Bed Bug Education

Jody Gangloff-Kaufmann, Allison Taisey Allen and Dini M. Miller

32.1 Introduction

Since the beginning of the modern bed bug resurgence, research-based information regarding bed bug biology, behavior, and control has grown exponentially in an attempt to meet the demand for public education. For those suffering from an infestation, the ultimate goal of education is to reduce their bed bug fear and to help them make financially wise decisions with regard to control efforts. For the pest management professional (PMP), education efforts are focused on helping them manage bed bug infestations more quickly and effectively.

The need for bed bug information and guidance has compelled entomologists and PMPs alike to engage new learners while tailoring their messages and the message delivery so as to reach as many people as possible. This chapter describes some of the education strategies that have emerged in the USA since the year 2000.

In the beginning of the bed bug resurgence there was no up-to-date scientific information regarding bed bug management. Written resources and research papers were limited to describing control techniques of a bygone era. These papers described antiquated strategies such as the burning of sulfur, hydrocyanic gas fumigation, or spraying the home with organochlorines, organophosphate, or pyrethrum insecticides (Usinger, 1966; see also Chapter 1). The information that was available described control strategies that were in direct conflict with the modern principles of integrated pest management. The new information that was needed by the public involved how to avoid bringing bed bugs into their homes, how to inspect their homes for bed bugs, and how to contribute to control efforts in the event of an infestation.

At the time when emerging bed bug experts were gathering new information, the options for delivering this new content were also changing. Internet technology and usage had undergone a vast expansion. The public now wanted information that was instant, engaging, shareable, and mobile-friendly. The use of websites, blogs, wikis, and social networks has greatly enhanced the rapid and low-cost exchange of all information (accurate and inaccurate) regarding bed bug management methods.

32.2 Strategies and Successes in Bed Bug Education

Since the turn of the 21st century, bed bug awareness has increased dramatically among the scientific community, as well as in government agencies and the pest management industry. For example, in 2002, the Fourth International Conference on Urban Pests was held in Charleston, South Carolina without a single scientific paper or poster presented on the topic of bed bugs (Jones *et al.*, 2002). However, a Google Trends search for the term "bed bugs" from the years 2004–2016 reveals a slow and steady increase in public interest in bed bugs.

Advances in the Biology and Management of Modern Bed Bugs, First Edition.
Edited by Stephen L. Doggett, Dini M. Miller, and Chow-Yang Lee.
© 2018 John Wiley & Sons Ltd. Published 2018 by John Wiley & Sons Ltd.

So who is responsible for delivering bed bug education to the public in the USA? In the 1800s, the US land grant university system established a public agricultural extension service. The purpose of the extension service was to deliver university research discoveries to people who could use it as part of their jobs (either in agriculture, engineering, or pest control). This cooperative extension service is still a primary mission of the US land grant university system. As expected, the cooperative extension response to bed bugs has grown rapidly since 2004 to include educational online videos, written fact sheets, bed bug websites, local volunteer task forces, and regional bed bug conferences. Bed bug education is unique in that it requires the inclusion of non-traditional audiences, such as social workers, hotel staff, and the medical community. It has to be recognized that groups other than the universities have been involved in delivering quality educational programs on bed bugs and their management in the USA. One particular individual (and his team) from the private sector has set an educational model for all others to follow.

32.3 Educational Programs Focusing on Bed Bugs

Perhaps the first PMP in the USA to begin lecturing on bed bugs was Dr Richard (Rick) A. Cooper of Cooper Pest Solutions, Lawrenceville, New Jersey. Cooper began speaking on bed bugs locally in 2003 to PMPs in New Jersey and Pennsylvania, and gave the first ever presentation at a national US conference on the modern bed bug resurgence. This was at the National Conference on Urban Entomology in 2004 in Phoenix, Arizona. Since then, he has been a highly sought-after speaker on bed bugs, and he produces numerous articles in pest management trade journals, as well as trade journals for other stakeholder groups. In 2007, he partnered with Larry Pinto and Sandra Kraft to develop the *Bed Bug Handbook: the Complete Guide to Bed Bugs and their Control* (Pinto *et al.*, 2007). This was the first textbook of the modern resurgence that aimed to educate PMPs and all stakeholders on bed bug management, but it also provided useful information on the history and impacts of bed bugs, as well as their biology.

In 2007, Cooper's company (along with his brother Phillip Cooper) developed the web site, BedBug Central (BBC; www.bedbugcentral.com), largely for the sale of bed-bug-related items. However, the direction of BBC changed in 2008, with the aim of educating the pest management industry and other stakeholders about bed bug management. This started with the creation of BBFREE (as in 'bed bug free'), a network of companies that promoted integrated pest management (IPM). The network still exists today and is over 160 companies strong in more than 40 US states, with representation in Canada as well. In 2008, Bed Bug TV was launched on the web site (hosted by Mr Jeff White) to post short videos on a wide variety of bed bug management topics. Over 80 videos have now been produced and all are free to download. Soon after, in 2009, BBC began a course that soon became known as "BedBug Boot Camp". This was (and still is) a three-and-a-half day training program covering classroom learning, field training, and business models for bed bugs. Over 300 companies have now attended this camp. In 2010, BBC held the first North American Bed Bug Summit in Chicago, which had over 500 attendees and included a number of international speakers. Since 2013, BBC has partnered with the US National Pest Management Association (NPMA) and the conference has been rebranded as the "Global Bed Bug Summit" (although the emphasis is still very much on the USA). The company ventured into product development in 2011, creating several bed bug monitors that are now widely employed. Since 2015, BBC has launched a series of e-learning modules on bed bugs.

In 2010, Cooper began his PhD under Dr Changlu Wang at Rutgers University, focusing on bed bug management in low-income housing. Both Cooper and Wang are highly productive and are the most published scientists in the field of bed bug research today. Their investigations have been translated back into practical control solutions, especially in the area of bed bug detection and monitoring (see Chapter 27).

Dr. Michael F. Potter, an urban entomologist at the University of Kentucky, began collaborating with various groups in 2007 to survey pest management companies (PMCs) on the bed bug resurgence. Initially this work was undertaken in association with *Pest Management Professional*, a trade publication (Potter, 2008). The survey collected information regarding the number of bed bug infestations the companies were treating annually, the products they were using, and the types of facilities they were treating. These surveys are still conducted today (now in association with the NPMA) and are the primary resource in the USA for tracking and quantifying the annual growth of the bed bug resurgence (Potter *et al.*, 2010, 2011, 2013, 2015). In 2007, Potter and Cooper, collaborated with another trade journal, *Pest Control Technology*, to create the *PCT Bed Bug Symposium*, which was presented at a number of localities across the USA. These workshops focused on educating PMPs about bed bug biology, behavior, and effective control methods.

The University of Kentucky has been continuously active in evaluating bed bug control methods. The results of these studies are presented regularly at pest management conferences, and published in pest management trade magazines so that the information is readily available to the industry and the public.

Virginia Tech is another land grant university that began focusing on bed bug issues early on in the resurgence. Dr Dini M. Miller is the Cooperative Extension Specialist – a role that combines elements of research and public education – responsible for urban pest management in the state of Virginia. In 2006, her laboratory began rearing bed bugs and conducting pesticide efficacy tests. The results of these studies were used to educate PMPs on how to improve their control efforts. In 2008, Miller began working with the Virginia Department of Agriculture and Consumer Services, Office of Pesticide Services (VDACS OPS), to present bed bug training programs tor hoteliers, apartment managers, social service agencies, and elder care facilities around the state of Virginia. Later in 2014, Miller and Molly Wilson established the Virginia Tech Bed Bug and Urban Pest Information Center (VTBBUPIC), funded by VDACS OPS. In 2015, the VTBBUPIC delivered 53 bed bug training programs at 51 different venues. Ninety-five percent of these programs were invited presentations, meaning that the training programs were requested by different state (Virginia), regional, national, and international stakeholder groups. International bed bug programs included training presentations for the pest management industry at the Expo de las Plagas in Mexico, the Structural Pest Management Association of British Columbia (Vancouver, Canada), and the Federation of Asian and Oceanic Pest Management Association in Penang (Malaysia). National and regional training invitations included the National Apartment Association, the Children's Law Center in Washington DC, the National Environmental Health Association, and the Global Bed Bug Summit. Thirty-four of these invited programs in 2015 were part of the VTBBUPIC mission, and were delivered exclusively within the state of Virginia.

One of the most novel and innovative outreach programs is directed by Dr Stephen Kells at the University of Minnesota. Kells is another bed bug expert with Cooperative Extension responsibilities. He began working with a multidisciplinary advisory committee in 2010 to address the need for providing quality bed bug information to the diverse and multicultural population in the state of Minnesota. After increasing reports of pesticide misapplications, bed bugs spreading to business offices, and bed bugs inhibiting the work of in-home service providers, the "Let's Beat the Bug" campaign was initiated. This campaign interacts with the public in three ways:

- via an information line
- by delivering in-person training to groups who are either directly affected by bed bugs, or who have clients affected by bed bugs
- via a website.

To provide stakeholders with the greatest access to bed bug information, the information line records telephone messages or emails that are answered by employees with bed bug expertise (via the telephone or email) within a specified period of time. "Let's beat the bug" brand in-person training is offered to community groups and professionals working within community organizations. A challenge with this type of training is that many people are unwilling to spend their time learning about bed bugs, unless they have encountered an actual infestation. To ensure good attendance, training events have been offered that carry continuing education credits for PMP staff, or that are delivered as part of larger community events, such as state fairs or information days for communities, or as part of meetings of landlords, health and safety committees, loss prevention seminars and so on.

Using in-person training data and Google Analytics, Shindelar and Kells (2017) have been able to track search terms and the pages visited by stakeholders using the website www.bedbugs.umn.edu. Interestingly, the most frequently accessed page on the website was "laundering items for bed bugs control". The second most frequently accessed page was "what not to do when you have bed bugs". The third most accessed page had to do with using freezing conditions to kill bed bugs. Interestingly, inquiries from clients contacting the information line demonstrated a similar theme of "how do I get rid of bed bugs on my own?" with 430 inquiries made between May 2012 and April 2016 on this topic alone. Overall, the frequency with which these particular pages (describing control methods) were accessed, and the questions received, indicated that those people using the information services believed that they already had active bed bugs in their homes.

Note that the second most common inquiry had to do with bed bug identification. People were quick to associate small foreign objects close to their bed as bed bugs, but the majority (77%) of samples sent to the University of Minnesota for identification were not actually bed bugs. Certainly, this finding illustrated the need for bed bug identification information. If people actually had bed bugs, most believed that they could solve the problem themselves. They were quite surprised at the level of detail and effort required for bed bug control. The priority of the information line was to encourage the use of PMCs as much as possible. If people were unable to afford pest management, the control recommendations they received depended on the situation.

The pest management industry also took an early leadership role in organizing bed bug educational events. In the late 2000s in the USA, the NPMA cooperated with bed bug researchers and extension educators to put together educational conferences, at which treatment products (insecticides) and methods (whole-home heat, and so on) were discussed in detail with regards to what was working and what was not. These conferences were open to the NPMA membership and any other interested parties.

In 2011 the NPMA also released the first edition of *Best Management Practices (BMPs) for Bed Bugs* (NPMA, 2011). This document outlined instructions for bed bug treatment preparation, bed bug detection methods, control methods, and insecticide use. There has also been a unique emphasis on client education contributing to successful bed bug control. The guidelines also contain specific information as to what clients should know about bed bugs, in order to improve their understanding and cooperation (see Chapter 24).

The US Environmental Protection Agency (EPA) also held "Bed Bug Summits" in 2009 and 2011. These meetings were designed as forums to solicit recommendations for addressing the bed bug resurgence in the USA. As a result of the EPA summits, the Federal Bed Bug Working Group was created as a cooperative effort among five US federal agencies that were specifically impacted by bed bugs. This group created the *Collaborative Strategy on Bed Bugs* (2015), which is a cooperative plan organized into four priority areas. These areas are: prevention, surveillance and IPM, education/communication, and research (FBBW, 2015).

In some ways, state and local governments within the USA have had an even greater impact on public education than has the federal government. For example, New York City was one of the first US cities to publicly suffer the physical and financial effects of bed bug infestations. In 2009, the New York City Council created the NYC Bed Bug Advisory Board, composed of city agency representatives, pest management experts, entomologists, and social service advocates. This group produced a report that was intended to guide the city's response to bed bug infestations (Bloom *et al.*, 2010). As a result of their work, city officials created an extensive "Bed Bug Portal" that made educational resources on controlling and preventing bed bugs available for diverse audiences in a wide variety of settings (NYC Health, 2016).

32.4 The Media

Fortunately, or unfortunately, both print and online news media have been instrumental in raising bed bug awareness. In 2010, a Yahoo.com article entitled "Bed bugs take bite out of the Big Apple" was viewed 35 million times in three months (Gangloff-Kaufmann, unpublished data). Indeed, 2010 may have been a peak in sensational bed bug news, as several popular retail stores in New York City were forced to close due to bed bug sightings (Bland, 2010). Today, bed bugs continue to be newsworthy, especially at a local level where human-interest stories about infested apartment buildings or evicted tenants appear daily. While most people are aware that bed bugs exist, the awareness gained from sensationalized news media only increases their bed bug fear. There is never any media attention given to successful control efforts or bed bug prevention. The NPMA has sought to change the media coverage of bed bug issues by offering a more "positive" perspective. The NPMA's "Bed Bug Awareness Week" has sought to provide science-based content to popular media outlets with the hope that the information would be widely distributed (see www.pestworld.org).

32.5 The Effect of Social Media on Bed Bug Education

In the past ten years, at least four "registry" websites have been established to collect and display customer reports of bed bug sightings in hotels, apartment buildings, and other types of businesses around the world (Harrison, 2014). Travel review sites are another venue for reporting bed bugs. These sites do not require validation or confirmation of the reports posted on them, which allows for false reports to be generated by otherwise unhappy customers or tenants. However, the benefits of social media greatly outweigh the risks and negative effects. The information exchange is instantaneous, interactive, and can foster supportive communities that offer good advice. In early 2006, the "Bedbugger" Yahoo Group was created as an online support group for people suffering from bed bugs. Bedbugger.com was independently established as a blog, discussion forum, and repository for educational resources, bed bug photos, videos, and control information. Over the last ten years, the Bedbugger.com discussion forum has amassed 16 213 members and 231 780 posts.

32.6 Identifying the Target Audience

Bed bug educational outreach must be targeted to the audiences that need it the most, and it must be provided in an easy-access format. To identify the target audience for bed bug education, the NYC Environmental and Health Portal uses a survey tool to gather data about city residents' health. In 2009, this survey started to include data on "adults reporting bed bugs in their home" (NYC DOHMH, 2016). The data from this survey documented that bed bug infestation rates were positively correlated with neighborhood poverty ratings (the higher the poverty rating, the more bed bug infestations present). In order to reach low-income communities, educators must know the audience and deliver the bed bug education materials that are appropriate in terms of the local language, level of literacy, and training expectations. This information must also be delivered in the most accessible format. A 2015 Pew Research Center report found that younger adults and non-whites with relatively low income and educational attainment levels are especially likely to be "smartphone-dependent" (Smith, 2015). Therefore, in 2017, information must be mobile-friendly to reach these intended audiences. Due to technology limitations in many low-income neighborhoods, face-to-face training sessions are often conducted. This is to ensure that everyone has the opportunity to ask questions about bed bugs, and receive accurate information.

32.7 Effective Adult Education

If education is a tool in the battle against bed bugs, we should seek to make our educational efforts as impactful as possible.

The term andragogy focuses on special needs of adult learners. Knowles (1980) identified six general assumptions about adult learning:

- need to know
- self-concept
- prior experience
- readiness to learn
- learning orientation
- motivation to learn." (Ota *et al.*, 2006).

Therefore, the most important element of an adult education program is the relevance of the information presented to the learner. People will be most receptive to the information if they need it to meet their current challenges. Learners will retain the most information when they can apply it in their day-to-day lives. Educators should inquire about the challenges the audience members are currently facing, and present information they can use immediately.

32.8 Measuring the Impacts of Bed Bug Education

In the USA, many funding agencies and university administration officials require that educators quantify the impacts of their education and training programs to justify the time that they spend delivering them. The impacts of bed bug education, such as changes in the learners' behavior after training, are notoriously difficult to measure when working with the general public. Impacts are somewhat easier to measure when working with specific audiences. For example, a US pest management industry survey conducted in 2005 revealed that just over 60% of respondents sometimes or always used aerosol foggers to control bed bugs (Gangloff-Kaufmann *et al.*, 2006). However, a research study conducted at Ohio State University in 2012 documented that aerosol foggers did not work for controlling bed bugs (Jones and Bryant, 2012). This study was presented at multiple regional and national pest management conferences, and was referenced in numerous pest management trade magazines. By 2015, the number of pest managers using aerosol foggers had plummeted to less than 1% (Potter *et al.*, 2015).

Similar to the pest management industry, university educators are encouraged to integrate surveys into their training programs so that data can be collected regarding the impact of training programs on learner knowledge and behavior. To collect data regarding their changes in knowledge, pre- and post-training surveys can be administered that either test the learners' knowledge directly, or ask them about their perceived knowledge on the topic before and after the training program.

Finally, an analysis of online searches for bed bug information provides indirect evidence regarding current educational needs and how educational outreach is working for the public. A study of Google search patterns for "bed bugs" and co-occurring terms between 2011 and 2014 found no evidence that the need for bed bug information had plateaued or declined. Therefore, even though we have direct evidence from in-person surveys that education is working, the number of online searches for bed bug control terms remain high. This suggests that new infestations continue to develop, and residents of newly infested homes and workplaces still need bed bug education (Sentana-Lledo *et al.*, 2015).

32.9 Conclusion

As bed bugs invade more and more businesses, homes, medical facilities, and other locations, the audience for bed bug education continues to increase. Educating all stakeholders about bed bug management methods remains an essential component in the prevention and control of this pest. Researchers and educators learn more and more about bed bugs every year. Therefore, the body of bed bug knowledge needs to be constantly refreshed. To be successful in training stakeholders to deal with their bed bug issues, experts need to be very adaptable. Educators must stay current with technology and social trends to ensure that even those stakeholders without bed bug concerns (today) can be taught how to avoid bed bug infestations in the future.

References

Bland, S. (2010) *Bed Bugs: Victoria's Secret Closure Points To A Bed Bug Comeback, Christian Science Monitor*, www.csmonitor.com/USA/2010/0720/Bed-bugs-Victoria-s-Secret-closure-points-to-a-bed-bug-comeback (accessed 30 May 2016).

Bloom, G.M., Cooper, R.A., Corea, R., Gangloff-Kaufmann, J.L. and Lopez, R. (2010) *Recommendations for the Management of Bed Bugs in New York City. New York City Bed Bug Advisory Board*, http://council.nyc.gov/downloads/pdf/ bed_bugs_report_2010.pdf (accessed 4 April 2011).

FBBW (2015) *Collaborative Strategy on Bed Bugs. US Federal Bed Bug Workgroup*, www.epa.gov/sites/production/files/2015-02/documents/fed-strategy-bedbug-2015.pdf (accessed 30 May 2016).

Gangloff-Kaufmann, J., Hollingsworth, C., Hahn, J., *et al*.(2006) Bed bugs in America: a pest management industry survey. *American Entomologist*, **52** (2), 105–106.

Harrison, R. (2014) *Trends in Bed Bug Management, Monitoring and Treatment*. Orkin Commercial Services. http://cdn.orkin.com/downloads/commercial/orkin-university-resources/trends-in-bed-bug-management_12-11-14.pdf (accessed 10 May 2016)

Jones, S.C., Zhai, J. and Robinson, W.H. (eds) (2002) *Proceedings of the 4th International Conference on Urban Pests*, Pocahontas Press, Inc. Blacksburg, VA.

Jones, S.C. and Bryant, J.L. (2012) Ineffectiveness of over-the-counter total-release foggers against the bed bug (Heteroptera: Cimicidae). *Journal of Economic Entomology*, **105** (3), 957–963.

Knowles, M.S. (1980) *The Modern Practice of Adult Education: From Pedagogy to Andragogy*, 2nd edn. Cambridge Book Co., New York, NY.

NYC DOHMH (2016) *New York City Department of Health and Mental Hygiene Environmental and Health Data Portal; Trends in Adults Reporting Bed Bugs in their Homes*. New York City Department of Health and Mental Hygiene. http://a816-dohbesp.nyc.gov/IndicatorPublic/VisualizationData.aspx?id=2030,719b87,92,ChartOverTime,Boroughs,Number (accessed 10 May 2016).

NYC DOHMH (2016) *New York City Department of Health and Mental Hygiene, Bed Bugs*, www1.nyc.gov/site/doh/health/health-topics/bedbugs.page (accessed 30 May 2016).

NPMA (2011) NPMA library update: Best management practices for bed bugs. *PestWorld*, **March/April**, I-XVI

NPMA (2017) PestWorld.org. "http://www.pestworld.org" www.pestworld.org (accessed 12 Oct 2017).

Ota, C., DiCarlo, C., Burts, D., Laird, R. and Gioe, C., (2006) Training and the needs of adult learners, *Journal of Extension*, **44** (6), https://joe.org/joe/2006december/tt5.php (accessed 10 October 2016).

Pinto, L.J., Cooper, R. and Kraft, S.K. (2007) *Bed Bug Handbook. The Complete Guide to Bed Bugs and Their Control*, Pinto & Associates, Mechanicsville, MD.

Potter, M.F. (2008) The business of bed bugs. *Pest Management Professional*, **76**, 24–25, 28–32, 34, 36–40.

Potter, M.F., Rosenberg, B. and Henriksen, M. (2010) Bugs without borders. Defining the global bed bug resurgence. *Pestworld*, **September/October**, 1–12.

Potter, M.F., Haynes, K.F., Rosenberg, B. and Henriksen, M. (2011) Bugs without borders. *Pestworld*, **November/December**, 4–15.

Potter, M.F., Haynes, K.F., Fredericks, J. and Henriksen, M. (2013) Bed bug nation are we making any progress? *Pestworld*, **September/October**, 4–11.

Potter, M.F., Fredericks, J. and Henricksen, M. (2015) Bed bugs across America. *Pestworld*, **November/December**, 4–14.

Shindelar, A.K. and Kells, S.A. (2017) The "Let's Beat the Bug! campaign". State-wide active public education against bed bugs in Minnesota. *Journal of Environmental Health*, **79** (7), 22–27.

Sentana-Lledo, D, Barbu, C.M., Ngo, M.N., *et al.* (2015) Seasons, searches, and intentions: what the internet can tell us about the bed bug (Hemiptera: Cimicidae) epidemic. *Journal of Medical Entomology*, **53** (1), 116–121.

Smith, A. (2015) *U.S. Smartphone Use in 2015*. www.pewinternet.org/2015/04/01/us-smartphone-use-in-2015 (accessed 30 May 2016).

Usinger, R.L. (1966) *Monograph of Cimicidae (Hemiptera – Heteroptera)*, Entomological Society of America, College Park.

Part VI

Bed Bug Control in Specific Situations

33

Low-income Housing

Richard Cooper and Changlu Wang

33.1 Introduction

Bed bugs are particularly difficult to eliminate in low-income housing communities, where the necessary financial resources and knowledge to cope with the expanding bed bug infestations are often lacking (Cooper *et al.*, 2015a). As a result, the prevalence of bed bugs is disproportionately high in low-income housing compared to other sectors of society (Rossi and Jennings, 2010; Doggett *et al.*, 2012; Wang *et al.*, 2016). In the USA, infestations have been limited to the Common bed bug, *Cimex lectularius* L., but in geographic regions within the 30° parallels, it is mostly the Tropical bed bug, *Cimex hemipterus* (F.) that is encountered (Doggett *et al.*, 2012). In Australia and the USA, the rapid escalation of bed bugs in low-income housing communities appears to have lagged behind the resurgence in other sectors of society (hotels, homes, and tourist locations) beginning around 2006 in Australia (Doggett *et al.*, 2011), and between 2008 and 2010 in the USA (Wong *et al.*, 2013; Wang *et al.*, 2016). Surveys also suggest a lag in the resurgence of bed bugs in low-income housing communities in Canada (Hwang *et al.*, 2005) as well as Malaysia and Singapore (How and Lee, 2010).

The impact of bed bugs in the lower socioeconomic sector has been most widely documented in the USA (Pinto *et al.*, 2007; Doggett *et al.*, 2012; Wong *et al.*, 2013; Stedfast and Miller, 2014; Bennett *et al.*, 2015; Cooper *et al.*, 2015a; Wang *et al.*, 2016), where the cost of managing bed bug infestations has been financially devastating to low-income housing communities (Wong *et al.*, 2013). Costs to residents can include replacement of infested furniture and personal belongings, purchase of pest control services or consumer pesticide products (which are often ineffective), as well as costs associated with preparation for treatment and medical treatment of bite symptoms (Doggett *et al.*, 2012; Cooper *et al.*, 2015a).

The reasons for the disproportionately high infestation rates in low-income housing communities include (Miller, 2010; Wong *et al.*, 2013; Bennett *et al.*, 2015; Cooper *et al.*, 2015a; Wang *et al.*, 2016):

- underreporting of bed bug infestations by residents
- reactive rather than proactive inspection of apartments
- low-bid pest control yielding poor-quality service
- lack of cooperation from property management and residents
- reduction rather than elimination of existing bed bug infestations
- lack of government funding to subsidize costs associated with bed bug management.

Ineffective bed bug management programs lead to chronic infestations and the spread of bed bugs within housing communities.

Advances in the Biology and Management of Modern Bed Bugs, First Edition.
Edited by Stephen L. Doggett, Dini M. Miller, and Chow-Yang Lee.
© 2018 John Wiley & Sons Ltd. Published 2018 by John Wiley & Sons Ltd.

33.2 Management of Bed Bugs in Low-income Housing

Integrated pest management (IPM), which emphasizes proactive inspection and involves a combination of non-chemical methods and the judicious use of pesticides, is the best approach for suppression of bed bug infestations (Pinto *et al.*, 2007; Bennett *et al.*, 2015). In multifamily settings, IPM requires a collaborative community- or building-wide effort involving residents, building staff, and pest control staff (Bennett *et al.*, 2015). Guidelines for implementation of bed bug IPM in affordable housing have been created by non-profit organizations (Taisey and Neltner, 2010; Maley *et al.*, 2014). In addition, "best management practices for bed bug management" have been developed by the US National Pest Management Association (NPMA, 2011; see also Chapter 24). Comprehensive recommendations for bed bug management using vetted methods of detection and control have also been created in Australia and the UK and are frequently updated to maintain their relevance (Doggett, 2013; BBF, 2013; see also Chapters 22 and 23).

Components of a successful building or complex-wide IPM program include:

- education of residents and housing staff
- proactive inspections to identify all apartments with existing bed bug activity and to identify new infestations
- affordable and effective treatment of apartments found to have bed bug activity
- procedures and policies that ensure that infestations are treated quickly and in the most appropriate manner, including follow-up visits to ensure elimination.

In the USA, four independent studies evaluating IPM for control of bed bugs in specific low-income apartments found that populations were reduced by more than 90%. However, elimination of the infestation was only achieved in approximately 30% of the apartments (Bennett *et al.*, 2015). Lack of cooperation from residents and the short-time frame (10–12 weeks) to implement the IPM program were listed among the reasons for the low elimination rates (Bennett *et al.*, 2015). In two separate long-term studies (1–2 years), building-wide bed bug IPM programs resulted in the elimination of at least 95% of treated apartments (Wang *et al.*, 2014a; Cooper *et al.*, 2015a). However, a small percentage of apartments were still infested in these long-term studies, due either to lack of resident cooperation or the identification of newly introduced infestations that had not yet been treated.

33.3 Components of a Successful Building- or Complex-wide IPM Program

33.3.1 Education

Building managers and pest management professionals (PMPs) are usually tasked with providing education to the entire housing community. Clearly identifying the roles and responsibilities of residents in the treatment effort is essential to the success of a bed bug management program (Bennett *et al.*, 2015). Education should include:

- dispelling common misconceptions about bed bugs
- recognizing signs of bed bug activity
- what to do if bed bug activity is suspected (adoption of effective non-chemical control procedures)
- what not to do (self-treatment and discarding of furniture).

Residents must be encouraged to report suspected bed bug activity and assured that there will be no negative consequences associated with doing so (Stedfast and Miller, 2014). In a study by Cooper *et al.* (2015a), delivering educational presentations to residents and providing them with factsheets was only partially effective in improving their understanding of bed bugs. Bennett *et al.* (2015) suggested that a more practical hands-on approach to training residents is necessary. For a more thorough discussion about the role of education in bed bug management see Chapter 32.

33.3.2 Identification of Apartments with Existing Bed Bug Activity

Achieving building-wide or community-wide control of bed bugs requires identification of all apartments with bed bug activity. Bed bugs can spread rapidly between apartments in multi-occupancy communities (Wang *et al.*, 2010; Booth *et al.*, 2012; Cooper *et al.*, 2015b). In fact, field studies have documented that over 50% of infestations occurred in apartment units that shared a common wall, floor, or ceiling with, or were located across the hall from, another infested unit (Doggett and Russell, 2008; Wang *et al.*, 2010). In addition to preventing the spread between units, bed bug infestations that were identified early were more easily controlled (Pinto *et al.*, 2007; Cooper *et al.*, 2015a,c).

Proactive inspections often identify many more bed bug infestations than reported by residents (Wang *et al.*, 2010; Cooper *et al.*, 2015a; Wang *et al.*, 2016), but proactive inspections that rely on visual inspection alone are less effective than those that use both visual inspection and effective monitoring (Wang *et al.*, 2010, Cooper *et al.*, 2015a). Community-wide proactive inspections (using a combination of visual inspection and interceptor traps below beds and furniture) in low-income housing communities revealed that residents had failed to report bed bugs in around 70% of the apartments where bed bug activity was discovered (Cooper *et al.*, 2015a; Wang *et al.*, 2016). In another study, in Australia, 50% of the bed bug infestations in a high-rise housing building for medical staff were not reported by residents (Doggett and Russell, 2008). An economic comparison of two effective building-wide inspection methods is discussed in Chapter 27 and in Wang *et al.* (2016).

Repeated building-wide inspections to identify newly introduced infestations (post-treatment) are just as important as the initial building-wide inspection (Cooper *et al.*, 2015a). At the start of a one-year study, bed bug activity was detected in 55 apartment units during the initial (pro-active) inspection of a 358-unit apartment community (Cooper *et al.*, 2015a). Of the 55 infested units, 71% had not been reported by residents and were unknown to the management. Over the next 12 months, 16 new apartments were identified with bed bug activity, of which only 4 had been reported by residents to property management. The additional 11 apartments were identified during the proactive inspections at 6 and 12 months (Cooper *et al.*, 2015a). The authors concluded that the initial, and then periodic, building-wide inspections were a key factor in the success of the bed bug IPM effort, which reduced the community infestation rate from 15% to 2% over the 12-month study period.

Infestation rates vary among buildings within a housing community as well as between housing communities (Wong *et al.*, 2013; Wang *et al.*, 2016). For example, inspection of 43 low-income apartment buildings in four different cities in New Jersey, USA, revealed infestation rates ranging from 3.8–29.5% (Wang *et al.*, 2016). Often the highest infestation rates occur in housing for elderly and disabled residents (Wong *et al.*, 2013; Cooper *et al.*, 2015a). In a low-income housing community in Virginia, USA, consisting of residents of all ages, 86% of 31 bed bug infestations existed in apartment buildings housing senior and disabled residents (Wong *et al.* 2013). For buildings with low infestation rates (<5% known infestations), the cost of conducting a building-wide proactive inspection may outweigh the benefits. Instead, an alternative approach can be used which limits proactive inspections to:

- apartments whose residents suspect bed bug activity
- apartments neighboring (above, below, adjacent, and across the hall) the unit with reported or confirmed activity
- apartments of new residents moving into the community.

Areas outside of apartments such as hallways and common areas must be inspected and/or monitored as well. Bed bugs can be easily transported from infested apartments to other areas on residents' belongings, such as walkers, wheelchairs, electric scooters, and laundry, or they can be introduced into buildings by visitors. The presence of bed bugs in offices, community or activity rooms, laundries, and so on can contribute to continued spread of bed bugs, so a regular inspection program of these areas should be implemented (see Chapter 27 for a discussion of detection and monitoring methods).

33.3.3 Preparation of Apartments for Treatment

Residents are often asked to carry out a variety of activities to prepare their unit to be treated for bed bugs. Preparations required by PMPs often involve removing, bagging, and laundering clothing in closets and dressers, bagging all items under or around beds and sofas, removing and laundering bed linen, and standing mattresses and box springs on end, in addition to numerous other preparations (Pinto *et al.*, 2007). Some residents, particularly those who are elderly or disabled, may be unable to carry out extensive preparations. Moreover, many of the preparations are burdensome, costly, and unwarranted based upon the severity or location of the infestation (Pinto *et al.*, 2007). Despite this, PMPs often refuse to treat the apartment unit if the preparations have not been carried out in full (Pinto *et al.*, 2007). Over half (61%) of pest management companies surveyed in the USA required preparations prior to treatment, of which 71% required all closets and dressers to be emptied (Potter *et al.*, 2015). Extensive preparation, as listed above, can potentially disrupt bed bugs, alter their distribution, and prevent the PMP from being able to properly assess the infestation during the initial service.

Rather than requiring extensive "one size fits all" preparations, requests for cooperation can be made based upon the conditions encountered as they relate to each infestation. For example, no preparations were requested of residents in two successful community-wide bed bug IPM programs housing the elderly and disabled residents (Wang *et al.*, 2014a; Cooper *et al.*, 2015a). In both studies, residents were encouraged to launder bed linen frequently and reduce clutter in their apartments, especially around sleeping areas. Assistance was provided to residents by housing staff as needed.

33.3.4 Treatment of Apartments with Bed Bug Activity

Treatment of apartments with bed bug activity must be aimed at elimination, not reduction, of infestations. Treatments must also employ affordable methods that have been proven to be effective. While pesticides remain an important part of a bed bug eradication program, non-chemical measures are also essential given the widespread resistance to most modern-day pesticides (Doggett *et al.*, 2012; Bennett *et al.*, 2015). Despite this known resistance, a survey by Wong *et al.* (2013) found that among 17 housing authorities that were currently being treating for bed bugs, 53% relied upon pesticides alone. However, Cooper *et al.* (2015a) were able to eliminate 95% of bed bug infestations using a treatment protocol that consisted primarily of non-chemical methods including:

- placement of interceptors under bed and furniture legs and in areas away from sleeping and resting areas
- encasement of mattresses and box springs
- applying steam to bed frames, head boards, foot boards, and upholstered furniture
- hot laundering and machine drying of bed linens, infested clothing, and other fabric items
- the use of a portable heat box to eliminate bed bugs from infested items that could not be laundered.

Similarly, Wang *et al.* (2014a), achieved greater than 95% elimination in an infested low-income housing community using a combination of chemical and non-chemical methods. For a review of effective bed bug control methods refer to Chapters 28 and 30.

33.3.5 Follow-up Service Visits

The number of service visits required to eliminate an infestation varies widely and is influenced by the infestation level (Singh *et al.*, 2013; Cooper *et al.*, 2015a), treatment history (Cooper *et al.*, 2015c), complexity of the furniture, degree of clutter, and level of resident cooperation (Wang *et al.*, 2014b; Bennett *et al.*, 2015; Cooper *et al.*, 2015a). Well-established infestations often require more service visits to eliminate the population than those infestations detected early on (Cooper *et al.*, 2015a). In a study by Cooper *et al.* (2015a), an average of 8.2 service visits were required to eliminate 54 established infestations that had been identified during an initial community-wide inspection. However, only 2.7 (average) service visits were required to eliminate newly introduced bed bug activity in 9 apartments where bed bugs were detected after the initial inspection.

Premature termination of treatment efforts can result in a population rebound, allowing bed bugs to spread from the infested apartments into others. Follow-up visits should continue until no more bed bug activity exists. However, determining when bed bugs have been eliminated can be very difficult. Cooper *et al.* (2015a) found resident satisfaction and visual inspection to be ineffective methods for confirming the elimination of infestations in treated apartments. The absence of bed bugs in interceptors under bed and furniture legs coupled with no bed bugs being found during visual inspections over three consecutive two-week follow-up visits (a total of six weeks) provided 97% confidence of elimination (Cooper *et al.*, 2015a,c).

33.3.6 Contracts and the Role of Property Management

Contracts for bed bug control in low-income housing are typically awarded on a low-bid basis, which does not promote the highest quality pest control (Wong *et al.*, 2013; Bennett *et al.*, 2015; Stedfast and Miller, 2014). PMPs are unlikely to include effective and sustainable IPM tactics in a low-bid process (Bennett *et al.*, 2015) unless the request for bid proposal (RFP) clearly specifies effective inspection methods, treatment methods, follow-up services, and measures for determining elimination. Housing agencies can control the quality of pest management by requiring companies to demonstrate their professionalism and success in similar accounts, describe their management plan, and have specific quality assurance measures in place. Examples of methods to include in an RFP include:

- proactive building-wide inspection (at least annually and more often with heavy infestations) using at least two proven inspection methods in buildings with estimated infestation rates greater than 5%
- inspection of neighboring apartments above, below, adjacent to, and across the hall from infested apartments in buildings with no building-wide inspection
- treatment of apartments with bed bug activity using methods proven to be effective, and not limited to the application of chemicals
- follow-up visits at two-week intervals until infestations have been eliminated.

An elimination protocol that consists of zero bed bug activity for a minimum of two follow-up visits, using at least two methods of inspection, has been proven to be effective.

Sustainable bed bug management also requires property management's commitment to, and support of, the bed bug management program. This includes ensuring access to all apartments for inspections and treatment, and working towards solutions in apartments whose residents are uncooperative or where exceptional challenges such as hoarding exist. In some low-income housing communities, in-house licensed pest management staff provide pest control services. Compared to contracted pest management services, the in-house pest control is more flexible, the staff are more familiar with the buildings and residents, and can more effectively address any obstacles in bed bug eradication. The involvement of housing staff contributed to the success of a community-wide bed bug IPM program implemented by Cooper *et al.* (2015a). In this study, a housing staff member licensed in the application of pesticides treated apartments infested with bed bugs. Assistance was also provided to residents whose apartments were being treated for bed bugs by subsidizing the cost of laundering, providing inexpensive metal bed frames to residents whose beds were sitting on the floor, and helping residents who were incapable of decluttering their apartment by contacting a family member to assist.

33.4 The Future

Robinson and Boase (2011) predicted that people from the middle and high socio-economic classes would eventually see a reduction in the number of new bed bug infestations because they could afford the high cost of control. However, people in the lower socio-economic classes will be unable to afford multiple technology-intensive treatment methods. Thus, to stem the spread of bed bugs throughout society, it is necessary to create programs that are effective, sustainable, and economically practical for low-income communities. These programs need to be supported by government agencies and the wider community in general (Robinson and Boase, 2011; Wang and Cooper, 2011; Doggett *et al.*, 2012; Koganemaru and Miller, 2013). However, the current high cost of bed bug IPM continues to be one of the greatest limitations to its adoption by housing agencies in many nations (Bennett *et al.*, 2015).

References

BBF (2013) *European Code of Practice, Bed Bug Management*, 2nd edn. Bed Bug Foundation, http://bedbugfoundation.org/ecop/ (accessed 15 April 2016).

Bennett, G.W., Gondhalekar, A.D., Wang, C., Buczkowski, G. and Gibb, T.J. (2015) Using research and education to implement practical bed bug control programs in multifamily housing. *Pest Management Science*, **72** (1), 8–14.

Booth, W., Saenz, V.L., Santangelo, R.G., *et al.* (2012) Molecular markers reveal infestation dynamics of the bed bug (Hemiptera: Cimicidae) within apartment buildings. *Journal of Medical Entomology*, **49** (3), 535–546.

Cooper, R., Wang, C. and Singh, N. (2015a) Evaluation of a model community-wide bed bug management program in affordable housing. *Pest Management Science*, **72** (1), 45–56.

Cooper, R., Wang, C. and Singh, N. (2015b) Mark-release-recapture reveals extensive movement of bed bugs (*Cimex lectularius* L.) within and between apartments. *PLoS ONE*, **10** (9), e013642.

Cooper, R., Wang, C. and Singh, N. (2015c) Effects of various interventions, including mass trapping with passive pitfall traps, on low-level bed bug populations in apartments. *Journal of Economic Entomology*, **109** (2), 762–769.

Doggett, S.L. (2013) *A Code of Practice for the Control of Bed Bug Infestations in Australia*, 4th edn, http://medent.usyd.edu.au/bedbug/cop_ed4.pdf (accessed 15 April 2016).

Doggett, S.L. and Russell, R.C. (2008) The resurgence of bed bugs, *Cimex* spp. (Hemiptera: Cimicidae) in Australia. In: *Proceedings of the Sixth International Conference on Urban Pests, July 13–16, 2008, Budapest, Hungary*. OOK-Press, Hungary, Veszprém.

Doggett, S.L., Orton, C.J., Lilly, D.G. and Russell, R.C. (2011) Bed bugs: the Australian response. *Insects*, **2** (2), 96–111.

Doggett, S.L., Dwyer, D.E., Peñas, P.F. and Russell, R.C. (2012) Bed bugs: Clinical relevance and control options. *Clinical Microbiology Reviews*, **25** (1), 164–185.

How, Y.F. and Lee, C.Y. (2010) Survey of bed bugs in infested premises in Malaysia and Singapore. *Journal of Vector Ecology*, **35** (1), 89–94.

Hwang, S.W., Svoboda, T.J., De Jong, I.J., Kabasele, K.J. and Gogosis, E. (2005) Bed bug infestations in an urban environment. *Journal of Emerging Infectious Diseases*, **11** (4), 533–538.

Koganemaru, R. and Miller, D.M. (2013) The bed bug problem: Past, present, and future control methods. *Pesticide Biochemistry and Physiology*, **106** (3), 177–189.

Maley, M., Taisey, A. and Koplinka-Loehr, C. (2014) *Integrated Pest Management: A Guide for Affordable Housing*, http://www.stoppests.org/what-is-ipm/guide/ (accessed 15 April 2016).

Miller, D.M. (2010) *Bed Bug Action Plan for Apartments*, http://www.vdacs.virginia.gov/pesticides/pdffiles/bb-apt1.pdf (accessed 15 April 2016).

NPMA (2011) *Best Management Practices for Bed Bugs*. National Pest Management Association, http://www.pestworld.org/media/3242/bed_bug_bmps_for_consumers_final.pdf (accessed on 15 April 2016).

Pinto, L.J., Cooper, R. and Kraft, S.K. (2007) *Bed Bug Handbook – the Complete Guide to Bed Bugs and Their Control*, Pinto & Associates, Inc., Mechanicsville, MD.

Potter, M.F., Haynes, K.F. and Fredericks, J. (2015) Bed bugs across America: the 2015 bugs without borders survey. *Pest World*, **November/December**, 4–14.

Robinson, W.H. and Boase, C.J. (2011) Bed bug (Hemiptera: Cimicidae) resurgence: Plotting the trajectory. In: *Proceedings of the Seventh International Conference on Urban Pests, August 7–10, 2011, Sao Paulo, Brazil*. Instituto Biologico, Ouro Preto, Brazil, pp. 315–318.

Rossi, L. and Jennings, S. (2010) Bed bugs: a public health problem in need of a collaborative solution. *Journal of Environmental Health*, **72** (8), 34–35.

Singh, N., Wang, C. and Cooper, R. (2013) Effectiveness of a reduced-risk insecticide based bed bug management program in low-income housing. *Insects*, **4** (4), 731–742.

Stedfast, M.L. and Miller, D.M. (2014) Development and evaluation of a proactive bed bug (Hemiptera: Cimicidae) suppression program for low-income multi-unit housing facilities. *Journal of Integrated Pest Management*, **5** (3), E1–E7.

Taisey, A.A. and Neltner, T. (2010) *What's working for Bed Bug Control in Multifamily Housing: Reconciling Best Practices with Research and the Realities of Implementation*, http://www.nchh.org/Portals/0/Contents/bedbug_report.pdf (accessed 15 April 2016).

Wang, C. and Cooper, R. (2011) Environmentally sound bed bug management solutions. In: *Urban Pest Management: An Environmental Perspective*, (ed P. Dhang), CABI International, Cambridge, pp. 44–63.

Wang, C., Saltzmann, K., Chin, E., Bennett, G.W. and Gibb, T. (2010) Characteristics of *Cimex lectularius* (Hemiptera: Cimicidae) infestation and dispersal in a high-rise apartment building. *Journal of Economic Entomology*, **103** (1), 172–177.

Wang, C., Saltzman, K., Gondhalekar, A., Gibb, T. and Bennett, G. (2014a) Building-wide bed bug management. *Pest Control Technology*, **42** (3), 70–74.

Wang, C., Singh, N. and Cooper, R. (2014b) What causes bed bug control failure? The resident factor. *Pest Control Technology*, **42** (8), 86–95.

Wang, C., Singh, N., Zha, C. and Cooper, R. (2016) Bed bugs: Prevalence in low-income communities, resident's reactions, and implementation of a low-cost inspection protocol. *Journal of Medical Entomology*, **53** (3), 639–646.

Wong, M., Vaidyanathan, N. and Vaidyanathan, R. (2013) Strategies for housing authorities and other lower-income housing providers to control bed bugs. *Journal of Housing and Community Development*, **70** (3), 20–28.

34

Multi-Unit Housing

Dini M. Miller

34.1 Introduction

Over the last several years, multi-unit housing has been the most prevalent and consistent source of Common bed bug, *Cimex lectularius* L., infestations in the USA (Potter *et al.*, 2011, 2015; Wang *et al.*, 2016). Bed bug control efforts in multi-unit housing facilities (including individually owned condominiums) can cost owners $100s to $100 000s of dollars for a single complex in a single year (Dini Miller, unpublished results). These costs have been quite a shock to the US housing industry because bed bugs have not been a problem for decades, and property management companies have never had bed bug remediation costs included in their budgets. So the question is, why is multi-unit housing so vulnerable to bed bug infestations? Is the proliferation of infestations simply the result of careless residents repeatedly bringing the bed bugs in, and then failing to cooperate with control efforts? Like most pest problems in the human environment, the answer is not that simple. It is only recently that bed bug researchers have begun to appreciate the wide range of factors that contribute to bed bug spread, and the subsequent control failures observed in multi-unit housing.

34.2 Challenges Unique to the Human Living Environment

The human home environment is relatively static from day to day, and this secure environment presents several advantages for bed bug population growth. The daily presence of the occupant(s) provides a regular food resource, which allows bed bug populations to feed, grow, and proliferate. When provided with a reliable host (regular blood meals) bed bug populations are capable of doubling or tripling in size within a few weeks (Polanco *et al.*, 2011). Secondly, some habits of the residents allow bed bugs to establish long-term aggregations in advantageous locations, such as:

- wheelchairs
- the toilet (Richard Cooper, Cooper Pest Solutions, New Jersey, unpublished results)
- the chair with the best view of the television.

In such locations, mates will be present and access to food requires minimal foraging effort. Finally, an abundance of stationary belongings (clutter) provides bed bugs with many harborage locations where they can hide, aggregate, and lay eggs. Apartment units with high levels of clutter provide bed bugs with thousands of hiding places, and this in turn makes treatment applications and other control efforts very difficult.

Advances in the Biology and Management of Modern Bed Bugs, First Edition.
Edited by Stephen L. Doggett, Dini M. Miller, and Chow-Yang Lee.
© 2018 John Wiley & Sons Ltd. Published 2018 by John Wiley & Sons Ltd.

34.3 Obstructions to Control Success

34.3.1 Challenges Unique to Multi-unit Housing

Recent research studies conducted in New Jersey (Cooper *et al.*, 2015a) and in the UK (Naylor, 2012) have documented that bed bugs can travel significant distances during their scotophase even if they are not actively foraging. Bed bug movement from an infested apartment unit into adjacent living spaces has been documented in several field studies (Potter *et al.*, 2013; Cooper *et al.*, 2015a), even when a host was present in their original location. Certainly, many of the construction features in multi-unit housing, such as drop ceilings, plumbing conduits, wall voids, and community hallways (Cooper *et al.*, 2015a), provide bed bugs with easy access to adjacent units. As a result, bed bug infestations can easily spread among units within a building (Wang *et al.*, 2010).

34.3.2 Human Host Behavior

Bed bugs are often brought into a home in the infested belongings of a visiting friend or family member. While this is also the case in a single-family home, the number of different residents in a multi-unit facility greatly increases the number of potential visitors to the building and thus the likelihood of a bed bug introduction. It is also fairly well known that bed bug infested furniture and belongings are often thrown away in communal trash depositories, where they are subsequently recovered by neighbors living within the same multi-unit community.

Another issue that has contributed to bed bug spread in multi-unit facilities is the failure of residents to detect and/or report bed bugs in their apartment units. Many elderly and disabled people live in multi-unit facilities. The elderly demographic is known to be particularly prone to bed bug infestation (Potter *et al.*, 2010a) because the elderly are often immunocompromised and have poor eyesight. Therefore, they do not always see the bed bugs or react to their bites (Potter *et al.*, 2010a). In fact, Cooper *et al.* (2015b) documented that 71% of the senior residents living in a multi-unit housing facility in New Jersey were unaware that they had bed bugs infesting their homes.

Even those renters who recognize that they have a bed bug infestation may be reluctant to report a bed bug problem to their housing managers. Many people still look at bed bugs as an indicator of poor hygiene. Residents do not want their apartment managers (or neighbors) to think they are "dirty", so they keep their knowledge of the infestation to themselves, often treating their bed bugs with insecticide products that they purchase from hardware stores or over the internet. In some cases, the renters have been dissatisfied with the pest control service that was provided by the property management and decide to take matters into their own hands. In other cases, renters may fear that they are going to be charged by the property management for bed bug control. All property managers/owners within the USA are required to provide (and pay for) a "safe and habitable" environment for their residents. Therefore, property managers/owners are financially responsible for eradicating all bed bug infestations. However, more and more often, property managers are rewriting their lease and rental agreements to make the resident financially responsible for bed bug control (note that these lease addendums have not been found to be legally binding). The threat of charging the residents has had the unforeseen consequence of renters choosing not to report bed bugs in their apartments to the management. This failure to report has resulted in property managers being completely unaware of the percentage of bed bug infested units in their buildings.

Finally, some renters may also fail to report bed bugs because they have illegal pets or people living in their apartment, or are engaging in illegal activities (such as drug sales). In short, the renters may be attempting to avoid financial penalties or eviction by keeping pest management professionals (PMPs) out of their units.

34.3.3 Financial Limitations for Multi-unit Housing Managers and Owners

Multi-unit housing managers and the owners have also unwittingly contributed to the bed bug spread within their own facilities in a number of ways. One of the major factors contributing to bed bug spread has been a lack of willingness on the part of the apartment management to proactively inspect their facilities, relying instead on resident complaints. The prevalent belief among multi-unit housing managers has been that bed bugs "are not part of my job" (Anonymous, New Orleans, LA, personal communication). However, as bed bugs continue to spread throughout multi-unit housing in the USA, housing owners (particularly those owning large properties, say of 4000 units) are now paying USD 100 000–500 000 annually for bed bug control. Arguably, bed bug management is now a health, safety, and budgeting factor for the housing industry.

However, even the conscientious manager or property owner has a difficult time mastering the management of a bed bug infestation. Bed bugs are still a relatively new pest for housing managers and they know very little about effective methods of treatment. In those states within the USA that require that the property owner to pay for bed bug remediation, property managers are eager to turn their bed bug problems over to the first low-priced pest management company (PMC) they can find. However, managers do not realize that most PMCs are also relatively new to bed bug control and rely heavily on liquid insecticide applications rather than a more integrated approach. While some PMCs have made it their business to become experts in bed bug control, many other companies are still relatively inexperienced, and underestimate the amount of time, number, and types of treatment necessary to control a bed bug infestation. This situation is compounded by bed bug resistance to insecticides.

34.3.4 Pest Management Limitations

Insecticide resistance in US bed bug populations has been documented repeatedly, particularly in regard to pyrethroid insecticides. Unfortunately, in the USA the vast majority of insecticides that are allowed to be used indoors are pyrethroids, with very few exceptions.

Liquid and aerosol spray formulation insecticides are generally the least expensive to purchase and apply (in terms of labor cost). Therefore, they are used by 96% (Potter *et al.*, 2015) of PMPs for bed bug control. When multi-unit housing managers hire a PMC, they seek out a competitive price and typically hire the lowest bidding contractor. Housing managers also expect their PMP to spray a liquid insecticide because they "know" insecticide spray kills insects. However, insecticide efficacy studies have repeatedly demonstrated that liquid insecticides have very little efficacy once the product is dry (Lilly *et al.*, 2009; Dini Miller, unpublished results).

For units that cannot be treated adequately with insecticide (usually due to extreme clutter), the use of heat has been widely marketed for bed bug control in multi-unit housing. Heat treating apartment units is still a relatively new method of pest control in multi-unit housing, and different heating systems can vary considerably in their efficacy. Most apartment managers believe that their choice for bed bug eradication is between heat or chemical treatments. Heat is appealing to apartment managers because it is often marketed as a nontoxic or "environmentally friendly" method of bed bug control. However, multi-unit housing owners rarely consider that not all heating systems are the same and that housing units of larger size or with high levels of clutter will often require more heat than many of the current heat systems are capable of producing.

The predominant issues that continue to hinder bed bug eradication in multi-unit housing can be summarized as follows:

- Due to the potential expense, property owners are reluctant to monitor/inspect for bed bugs proactively within their units.
- When bed bugs are found, property owners want the least expensive treatment possible.
- Property managers dread bed bug complaints and want to turn all bed bug issues over to a PMC. Rarely do they make any effort to evaluate (inspect) the infestation or treatment results themselves.
- Property managers and owners often have little or no ability to evaluate whether or not a proposed bed bug treatment (chemical or heat) has the potential to work in their buildings.
- Both apartment managers and PMPs are quick to accuse the apartment residents of bringing in more bed bugs if any bed bugs are found in the unit after a treatment. Bed bug discovery after treatment is almost never blamed on inadequate treatment or treatment failure (say due to pesticide resistance).

Compounding the situation further, most pest management contracts in the USA contain very vague language, which leaves the multi-unit property managers/owners with limited knowledge of how their buildings are being treated. Below are examples of bed bug contract language:

- *"Treatment preparation instructions shall be given to all residents"* (from Norfolk, VA 2016, Dini Miller, unpublished results). This pest management contract was written for a facility housing elderly disabled residents. The contract also does not mention that the preparation instructions require residents to empty all closets, vacuum all furniture and surfaces in the unit, wash and dry all fabrics, move all belongings to the living room, and move all furniture away from the walls. Many residents are physically incapable of performing these required preparations. Further, these extensive preparations are potentially counterproductive, in that infested items are very likely moved, or disturbed, thus scattering bed bug aggregations.
- *"A crack and crevice liquid treatment will be performed under all baseboards* [a narrow wooden board or vinyl strip running along the base of an interior wall] *in the unit"* (Myrtle Beach SC, USA 2015). Bed bug insecticide resistance aside, this statement does not even indicate that the "liquid" is an insecticide).
- *"Ambient heat treatment shall be used for all treatments"* (Richmond, VA, USA 2013). Ambient treatment implies that the air will be treated for bed bugs. Unfortunately, this was indeed the case. The PMPs operated three heaters in a two-story, three-bedroom apartment, and left the heaters in place for 4 h. No temperature measurements (ambient or otherwise) were taken during the treatment.
- Several contracts also specify that *"a bed bug sterilant will be used"*. These contracts are referring to products that contain an insect growth regulator, and specifically a juvenile hormone analogue (JHA). Several studies have shown that JHAs have minimal effects on bed bugs (Goodman *et al.*, 2012; David Moore and Dini Miller, Virginia Tech, Blacksburg, unpublished results).

In addition to these contractual issues, few, if any, current US bed bug management contracts provide the property owner with a monitoring record of the number of bed bugs found in an apartment (estimated or relative, for example "fewer than 10" or "more than 600") prior to treatment. Nor do pest management contracts typically require that treatment effectiveness (percent reduction in bed bugs) be reported. So most multi-unit housing managers or owners have no idea whether or not the treatment was effective. Having said that, many companies do provide a warranty that states that they will come back and retreat for free if bed bugs are found within 30 days of the treatment regimen (usually three applications made at two-week intervals) being completed. However, if bed bugs are found after treatment, both managers and PMPs tend to fall back on the argument that the resident brought new bed bugs into the unit, rather than suggesting that the treatment was incomplete.

Another aspect of pest management practices in the USA that hinders bed bug control in multi-unit housing is that most PMPs offer only a single protocol for bed bug treatment. In other words, the same protocol is used for all units, regardless of the unit size, level of clutter, or size of the bed bug infestation. This "one-size-fits-all" approach has resulted in both "overkill" (say, two technicians working a heat treatment for 7 h to kill 20 bed bugs) and "underkill", where one technician takes 20 min to apply a liquid insecticide in a unit infested with hundreds of bed bugs. While overkill with heat may be successful, the expense is excessive. However, underkill is more common and more problematic. The author has spoken to many apartment managers who complain that single units having to be re-treated eight or nine times. However, as stated above, the resident is often blamed for bed bug re-introduction, rather than treatment failure being suggested. The ultimate result of these multiple treatments is a cost to multi-unit housing owners, particularly if they own multiple complexes, ranging from USD 100 000–1 million every year (Dini Miller, unpublished data; Karen Vail, University of Tennessee, Knoxville, unpublished data; see also Chapter 15).

34.4 Future Prospects for Success

At present, Integrated Pest Management has not been widely implemented in multi-unit housing, but there is hope for the future. As apartment managers and owners spend more and more money on ineffectual control efforts, they become more willing to educate themselves about effective bed bug management methods. In addition, every year more field research is being conducted to evaluate bed bug control techniques. Consequently, we have recently increased our knowledge regarding specific challenges that hinder bed bug control efforts. Examples include the difficulties of working with elderly people who cannot prepare for treatment, or bed bug resistance to residual insecticides, or the possibility that bed bugs put into bags as part of treatment preparation will be let out again. This new knowledge will help us understand how these challenges might be overcome. Field studies have also educated researchers so that they make effective treatment recommendations, particularly monitoring, vacuuming, or use of desiccant dusts, as opposed to recommendations that may be ineffective, such as throwing belongings away, making preventative insecticide applications, or "cleaning". Finally, there are many PMCs that have decided to make bed bug remediation a specific focus in their service offerings. These companies are going to great lengths to become experts in the field of bed bug remediation, and have developed multiple management protocols so that they can effectively address bed bug infestations in a variety of challenging field situations.

References

Cooper, R.A., Wang, C. and Singh, N. (2015a) Mark-release-recapture reveals extensive movement of bed bugs (*Cimex lectularius* L.) within and between apartments. *PloS ONE*, **10** (9), e0136462.

Cooper, R.A., Wang, C. and Singh, N. (2015b) Evaluation of a model community-wide bed bug management program in affordable housing. *Pest Management Science*, **72** (1), 45–56.

Goodman, M.H., Potter, M.F. and Haynes, K.F. (2012). Shedding light on IGRs and bed bugs. *Pest Control Technology*, **40** (8), 38, 40–43, 44–46.

Lilly, D.G., Doggett, S.L., Orton, C.J. and Russell, R.C. (2009) Bed bug product efficacy under the spotlight – Part 2. *Professional Pest Manager*, **Apr/May**, 14–15, 18.

Naylor, R. (2012) Ecology and dispersal of the bedbug. PhD thesis, University of Sheffield.

Polanco, A., Brewster, C.C. and Miller, D.M. (2011) Population growth potential of the bed bug *Cimex lectularius* L.: a life table analysis. *Insects*, **2** (2), 173–185.

Potter, M.F., Haynes, K.F., Connelly, K., *et al.* (2010a) The sensitivity spectrum: human reactions to bed bug bites. *Pest Control Technology*, **38** (2), 70–74.

Potter, M.F., Rosenberg, B. and Henriksen, M. (2010b) Bugs without borders: defining the global bed bug resurgence. *Pest World*, **Sept/Oct**, 8–20.

Potter, M.F., Haynes, K.F., Rosenberg, B. and Henriksen, M. (2011) Bugs without borders. *Pest World*, **Nov/Dec**, 4–15.

Potter, M.F., Gordon, J.R., Goodman, M.H. and Hardin, T. (2013) Mapping bed bug mobility. *Pest Control Technology*, **41** (6), 72–74, 76, 78, 80.

Potter, M.F. Fredericks, J. and Henriksen, M. (2015) *Bed Bugs Without Borders Executive Summary*, http://www.pestworld.org/news-hub/pest-articles/2015-bugs-without-borders-executive-summary/ (accessed 1 June 2016).

Wang, C., Saltzmann, K., Chin, E., Bennett, G.W. and Gibb, T. (2010). Characteristics of *Cimex lectularius* (Hemiptera: Cimicidae) infestation and dispersal in a high-rise apartment building. *Journal of Economic Entomology*, **103** (1), 172–177.

Wang, C., Singh, N., Zha, C. and Cooper, R. (2016). Bed bugs: prevalence in low-income communities, resident's reactions, and implementation of a low-cost inspection protocol. *Journal of Medical Entomology*, **53** (3), 639–646.

35

Shelters

Molly S. Wilson

35.1 Introduction

Throughout the world, there are people without permanent homes who turn to shelters to avoid the nightly outdoor elements. The purpose of a shelter is to provide a safe, protected place in which people can avoid exposure to domestic violence, drugs and alcohol, and the dangers of sleeping in a public place. Unfortunately, shelters have also become a haven for bed bugs.

There are many types of shelters, including homeless shelters, emergency shelters, group homes, and transitional living centers, which provide housing, food, and overnight lodging for varying periods (SAMHSA, 2016). In some situations, shelter facilities will also provide educational, counseling, and vocational services to individuals and families, with the goal of preparing these people to find and maintain a stable living environment. Each type of shelter will present unique challenges when it comes to bed bug management.

35.2 Challenges in Shelters

All shelters are unique when it comes to their management (government, religious, community), patronage (age, gender, sexual orientation, families), resources (education, child care, vocational support), and capacity. However, there is a commonality among shelters in that multiple people utilize a shelter at once, and that the population can change daily, with new people arriving and former residents departing all within a matter of 24 h.

It is this commonality within shelters that presents the greatest challenge to bed bug management. With multiple and different people entering a particular shelter on a daily basis, bed bugs have the potential to be brought in frequently and repeatedly. Generally, shelter management personnel have no control over where patrons and their belongings have previously been located or stored. Similarly, patrons often cannot ensure that their belongings are bed bug-free prior to entering a shelter.

An additional challenge with regard to bed bugs in shelters is that homeless people (frequent patrons) have a three-fold greater risk of being exposed to insect bites (including but not limited to bed bugs, lice, and mosquitoes) and skin infestations (such as scabies) than the housed population (Wright, 1990). These bites and infestations are thought to be the result of living in crowded or outdoor environments, combined with the inability of homeless people to regularly bathe and launder their clothing, and their potentially poor physical and mental health (Hale *et al*., 2005). Ultimately, the increased probability of bites and skin infestations on patrons correlates with an increased probability of those patrons bringing parasites into shelter facilities.

Advances in the Biology and Management of Modern Bed Bugs, First Edition.
Edited by Stephen L. Doggett, Dini M. Miller, and Chow-Yang Lee.
© 2018 John Wiley & Sons Ltd. Published 2018 by John Wiley & Sons Ltd.

35.3 Obstacles to Successful Control

In addition to the unique challenges, shelters also face multiple limitations when controlling and managing bed bugs. The first of these limitations is financial. The local governments and communities that support shelters are often challenged by funding limitations. Facility maintenance (which often includes pest management) remains a high priority in shelters throughout the world. However, the fact remains that many facilities do not currently, and will not in the future, have the funds to repeatedly pay for bed bug control (Hottel et al., 2014). The fear of budget reductions is a prevalent concern for shelters throughout the world. An example from the USA is the New York City Department of Homeless Services (NYC DHS), which operates 73 single adult shelters, 22 facilities for adult families, 95 facilities for families with children, 49 hotels, and 17 cluster (group home) sites throughout the city. In 2016, the NYC DHS was subject to a budget cut of approximately USD 192 million for the 2017 fiscal year (Sompura, 2016). The reason for the budget cut was an anticipated decline in the shelter population because of incoming homeless population reduction initiatives, despite ongoing high levels of homelessness in the city (Stefanski, 2016). Budget cuts such as those in New York are highly problematic for shelter facilities that already have stressed budgets.

Another consequence of the financial limitations in shelters is the lack of space. The facilities that house shelters are often either used for other purposes during the daytime (churches, schools, and so on), or have been used for other purposes in the past (old schools, hotels, hospitals). In cases where these buildings serve multiple purposes and may be continuously occupied, shelters do not have the physical space to alter the facility in order to incorporate bed bug management strategies, such as introducing additional bed bug inspection areas into their check-in processes, or space for storing sleeping furniture or mats that are less conducive to bed bugs (Doggett, 2013).

While bed bug prevention methods can be conceptualized and promoted – for example installation of monitoring devices and mattress encasements, facility inspection protocols, purchase of appropriate sleeping furniture or mats, increased personal hygiene – many shelters are so resource-limited that such steps are rarely implemented. When bed bug problems arise, it is prudent for shelters to have bed bug policies in place, which staff and volunteers can refer to, but shelters can often be limited further by a lack of communication between patrons and management (Hersberger, 2003; Stennett et al., 2012).

35.4 Methods of Control in Shelters

Central to bed bug control in shelters is the implementation of a management policy. The implementation of these policies must be reinforced by providing education to patrons and staff to ensure that the policies are well-known and adhered to. In the USA, there are academic and professional organizations that have produced guidelines to assist shelter staff and management in developing their own bed bug policies for use in their facilities. The New York State IPM Program published a 40-page set of guidelines specifically for the prevention and management of bed bugs in shelters and group living facilities (Gangloff-Kaufmann and Pichler, 2008). These guidelines are tailored toward both public and private facilities and thoroughly cover bed bug identification, best management practices for bed bug prevention and control, shelter intake procedures, and personal protection for patrons and staff. The Michigan Department of Community Health and the Michigan Bed Bug Working Group produced a 118-page, *Manual for the Prevention and Control of Bed Bugs* to provide "comprehensive guidance to identify, prevent, and manage bed bugs" (Michigan Department of Community Health, 2010). The Michigan Department of Community Health manual contains a guide for shelter intake and assessment procedures, which are intended to be used when screening new shelter patrons. The manual recommends that bed bug screening be incorporated into medical queries and be phrased in compassionate language, such as, "Bed bugs, lice, scabies, and other things can cause medical concerns. Do you need help with a bed bug, or any other pest issue?". The manual also suggests that shelters create a "Bed Bug Treatment Agreement" in addition to their other bed bug preparedness policies. This agreement is intended to facilitate and encourage cooperation between shelter patrons and management.

35.5 Key Elements to Successful Control

Ideally, shelters and other temporary housing facilities should have a pest management plan in place to appropriately deal with bed bugs before they become a problem. Christie-Smith and Lassiter (2012) implemented a successful multidisciplinary approach for bed bug elimination in a 30-bed homeless shelter in northern Texas. They stated that the two key components to the plan's success were that the plan was long-term and included "ongoing routine inspection…at frequent consistent intervals". A long-term plan for bed bug management in shelters should include intake instructions for patrons' belongings, where the items are inspected and either treated or quarantined if bed bug evidence is found (Gangloff-Kaufmann and Pichler, 2008). Long-term or ongoing plans should also include continual monitoring of the sleeping and recreation areas within a shelter. Long-term and ongoing recordkeeping will also help to identify those areas or situations that are most conducive to bed bug infestation (Koehler, 2013).

Education for both staff and residents is paramount when attempting to manage bed bugs in shelters. For example, a 2008 survey of homeless shelters in Hawai'i found that shelter staff and management personnel obtained most of their bed bug knowledge from the internet, and would only begin looking for bed bugs and their evidence after residents complained of bites (Fickle *et al.*, 2008). Had the staff been educated about bed bugs, they would have implemented preventative intake procedures and bed bug management efforts prior to receiving bite complaints from the shelters' patrons.

A unique program that has been applied in Canadian shelters is the "Bug and Scrub" program, which was developed by the Toronto Public Health Department (Shum *et al.*, 2012). The Bug and Scrub program provided real employment opportunities to homeless people, by teaching them proper and efficient bed bug management techniques, such as inspection, steam-cleaning, laundering, and the handling of infested personal items. In locations where this program was implemented, both shelter staff and patrons had a functional knowledge of bed bug management procedures and were able to effectively avoid further infestations (Shum *et al.*, 2012).

Overall, bed bugs will continue to be a problem in shelters because shelter patrons come into contact with bed bugs frequently in their transient lifestyle. Shelters also face physical and financial resource limitations when trying to control bed bugs in their facilities. With the implementation of both proactive and reactive bed bug management policies as well as educational efforts aimed at shelter staff and patrons, shelters will be more prepared to appropriately manage bed bug problems.

References

Christie-Smith, A. and Lassiter, A.D. (2012) A multidisciplinary approach toward successful bed bug elimination in a homeless domiciliary setting. *American Journal of Infection Control*, **40** (5), e111.

Doggett, S.L. (2013) *A Code of Practice for the Control of Bed Bug Infestations in Australia*, 4th edn, Department of Medical Entomology and The Australian Environmental Pest Managers Association, Sydney.

Fickle, V.J., Yang, P.J. and Olmsted, G.K. (2008) Examination of bed bug (*Cimex lectularius* Linnaeus) infestations on the island of Oahu, Hawai'i. *Hawaii Journal of Public Health*, **1** (1), 36–39.

Gangloff-Kaufmann, J.L. and Pichler, C. (2008) *Guidelines for Prevention and Management of Bed Bugs in Shelters and Group Living Facilities*. New York State IPM Program, Cornell University Cooperative Extension, IPM No. 618.

Hale, A., Allen, J., Caughlan, J., *et al.* (2005) Bugs that bite: Helping homeless people and shelter staff cope. *Healing Hands*, **9** (1), 1–4.

Hersberger, J. (2003) A qualitative approach to examining information transfer via social networks among homeless populations. *New Review of Information Behaviour Research*, **4** (1), 95–108.

Hottel, B., Pereira, R. and Koehler, P. (2014) Helping those in need. *Pest Control Technology*, **42** (12), 92–94.

Koehler, P.G. (2013) Practical research on bed bug inspection, detection, & monitoring. Speech presented at the Global Bed Bug Summit in Denver, Colorado, December 5, 2013.

Michigan Department of Community Health (2010) *Getting the Bed Bugs Out: A Guide to Controlling Bed Bugs in Your Home*, https://www.michigan.gov/documents/emergingdiseases/Getting_the_Bed_Bugs_Out_Guide_442175_7.pdf (accessed 2 June 2016).

SAMHSA (2016) *Homelessness and Housing*. Substance Abuse and Mental Health Services Administration, http://www.samhsa.gov/homelessness-housing (accessed 18 May 2016).

Shum, M., Comack, E., Stuart, T., *et al.* (2012) Bed bugs and public health: new approaches for an old scourge. *Canadian Journal of Public Health*, **103** (6), e399–403.

Sompura, D. (2016) *Report of the Fiscal 2017 Preliminary Budget and the Fiscal 2016 Preliminary Mayor's Management Report*. Department of Homeless Services, the Council of the City of New York. New York. 15 March 2016, http://council.nyc.gov/html/budget/2017/pre/071%20Department%20of%20Homeless%20Services.pdf (accessed 2 June 2016).

Stefanski, S. (2016) *Homeless Shelter Spending Increased To Record High This Year, Yet Next Year Remains Underfunded. New York City Independent Budget Office*, http://www.ibo.nyc.ny.us/iboreports/homeless-shelter-spending-increased-to-record-hight-this-year-yet-next-year-reamins-underfunded-march-2016.pdf (accessed 4 April 2017).

Stennett, C.R., Weissenborn, M.R., Fisher, G.D. and Cook, R.L. (2012) Identifying an effective way to communicate with homeless populations. *Public Health*, **126** (1), 54–56.

Wright, J.D. (1990) Health care for homeless people: evidence from the national health care for the Homeless Program. In: *Under the Safety Net* (ed. P.W. Brickner *et al.*), WW Norton & Co., New York, pp. 15–31.

36

Hotels

David Cain

36.1 Introduction

Bed bugs have the potential to create significant crisis events for any provider in the hospitality industry (Liu *et al.*, 2015; Penn *et al.*, 2015, 2017). As infestations continue to spread around the globe (Doggett, 2006), a high occupant turnover makes hotels and the hospitality sector much more susceptible to new infestations than domestic settings. At the same time, the past tendency for many in the industry to "rather not think about bed bugs" has provided the insect with the perfect opportunity to quietly and quickly spread.

The hospitality sector is especially vulnerable to the fiscal impacts of bed bugs. Infestations are costly to treat and, if not detected early or improperly treated, bed bugs can quickly spread from the original source to infest other rooms. However, treatment costs can be minimal compared with the loss of business due to brand damage, resulting from on-line reviews warning of the presence of bed bugs within a hotel. Furthermore, there is the potential for huge payouts associated with litigation.

36.2 Bed Bugs in Hotels: The Challenges

One of the greatest challenges unique to the accommodation industry regarding bed bugs is the reality that either the current or a previous guest is most likely to have been the source of the infestation. Most people do not wish to encounter bed bugs (Meek, 2007), but keeping a hotel free of bed bugs can only be achieved by eliminating guests. This is clearly not feasible (O'Neill, 2012). The alternative is to screen all incoming guests for bed bugs, which presents ethical and logistical issues.

The financial impact of bed bug infestations can be very high and may include:

- cost of guest compensation (including the reimbursement of the stay, compensation to restore good will, and medical expenses)
- treatment costs
- loss of revenue with closure of infested rooms for treatment
- preparation costs and repairs (in some cases various fixtures need to be removed for treatment)
- replacement of furnishings (stained by bed bug feces) and cleaning costs to remove bed bug signs
- loss of future business
- public relations expenses to protect brand reputation
- potential legal expenses.

Advances in the Biology and Management of Modern Bed Bugs, First Edition.
Edited by Stephen L. Doggett, Dini M. Miller, and Chow-Yang Lee.
© 2018 John Wiley & Sons Ltd. Published 2018 by John Wiley & Sons Ltd.

Table 36.1 A comparison of known number of rooms infested with bed bugs in three London hotels, pre- and post-inspection, and rate of new infestations.

Hotel type	No. beds	Pre-inspection Number of rooms	Post-inspection	New infestations Rooms/month
Business	300	5	38	1
General	260	2	8	0.8
Budget	65	10	65	0.5

Source: David Cain, unpublished data.

In the USA, there have been high profile legal cases with million-dollar headlines (Higgins and Shilling, 2013) and hundred-thousand-dollar pay outs (Hedgpeth, 2015; see also Chapter 41). Fortunately, in most countries such payouts are not the norm and many cases are now settled out of court for much lower sums (see Part 7 of this book).

A major issue has been that hotels have tended to take a reactive approach in which a guest, member of staff, or canine pest inspection team identifies an infestation, which is then treated. The problem with this approach is that it fails to detect the problem before it has already become an issue to the organization. For hotels, the avoidance of guest complaints should be the ultimate goal in order to protect the brand. Quarterly or six-monthly bed bug inspections (a common approach and often considered good practice) may mean that infestations can be present for up to 3–6 months before detection. This interval is simply too long and increases the risk that an initial introduction could become established and spread to other rooms. A reactive approach to bed bug management means that it is not uncommon to underestimate the true number of infestations during the initial investigation, as evident from Table 36.1. Therefore a more regular inspection regimen appears justified.

The hotel environment has additional obstructions when compared to the domestic setting. These bring further complexities to the process of achieving successful eradication. They are detailed in Table 36.2.

As noted in Table 36.2, many hotel managers often select pest services on price alone. Often companies offering low-priced quotes do not have any practical experience in bed bug management and do not appreciate the time involved in achieving successful eradication. Due to the limited budget, they are forced to undertake a quick insecticide application only. As a result, treatment failures are common, and this can be due to:

- the use of aerosol and fogger based insecticides, which do not reach bed bug harborages (Jones and Bryant, 2012)
- inappropriate insecticides and applications, with no attention to resistance issues
- over application of insecticides, notably the pyrethroids, which can result in insect dispersal
- over-application of organic desiccants, most notably diatomaceous earth (DE), or the application of incorrect forms of DE.

Treatment failures can also be the result of a failure to undertake a follow-up inspection. The consequence of a lack of post-treatment inspection is that unsuccessful eradication attempts are not identified and consequently bed bugs can often disperse to infest other locations. It can take some time before senior management or the business owner steps in to seek a more complete and proactive solution.

36.3 Successful Bed Bug Management in Hotels

In developing a bed bug service solution for hotels, there are number of key elements, perhaps the most critical being early bed bug detection. It is important to detect infestations before guests became aware of the issue and prior to infestations having an opportunity to establish. Early detection means that the infestation

Table 36.2 Challenges to achieve bed bug eradication in hotels.

Factor	Challenge
Fixed décor, with no spares	Few options to remove and replace fittings if spares not available.
Pressure to reoccupy room	Loss of revenue on a tight margin can quickly result in significant profit losses.
Egg hatching time	Time taken for eggs to hatch reduces viability of chemical treatments; multiple treatments needed for many products.
Fitted construction	Modern hotel room design with fitted head boards and sideboard units can make access for treatment difficult. They also provide protected refugia, resulting in treatment failures if ignored.
Disruption of operations	Modern hotels are expected to reset rooms between guests within the hours of 8 am and 3 pm, so treatment visits are disruptive to scheduled tasks.
Internal culture	In many medium-to-large organizations there can be a lack of integration between management, front of house, housekeeping, and maintenance. Some organizations focus on blame rather than working on issues in a cooperative sense.
Lack of spare rooms	If hotel is fully booked and a guest raises the alarm about bed bugs in the middle of the night, there may be no accommodation available to transfer them into.
Lack of communication with peers	Stigma incorrectly associated with bed bugs means hotel staff less likely to seek advice from colleagues.
Lack of professional training	Training of hotel staff has not kept up with the bed bug resurgence.
Operational time pressures	The industry in the UK aims to service a room in in about 12 min, leaving little time for bed bugs inspections, which can add significantly to costs.
Cost of treatment	Although cost of treatment is a small percentage of overall potential cost of an infestation, many will select cheapest quotation for control, leading to treatment failure and higher costs in the long term.

Source: David Cain, unpublished results.

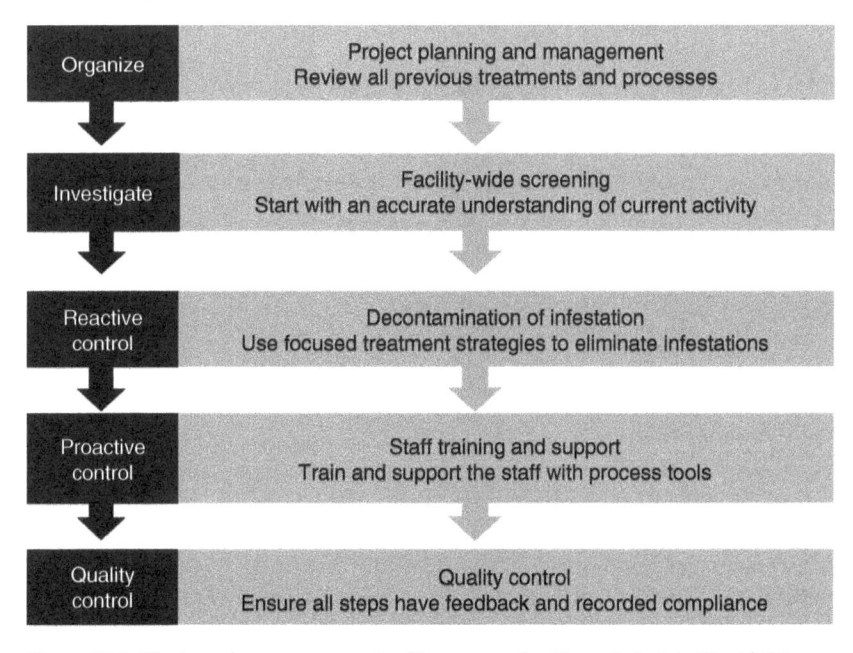

Organize — Project planning and management. Review all previous treatments and processes

Investigate — Facility-wide screening. Start with an accurate understanding of current activity

Reactive control — Decontamination of infestation. Use focused treatment strategies to eliminate infestations

Proactive control — Staff training and support. Train and support the staff with process tools

Quality control — Quality control. Ensure all steps have feedback and recorded compliance

Figure 36.1 The broad processes required to manage bed bugs in hotels (David Cain, unpublished results and after various sources, see also www.bed-bugs.co.uk/hotels).

can be eradicated in a matter of minutes rather than hours, even without room closure. Figure 36.1 outlines the processes for bed bug management in hotels and this proactive system is based around defining procedures, weekly inspections, and a simple treatment protocol for light infestations. A critical step in minimizing bed bug impacts in hotels involves compliance and quality control. It is necessary to ensure the inspections are being completed as scheduled, because human error and task fatigue can become an issue.

It is important that hotels have a bed bug management policy in place (see, for example, Doggett, 2011). Such a policy should outline the procedures that ensure that everyone in the organization understands the processes and the roles they play in resolving bed bug issues. Clearly demarcated responsibilities will aid staff in working together rather than passing the blame and responsibility onto others.

As noted above, early bed bug detection is the key to minimizing bed bug impacts. However there are challenges in using monitors in the hotel environment. While various studies have demonstrated that pitfall monitors can be effectively employed to detect bed bugs (see Chapter 27), unfortunately this type of monitor is not acceptable for the hotel industry. Pitfall monitors are obvious and not aesthetically appealing, and almost advertise that there is a bed bug issue, as well as adding maintenance tasks. Instead, a more discrete monitoring system is required. Bed Bugs Limited (the company set up by the author) developed a simple passive monitor technology (GB2470307B and GB2463953) (Intellectual Property Office, 2009a, 2009b) that has a white skirt, which bed bugs defecate on. When checked weekly as part of a routine inspection regimen of the beds, these monitors can assist in the detection of bed bugs before the breeding cycle starts.

For treating light infestations, steam can be employed, and this method can be conducted by the pest management professional or by the client themselves, if they are trained in how to use the equipment properly. The author's company employs the super-heated steamer Cimex Eradicator™ (Polti, 2014), which enables the room to be returned to normal operations within 45 min, ensuring minimal disruption to the facility.

In terms of reactive bed bug management within hotels, an IPM approach is always recommended, as outlined in Part 5 of this book (which should be consulted for more details on control). By using a convective heater to raise the temperature of the room to above 50 °C, it is possible to achieve 100% eradication of bed bugs and their eggs, so long as the air flow does not permit cool spots, or heat sinks (such as tiles or brickwork) do not significantly reduce temperatures (see Chapter 28). Heat alone is not without its risks, and failures, from damage to furniture, accidental fires, triggering of sprinkler systems, and so on means that it is far from a robust methodology for inexperienced personnel. Heat systems also have significant start-up costs.

In the case of hotels, it is often only after senior management realize that a proactive bed bug management program costs significantly less than reactive pest management in the long term that such a program is implemented. However, even if a hotel adopts an optimal pest management plan, the presence of bed bugs in the hotel has the ability to ruin the reputation and profitability of a hotel, if not the entire brand.

References

Doggett, S.L. (2006) Bed bugs: the unwanted guest. *Executive Housekeeper*, **10** (1), 43–45.

Doggett, S.L. (2011) *A Bed Bug Management Policy and Procedural Guide*, Department of Medical Entomology, Westmead Hospital, Westmead.

Hedgpeth, D. (2015) *Jury in Prince George's County Awards Woman $100,000 in Bedbug Case*, https://www.washingtonpost.com/local/crime/jury-in-prince-georges-county-awards-woman-100000-in-bedbug-case/2015/09/18/25b83036-5e26-11e5-8e9e-dce8a2a2a679_story.html (accessed 25 July 2016).

Higgins, L. and Shilling, E. (2013) *N.Y. Woman Sues Hotel For $7 Million in Bedbug Attack*, http://www.usatoday.com/story/news/nation/2013/01/10/women-sues-hotel-over-bedbugs/1566206/ (accessed 25 July 2016).

Intellectual Property Office (2009a) *GB2463953 – A Bed Bug Detecting Device and method of detection*, https://www.ipo.gov.uk/p-ipsum/Case/PublicationNumber/GB2463953 (accessed 25 July 2016).

Intellectual Property Office (2009b) *GB2470307 – A Bed Bug Detecting Device and Method of Detection*, https://www.ipo.gov.uk/p-ipsum/Case/PublicationNumber/GB2470307 (accessed 25 July 2016).

Jones, S.C. and Bryant, J.L. (2012) Ineffectiveness of over-the-counter total-release foggers against the bed bug (Heteroptera: Cimicidae). *Journal of Economic Entomology*, **105** (3), 957–963.

Liu, B.J., Pennington-Gray, L. and Klemmer, L. (2015) Using social media in hotel crisis management: the case of bed bugs. *Journal of Hospitality and Tourism Technology*, **6** (2), 102–112.

Meek, F. (2007) Guests consider bed bugs. *Lodging Hospitality*, **63** (16), 94–95.

O'Neill, L. (2012) The bedbug menace. *Caterer and Hotelkeeper*, **27-April**, 30–31.

Penn, J., Hu, W. and Potter, M.F. (2015) Disturbed! *Pest Control Technology*, **43** (11), 24–25, 26, 28, 30, 32.

Penn, J.M., Penn, H.J., Potter, M.F. and Hu, W. (2017) Bed bugs and hotels: traveler insights and implications for the industry. *American Entomologist*, **63** (2), 79–88.

Polti (2014) *Cimex Eradicator*, http://www.polti.com/en/catalog/consumer/212/398 (accessed 25 July 2016).

37

Healthcare Facilities

Stephen L. Doggett

37.1 Introduction

Bed bugs in healthcare facilities (HCFs) have posed a problem for almost as long as humans have placed the sick in facilities to receive medical treatment. Unfortunately, with the global resurgence, bed bugs in HCFs have again become a genuine issue (Reynolds, 2008; Anonymous, 2010a,b; Henner, 2011; Hurst and Humphreys, 2011; Masterson, 2011; Doggett *et al.*, 2012; Zipple *et al.*, 2012; Laliberté *et al.*, 2013; Williams, 2013; Hussain *et al.*, 2014; Sfeir and Munoz-Price, 2014; Bandyopadhyay *et al.*, 2015; Potter *et al.*, 2015; Sheele *et al.*, 2016, 2017a,b). For example, in a 2015 survey of pest management professionals (PMPs) in the USA, 36% of respondents reported treating bed bugs in hospitals, an increase from 33% in 2013 and 12% in 2010 (Potter *et al.*, 2015). Furthermore, the same survey revealed that 58% of PMPs had treated bed bugs in nursing homes and 33% in doctor/outpatient clinics. This chapter examines the challenges and history of bed bugs in HCFs, the treatment processes employed, and the management protocols used to eradicate bed bugs therein.

37.2 The Challenges Bed Bugs Pose to Healthcare Facilities

HCFs present many challenges to the management of bed bugs. Patients are ill, possibly immunocompromised, and may be more sensitive to the injurious effects of chemicals. Often wards have high occupancy rates and the opportunity to close sections for treatment can be limited. There are high staff and visitor numbers, meaning there are ample opportunities for bed bugs to be introduced or passed onto others (Dana, 2012). Beds and other equipment in wards are often not static and are moved around between rooms, which can facilitate bed bug spread. Headlines on bed bugs in HCFs are alarming for both patients and staff (Leininger-Hogan, 2011), while infestations in HCFs can be expensive. One emergency department in the USA spent over USD 12 000 on bed bug remediation over a 22-week period (Totten *et al.*, 2016).

It is not uncommon for patients to present to medical practitioners with signs of a bed bug infestation (Podczervinski *et al.*, 2014). In Anderson, Ohio, staff reporting seeing bed bugs on patients on a weekly basis (Erdogan *et al.*, 2010; Aultman, 2013). A survey from 2014 revealed that bed bug related visits to emergency departments across the USA over a ten-year period increased by a factor of more than 750, from 21 visits in 2001 to 15 945 in 2010 (Langley *et al.*, 2014). In Cleveland, Ohio, between 1 August 2014 and 31 August 2015, a bed bug "event'" was recorded every 2.2 days in a tertiary care medical center (Sheele *et al.*, 2017a) and patients presenting to the emergency department with bed bugs were more likely to be older males requiring hospitalization (Sheele *et al.*, 2017b). Presumably, with more patients transporting bed bugs to HCFs, the risk of the facility becoming infested also increases. An unfortunate consequence of this is that many patients have

Advances in the Biology and Management of Modern Bed Bugs, First Edition.
Edited by Stephen L. Doggett, Dini M. Miller, and Chow-Yang Lee.
© 2018 John Wiley & Sons Ltd. Published 2018 by John Wiley & Sons Ltd.

been refused medical treatment (Doggett *et al.*, 2012; Laliberté *et al.*, 2013), which poses serious ethical issues as there is little justification to exclude patients from essential healthcare (Aultman, 2013; Laliberté *et al.*, 2013).

The risk of an infestation in HCFs has yet to be quantified but is likely to be greater in locations with higher bed bug incidence such as inner cities and facilities that treat the socially disadvantaged. Bed bugs are probably introduced into hospitals much as they are transferred in other situations, namely via human belongings. In Austria, the Common bed bug, *Cimex lectularius* L., was brought into a hospital in the leg prosthesis of a patient admitted to the intermediate care unit (Paulke-Korinek *et al.*, 2012). In Canada, a patient with Alzheimer's disease was admitted with infested clothing (Sabou *et al.*, 2013). Wheelchairs have been reported as a harborage location for bed bugs and a means by which they can be spread (Anderlonis, 2015). Furthermore, ambulances have also been found infested with bed bugs (Potter *et al.*, 2015); such infestations could lead to the widespread dispersal of the insect across multiple HCFs.

37.3 The History of Bed Bugs in Healthcare Facilities

Probably the earliest record of bed bugs in HCFs goes back to 1765, when the English surgeon Robert Sharp observed the use of iron bedsteads in Italy (Sarasohn, 2013). He proposed that they should be adopted for English hospitals to minimize the suffering of "so many thousands of miserable wretches, that are tormented sometimes to death, by these nauseous vermin" (Sarasohn, 2013). Even back in the 1700s, it was recognized that harborage reduction is an important means of bed bug management.

In 1904, H.E. Durham reported capturing bed bugs in the "Lunatic Asylum and General Hospital...at Kuala Lumpur" for his research into the cause of beriberi (Durham, 1904). A Dr. Campbell in 1925 described the results of his experiments on bed bug movement in hospitals (Campbell, 1925). He placed marked insects in an unoccupied cot one evening and found them in an occupied cot at the other end of the ward the next morning. In India during the 1930s, a weekly regimen of treating bedsteads and mosquito nets with boiling water and steaming the mattresses was introduced to control bed bugs in a hospital (Pachecho, 1935). In addition, kerosene spray along with coconut oil was applied to the crevices.

It appears that bed bugs in HCFs were such a common phenomenon that they, along with hotels, were constructed in ways that were intended to minimize potential harborages (Hartnack, 1939). This included minimizing woodwork, installing no picture moldings, having steel windows and doors, using no wallpaper, and enclosing pipework (Hartnack, 1939).

In Australia, bed bugs also occurred in HCFs during the 1930s and 1940s. One nurse during these years recalls that beds were steel framed and had to be regularly dismantled and painted liberally with kerosene. Bedding was boiled in soapy water and bed legs were placed into cut-down tin cans filled with motor oil (Christopher Orton, University of NSW, Sydney, unpublished data; see also Chapter 7).

By the 1950s, DDT was regularly being used for the control of bed bugs in hospitals around the world (Busvine, 1957) and the bed bug problem declined in the developed world about that time.

37.4 Bed Bugs in Healthcare Facilities with the Modern Resurgence

Since the onset of the modern bed bug resurgence, there have been a number of reports that describe bed bugs in HCFs and the measures undertaken to control them. In 2009, an infestation occurred in a maternity unit in New York; bed bugs were observed in a sleeper chair (Adeyeye *et al.*, 2010). Subsequently, a canine was used for bed bug detection in the unit, and the dog identified eleven additional rooms that were also infested (although the publication did not state whether or not bed bugs were then visually confirmed by the dog handler). Eventually the infestation spread to encompass offices and other units. Control measures included the implementation of quarterly inspections and the development of bed bug guidelines.

In New Delhi, India, there was an outbreak of the Tropical bed bug, *Cimex hemipterus* (F.) in a neonatal unit during late 2013 (Bandyopadhyay et al., 2015). It is thought that the bed bugs were brought in on patient blankets. Itchy papular eruptions were first reported in patients, and then staff, and the presence of bed bugs was subsequently confirmed. Overall, 39 individuals were affected in the outbreak. In response, the infested unit was closed and patients were told to bag, remove, and hot launder all belongings. Baby mattresses were discarded and cots disinfected. Cracks and crevices were sealed with plaster of Paris and the rooms treated with insecticide. Staff received education on bed bugs and the infestation was declared eradicated after two weeks of treatment. Subsequent inspections over the following six months revealed no further infestations.

The author of this chapter has been involved in a number of infestations in HCFs. In a case in a doctor's treatment rooms, the chairs in the waiting area became infested. Replacement of the infested chairs quickly solved the problem. One hospital in south-east Sydney, Australia, had a massive problem in the staff quarters between May 2003 and June 2005, with 20% of the 320 rooms in the accommodation complex infested (Doggett and Russell, 2008). Some of the apartments were reserved for visiting families of sick children. Family members would also sleep in the child's hospital room and, in the process, bed bugs (*C. lectularius*) were transferred from the accommodation complex to the ward. Eventually the infestation was controlled using a combination of vacuuming and insecticides, and no further infestation has since occurred. A key element to the success in this case was that every room in the staff quarters was inspected. Interestingly, half of the rooms that were found to be infested had not previously been recognized as such.

Another hospital in western Sydney with a children's ward had 18 rooms that became infested during 2007 (Doggett and Russell, 2008). Somewhat surprisingly, only one bed was found positive for bed bugs. Small numbers of bed bugs were detected in bedside furniture, while the heaviest infestations were in the lounge/bed chairs where parents often slept overnight. After each patient, the mattress and bed frame were sterilized with a disinfectant that was found to have a high level of topical efficacy for controlling bed bugs (Stephen Doggett, unpublished). However, the lounge chairs were not routinely treated with the disinfectant. The infestation was ultimately controlled through the use of steam, vacuuming, and insecticides. The hospital developed a bed bug management policy and no further infestation has since occurred.

37.5 Bed Bug Management in Healthcare Facilities

Many authors recommend that hospitals should have procedural guidelines that address bed bugs and that they should follow an IPM approach (Pinto et al., 2007; Ruckriegel et al., 2011; Munoz-Price et al., 2012; Zipple et al., 2012; AHS, 2013; Barnes and Murray, 2013; Winegar et al., 2013; Podczervinski et al., 2014). Also, any patient exposed to bed bugs in the facility should be informed of the risk (Henner, 2011). Broadly speaking, guidelines can be divided into reducing the risk of contracting bed bugs, early detection, control measures, follow up inspections, and the review of any infestation. It is vital to undertake control measures in a timely manner and to document the steps taken. The education of all staff who may be involved in the recognition or management of the infestation is also essential (Barnes and Murray, 2013).

Bed bug risk reduction involves the use of furniture that provides few harborages, such as hard plastic (rather than fabric chairs), and ensuring that there are minimal cracks and crevices in the room. Patients who may present with bite marks need to be investigated in case their belongings are infested (Munoz-Price et al., 2012). Any suspicion of bed bugs should result in the quarantining and treatment of the patient's belongings using heat treatments or freezing (Munoz-Price et al., 2012). If possible, patients should be discouraged from bringing personal belongings into the facility, especially bed linen and blankets (AHS, 2013; Bandyopadhyay et al., 2015). Areas prone to bed bugs should be regularly vacuumed and/or steam cleaned (AHS, 2013). Equipment that has been exposed to or infested with bed bugs should be labeled and confined to a particular room or area, to avoid possible spread of an infestation. Staff should be trained in bed bug recognition and bed bug information sheets should made available to patients (AHS, 2013).

Early detection involves regular inspections, especially of high risk areas (such as beds for families in wards), possibly with the aid of canine detection (Harrison and Lawrence, 2009; Munoz-Price *et al.*, 2012; Harrison, 2014). Several bed bug researchers recommend the use of bed bug traps, but the effectiveness of such devices in HCFs has largely yet to independently assessed. One study examined the ability of the Verifi (FMC, Philadelphia, PA, USA; a trap that is no longer marketed) to detect bed bugs in an emergency department and it was found inefficient (Sheele *et al.*, 2016). However, few traps were employed in the study and most were positioned outside of the rooms where the bed bug activity occurred.

Any unexplained bites and rashes on patients while in the HCF should be investigated. If bed bugs are found, specimens should be captured and their identity confirmed by a pest management professional with experience in bed bug management (AHS, 2013). Once an infested room is positively identified, it should be closed and quarantined, and all potentially infested items bagged before removal from the room prior to treatment (Munoz-Price *et al.*, 2012). Non-chemical means of control should be undertaken first, including the use of vacuuming, steam, and portable heat chambers. Insecticides should only be used if the non-chemical measures fail to achieve control. The decision to use an insecticide should take into consideration the condition of any patient that may inhabit the room post-treatment in case of chemical sensitivities. Follow-up inspections should be undertaken to ensure that the infestation is controlled and not spread to other areas of the facility (Bandyopadhyay *et al.*, 2015). A debrief following the eradication of the infestation should be undertaken to review and enhance bed bug policies and procedures (Podczervinski *et al.*, 2014). In all cases, bed bug management should follow best practice, as outlined in one of the industry standards (BBF, 2013; Doggett, 2011, 2013; NPMA, 2013; see also Chapters 22–24).

While the modern global resurgence continues, bed bugs will pose an ongoing threat to HCFs. For the most part, such locations are modestly furnished, with few locations where bed bugs can hide, and bed bug eradication is not overly challenging. However, having policy and procedural guidelines in place in the event of an infestation will ensure that bed bugs are managed in a timely and appropriate fashion, without further compromising the health of patients.

References

Adeyeye, A., Adams, A., Herring, L. and Currie, B.P. (2010) Bed bug infestation on a maternity unit in a tertiary care center. *American Journal of Infection Control*, **38** (5), e82–e83.

AHS (2013) *Bed Bug Management Protocols for Health Care Workers*. Alberta Health Services, http://www.albertahealthservices.ca/ipc/hi-ipc-bedbug-management-protocol-hcw.pdf (accessed 19 May 2016).

Anderlonis, F.M.J. (2015) Don't let the bedbugs bite! *Nursing*, **45** (7), 46–47.

Anonymous (2010a) EDs trying not to let the bed bugs bite. *ED Management*, **22** (9), 100–101.

Anonymous (2010b) PA hospital had bed bugs. *Star Phoenix* (6 November), p. A4.

Aultman, J.M. (2013) Don't let the bedbugs bite: the Cimicidae debacle and the denial of healthcare and social justice. *Medicine, Health Care and Philosophy*, **16** (3), 417–27.

Bandyopadhyay, T., Kumar, A. and Saili, A. (2015) Bed bug outbreak in a neonatal unit. *Epidemiology and Infection*, **143** (13), 2865–2870.

Barnes, E.R. and Murray, B.S. (2013) Bedbugs: what nurses need to know. *American Journal of Nursing*, **113** (10), 58–62.

BBF (2013) *European Code of Practice for Bedbug Management*, 2nd edn. Bed Bug Foundation, London.

Busvine, J.R. (1957) Recent progress in the eradication of bed bugs. *The Sanitarian*, **May**, 365–369.

Campbell, C.A.R. (1925) *Bats, Mosquitoes and Dollars*, Stratford Company, Boston.

Dana, R. (2012) Control with care. Infestations in health-care facilities present PMPs with obstacles and opportunities. *Pest Management Professional*, **80** (9), S6, S9–S10.

Doggett, S.L. (2011) *A Bed Bug Management Policy & Procedural Guide for Accommodation Providers*, Westmead Hospital, Sydney.

Doggett, S.L. (2013) *A Code of Practice for the Control of Bed Bug Infestations in Australia*, 4th edn, Department of Medical Entomology and The Australian Environmental Pest Managers Association, Sydney.

Doggett, S.L. and Russell, R.C. (2008) The resurgence of bed bugs, *Cimex* spp. (Hemiptera: Cimicidae) in Australia. In: *Proceedings of the Sixth International Conference on Urban Pests, July 13–16, 2008, Budapest, Hungary*. OOK-Press, Veszprém.

Doggett, S.L., Dwyer, D.E., Peñas, P.F. and Russell, R.C. (2012) Bed bugs: clinical relevance and control options. *Clinical Microbiology Reviews*, **25** (1), 164–192.

Durham, H.E. (1904) Notes on beriberi in the Malay Peninsula and on Christmas Island (Indian Ocean). *The Journal of Hygiene*, **4** (1), 112–155.

Erdogan, J., Martin, T. and Payne, R. (2010) EDs trying not to let the bed bugs bite. *ED Management*, **22** (9), 100–101.

Harrison, R. (2014) Lifting the sheets. Preventing and managing health care bed bug problems. *Health Facilities Management*, **27** (2), 49–51.

Harrison, R. and Lawrence, B. (2009) *Pulling Back The Sheets on The Bed Bug Controversy*, http://www.ahe.org/ahe/content/orkin/ahe-bedbug-white-paper.pdf (accessed 20 July 2016).

Hartnack, H. (1939) *202 Common Household Pests of North America*, Hartnack Publ. Co., Chicago.

Henner, K.A. (2011) What are the ethical ramifications and suggested guidelines when a dental office is confirmed as having an active bedbug infestation? *Journal of the American Dental Association*, **142** (12), 1398–1399.

Hurst, S. and Humphreys, M. (2011) Bedbugs: not back by popular demand. *Dimensions of Critical Care Nursing*, **30** (2), 94–96.

Hussain, M., Khan, M.S., Wasim, A., *et al.* (2014) Inpatient satisfaction at tertiary care public hospitals of a metropolitan city of Pakistan. *Journal of the Pakistan Medical Association*, **64** (12), 1392–1397.

Laliberté, M., Hunt, M., Williams-Jones, B. and Feldman, D.E. (2013) Health care professionals and bedbugs: an ethical analysis of a resurgent scourge. *HealthCare Ethics Committee Forum*, **25**, 245–255.

Langley, R., Mack, K., Haileyesus, T., Proescholdbell, S. and Annest, J.L. (2014) National estimates of noncanine bite and sting injuries treated in US hospital emergency departments, 2001–2010. *Wilderness and Environmental Medicine*, **25** (1), 14–23.

Leininger-Hogan, S. (2011) Bedbugs in the intensive care unit. A risk you cannot afford. *Critical Care Nursing Quarterly*, **34** (2), 150–153.

Masterson, D. (2011) Preventing uninvited visitors to radiology: bedbugs. *Journal of Radiology Nursing*, **30** (2), 67–69.

Munoz-Price, L.S., Safdar, N., Beier, J.C. and Doggett, S.L. (2012) Bed bugs in healthcare settings. *Infection Control and Hospital Epidemiology*, **33** (11), 1137–1142.

NPMA (2013) *Best Management Practices for Bed Bugs – Consumer Edition*. National Pest Management Association, http://www.pestworld.org/media/560191/bed_bug_bmps_for_consumers_final.pdf (accessed 19 May 2016).

Pachecho, J.N. (1935) A simple method of bug destruction. *Indian Medical Gazette*, **70** (2), 75–76.

Paulke-Korinek, M., Szell, M., Laferl, H., Auer, H. and Wenisch, C. (2012) Bed bugs can cause severe anaemia in adults. *Parasitology Research*, **110** (6), 2577–2579.

Pinto, L.J., Cooper, R. and Kraft, S.K. (2007) *Bed Bug Handbook-the Complete Guide to Bed Bugs and Their Control*, Pinto & Associates, Inc., Mechanicsville, MD.

Podczervinski, S., Fauver, J., Helbert, L., *et al.* (2014) A multidisciplinary approach to minimize exposures to bed bugs (*Cimex lectularis*) in a large ambulatory cancer center. *American Journal of Infection Control*, **42** (6), S85–S86.

Potter, M.F., Haynes, K.F. and Fredericks, J. (2015) Bed bugs across America. *Pestworld*, **November/December**, 4–14.

Reynolds, E. (2008) Patients flee hospital infested by bed bugs. *The Daily Mirror* (26 September), p. 21.

Ruckriegel, C., Pascarella, L., Harmon, P., *et al.* (2011) Bed bugs? Preventing mayhem with a protocol. *American Journal of Infection Control*, **39** (5), E65.

Sabou, M., Imperiale, D.G., Andres, E., *et al.* (2013) Bed bugs reproductive life cycle in the clothes of a patient suffering from Alzheimer's disease results in iron deficiency anemia. *Parasite*, **20**, 1–5.

Sarasohn, L.T. (2013) "That nauseous venomous insect:" bedbugs in early modern England. *Eighteenth-Century Studies*, **46** (4), 513–530.

Sfeir, M. and Munoz-Price, L.S. (2014) Scabies and bedbugs in hospital outbreaks. *Current Infectious Disease Reports*, **16** (8), 412.

Sheele, J.M., Mallipeddi, N., Chetverikova, M., Mothkur, S. and Caiola, C. (2016) FMC Verifi traps are not effective for quantifying the burden of bed bugs in an emergency department. *American Journal of Infection Control*, **44** (9), 1078–1080.

Sheele, J.M., Barrett, E., Farhan, O. and Morris, N. (2017a) Analysis of bed bug (*Cimex lectularius*) introductions into an academic medical center. *Infection Control and Hospital Epidemiology*, **38** (5), 623–624.

Sheele, J.M., Gaines, S.L., Maurer, N., *et al.* (2017b) A survey of patients with bed bugs in the emergency department. *American Journal of Emergency Medicine*, **35** (5), 697–698.

Totten, V., Charbonneau, H., Hoch, W., Shah, S. and Sheele, J.M. (2016) The cost of decontaminating an ED after finding a bed bug: results from a single academic medical center. *American Journal of Emergency Medicine*, **34** (3), 649.

Williams, J. (2013) Bed bugs in hospitals: more than just a nuisance. *Canadian Medical Association Journal*, **185** (11), E524–E524.

Winegar, R.D., Rick, S. and Johnson, A. (2013) Bed bugs and beyond: a call to action for advanced practice registered nurses. *The Journal for Nurse Practitioners*, **9** (8), 536–540.

Zipple, A.M., Batscha, C.L., Flaherty, P. and Reynolds, J.L. (2012) Don't get bugged. Practical strategies for managing bedbug infestations in psychiatric rehabilitation programs. *Journal of Psychosocial Nursing*, **50** (7), 22–26.

38

Aircraft

Adam Juson and Catherine Juson

38.1 Introduction

The management of bed bugs onboard aircraft is one of the most challenging issues facing pest management professionals (PMPs) around the world. Bed bugs can easily be introduced onto aircraft by passengers, luggage, crew, and ground personnel. Both the Common bed bug, *Cimex lectularius* L., and the Tropical bed bug, *Cimex hemipterus* (F.), have been detected on commercial aircraft, although *C. lectularius* is by far the most prevalent species (Adam Juson, unpublished results). PMPs face a diverse range of infested environments that have little in common except their vulnerability to bed bugs. The body of an aircraft presents a unique challenge to any PMP charged with eradicating an infestation. This chapter focuses on bed bug management on aircraft and reviews the challenges unique to this situation, as well as the protocols required to achieve successful eradication.

38.2 Aviation Entomology – a Brief History

Kisluik (1929) initiated the field of research investigating insect transport and infestation potential on aircraft, when he inspected the Graf Zeppelin in 1928. He found ten species of insect pests on-board, including one "bug". Since that time, concerns have been raised regarding various biting insects being transported on aircraft (Griffitts and Griffitts, 1931). Swain (1952) predicted that aircraft would be a major distributor of insect species, and Sullivan *et al.* (1958) reported on how insects might survive on aircraft. Public health pests have received the most attention due to their potential to spread agents of disease (Evans *et al.*, 1963; Basio *et al.*, 1970; Otaga *et al.*, 1974; Russell *et al.*, 1984; Goh *et al.*, 1985). Disinsection (a term indicating the process of treating of an aircraft so as to eliminate insects) of aircraft cabin environments to control mosquito species on international flights is now required under the World Health Organization, *International Health Regulations* (WHO, 2005).

The resurgence of bed bugs around the world (Robinson and Boase, 2011) has included their introduction onto commercial aircraft (Haiken, 2011). The efficacy of a limited number of detection and control devices for bed bug management on commercial aircraft was reviewed by Juson (2014). Little is known about the prevalence of bed bugs or their control on private, medical, or military aircraft. However, NATO's advisory group for aerospace research and development acknowledges that bed bugs are carried onto aircraft, and are known to be exchanged between passengers (Ellis, 1996).

Advances in the Biology and Management of Modern Bed Bugs, First Edition.
Edited by Stephen L. Doggett, Dini M. Miller, and Chow-Yang Lee.
© 2018 John Wiley & Sons Ltd. Published 2018 by John Wiley & Sons Ltd.

38.3 Bed Bug Management on Aircraft: The Challenges

One of the challenges facing PMPs when working on aircraft is access. An aircraft's duty cycle limits opportunities for non-critical inputs, such as bed bug inspections. Long-haul aircraft spend approximately 16 h/day in flight, giving bed bug populations virtually unrestricted feeding opportunities. The presence of sedentary hosts, coupled with a warm and stable air temperature, provide near optimal conditions for bed bug population establishment and growth.

The cost of bed bug management on aircraft is quite variable depending on how the aircraft is employed. Private (luxury) aircraft, commercial aircraft, and military aircraft have very few financial restrictions with regard to the cost of bed bug control. This is because the cost of bed bug management in these aircraft is minimal when compared to the expense associated with even a short flight delay. However, domestic medical aviation is often charitably funded and, as a result, very cost conscious. Air ambulance infestations are not common, but they are amongst the most time consuming and costly to resolve. This is due to the complex interior structure of the aircraft, which is designed to house medical equipment (Adam Juson, unpublished results).

The complex nature of the aircraft interior limits both visual inspection and treatment ability. In many cases it is not possible to access deep harborages on the airframe without dismantling the aircraft interior. The widespread use of light honeycomb structures in both aircraft seating and cabin construction allows bed bugs to harbor in protected microclimates that are in close proximity to the hosts. These ideal harborages allow infestations to go unnoticed for considerable periods of time and make treatment efforts very challenging.

Aircraft engineers and other aircraft employees are typically reluctant to work alongside PMPs onboard an infested aircraft. Widespread fear of bed bug exposure, coupled with the perceived risk of infesting their own homes, limits employee cooperation and leaves the PMP to inspect and locate sites of infestation on their own.

Another challenge for PMPs is that aviation is one of the world's most highly regulated industries. Not surprisingly, the pest management options that are acceptable for use on an aircraft are quite limited and these options must fit within the aviation regulatory framework. In addition, all pesticides used for insect control must meet the AMS1450a specifications (SAE International, 1995) and carry cabin approval (further local restrictions may also apply). This limits the currently available actives to a few pyrethroids and carbamates. Consequently, any new or innovative control technologies that might be of use on aircraft must be reviewed, evaluated, and go through an often lengthy approval process before they can be applied.

Another challenge for aircraft disinfestation is that any work carried out on board the aircraft, including pest control, has to be overseen by a licensed aircraft engineer. This necessitates that engineers be present during the strip-down and re-cover of passenger seating before and after chemical treatment. Engineers must also be present for the removal and reinstatement of heat-sensitive items, such as life vests and emergency exit slides, during thermal disinsection (heat treatment). The requirement that an engineer be present increases the cost of bed bug remediation and often puts pressure on the pest management crew to complete their treatment as quickly as possible.

As is often the case in any insect infestation, a small number of unauthorized (and unproven) treatments may be undertaken by the cabin crew. Crew members will often use disinsection aerosols (pyrethroid-based insecticides intended to control mosquitoes and agricultural pests that may pose a health or economic threat to various nations) when bed bugs are detected in flight. This application often results in flushing the population from the harborages into the passenger cabin. Passengers are often moved from infested seats to open seats elsewhere in the aircraft, potentially spreading the infestation.

Few academic studies have focused on evaluating best practices for bed bug remediation on aircraft. Those methods that have proved successful in aviation have come from research studies conducted in other environments.

38.4 Bed Bug Management on Aircraft

Bed bug management techniques focus on detection first, and then eradication. The application of these techniques must fit into the airlines' flying schedules. The severity and spread of bed bug activity is obviously directly correlated with the elapsed time since the inoculating event, or the failure of the previous treatment to eradicate the problem. As preventing a bed bug introduction is not possible, the early detection of infestations is vital. The complex nature of aircraft seating products severely limits the detection systems that can be used on board aircraft.

Refuge monitors, lure monitors, electronic air sampling detectors, and scent detection dogs have all been evaluated on aircraft to determine their utility and practicality (Juson, 2014). Harborage and lure-based monitors were quickly determined to be unviable, as these products require not only an initial set up, but regular follow-up visits to check for activity. The monitor maintenance alone made them unfavorable with airline operators. Furthermore, these monitoring devices could not compete with the attractiveness of the airline seats for bed bug harborage.

Alternative monitoring methods, such as the use of air sampling devices, were promoted extensively by some commercial operators. One particular device contained an infrared absorption cell calibrated to recognize the spectrum of gases resulting from bed bug respiration. However, this device was found to be very time consuming to use, and resulted in a large number of false positives, due to warm air pockets in the airline seats.

Although somewhat controversial, the most accurate and time-efficient method of bed bug detection on aircraft was the use of scent detection dogs. The ability of scent detection dogs to locate live bed bug infestations in a confined environment is well-documented (Pfiester *et al.*, 2008). However, the reported accuracy among different dog teams (inspecting multi-unit housing) has been shown to be quite variable (accuracy ranging from 15–98%, and a mean false positive rate of 15%; Cooper *et al.*, 2014; see also Chapter 27). However, the two studies that have been conducted onboard commercial aircraft found the canine accuracy to be 95% (Juson, 2014), with a mean false positive rate of 16% (Adam Juson, unpublished results). In heavily infested aircraft, dogs have difficulty detecting the periphery of the infestation, and will display scent saturation behavior in a closed, infested aircraft.

Many bed bug control methods have been attempted to manage active infestations. Some airlines attempted to use modified methods based around the WHO guidelines for aircraft disinsection (WHO, 2013). In essence, disinsection treatments involve either the application of "residual" pyrethroids every 6–8 weeks, or the use of pyrethroid-based total release aerosols. Both of these pyrethroid applications have been observed to agitate and disperse bed bug populations. This complicates control efforts further by spreading the infestation. In addition, it has been well documented that total release aerosols are ineffective for controlling bed bugs. This is because the chemical droplets do not penetrate harborages, and the bed bugs that contact the treated surfaces are highly resistant (Romero *et al.*, 2007) so the product has no residual activity (Jones and Bryant, 2012).

In summary, the chemical treatments currently applied by PMPs on aircraft have produced disappointing results (Juson, 2014). The complex construction of aircraft seating and the restrictions on dismantling seats does not allow PMPs enough access to make an adequate insecticide application. Additionally, the aircraft cabin environment rapidly degrades chemical insecticides so that the products have no residual activity for controlling resistant bed bugs.

38.5 Improving the Pest Management Protocol

Current pest management methods on aircraft need to be improved. As bed bug populations continue to proliferate throughout the world, consumer complaints will no doubt drive efforts toward improving eradication methods. Thorough inspections of the aircraft will be necessary to locate the areas of infestation in the cabin. Insecticide applications, and other treatment methods, will need to be applied and evaluated for efficiency. The treatment processes will need to be flexible so that they can be scaled to adequately address the size and distribution of the infestation on the type of airframe being treated.

Most recently, a four-phase approach has proven to be the most effective for treating bed bugs on a variety of aircraft. This sequential protocol involves using a vacuum to reduce the number of bed bugs immediately, then using steaming to kill additional bed bugs and their eggs. Steaming is followed by the application of a desiccant dust to provide residual activity. The success of these control measures is determined by re-inspection of the aircraft for survivors at weekly intervals following the initial treatment (Adam Juson, unpublished results). The specific elements of this treatment protocol are described below.

The information gathered from the initial inspection (the relative number of bed bugs and their specific locations) is used to inform the PMP where to vacuum. The vacuuming physically removes both live and dead bed bugs, and their cast skins. This initial removal of bed bugs and their debris does a lot to improve the efficacy of subsequent treatments. Vacuuming the major bed bug harborages found within aircraft seating has been found to reduce living bed bug biomass by ~70%.

Topical heat treatment using steam is highly effective at killing bed bug eggs left behind after vacuuming. Steam is also good for penetrating the airline seating. Steam has been found most effective for treating light-to-moderate infestations. Treating airline seating with steam requires the removal of the seat cover by airline personnel. The cover is then thoroughly vacuumed, and steam applied to all seams. The steamer is used to treat the seat structure and cushioning. Steam should be applied methodically, starting at the head rest, moving down the seat back, and finally onto the base of the seat and the foot rests. The treatment should be focused on, but not limited to, bed bug harborages that would be located near the passenger's shoulders and waist. It is very important that the steam velocity exiting the steamer head be limited to minimize the risk of blowing live bed bugs around inside the aircraft.

In the rare case of a high-level infestation, localized or full cabin heat treatment should be considered. If the infestation is restricted to the seating cabin, the use of a tented heat treatment will provide satisfactory results, providing that all potential harborage locations reach 55 °C, and there are no cold spots. If the infestation extends behind the cabin wall paneling, or into the floor runners or electrical runs, it will be necessary to pre-heat the exterior of passenger aircraft before heating the interior cabin. The exterior heating will prevent bed bugs from migrating behind the aircraft's thermal blanketing.

Residual treatment is applied to key bed bug harborage locations using a desiccant dust (which must have AMS1450a approval). This application will prevent re-infestation of harborages in passenger seating and other hard-to-treat locations. Desiccant dusts will not only kill bed bugs exposed to the dust but will act in a prophylactic manner to reduce the ability of newly introduced bed bugs from establishing an infestation.

As with any bed bug treatment, the infested treated area needs to be inspected post-treatment to evaluate the results. Re-inspection should be undertaken at weekly intervals to determine treatment efficacy. Any remaining bed bugs should receive focused remedial treatment until complete control is achieved.

In summary, the management of bed bugs on aircraft poses many logistical and engineering challenges. Furthermore, the industry is so highly regulated that significant constraints are imposed as to the range of control options that are available to the PMP. Similar to managing bed bug infestations in other locations, only an integrated and sustained approach will ensure bed bug eradication on aircraft.

References

Basio, R.G., Prudencio, M.J. and Chanco, I.E. (1970) Notes on the aerial transportation of mosquitoes and other insects at the Manila International Airport. *Philippine Entomologist*, **1** (5), 407–408.

Cooper, R., Wang, C. and Narinderpal, S. (2014). Accuracy of trained canines for detecting bed bugs (Hemiptera: Cimicidae). *Journal of Economic Entomology*, **107** (6), 2171–2181.

Ellis, R.A. (1996) *Aircraft Disinsection: A Guide for Military & Civilian Air Carriers*, http://citeseerx.ist.psu.edu/viewdoc/download?doi=10.1.1.215.4945&rep=rep1&type=pdf (accessed 28 July 2016).

Evans, B.R., Joyce, C.R. and Porter, J.E. (1963) Mosquitoes and other arthropods found in baggage compartments of international aircraft. *Mosquito News*, **23**, 9–12.

Goh, K.T., Ng, S.K. and Kumarapathy, S. (1985) Disease-bearing insects brought in by international aircraft into Singapore. *Southeast Asian Journal of Tropical Medicine and Public Health*, **16** (1), 49–53.

Griffitts, T.H.D. and Griffitts, J.J. (1931) Mosquitoes transported by airplanes. *U.S. Public Health Reports*, **46**, 2775–2782.

Haiken, M. (2011) *Bed Bugs on Airplanes*, http://www.forbes.com/sites/melaniehaiken/2011/11/21/bed-bugs-on-airplanes-how-to-fly-bed-bug-free/#7e1a2e167dd5 (accessed 27 July 2016).

Jones, S.C. and Bryant, J.L. (2012) Ineffectiveness of over-the-counter total-release foggers against the bed bug (Heteroptera: Cimicidae). *Journal of Economic Entomology*, **105** (3), 957–963.

Juson, A. (2014) Management of bed bugs on commercial aircraft. In: *Proceedings of the Eighth International Conference on Urban Pests, 2014, July 20–23, Zurich, Switzerland*. OOK-Press, Veszprém.

Kisliuk, M. (1929). Air routes, German dirigible "Graf Zeppelin" and quarantines. *Entomological News*, **40** (6), 196–197.

Otaga, K., Tanaka, I., Ito, Y. and Morii, S. (1974) Survey of the medically important insects carried by international aircraft to Tokyo International Airport. *Japanese Journal of Sanitary Zoology*, **25**, 177–184.

Pfiester, M., Koehler, P. and Pereira, R. (2008) Ability of bed bug detecting canines to locate live bed bugs and viable bed bug eggs. *Journal of Economic Entomology*, **101** (4), 1389–1396.

Robinson, W.H. and Boase, C.J. (2011) Bed bug resurgence: plotting the trajectory. In: *Proceedings of the Ninth International Conference on Urban Pests, 2011, August 7–10, Ouro Preto, Brazil*. Instituto Biológico, São Paulo.

Romero, A., Potter, M., Potter, D. and Haynes, K. (2007) Insecticide resistance in the bed bug: A factor in the pests sudden resurgence? *Journal of Medical Entomology*, **44** (2), 175–178.

Russell, R.C., Rajapaksa, N., Whelan, P. and Langsford, W.A. (1984) Mosquito and other insect introductions to Australia aboard international aircraft, and the monitoring of disinsection procedures. In: *Commerce and the Spread of Pests and Disease Vectors*, (ed M. Laird), Praeger, New York, pp. 109–142.

SAE International (1995) *AMS1450A: Aircraft Disinfectant (Insecticide)*, SAE International, Warrendale, PA.

Sullivan, W.N., du Chanois, F.R. and Hayden, D.L. (1958) Insect survival in jet aircraft. *Journal of Economic Entomology*, **51** (2), 239–241.

Swain, R.B. (1952) How insects gain entry. In: *Insects: the Yearbook of Agriculture, 1952*. (eds F.C. Bishop *et al.*), US Government Printing Office, Washington, DC, pp. 350–355.

WHO (2005) *International Health Regulations*, 2nd edn, World Health Organization, Geneva.

WHO (2013) *Aircraft Disinsection Insecticides*. World Health Organization, Geneva, http://www.who.int/ipcs/publications/ehc/ehc243.pdf (accessed 18 October 2016).

39

Cruise Ships and Trains

David G. Lilly and Garry Jones

39.1 Introduction

While much of the attention to the bed bug resurgence has been focused on domestic settings and the hospitality sector, peripheral effects have nonetheless been felt in a range of industries, including various forms of mass transportation. This chapter will review the incidence of bed bugs on two forms of mass transport: cruise ships and trains. The challenges presented by an infestation in these specific situations will be discussed, as well as some of the treatment methods employed to achieve eradication.

39.2 Cruise Ships and Ferries

The industry in leisure cruising has been one of the strongest global tourism performers in recent years (Anonymous, 2012a) and is expected to continue to grow. However, as with many other sectors of society, cruise ships, ocean liners, luxury vessels, and even house boats have not been immune to the scourge of bed bug infestations (Silverstein, 2011). Two potential sources for on-board bed bug introductions are the passengers and the crew. This is because both groups will stay in potentially infested accommodation prior to boarding the ship. Housekeeping staff trained in bed bug recognition represent a critical component of any pest management program as they regularly service guest rooms. Interestingly however, cruise ship staff rooms are not serviced by housekeeping and this can delay the recognition of a bed bug infestation (David Lilly and Garry Jones, unpublished results).

The management and eradication of bed bugs on cruise ships can be an extremely challenging process. Many factors, such as the complex physical structure of the interior environment (Mouchtouri *et al.*, 2008), the inherent limitations on passenger cooperation (and spare cabin availability), and the transient nature of cruise ship itineraries, which limit potential pest management professional (PMP) assistance (David Lilly and Garry Jones, unpublished results), can all challenge eradication efforts.

Bed bugs infesting ships is not a new phenomenon; in fact it is thought that the movement of wooden sailing vessels around the world was the means for the early spread of bed bugs globally (see Chapters 1 and 7). Records of infestations on naval ships date back to the early 20th century (Phelps, 1924; Baker, 1934; Anonymous, 1939) where bed bugs were recognized as "a worse nuisance on board than they were ashore". However, in those days they were considered "easy to prevent…but exceedingly difficult to exterminate" (Anonymous, 1939) with treatments of sulphur dioxide, hydrogen cyanide, and/or heat necessary for eradication (Baker, 1934; Anonymous, 1939).

In the 21st century however, cruise ships and ferries have become dramatically larger and more structurally complex. In 2015, the global average passenger occupancy on ocean liners reached 3000 guests (Chanev,

---

Advances in the Biology and Management of Modern Bed Bugs, First Edition.
Edited by Stephen L. Doggett, Dini M. Miller, and Chow-Yang Lee.
© 2018 John Wiley & Sons Ltd. Published 2018 by John Wiley & Sons Ltd.

2015). This was a milestone that also corresponded with one of the strongest periods of growth in leisure cruise ship patronage (Anonymous, 2012a; Anonymous, 2015). Consequently, when combined with the resurgence of bed bugs over the last two decades, the potential for bed bugs to once again pose a risk to the leisure maritime industry is tangible.

Nonetheless, the overall incidence of bed bugs on ships appears sparse or, for obvious reputational reasons, under-reported. As a result, only three peer-reviewed studies directly dealing with bed bugs on ships have been published since the modern resurgence began. The first was a survey of passenger ships entering Greece, where complaints of bed bugs (species not recorded) was reported on three ships in 2003 (Mouchtouri et al., 2008), with "several" bugs collected from lounge and cabin areas. No details on the control regimen used in these incidences were provided (Mouchtouri et al., 2008).

A similar survey during 2011 of an infested "entry ship" from China collected over 300 specimens of the Tropical bed bug, *Cimex hemipterus* (F.), from crew cabins. It was determined that the calculated incidence rate and bed bug density was 180 bed bugs per person-hour, and 300 bed bugs per square meter respectively (Bo et al., 2013). Despite the apparent high density of bed bugs, complete eradication was achieved using a 0.5% malathion solution and applied to all residual surfaces and cracks and crevices (Bo et al., 2013).

A case study of fumigation of a *C. hemipterus* infestation on a ship in China was also documented in 2014. Sulfuryl fluoride was successfully used to eradicate bed bugs that had infested the on-board living areas (You et al., 2014). An earlier organophosphate-based insecticide treatment had proven unsuccessful, thus resulting in the need for fumigation.

Media reports and complaints from passengers on online travel review sites are perhaps the best proxy for gauging infestation rates on cruise ships. These sites typically indicate an infrequent yet persistent issue (Doheny, 2005; Silverstein, 2011; Anonymous, 2012b). Some of the earliest reports of bed bugs on shipping vessels during the modern resurgence involved two large passenger cruise ships (Doheny, 2005). A further, cursory, search of web-based review sites (such as TripAdvisor.com and cruisereviews.com) reveals similar unsubstantiated instances across multiple cruise ships and lines. Despite the lack of documented reports of bed bugs on cruise ships in recent years, many PMPs will casually mention that such treatments are not uncommon, and the authors have been involved in the management of several infestations on cruise ships.

Eradication strategies for bed bug infestations on modern cruise ships are limited for several reasons. The major factor is that fast, effective control is a complex logistical process. When a bed bug infestation has been discovered, arrangements are typically made for the PMP to either board the ship at the next available port, or to review a report of the infestation circumstances (David Lilly and Garry Jones, unpublished results). Nonetheless, most infestations will be treated by trained crew (who are not PMPs) according to pre-determined protocols (which are usually developed in conjunction with a PMP), employing a selection of insecticides kept on the ships. Three key components are required for this process to be successful, namely the appropriate management of the passengers and their belongings, pre-treatment disinfestation procedures, and the adherence to treatment protocols.

The inappropriate handling of passengers exposed to bed bugs, and their belongings, can lead to the spread of an infestation within the ship, and a distinctly negative customer experience. Reassignment of passengers to a new, uninfested cabin may not always be possible during the cruise, since most ships often run at 100% occupancy (Anonymous, 2012a; Mark, 2016). Therefore it is not uncommon for the treatment of an infestation to be conducted whilst the cabin is "occupied". Regardless, the fact that passengers may either be moved to an uninfested cabin, or returned to a treated cabin at night, predicates strict guidelines with respect to moving their belongings (such as clothes and luggage) and preparing the cabin for treatment. The authors have been involved with several situations where infestations have spread due to the movement of passengers from an initially infested cabin to another (David Lilly and Garry Jones, unpublished results). Consequently, if the passenger is being relocated (and often, even if they are not), personal belongings including all clothes, are typically bagged and sent for complimentary laundering. The luggage is usually left in the infested room until it can be inspected and treated (if necessary). Freezers may be occasionally used for the disinfestation of luggage, although it is more commonly inspected and treated within the room. Alginate (dissolvable) bags are often employed for transporting and laundering passenger clothing.

In instances where a lack of spare cabin availability precludes relocation, passengers are typically advised that maintenance is necessary and that they must vacate the cabin for the day. Once vacated, the cabin is stripped in order to expose the deck, deck-head, and bulk-heads. This can be a particularly laborious task as all furnishings and walls/flooring are securely fixed to prevent movement during any heavy seas, and typically requires multi-departmental coordination (say between the accommodation and maintenance crews). However, it is important that this room preparation is undertaken at the earliest available opportunity. This is because bed bugs are commonly found in the voids behind decorative walls and fixed furnishings. Removing these fixtures will provide time for a treatment to proceed before the passenger(s) return (David Lilly and Garry Jones, unpublished results). Once preparation is complete, most floor coverings (such as carpet skirting) and soft furnishing materials (such as bed valances) are bagged and, depending on the ship or cruise line, sent for hot washing or incineration. Incineration is commonly used because it can dramatically reduce the scope of a treatment, and ships routinely maintain a stock of all cabin furnishings and fittings, allowing for rapid cabin refurbishment at sea (David Lilly and Garry Jones, unpublished results).

Even with all of the preparations described above, treatment of an infested cabin is still a nuanced affair that requires attention to detail, adherence to treatment protocols, and use of the right tools and insecticides. It is a requirement of cruise companies to conduct pest management (for various pests) whilst isolated from land. While this may not be true for all cruise lines, most of the products selected for use are low-toxicity, "ready-to-use'" aerosol sprays. These insecticides are closer to consumer-use products than professional-grade formulations. Even selecting products that are low in odor must be considered, in an effort to reduce passenger perception of exposure, as ships have complex ventilation systems that may carry strong odors from one cabin to other areas. Consequently, steam and/or high-efficiency particulate air (HEPA) vacuums are heavily relied upon. Despite its advantages of leaving no residue and causing an immediate kill, steam is time-consuming and attention to detail is required. Nonetheless, it is recommended that steam is used throughout the cabin and on all bedding frames and furniture, followed by insecticide sprays (low odor) on solid furniture. Voids can be treated with insecticidal dust or diatomaceous earth products, before the cabin is reconstructed and all surfaces thoroughly cleaned. Passengers would normally be permitted to re-enter after a 4–6 h clearance period, or as long as the cabin can be ventilated through opening of portholes and/or balcony doors.

Finally, education of new crew is essential to ensure that:

- protocols are adhered to
- ongoing monitoring ensures that past infestations have been eradicated and that new infestations are detected early.

Given the complexity of some ships (with over 2000 cabins) schematic maps of the ship are essential for infestation tracking and risk identification. "Certification'" training is provided to a small selection of crew who are directly responsible for undertaking treatments, with broader training provided to all crew members about bed bug awareness and the protocols for reporting a suspected infestation.

39.3 Trains

Infestations of bed bugs on trains appear to be a more frequently reported occurrence than on cruise lines. Reports have come from around the world including Australia (Anonymous, 1916), Asia (Zhang, L.W., *et al.*, 2010; Li *et al.*, 2011; Chen *et al.*, 2012; Chen and Huo, 2013; Yu *et al.*, 2013; Qi *et al.*, 2014; Zhu *et al.*, 2014; Zheng *et al.*, 2015), the Indian subcontinent (Webster, 1934), Europe (Vermeer and Vroege, 1949; Anders *et al.*, 2010; Anonymous, 2010a), and the USA (Potter, 2008; Anonymous, 2010b; Potter *et al.*, 2010; Wisniewski, 2016). It is worth noting that most of these reports relate to the modern resurgence.

The itinerant nature of trains also presents bed bug management challenges. Detection is arguably harder to achieve due to passengers spending only minutes to hours on the train instead of days. Therefore, there is a high chance that the affected bitten person will have stood up, moved, or departed before realizing a bite has occurred. In addition, as with planes and ships, trains can be inherently complex structures with intricate seat construction and with various voids for storage and the routing of wiring.

China has been at the forefront of reports of bed bugs on trains, with infestations identified across a number of railway districts (Zheng *et al.*, 2015). Several studies looking specifically at the incidence of infestations in sleeper-car trains found bed bugs around couchettes (100% occurrence), on middle or low berths (48 and 37–67% occurrence respectively), sleeper compartments (16% occurrence), and in voids behind boards (Li *et al.*, 2011; Chen and Huo, 2013; Yu *et al.*, 2013; Qi *et al.*, 2014). All of these features make control difficult (but still achievable). One factor that perhaps made control easier in the above examples was the availability of insecticides in China that have been withdrawn from other markets. For instance, treatments consisting of propoxur, propoxur and fenthion, cypermethrin and fenthion, and chlorpyrifos and cyfluthrin were reportedly employed, and it was claimed that 100% control was achieved in all cases (Li *et al.*, 2011; Zhang *et al.*, 2014; Zhu *et al.*, 2014; Liu and Fan, 2015).

Given the constant supply of hosts and available habitat, it is perhaps not surprising that the incidence of bed bugs is increasing on trains and other forms of public transport. A 2010 survey of PMPs by the US National Pest Management Association (NPMA) found 9% had encountered bed bugs on public transport (Anonymous, 2010b), an increase from 4% from three years prior (Potter, 2008). Whilst not specifically stated, it is presumed some of this "transport" included trains. In China, the abovementioned surveys similarly found a 2.6% infestation rate in 2013 (Yu *et al.*, 2013), and an increase from 2.5% to 14.1% from 2011 to 2012 (Qi *et al.*, 2014).

Anecdotally, the authors have experienced multiple infestations in trains and, to date, only one has involved a single carriage, in which nearly 80% of the carriage seats were infested (David Lilly and Garry Jones, unpublished results). More typically, infestations have involved 2–3 carriages (maximum of 6) in a travelling set. It is axiomatic, therefore, that by the time an infestation has been discovered, it has typically advanced to a moderate or severe level, making control more challenging and the risk of dispersal greater. Indeed, it has been hypothesized that public transport within cities may be a mechanism of bed bug dispersal (Anonymous, 2009; Anonymous, 2010a). However, the evidence to date of high genetic diversity between infestations (Saenz *et al.*, 2012; Fountain *et al.*, 2014) suggests that any effect from a public-transport-derived founder population (or populations) is negligible, or due to other sources (Juson, 2014).

Bed bug control on trains also presents challenges. As mentioned above, infestations often involve multiple carriages. This leads to a significant disruption in service when carriages have to be treated. There is also great pressure to return them to service at the earliest opportunity. Given the complexity of carriage structure, fumigation would normally be regarded as the most efficacious and cost-effective treatment option. However, due to the need to undertake such treatments in close proximity to other maintenance facilities and human workers, fumigation has largely been excluded as an option in Australia (David Lilly and Garry Jones, unpublished results). In such circumstances, labor-intensive dismantling of couchettes, seats, paneling, and other fixtures may be required. The required steps may also include treating voids below the floor and the interior wall of carriages. Consequently, treatment options involve a detailed pattern of inspection, treatment, and an intervening period of quarantine, before the process is repeated and observed until no further bed bugs are evident.

39.4 Conclusion

Bed bugs present a risk to various forms of mass transport, including trains and the leisure cruise industry. However, based on the few publications to date, it is difficult to ascertain the true extent of infestations and thus this is an area that warrants further investigation. Nonetheless, the inherent structure of trains and ships, and the complex itinerant nature of both the travelling guest and the vessel or carriage, make bed bug detection and control a challenging and expensive prospect.

References

Anders, D., Broecker, E.B. and Hamm, H. (2010) *Cimex lectularius* – an unwelcome train attendant. *European Journal of Dermatology*, **20** (2), 239–240.

Anonymous (1916) Bug inspector required. *Canowindra Star and Eugowra News* (7 April) p. 3.

Anonymous (1939) Bedbugs in ships. *Lancet*, **234** (6047), 209–210.

Anonymous (2009) A passion for bedbugs. *Pest*, **1**, 22.

Anonymous (2010a) Unlocking bedbug population dynamics. Has public transport paved the way for their return? *Pest*, **8**, 11–13.

Anonymous (2010b) Bugs without borders. *Pest*, **10**, 14.

Anonymous (2012a) *The Cruise Industry in Australia 2012*, http://www.tourism.australia.com/the-cruise-industry.aspx (accessed 27 July 2016).

Anonymous (2012b) *Bed Bugs At Sea: How Serious A Problem?*, http://www.traveltruth.com/2012/04/17/will-we-be-sharing-our-cruise-accommodations-with-strangers/ (accessed 18 July 2016).

Anonymous (2015) *Cruise Industry Overview – 2015*. Florida-Caribbean Cruise Association, http://www.f-cca.com/research.html (accessed 27 July 2016).

Baker, R.E. (1934) Extermination of bedbugs on board ship. *Naval Medical Bulletin*, **32** (2), 193–194.

Bo, J.X., Zhu, S.Y., Yang, C.G., et al. (2013) Investigation and disposal of bedbugs carried by an entry ship. *Zhongguo Meijie Shengwuxue ji Kongzhi Zazhi (Chinese Journal of Vector Biology and Control)*, **24** (5), 462–463, 466.

Chanev, C. (2015) *Cruise Ship Passenger Capacity*, http://www.cruisemapper.com/wiki/761-cruise-ship-passenger-capacity-ratings (accessed 26 November 2015).

Chen, B.S. and Huo, W. (2013) Investigation of bed bug outbreak in a passenger train number 1524 at Shijiazhuang railway passenger train district. *Yi Xue Dong Wu Fang Zhi (Chinese Journal of Medical Pest Control)*, **29**, 1127–1128.

Chen, S.M., Tang, S.X., Zheng, J.M., Wan, Y.H., Jiang, X. and Wang, G. (2012) Study on the roosting habit of bedbug and the effect of control measure in the passenger train. *Zhongguo Meijie Shengwuxue ji Kongzhi Zazhi (Chinese Journal of Vector Biology and Control)*, **23** (1), 86–87.

Doheny, K. (2005) *Don't Let The Bedbugs Bite? That's Easier Said Than Done Lately*, http://articles.latimes.com/2005/may/22/travel/tr-healthy22 (accessed 6 October 2016).

Fountain, T., Duvaux, L., Horsburgh, G., Reinhardt, K. and Butlin, R.K. (2014) Human-facilitated metapopulation dynamics in an emerging pest species, *Cimex lectularius*. *Molecular Ecology*, **23** (5), 1071–1084.

Juson, A. (2014). Bed bugs "sharing the love". *Professional Pest Controller*, **75** (5), 33–34.

Li, G.M., Wang, X.L. and Zhang, S.M. (2011) Investigation and study of the bugs in the sleeping cars of passenger trains. *Yi Xue Dong Wu Fang Zhi (Chinese Journal of Medical Pest Control)*, **27**, 813–814.

Liu, P.T. and Fan, C.Y. (2015) Investigation and control of bedbug infestation on a passenger train. *Zhongguo Meijie Shengwuxue ji Kongzhi Zazhi (Chinese Journal of Vector Biology and Control)*, **26**, 536.

Mark, J. (2016). *Buying Unsold Cruise Cabins and What Happens to Them*, http://cruisefever.net/0115-buying-unsold-cruise-cabins-and-what-happens-to-them/ (accessed 17 August 2016).

Mouchtouri, V.A., Anagnostopoulou, R., Samanidou-Voyadjoglou, A., et al. (2008) Surveillance study of vector species on board passenger ships, risk factors related to infestations. *BMC Public Health*, **8**, 1–8.

Phelps, J.R. (1924) Eradication of vermin on board ship. *Naval Medical Bulletin*, **20** (2), 247–268.

Potter, M.F. (2008) The business of bed bugs. *Pest Management Professional*, **76** (1), 24–25, 28–32, 34, 36–40.

Potter, M.F., Rosenberg, B. and Henriksen, M. (2010) Bugs without borders. Defining the global bed bug resurgence. *Pestworld*, **Sep/Oct**, 1–12.

Qi, J.X., Zhu, J.H., Fan, H.L., Yao, W.Y. and He, J.N. (2014) Investigation on bedbugs in sleeper compartments of passenger trains. *Zhonghua Wei Sheng Sha Chong Yao Xie (Chinese Journal of Hygienic Insecticides and Equipments)*, **20** (5), 490–492.

Saenz, V.L., Booth, W., Schal, C. and Vargo, E.L. (2012) Genetic analysis of bed bug populations reveals small propagule size within individual infestations but high genetic diversity across infestations from the eastern United States. *Journal of Medical Entomology*, **49** (4), 865–875.

Silverstein, E. (2011) Itchy and scratchy: Are there bed bugs on your cruise ship? http://www.cruisecritic.com.au/articles.cfm?ID=1211 (accessed 27 June 2016).

Vermeer, D.J. and Vroege, C. (1949) Een epidemie van bulleuze dermatitis aan de benen bij trampassagiers, veroorzaakt door *Cimex lectularius* [The epidemiology of leg bullous dermatoses in train passengers, caused by *Cimex lectularius*]. *Nederlands Tijdschrift voor Geneeskunde*, **93** (5), 4161–4167.

Webster, W.J. (1934) Bed bugs in rarified air. *Indian Journal of Medical Research*, **21** (3), 523.

Wisniewski, M. (2016) *Chicago Tribute, CTA Pulls Red Line Car After Bedbug Report*, http://www.chicagotribune.com/news/ct-cta-red-line-bed-bugs-20160927-story.html (accessed 6 October 2016).

You, M.C., Dong, S.D., Qiu, J.L., *et al.* (2014) Fumigation treatment with sulfuryl fluoride in an old vessel seriously infested with bedbugs. *Zhonghua Wei Sheng Sha Chong Yao Xie (Chinese Journal of Hygienic Insecticides and Equipments)*, **20** (1), 93–94.

Yu, H.Y., Yang, J., Ren, C., *et al.* (2013) Investigation and emergency disposal of bedbug infestation in passenger train and observation of control effects. *Zhongguo Meijie Shengwuxue ji Kongzhi Zazhi (Chinese Journal of Vector Biology and Control)*, **24** (1), 69–71.

Zhang, G.P., Chi, D., Liang, X.D. and Sun, K. (2014) Management and advices about an encroachment of bedbugs on passenger train. *Zhonghua Wei Sheng Sha Chong Yao Xie (Chinese Journal of Hygienic Insecticides and Equipments)*, **20**, 612.

Zhang, L.W., Qian, J.X. and Shu, H.C. (2010) Survey and control measures of bedbug infestation in passenger trains. *Zhongguo Meijie Shengwuxue ji Kongzhi Zazhi (Chinese Journal of Vector Biology and Control)*, **21** (5), 523.

Zheng, J.M., Chen, S.M., Jiang, X. and Huang, D.Y. (2015) Passenger trains bedbug prevention and control. *Zhonghua Wei Sheng Sha Chong Yao Xie (Chinese Journal of Hygienic Insecticides and Equipments)*, **31**, 61–65.

Zhu, J.H., Qi, J.X., Fan, H.L., Yao, W.Y. and He, J.N. (2014) The effect of propoxur and fenthion applied in passenger trains for bedbug control. *Zhonghua Wei Sheng Sha Chong Yao Xie (Chinese Journal of Hygienic Insecticides and Equipments)*, **20** (4), 363–364.

40

Poultry Industry
Allen Szalanski

40.1 History – Cimicids and Poultry

Bed bugs can be important pests in commercial poultry operations. This fact has been often overlooked by entomologists working on bed bugs in the urban environment. Cimicids, and particularly the Common bed bug, *Cimex lectularius* L., can infest poultry houses and cause severe cutaneous reactions in the birds as well as other clinical effects (Dulceanu *et al.*, 1975; Monov and Topalski, 1980; De Vaney, 1986). It is well-documented that *C. lectularius* attacks poultry in North America, Europe, and the former Soviet Union (Frolov, 1966; Monov and Topalski, 1980; Steelman, 2000; Tabler *et al.*, 2015). However, other cimicids have been documented as pests of poultry, including:

- the Mexican chicken bug, *Haematosiphon inodorus* (Duges)
- the Brazilian chicken bug, *Ornithocoris toledoi* Pinto
- the related species, *Ornithcorus pallidus* Usinger;
- and the Tropical bed bug, *Cimex hemipterus* (F.).

These cimicids find harborage in avian nests and rearing facilities, and can cause economic damage to poultry by feeding on their legs and exposed skin, causing stress and weight reduction.

40.2 Mexican Chicken Bug, *Haematosiphon inodorus*

Haematosiphon inodorus was first described from Mexico in the 1890s. It has since been found in Central America, and historically in western North America and Florida (Townsend, 1894; Blatchley, 1928; Usinger, 1966). Hosts of *H. inodorus* include the golden eagle, *Aquila chrysaetos* L., and domestic chicken, *Gallus gallus* L., (Lee, 1954; Usinger, 1966). *Haematosiphon inodorus* were collected in California during 1939 from nest caves of the California condor, *Gymnogyps californianus* Shaw. *Haematosiphon inodorus* were also collected in the nests of great horned owls, *Bubo virginianus* Gmelin, in Oklahoma in 1941 (Usinger, 1947). This species is also known to be an ectoparasite of barn owls. Lee (1959) observed as many as 1778 *H. inodorus* in a single barn owl nest. Grubb *et al.* (1986) observed up to 30 000 *H. inodorus* in a bald eagle nest in Arizona. Other hosts of the Mexican chicken bug include the domestic turkey, *Meleagris gallopavo* L., (Lee 1955), the red-tailed hawk, *Buteo jamaicensis* Gmelin, the prairie falcon, *Falco mexicanus* Schlegel, (Platt 1975), and the turkey vulture, *Cathartes aura* L., (Wilson and Oliver, 1978).

There are several theories as to how *H. inodorus* became a pest of domestic poultry in the USA and Mexico. Usinger (1966) proposed that raptors may have carried *H. inodorus* to domestic chicken yards while preying or scavenging on the birds there. Another possibility is that *H. inodorus* was transmitted to domestic poultry by raptors nesting in barns and other farm buildings (Wilson and Oliver, 1978).

Townsend (1894) documented *H. inodorus* as being an important pest of poultry in southern New Mexico when he observed them "swarming in great numbers in the hen-houses, infesting the inmates and roosts, and covering the eggs with the black specks of its excrement." Besides many avian hosts, the Mexican chicken bug can also be a pest of humans. Townsend (1894) reported that *H. inodorus* "also spreads from the hen roosts to the (human) dwelling houses, where it proves to be more formidable than the common bed-bug." Lee (1955) reported that *H. inodorus* had a natural, though rare, association with people and "has no reluctance to feed on human subjects." Harwood and James (1976) stated that *H. inodorus* infesting human dwellings was a rare occurrence, but that the infestation may be severe. Reinhardt (2012) proposed that *H. inodorus* has had a long association with humans in North America and may have been a pest of the Hopi people in Arizona. *Haematosiphon inodorus* was regarded as an important pest of poultry in facilities in southwestern USA into the 1940s and 1950s (Back and Bishopp, 1942; Usinger, 1947; Hall and Wehr, 1953). In the southeastern USA, *H. inodorus* was recorded in Lakeland, Florida by Blatchley (1928).

It is interesting that *H. inodorus* is no longer considered a pest of poultry in the USA and Mexico. Although *H. inodorus* was well documented in nests or on raptors in southwestern USA from the 1970s into the mid-1980s (Grubb *et al.* 1986; Platt, 1975, Wilson and Oliver, 1978), there appears to be no published record of it being associated with poultry facilities in the USA since the late 1940s (Usinger, 1947). One possible explanation for this was the widespread use of DDT and other organochlorines in poultry facilities in the 1940s (Kulash and Maxwell, 1945; Kulash, 1947). This DDT exposure resulted in populations of *C. lectularius* becoming highly resistant to DDT in poultry facilities in the USA (Steelman *et al.*, 2008). However, if *H. inodorus* in those poultry facilities did not have the resistant genes for organochlorides, it is possible that they were eradicated by DDT.

40.3 Brazilian Chicken Bug, *Ornithocoris toledoi*

Ornithocoris toledoi was first described in poultry facilities from the vicinity of Sao Paulo, Brazil (Pinto, 1927). Snipes *et al.* (1941) stated that this cimicid was a serious pest of poultry in Brazil and could occur in human dwellings where chickens were roosting. Interestingly, Snipes *et al.* (1941) stated that the human occupants of these dwellings did not report any feeding of *O. toledoi* on them, which may indicate that this species does not feed on humans. Harwood and James (1979) reported that *O. toledoi* was once a poultry pest of considerable importance, but was brought under control using residual insecticides. Another species of the genus *Ornithocoris*, *O. pallidus*, was also described from Brazil and found in the nests of the blue and white swallow, *Notiochelidon cyanoleuca* (Viellot) (Usinger, 1959). Usinger reported that *O. pallidus* was associated with poultry facilities in Florida and Georgia, USA (Usinger, 1959, 1966), where it was probably introduced from Brazil (Harwood and James, 1979). In 1998, *O. pallidus* was observed living in human dwellings in Louisiana, USA (Carlton and Story, 1998).

40.4 Tropical Bed Bug, *Cimex hemipterus*

The Tropical bed bug, *C. hemipterus*, is found in Africa, Asia, Australia, and South America, where it can be a pest of both humans and poultry (Rosen *et al.*, 1987), although *C. hemipterus* (and *C. lectularius*) have not been reported affecting poultry in Australia (Stephen Doggett, unpublished data). Rosen *et al.* (1987) observed that *C. hemipterus* were found in the barns and in the houses of poultry workers in 1981 in Israel. Owen (2005) reported finding *C. lectularius* in chicken houses in Papua New Guinea, but more than likely this was *C. hemipterus* because *C. lectularius* has not been reported from this region.

40.5 The Common Bed Bug, *Cimex lectularius*

Reports of *C. lectularius* being found in poultry facilities in the USA date as far back as the early 1900s (Rucker, 1912). Back and Bishopp (1942) observed that *C. lectularius* frequently infested poultry and pigeon houses in the USA. They also reported that log houses, which were often used for poultry in farms in the USA during that era, were especially subject to bed bug infestations due to the cracks and holes made by wood boring insects, which made excellent refuges for bed bugs. Several studies were published on the occurrence of *C. lectularius* in poultry facilities located in North Carolina and Delaware during the 1940s and 1950s. Kulash and Maxwell (1945) reported a severe infestation of *C. lectularius* in poultry facilities located in North Carolina, USA. They stated that three chicken houses that had been in use for 15 years had bed bugs in them for most of those years. MacCreary and Catts (1954) examined 322 flocks of chickens in 1954, but found *C. lectularius* in only one Delaware poultry facility. Very little was published regarding *C. lectularius* in USA poultry facilities after the 1950s (Fletcher and Axtell, 1993; Steelman, 2000). However, since the early 2000s there have been several publications on bed bugs associated with poultry in Arkansas (Steelman *et al.*, 2008) and Mississippi (Carter *et al.*, 2011; Goddard, 2012; Tabler *et al.*, 2015). In Eastern Europe, Frolov (1966, 1971) reported the presence of *C. lectularius* in wooden poultry houses located in the Soviet Union in 1966. Similarly, *C. lectularius* was reported to be living in poultry facilities in Bulgaria during the 1970s (Monov and Topalski, 1980).

During blood feeding on avian hosts, bed bugs will crawl to the host, engorge rapidly, usually in 10 min or less, and then return to harborage locations. Due to the nocturnal feeding of *C. lectularius* on its host, it is rarely seen on the birds, and can only be monitored by examining potential hiding places, evidence of fecal spots (on posts, nest boxes, and other surfaces), and by the presence of lesions on the breast and legs of the birds (Arends and Robertson, 1986).

40.6 Biology and Impact of *Cimex lectularius* on Poultry

It is well-documented that *C. lectularius* can be an important pest in poultry facilities in the USA (Hall and Wehr, 1953; Usinger, 1966; Axtell and Arends, 1990; Steelman, 2000; Mullen and Durden, 2002). Bed bugs may occur occasionally in poultry production facilities, but most often, it infests broiler breeder facilities (that is, birds mass reared for meat consumption). The feeding of the bed bugs results in reduced egg production and anemia in the birds, making the presence of bed bugs an economic problem (Axtell, 1999; Carter *et al.*, 2011). In a typical breeder facility, thousands of hens and roosters, are maintained (at a typical ratio of 10 hens to one rooster) to yield hatching eggs for the production of broilers. As noted, broiler breeder facilities are the primary type of poultry operation that is infested with bed bugs (Goddard, 2012) and there are two main explanations for this. Firstly, each side of a broiler-breeder facility usually has a slatted wooden platform. These wooden slats provide an optimal environment for bed bugs (Fletcher and Axtell, 1993). Secondly, compared to broiler operations in which the facility is cleaned out after each 50-day grow out, the broiler-breeder egg production cycle is usually 265 days long (Steelman, 2000). This allows bed bug populations to reach high numbers before the egg production cycle ends and the facility is emptied for cleaning.

Compounding the situation further, bed bugs can survive for a long period of time without a blood meal. For example, nymphs may survive without a blood meal for up to 70 days, and adults can live without a blood meal for up to a year depending on the temperature (Axtell, 1999). This ability to withstand starvation allows *C. lectularius* in broiler breeder facilities to survive periods when birds are not present in the structure (Steelman, 2000).

Carter *et al.* (2011) reported that a severe bed bug infestation in two broiler breeder houses on a single farm in southeastern USA caused approximately 10% reduction in egg production. Blood samples taken from chickens in this facility indicated that the birds had anemia. In addition to anemia, cimicid feeding has been known to cause mortality in avian species. The Mexican chicken bug, *H. inodorus* was implicated in the death of bald eagle *H. leucocephalus* chicks when 21 000–31 000 Mexican chicken bugs were found inhabiting the nest (Grubb *et al.*, 1986).

40.7 Dispersal

Bed bugs of various life stages (eggs, nymphs, or adults) may be introduced into poultry facilities in boxes, clothing, cages, manure removal equipment, or any other equipment brought from another infested poultry house (Steelman, 2000). For example, as many as 2500 bed bugs were removed from the cracks of two wooden crates taken from one poultry market in the USA (Back and Bishopp, 1942). These crates are often used for shipping poultry or for containing poultry at the market. MacCreary and Catts (1954) suggested that the practice of moving bagged, unused feed from one farm to another might also contribute to the spread of bed bugs in poultry houses. If precautions are not taken, pullet farms may not only supply 22-week-old hens to breeder farms but also bed bugs (Steelman, 2000). Jacobs (2015) noted that poultry workers can carry bed bugs to their residences from their places of work. However, this observation was not new (Hall and Wehr, 1953; Axtell, 1999; Campbell, 1996). Rosen *et al.* (1987) in Israel also observed that *C. hemipterus* were found in the barns and houses of poultry workers in 1981.

Another means of bed bug dispersal lies in the ability of cimicids to feed from alternate hosts. As previously stated, Usinger (1966) proposed that raptors could carry the Mexican chicken bug, *H. inodorus*, to domestic fowl in chicken yards while preying on birds. Another possible method of transportation of *H. inodorus* to domestic fowl is the raptor host choosing to nest in barns and other farm buildings that also house chickens (Wilson and Oliver, 1978).

Feral pigeons, *Columbia livia* Gmelin, are also known to be hosts of *C. lectularius* (Döhring, 1958; Teschner, 1964; Usinger, 1966). These birds are also known to nest in poultry-rearing facilities, and could thus function as transporters of bed bugs to poultry.

A population genetics study conducted on 22 populations of *C. lectularius* from human dwellings and poultry facilities in the USA, Australia (human dwellings only), and Canada used mitochondrial DNA 16S gene markers to reveal that haplotypes were shared between bed bugs found in human dwellings and poultry facilities (Szalanski *et al.*, 2008). A more recent population genetics study (Balvin *et al.*, 2012) examining bed bugs collected from humans and bats found that bed bugs populations from these two hosts were genetically isolated. Three of the four bed bug 16S haplotypes collected from humans in the Balvin *et al.* (2012) study were also found in the study by Szalanski *et al.* (2008). This haplotype co-incidence provides evidence that the European bed bug diversity may be related to American bed bug diversity.

40.8 Bed Bug Control in Poultry Facilities

Rucker (1912) suggested that benzene or kerosene be injected behind wainscoting (wooden paneling) and applied to floor cracks in poultry facilities for *C. lectularius* control. Other historical methods have included the application of turpentine or boiling water to structures inside the poultry facility. Fumigation by burning sulfur inside the barn has also been recommended for bed bug control. Due to changes in regulations, and health and safety concerns, none of these historical solutions are recommended today. More recently, Frolov (1971) recommended the use of the predacious mite *Cheyletus polymorphus* Volgin for controlling *C. lectularius* in wooden poultry facilities in the Soviet Union. Frolov (1971) found that *C. polymorphus* would feed on first instar nymphs, but could not prey on late instar nymphs or bed bug adults. This was most likely due to the mites' inability to bite through the thicker integument of later instars.

In 1945, Kulash and Maxwell found that a solution of 5% DDT in kerosene was effective for controlling bed bugs in poultry houses in North Carolina. Two months after treatment, no living adults or nymphs were found. Hall and Wehr (1953) reported that *C. lectularius* could be controlled in poultry houses using a single application of DDT solution applied at a 50% concentration. Hall and Wehr (1953) also noted that the DDT solution was equally effective for controlling the Mexican chicken bug. Rosen *et al.* (1987) found that treatment of poultry barns in Israel with 2% malathion controlled *C. hemipterus*. Historically, both malathion and carbaryl have been recommended for bed bug control, particularly after bed bugs developed resistance to DDT (Steelman, 2000).

In the USA, insecticides that can be applied in poultry facilities (when the birds are present) include permethrin formulations, tetrachlorvinphos and dichlorvos formulations, and tetrachlorvinphos (Studebaker, 2015; Tabler et al., 2015). Residual insecticides that can be applied when birds are not present in the facility include lambda-cyhalothrin, cyfluthrin, and beta-cyfluthrin (Studebaker, 2015; Tabler et al., 2015).

Heat treatments can also be used to control bed bugs when the poultry house is empty. Heating of the interior of the poultry facility until all bed bug harborages reach 55 °C will kill bed bug adults, nymphs, and eggs (Tabler et al., 2015).

40.9 Insecticide Assays

Fletcher and Axtell (1993) published a study on the susceptibility of *C. lectularius* collected from a broiler breeder chicken house in North Carolina, USA, to nine different insecticides. LC_{50} assays were conducted by placing treated filter papers inside Petri dishes and exposing ten adult bed bugs to the treated filter paper to determine lethal concentrations. Four replicates per concentration were tested on four different days ($n = 960$ bed bugs for each insecticide). After a 24-h exposure period, it was found that the bed bugs were susceptible to acetylcholinesterase inhibitors such as dichlorvos (organophosphate), and bendiocarb (carbamate), and also, pirimiphos methyl (organophosphate). The LC_{50} and LC_{90} values generated when tested for the control of bed bugs were <100 ppm. Tests using other insecticides like permethrin (pyrethroid), malathion (organophosphate), lambda-cyhalothrin (pyrethroid), carbaryl (carbamate), and tetrachlorvinphos (organophosphate), were moderately toxic, recorded LC_{90} values ranging from 200 to 500 ppm. Fenvalerate (pyrethroid) demonstrated a very low toxicity, with a calculated LC_{90} value of >41 000 ppm.

Goddard (2013) in 2011 evaluated a group of *C. lectularius* collected from a poultry facility in Mississippi, USA, to determine their susceptibility to various insecticides. Individual adult bed bugs were placed in Petri dishes containing filter paper, which had been dosed with insecticide at the label rate. One bed bug was placed in each Petri dish, and each product had ten replicates. Propoxur killed 100% of the bed bugs exposed to filter paper within 1 h. The imidacloprid-betacyfluthrin mixture killed 100% of bed bugs within 24 h. In contrast, deltamethrin killed 50% of the bed bugs within 24 h. Similarly, chlorfenapyr and Swepe-tite (herbal plant oil extract) killed 40% of the bed bugs within 24 h.

Steelman et al. (2008) evaluated the susceptibility of *C. lectularius* collected from three broiler breeder poultry facilities in Arkansas, USA, to 12 insecticide formulations. The active ingredients were dissolved in acetone (3–4 different concentrations) and applied to glass vials as per Plapp (1971). Each bioassay of ten bed bugs was replicated three times for each insecticide concentration. The order of toxicity determined for each active ingredient (in order of most toxic to least toxic) based on LC_{50} values were: lamda-cyhalothrin, bifenthrin, carbaryl, imidacloprid, fipronil, permethrin, diazinon, spinosyn, dichlorvos, chlorfenapyr, and DDT. Two of the three bed bug populations had a history of repeated exposure to organophosphate insecticides over a ten-year period. All three bed bug populations were highly resistant to DDT, with LC_{50} values of >100 000 ppm.

Most recently, Goddard and Mascheck (2015) conducted a study evaluating eight insecticides for the control of *C. lectularius* collected from a broiler breeder operation in Mississippi, USA. The insecticides were applied to ceramic tiles at the required label rate. The treated tiles were placed inside Petri dishes containing 20 bed bugs per dish. Three replicates of each insecticide were assayed and bed bug mortality was recorded after 24 h. Goddard and Mascheck (2015) found that formulations of propoxur, tetrachlorvinphos-dichlorvos mixture, and silica gel produced 100% mortality. The organophosphate chlorpyrifos killed 96% of the bed bugs in 24 h, but the pyrethroids beta-cyfluthrin and deltamethrin killed only 73% and 93%, respectively. The "natural" products tested included diatomaceous earth formulated with dinotefuran, and phenethyl propionate formulated with soybean oil and clove oil, with mortality rates of 46 and 45% respectively, at 24 h post-treatment.

The relatively low level of pyrethroid efficacy observed by Fletcher and Axtell (1993) for the control of *C. lectularius* contrasts with the observations made by Steelman et al. (2008) and Goddard and Maschek (2015). While these conflicting studies may suggest that the protocol used for evaluating bed bug susceptibility to a specific insecticide may have influenced the results, it is even more likely that bed bugs from different geographic locations may have different susceptibility to different products. This would of course depend on the insecticide treatment histories of the respective poultry facilities from which the bed bugs were collected. Interestingly, there have been no reports of high levels of pyrethroid resistance among bed bug populations collected from poultry facilities (Davies *et al.*, 2012). However, high levels of pyrethroid resistance have been observed in other arthropod pests found in poultry facilities, including the lesser mealworm (Hamm *et al.*, 2006), the red poultry mite (Beugnet *et al.*, 1997), the northern foul mite (Mullens *et al.*, 2004), and in filth flies (Scott *et al.*, 2000). Based on this lack of resistance data, it cannot be assumed that local poultry facilities are the source of pyrethroid-resistant bed bugs found in nearby human dwellings (Davies *et al.*, 2012).

References

Arends, J.J. and Robertson, S.H. (1986) Integrated pest management for poultry production: Implementation through integrated poultry companies. *Poultry Science*, **65** (4), 675–682.

Axtell, R.C. (1999) Poultry integrated pest management: status and future. *Integrated Pest Management Review*, **4** (1), 53–73.

Axtell, R.C. and Arends, J.J. (1990) Ecology and management of arthropod pests of poultry. *Annual Review of Entomology*, **35**, 101–126.

Back, E.A. and Bishopp, F.C. (1942) Bedbugs as pests of poultry. In: *Keeping Livestock Healthy Yearbook of Agriculture, 1942*, (eds G. Hambidge and M.J. Drown) United States Department of Agriculture, Washington, D.C., pp. 1068–1071.

Balvin, O., Munclinger P., Kratochvil, L. and Vilimova J. (2012) Mitochondrial DNA and morphology show independent evolutionary histories of bedbug *Cimex lectularius* (Heteroptera: Cimicidae) on bats and humans. *Parasitology Research*, **111** (1), 457–469.

Beugnet, F., Chauve, C., Gauthey, M. and Beert, L. (1997) Resistance of the red poultry mite to pyrethroids in France. *Veterinary Record*, **140** (22), 577–579.

Blatchley, W.S. (1928) The Mexican chicken bug in Florida. *Florida Entomologist*, **12** (3), 43–44.

Campbell, J.B. (1996) G89-954 A guide for managing poultry insects (revised April 1996). Historical materials from University of Nebraska-Lincoln Extension. Paper 1147.

Carlton, C.E. and Story, R.N. (1998) *A New Record of the Bird-Feeding Cimicid Ornithocoris pallidus Usinger in Louisiana* (Hemiptera: Cimicidae), http://lsuinsects.org/research/ornithocoris_pallidus/ (accessed 17 May 2016).

Carter, J., Magee, D., Hubbard, S.A., Edwards, K.T. and Goddard, J. (2011) Severe infestation of bed bugs in a poultry breeder house. *Journal of the American Veterinary Medical Association*, **239** (7), 919.

Davies, T.G.E., Field, L.M. and Williamson, M.S. (2012) The re-emergence of the bed bug as a nuisance pest: implications of resistance to the pyrethroid insecticides. *Medical and Veterinary Entomology*, **26** (3), 241–254.

De Vaney, J.A. (1986) Ectoparasites. *Poultry Science*, **65** (4), 649–656.

Dulceanu, N., Dascalu, A., Clipa, V. and Sasu, E. (1975) Observations on attacks by *Cimex lectularius* on fowls. *Cercetari Agronómico in Moldova*, **8**, 121–124.

Döhring, E. (1958) Plagen durch verwilderte Haustauben. *Ornithologische Mitteilungen*, **3**, 41–46.

Fletcher, M.G. and Axtell, R.C. (1993) Susceptibility of the bedbug, *Cimex lectularius*, to selected insecticides and various treated surfaces. *Medical and Veterinary Entomology*, **7** (1), 69–72.

Frolov, B.A. (1966) The prevention and eradication of ectoparasites of birds. *Veterinariya*, **43** (3), 106–108.

Frolov, B.A. (1971) Predacious mites as natural enemies of insects that parasitise birds. *Vestnik sel'sko-khozyaistvennoi Nauki*, **16**, 95–97.

Goddard, J. (2012) What to do about bed bugs in poultry houses. Mississippi State Extension Service publication POD-05-12.

Goddard, J. (2013) Laboratory assays of various insecticides against bed bugs (Hemiptera: Cimicidae) and their eggs. *Journal of Entomological Science*, **48** (1), 65–69.

Goddard, J. and Maschek, K. (2015) Laboratory assays with various insecticides against bed bugs taken from a poultry house in Mississippi. *Midsouth Entomologist*, **8**, 10–15.

Grubb, T.G., Eakle, W.L. and Tuggle, B.N. (1986) *Haematosiphon inodorus* (Hemiptera: Cimicidae) in a nest of the bald eagle (*Haliaeetus leucocephalus*) in Arizona. *Journal of Wildlife Diseases*, **22** (1), 125–127.

Hall, W.J. and Wehr, E.E. (1953) *Diseases and Parasites of Poultry*, USDA Farmers' Bulletin No. 1652.

Hamm, R.L., Kaufman, P.E., Reasor, C.A., Rutz, D.A. and Scott, J.G. (2006) Resistance to cyfluthrin and tetrachlorvinphos in the lesser mealworm, *Alphitobius diaperinus*, collected from the eastern United States. *Pest Management Science*, **62**, 673–677.

Harwood, R. F. and James, M.T. (1979) *Entomology in Human and Animal Health*, Macmillan Pub. Co., New York.

Jacobs, S.B. (2015) Bed bugs, *Cimex lectularius* L. The Pennsylvania State University Cooperative Extension Entomological Notes PH-1.

Kulash, W.M. (1947) DDT for control of bedbugs in poultry houses. *Poultry Science*, **26** (1), 44–47.

Kulash, W.M. and Maxwell, J.M. (1945) DDT and bedbugs in chicken houses. *Journal of Economic Entomology*, **38** (5), 606.

Lee, E.D. (1954) First report of the poultry bug from a Golden Eagle's nest, with new locality records. *Journal of Economic Entomology*, **47** (6), 1144.

Lee, R.D. (1955) The biology of the Mexican chicken bug, *Haematosiphon inodorus* (Duges) (Hemiptera: Cimicidae). *Pan-Pacific Entomologist*, **31**, 47–61.

Lee, R.D. (1959) Some insect parasites of birds. *Audubon Magazine*, **61**, 214–215, 224–225.

MacCreary, D. and Catts, E.P. (1954) *Ectoparasites of Delaware Poultry Including a Study of Litter Fauna*, University of Delaware Agricultural Experiment Station Bulletin No. 307: 4–16.

Monov M. and Topalski, E. (1980). Distribution and seasonal dynamics of bed-bugs in poultry farms. *Veterinarna Sbirka*, **78** (2), 21–22.

Mullen, G. L. and Durden, L.A. (eds) (2002) *Medical and Veterinary Entomology*, Academic Press, New York.

Mullens, B.A., Velten, R.K., Hinkle, N.C., Kuney, D.R. and Szijj, C.E. (2004) Acaricide resistance in northern fowl mite (*Ornithonyssus sylviarum*) populations on caged layer operations in southern California. *Poultry Science*, **83** (3), 365–374.

Owen, I.L. (2005) Parasitic zoonoses in Papua New Guinea. *Journal of Helminthology*, **79**, 1–14.

Pinto, C. (1927) *Ornithocoris toledoi*, a new genus and species of fowl's bug. *Revista de Biologia e Hygiene*, **1927**, 17–22.

Plapp, E. W. (1971) Insecticide resistance in *Heliothis*: tolerance in larvae of *H. virescens* as compared with *H. zea* to organophosphate insecticides. *Journal of Economic Entomology*, **64** (5), 999–1002.

Platt, S.W. (1975) The Mexican chicken bug as a source of raptor mortality. *Wilson Bulletin*, **87**, 557.

Reinhardt, K. (2012) Pesets'ola: which bed bug did the Hopi know? *American Entomologist*, **58** (1), 58–59.

Rosen, S., Hadani, A., Gur Lavi, A., *et al*. (1987) The occurrence of the tropical bedbug (*Cimex hemipterus*, Fabricius) in poultry barns in Israel. *Avian Pathology*, **16** (2), 339–342.

Rucker, W.C. (1912) The bedbug. *Public Health Reports*, **27**, 1854–1856.

Scott, J.G., Alefantis, T.G., Kaufman, P. and Rutz, D.A. (2000) Insecticide resistance in house flies from caged-layer poultry facilities. *Pest Management Science*, **56**, 147–153.

Snipes, B.T., Carvalho, J.C.M. and Tauber, O.E. (1941) Biological studies of *Ornithocoris toledoi* Pinto, the Brazilian chicken bug. *Iowa State College Journal of Science*, **15** (1), 27–37.

Steelman, F.D. (2000) Biology and control of bed bugs (*Cimex lectularius* L.) in poultry houses. *Avian Advice*, **2** (2), 10–15.

Steelman, C.D., Szalanski, A.L., Trout, R., *et al*. (2008) Susceptibility of the bed bug *Cimex lectularius* L. (Heteroptera: Cimicidae) collected in poultry production facilities to selected insecticides. *Journal of Agricultural and Urban Entomology*, **25** (1), 41–51.

Studebaker, G. (2015) *Insecticide Recommendations for Arkansas*, Division of Agriculture, University of Arkansas. 304 pp.

Szalanski, A.L., Austin, J.W., McKern, J.A., Steelman, C.D. and Gold, R.E. (2008) Mitochondrial and ribosomal internal transcribed spacer 1 diversity of *Cimex lectularius* (Hemiptera: Cimicidae). *Journal of Medical Entomology*, **45** (2), 229–236.

Tabler, T., Loftin, K.M., Farnell, M., Wells, J. and Yakout, H.M. (2015) Bed bugs: Difficult pests to control in poultry breeder flocks. Mississippi State Extension Service, Publication 2881.

Teschner D. (1964) Die Bedeutung der Nester verwilderter Tauben in Grossstadten. *Anzeiger fur Schadlingskunde, Berlin*, **37**, 40–43.

Townsend, C.H.T. (1894) Note on the curoco, a hemipterus insect which infests poultry in southern New Mexico. *Proceedings of the Entomological Society of Washington*, **3**, 40–42.

Usinger, R.L. (1947) Native hosts of the Mexican chicken bug, *Haematosiphon inodora* (Duges) (Hemiptera, Cimicidae). *Pan-Pacific Entomologist*, **23** (3), 140.

Usinger, R.L. (1959) New species of Cimicidae (Hemiptera). *The Entomologist*, **92**, 218–222.

Usinger, R.L. (1966) Distributional patterns of Cimicidae in western North America. *Bulletin of the Entomological Society of America*, **12** (2), 112.

Wilson, N. and Oliver, Jr. G. (1978) Noteworthy records of two ectoparasites (Cimididae and Hippoboscidae) from the turkey vulture in Texas. *The Southwestern Naturalist*, **23** (2), 305–307.

Part VII

Legal Issues

41

Bed Bugs and the Law in the USA

Jeffrey Lipman and Dini M. Miller

41.1 Introduction

The bed bug resurgence in the USA has brought about a new era of legal conflict between owners of infested locations (hotels and apartments) and unknowing patrons who stay in the them. These patrons have typically been bitten, or worse, have transported the bed bugs back to their own homes. The financial compensation sought by these bed bug victims for their physical and mental suffering and their loss of infested belongings has opened a whole new field of legal research. This research focuses on determining the responsibilities of the property owner with regards to the habitability of their facilities as well as how much financial compensation (if any) a victim is entitled too. Deciding on compensation can be very tricky considering that it can almost always be argued that the "victim", particularly if it is an apartment resident, might be responsible for bringing the bed bugs into the facility in the first place.

Pest management efforts further complicate the situation. Bed bug control is a long and tedious process, and these pests have proven to be very difficult to eradicate. Finding a bed bug after a pest control treatment raises the question as to who is responsible for that bed bug. Is it a bed bug that the pest management company failed to kill? Or did the resident bring in another bug after the treatment was completed? In which situation is the property owner responsible for paying for additional treatments? It is best to consult with a lawyer.

The USA has a reputation for being one of the most litigious countries in the world. In a recent report produced by Clements Worldwide (Anonymous, 2017) according to the American Bar Association, the USA has the most lawyers per capita (1 lawyer for every 300 people). In the same report, it is stated that the USA is ranked number five in the world for the number of litigations (74.5) per 1000 people. The Clements Worldwide report also notes that the USA ranks very high on the Norton Rose Fulbright Litigation Trends Annual Survey, with 55% of businesses having more than five lawsuits filed against them in the previous 12 months. This is opposed to 23% in the UK and 22% in Australia (Anonymous, 2017). Businesses and citizens within the USA are very familiar with the use of litigation for resolving their disputes. No doubt this contributes to the abundance of lawyers in the USA. Not surprisingly, since the onset of the bed bug resurgence, the USA now has litigators who specialize in bed bug-related cases. These individuals have intensely studied the residents' rights and responsibilities as well as housing and accommodation laws as they apply to the control of "vermin" on public and privately owned property.

This chapter will review the legal issues involved when attempting to determine who is at fault in human/bed bug interactions, and who is financially responsible for mitigating bed bug infestations. Special attention is devoted to the legal interplay between property owners, pest management professionals (PMPs), and consumers inhabiting multifamily housing (both private and government subsidized), or hotels. The legal implications of being a property owner who implements an inspection-based "proactive" approach to bed bug prevention will be compared with those of being an owner who is only "reactive", or complaint motivated, in addressing their bed bug infestations. Finally, this chapter examines the litigation landscape both from the PMP and consumer perspectives.

41.2 Registration of Pesticides

To put this chapter into context it is important to note that all pesticide formulations that are sold in the USA must be registered with the US Environmental Protection Agency (EPA). The Food Quality Protection Act of 1996 requires that all pesticide active ingredients go through rigorous safety testing to ensure that they will have no harmful effects on humans (or pets) who may be exposed to a variety of pesticide residues in their food, drinking water, homes, and other specific locations over the course of their lifetime. It is from this extensive safety testing that the pesticide product labels are generated. These labels dictate how and where these active ingredients are to be applied. For example, the active ingredients must be formulated at a specific concentration (say 0.06%), then applied at a specific volume per square meter (say $10\,\text{ml/m}^2$), then applied in a specific location (on a corn crop, or in cracks and crevices in a home), or applied on surfaces (mattress seams and tufts). These label requirements are specified so the pesticide exposure risk to humans is minimized. Because the pesticide label is Federal law, the pesticide applicators must be familiar with the label directions of every product they use, and follow the label directions to the letter.

41.3 Legal Requirements Regarding Who Can Apply Pesticides in the USA

While US Federal law requires that all pesticide applicators apply each pesticide according to its label directions, the law governing who is allowed to apply pesticides on properties can vary from state to state. In all states, pest control business owners must have at least one certified applicator on staff. Being certified means that the individual has had at least one year of on-the-job pest management training, has passed the state core examination (covering pesticide application and environmental safety), has passed an examination in at least one of the specific pesticide application categories (for example, general household pests, wood destroying insects, lawn and ornamental), and can now supervise pest management technicians. A pest management technician may also apply pesticides. However, a technician is different from a certified applicator in that they (typically) have not yet had a year of pest management experience, and they have not taken the certification examination in any specific category. The pest management technician is always working under the direct supervision (visual or via telephone) of a certified applicator, although that applicator may not be on-site. The training requirements for pest control technicians vary from state to state. In certain states, technicians are required to take the core examination (pesticide applicator safety) after their probationary period. In other states, the technicians are not required to take any examination but function completely under the direction of the certified applicator, who is ultimately responsible for the technician's ability to follow the pesticide label directions (Jeffrey Rogers, Virginia Department of Agriculture and Consumer Services, Office of Pesticide Services, personal communication).

Yet, even within a state, local ordinances with respect to who can apply pesticides may differ from state statutes. It is always critical to remember that a local ordinance can often be more protective of consumers than the minimum protections provided by state laws. For example, in California, a resident may apply pesticides to his own property. This principle applies not only to homeowners but to owners of apartment complexes, and even super structures. However, this is not the case in structures that receive federal funding or rent vouchers from the US government (known as Section 8 housing). Also, in the City of San Francisco, housing owners are not allowed to treat bed bug infestations, even on their own properties. San Francisco, in Section 4.2 of its bed bug ordinance, requires that only PMPs with a Branch 2 license (equivalent to the certified applicator described above; a Branch 2 means the operator is licensed to control non-wood destroying pests, including bed bugs, and has taken an exam qualifying him/her to apply pesticides indoors) issued from the State Structural Pest Control Board, may treat bed bug infestations. Section 4.2 applies to "Owners, Operators, Property Managers, building employees or tenants" who might be interested in attempting their own pest control.

41.4 Legal Requirements for PMPs Regarding the Standard of Care

In San Francisco, CA, only certified PMPs or their trained technicians are allowed to perform bed bug monitoring and treatment. However, this is not true for privately owned facilities in the rest of California or in other states, where the owner or a facilities employee may perform the on-site pest control. In addition, heat treatment for bed bug control is not regulated in many states, allowing people who are not PMPs to perform bed bug treatments as long as they do not apply any pesticides. It is unfortunate that most states, including California, still lack standard requirements for bed bug treatments. The US Housing and Urban Development (hereinafter "HUD") Notice H 2012-5 for HUD subsidized housing and Notice PIH 2012-17 for Public and Indian Housing, does require inspections to be conducted by a qualified third person (how they are qualified is not specified). However, pesticide applications can only be made by a certified applicator or pest control technician. Yet even in HUD facilities, "heat only" companies (who do not apply pesticides and are therefore not required to be certified applicators in most states) are often hired to engage in bed bug control activities that should be reserved for only the most qualified and experienced PMP. In 2015, HUD published Notice H 2012-15: Guidelines on Addressing Infestations in HUD-Insured and Assisted Multifamily Housing (HUD, 2012). These guidelines were addressed to multi-family housing owners and management agents (O/As) with the intention of ensuring that bed bug infestations would be addressed promptly (before spreading), and that residents would not be penalized for reporting bed bug incidents. In this document, HUD "encourages Multifamily O/As to develop an Integrated Pest Management Plan (IPM) to focus on preventing infestations." Such plans describe the "ongoing efforts the property management will take to prevent and respond to pests". Also in this document, HUD provides a bulleted list of the "Principles of IPM" according to the EPA. Sadly, the EPA makes absolutely no mention of monitoring for bed bugs. Monitoring is the foundation of any IPM program, and is also the principal method of early detection, which is essential for preventing bed bug spread. There is no question that US Federal agencies are still learning how to appropriately address bed bug infestations.

41.5 Public Health Acts Regarding Bed Bugs

Once again, San Francisco, CA appears to have the most advanced bed bug policies and regulations within the USA. The *San Francisco Department of Health Director's Rules and Regulations for Prevention and Control of Bed Bugs* were adopted by the City of San Francisco in 2012 (SFDPH, 2012). Although these rules and regulations do not specifically require the use of IPM, the regulations do employ a standard of care that reflects the basic practices of that approach. These practices include monitoring with bed bug interceptors (traps), providing resident education, inspection and treatment of housing units adjacent to the infested unit, and documentation of all bed bug complaints, treatments, and inspections. Note that if there is a lack of records specifically documenting that the Director's Rules and Regulations were followed (including the fact that a PMP with a Branch 2 license was employed), the building owner and management staff would be subject to legal liability.

41.6 Bed Bug-related Statutes, Laws, and Ordinances

While some states have enacted specific bed bug laws, all states have enacted some version of the Uniform Landlord Tenant Act (ULTA). Under the ULTA, the building owner is responsible for "tendering habitable premises" to the tenant and paying for routine maintenance. This would include pest control. Note that ULTA coupled with the implied "Warranty of Habitability" places a heavy burden on building owners to engage in aggressive bed bug management and control. The implied "Warranty of Habitability" warrants that a building and its living spaces will be habitable and free from such conditions as pest infestations from the time the tenant's lease begins until the tenant moves out. Implicit in this standard is the requirement that the building owner, on some occasions, conduct a reasonable inspection to search out defects on the property. Bed bugs qualify as a "defect".

Most state laws including the ULTA would preclude shifting the costs of pest control from the landlord to the tenant based on the landlord being required to tender the habitable premises. However, some states have allowed some cost shifting under limited situations ("the resident brought the bugs in again"). While some shifting may be allowed, the landlord must still provide the pest control for the building under all circumstances; the "blame game" only occurs if the bed bugs have been remediated, but then reappear.

Other common law legal theories that can be used in bed bug litigation cases would include premises liability. For instance, the Restatement[1] 2d of Torts § 342, (a tort is a civil wrongful act that causes someone to suffer loss or harm, leading to civil legal liability) states a possessor of land is liable if he or she has reason to know of a "injurious" condition, and realizes it can cause harm to those who may occupy the premises, and fails to notify those who may enter or occupy the land or if he or she fails to use reasonable care to remedy the condition (Torts 2d § 342)

In common law jurisdictions, the implied Warranty of Habitability can be used as a basis for building owner–tenant lawsuits. Restatement 2d of Property: Landlord & Tenant, § 17.6 states a building owner is subject to liability for harm caused to the tenant on the property by a condition existing before, or arising after, the tenant has possession (of the home) if he or she (the owner) has not taken reasonable care to repair the dangerous condition. In this instance, if a tenant reports a bed bug infestation to the building owner, and the building owner fails to take proper steps to eradicate the bed bugs, a tenant could sue under the implied warranty of habitability. This is because the implied warranty requires the building owner to maintain the property in a habitable condition. A violation of that warranty results in the liability of the landlord for any harm that arises out of the breach (Property 2d: Landlord & Tenant § 17.6). Again, with regards to bed bugs, this warranty implies the building owner will conduct reasonable inspections to search for bed bug introductions or infestations on a regular basis.

Another common law theory of liability for bed bug lawsuits would be unjust enrichment. Under Restatement 3d of Restitution § 1, unjust enrichment occurs when one side of a transaction receives a benefit, which would result in a loss on the other side of the transaction (Restitution 3d § 1). Referring to bed bug cases, a victim of a bed bug infestation may sue under unjust enrichment when he or she has paid the landlord/hotel for the occupancy of the premises but the building owner has not made reasonable efforts to keep the premises safe. Thus, there is a resulting benefit to the building owner (receiving rent), while there is a loss to the victim (becoming infested). In cases such as these, the building owner would be liable to pay restitution (Restitution 1st § 116). This restitution would include abatement of rent or hotel guest room charges.

Victims of bed bug infestations may also sue under the Unfair and Deceptive Acts and Practices (UDAP) theory. In Iowa, Iowa Code section 714H.3, defines unfair and deceptive acts, stating;

> "A person shall not engage in a practice or act that the person knows, or reasonably should know, is an unfair practice, deception, fraud, false pretense, or false promise; or the misrepresentation, concealment, suppression, or omission of a material fact, with the intent that others rely upon..." (the deception or false pretense in this case, is an owner of a residence acting as if there are no bed bugs on the premises, when they know there are).

---

1 Note that a "restatement" refers to a judge interpreting a law or ordinance and translating it into more understandable terms (laymen's terms) under which it will be applied. *Koenig v. Koenig* (a 2009 court case that set a legal precedent) established one criterion that local court systems use in determining premises liability. That criterion is to determine how reasonable (the cost of) an inspection, repair or warning would be compared to the burden it would put on the owner of the premises. This standard places a duty on the landowners to conduct reasonable inspections to search for latent defects (*Koenig v. Koenig* 2009).

Similarly, the Iowa Code § 714H.3 is applicable when a building owner rents a room or part of their premises while knowing there is a bed bug infestation. The owner's intent is that the guests will stay because they rely on the owner's omission (owner did not mention bed bugs at the time of check in) to mean there is no infestation. Similar to Iowa, Illinois's Uniform Deceptive Trade Practices (UDAP) Act states that a person engages in a deceptive trade practice when they "…represent that goods or services are of a particular standard, quality, or grade (in other words, that theirs is a bed bug-free hotel) or that the goods are a particular style or model, but they are actually of another (in other words, that there *are* bed bugs in the hotel; 815 Ill. Comp. Stat. Ann. 510/2(7)). Additionally, in Missouri, merchants may be liable for unfair and deceptive acts if they knowingly conceal, suppress, or omit evidence in connection to the sale of goods or services (MO. Ann. Stat. § 407.020). Remedies for a UDAP violation may include abatement of rent for the victim, or even treble damages (payment of three times the actual costs) being paid for ascertainable losses and attorney fees.

41.7 Laws Addressing Bed Bug Remediation

Legislatures and local governmental authorities have struggled when enacting bed bug laws to balance the rights of consumers with the pragmatic realities facing building owners who sometimes face catastrophic costs when attempting to control bed bugs. The trend clearly favors protecting the rights of tenant consumers. Again, the most progressive of these laws can be found in San Francisco, CA, where moderate income families pay very expensive monthly rents.

Most prevalent is the remedial provisions found in San Francisco's Department of Public Health Director's Rules and Regulations for Prevention and Control of Bed Bugs (SFDRR); Rent Board Section 37.10B. (San Francisco Municipal Admin. Code § 37.10B). This remedial ordinance prohibits building owners, agents, contractors, subcontractors or employees of the building owners from acting in bad faith in performing the repairs or maintenance required by state, county, or local housing laws. The SFDRR prescribe specific time limits with regards to the number of days it takes to inspect and treat an apartment unit once a bed bug complaint has been made. A follow-up inspection schedule is also required once a bed bug is found (inspections are required to be made post-treatment by a PMP in both the infested units and units adjacent to infested units). The SFDRR also create an affirmative duty for owners and managers to assist mobility-impaired and elderly residents with unit preparation. Finally, all bed bug inspections and treatments must be recorded within a documentation system.

Failure to comply with the SFDRR will result in financial penalties. The penalties found in Section 37.10(B) include:

- reduction in the victim's rent
- fines to the owner up to USD 1000 plus six months in jail
- defense to an eviction (legal prevention of an eviction)
- injunction brought by either the City attorney or the tenant to force remediation
- paying three times the actual cost for damages (including retribution for emotional distress, and attorney fees).

Below are some of the operative provisions that specifically apply to bed bug management and control requirements (San Francisco ordinance 229-12, effective December 14, 2012). These include:

- developing a written bed bug prevention and control plan
- tenant education
- development of building maintenance procedures to prevent bed bug harborage
- use of monitoring devices on a proactive basis
- routine monitoring and inspections by trained employees and PMPs
- sealing cracks and crevices between adjacent apartment/hotel units

- providing instructions for bed bug treatment preparation
- recording responses to tenant complaints using an official complaint response log
- retention of a Branch 2 Licensed PMP to verify tenant bed bug complaint within two working days
- inspection of all adjacent units (above, below, next to, and across from)
- disclosure of bed bug inspection findings by PMP to all tenants
- providing durable and sealable bags for pre- and post-laundering of personal items
- assisting tenants that have disabilities with preparation and accommodations
- treatments only by a Branch 2 Licensed PMP (See generally, Directors' Rules and Regulations for Prevention and Control of Bed bugs).

While San Francisco, CA, is exceptional in its detailed requirements, other states and cities do have laws regarding how infestations are to be addressed, although most are far less specific. For example, the state of Alabama, Section 420-3-11-.12 requires hotels to immediately close the rooms where bed bugs have been sighted, and/or treated until the Health Office has declared the bed bugs eliminated (Ala. Admin. Code r. 420-3-11-.12). Arizona has declared bed bugs to be a public nuisance that are dangerous to public health. Therefore, the Arizona Code Section 33-1319 requires building owners to provide tenants with bed bug education. The Arizona code also requires tenants to comply with bed bug prevention practices, such as not moving materials (furniture and so on) into a dwelling that is infested with bed bugs (A.R.S. § 33-1319). In addition, the Florida Revised Code Section 85.51 places a duty on the building owners to maintain habitable premises. This code specifically includes a duty on the part of a landlord to exterminate bed bugs (F.R.C. § 85.51).

Maine has a progressive statute in Section 6021-A. Maine's law states that the building owner must respond to a bed bug complaint in the following way:

- inspecting for bed bugs within five days of a verbal tenant complaint
- hiring a PMP within 10 days of a bed bug finding
- disclosure to adjacent units of a confirmed bed bug finding, as well disclosure to prospective tenants of the infested and adjacent units
- refraining from renting a unit suspected of having bed bugs
- providing assistance, including financial assistance, to tenants who have difficulties with bed bug inspection and preparation (14 M.R.S § 6021-A).

If the building owner fails to act reasonably, and take prompt, effective steps to treat bed bugs, the owner may be liable for USD 250, or actual damages plus attorney fees. Note that Maine's statute also placed a duty on tenants to cooperate with bed bug control measures, including granting the building owner access to a unit, and allowing the building owner to shift the bed bug management costs to tenants who fail to provide access for the building owner to inspect and treat.

Newly enacted State laws such as California Code Section 1954.602 (2017) and South Dakota Administrative Code Section 44:02:05:08 follow a trend requiring exterminators who perform bed bug inspections, remediation, and post follow up treatments and inspections, to carry a Branch 2 license. While these new laws do not provide the same protections as some local ordinances such as San Francisco's the trend of enacting laws favoring tenant protection is clear.

The enactment of bed bug laws and ordinances are somewhat fluid in several states, and local governments are considering ways of protecting consumers, while at the same time being sensitive to the economic realities that face property owners. However, in the end, the state ULTA places the burden of tendering habitable premises on the building owner. Tenants lack the financial resources to handle a bed bug problem. So pragmatically, if the building owner fails to take steps to remedy the infestation, the bed bugs will progress through the structure creating a clear and present danger to other tenants and employees of the building. This will result in the landlord being liable for damages. Hence, regardless of specific cost-shifting aspects of some laws, building owners ultimately own the bed bugs once the insects are inside their building. Therefore, owners can be held liable for failing to address the infestation appropriately. An excellent source of state bed-bug-specific laws (as of November 2016), can be found at the National Pest Management Association web site (NPMA, 2016).

41.8 Tenants and Public Housing

Bed bug laws in public housing (provided or subsidized by the US Federal government for low-income individuals) and laws relating to those who receive public assistance are very complicated. HUD issues what are called "Notices". It is not clear what legally binding effect these notices have on the owner–tenant relationship. What is clear is that these notices prescribe some interpretation of integrated pest management (IPM) and a threshold standard of care in those buildings owned and operated by governmental authorities, and those buildings accepting tenants on government assistance.

The Environmental Protection Agency (EPA) and the Centers for Disease Control and Prevention (CDC) declared bed bugs as a pest of significant public health importance in 2010 (Anonymous, 2010). In response to the 2010 statement, HUD issued bed bug IPM guidelines for public housing, and Section 8 housing in Notice PIH-2012-17. Public housing properties were/are strongly encouraged to develop an IPM plan that:

- raises bed bug awareness through education
- suggests the inspection of infested units and adjacent living spaces
- the inspection of luggage when one returns from a trip
- the inspection of second hand items before bringing those items into the building
- correctly identifying the suspected pest as a bed bug
- documenting dates and locations where bed bugs are found
- cleaning items in bed bug infested areas
- reducing clutter where bed bugs can hide
- eliminating bed bug habitats
- physically removing bed bugs
- all pesticides are applied by a certified applicator and according to label instructions
- following up bed bug treatments with additional inspections.

The HUD notice also encourages training the housing staff to identify bed bugs, responding to bed bug complaints quickly, and performing ongoing prevention actions (2012 HUD PIH LEXIS 109 Notice PIH 2012-17). This notice also prohibits HUD housing from denying tenancy to a resident based on the tenant having experienced a prior bed bug infestation. HUD facilities also cannot charge a tenant (cost shift) for the price of the bed bug treatment. However, tenants are expected to cooperate with treatment and inspection efforts under the notice.

Note that HUD housing also encompasses government subsidized housing. These housing types are owned by private entities, who receive money from HUD to subsidize rents. These entities are governed by HUD Notice H 2012-5. The HUD notice references 24 CFR Part 5, Subpart G, which requires HUD housing to be decent, safe, sanitary, and in good repair (2012 HUD PIH LEXIS 109 Notice H 2012-5). This HUD notice applies to:

- properties assisted with Section 8 Project Based Rent Assistance, Rent Supplement, or Rental Assistant Payment (RAP) contracts
- properties with active Section 202 Direct Loans, Section 202/162, Section 202 and 811 Capital Advances, and Section 202 Senior Preservation Rental Assistance Contracts, or Section 811 Project Rental Assistance demonstration funding
- properties with active FHA insurance first mortgages under Section 207 pursuant to 223(f), 221(d)(3, 221(d)(4), 221 (d)(5), 231 and 236.

HUD Notice H 2012-5 includes guidance aimed at preventing and addressing bed bug infestations in these subsidized facilities.

HUD Notice H 2012-5 (2012) list the same (inadequate) *EPA Principles of IPM* that are used in the guidelines for government or public housing. There are slight differences with respect to the ability to cost shift, which may be allowed, if and only if, the tenant refuses entry to the unit, or fails to cooperate in some other way that directly hinders control efforts. Of specific interest to subsidized housing owners that are governed by this notice is the ability of an owner or agent to contact HUD and apply for project resources (funding) to move (disabled) residents to another unit during treatments.

Notice H 2012-5 also states that assisted housing facilities may request resources to help pay for bed bug control costs (if funds are available), or that HUD may consider use of rental assistance funds to pay reasonable and necessary expenses for bed bug control.

HUD Notice H 2012-5 also provides guidance for recurring infestations. Under the notice, an owner or agent may be proactive in offering inspections of the unit as well as the new tenant's belongings prior to move-in. Unlike the public housing counterpart, this notice allows the building owner to require treatment of infested furniture before moving in. However, this treatment will be conducted at the housing owner's expense. While the governmental housing notice prohibits denial of tenancy prior to bed bug experience, HUD Notice 2012-5 will allow for eviction and denial of a tenancy where the owner follows the HUD notice and the tenant causes the infestation, or interferes with bed bug management and control (2012 HUD PIH LEXIS 109 Notice H 2012-5). While these HUD notices are not statutes or regulations, they both prescribe a minimum standard of care that, if not followed, will impose liability upon the property owner.

41.9 Legal Standing Clients Encountering Bed Bugs in Temporary Occupancies

Guests or tenants may have legal standing to sue if a hotel, or other temporary occupancy, knowingly rents or leases them a unit that is infested with bed bugs. Under most state laws, "knowing" can be established when the building owner or hotel owner has reason to know or should have known about the existence of bed bugs. The Arizona Revised Statutes Section 33-1319 provides standing for a tenant to make a claim against a landlord when the building owner entered into a lease agreement with the tenant for a bed bug infested dwelling (Ariz. Rev. Stat. § 33-1319). Similarly, the state of Kansas has a regulation where a hotel guest can sue under the Kansas Administrative Regulation Section 4-27-9 if there is a known bed bug infestation in a hotel room, about which the hotel failed to inform the victim (Kan. Admin. Reg. § 4-27-9). Another situation which provides tenants or patrons with the legal standing to sue a landlord or hotel is if the management of the facility has received a report of an infestation and has failed to investigate the report, or properly remedy the infestation. Most states require hotel rooms and other occupancies be properly inspected if there is a bed bug complaint. For example, New Hampshire Revised Statutes Annotated Section 540-A:3 requires a landlord to investigate a tenant's report of an infestation (N.H. Rev. Stat. Ann. § 540-A:3). Section 48-A:11 further requires the landlord or hotel to conduct a reasonable inspection before renting the unit to a tenant or guest (N.H. Rev. Stat. Ann § 48-A:11).

41.10 Bed Bug Lawsuit Landscape

Bed bug lawsuits are prevalent throughout the USA. Perhaps the most well-known case is *Mathias v. Accor Econ. Lodging, Inc.* (Mathias v. Accor. Econ. Lodging, Inc., 2003). The Mathias case involved a hotel renting a room to a guest with the full knowledge that the room had a bed bug infestation. Therefore this case really stood out for the proposition that there can exist conduct so outrageous that it justifies mammoth punitive damages or punishment. Where most courts might limit a punitive damages award to triple the compensatory damages, in this case the court let stand USD 186 000 in punitive damages to each of two plaintiffs who had received only USD 5000 in compensatory damages. The facts of this case arose in 1998 – well before bed bugs resurfaced as a global concern. A year later, a New York court decided *Ludlow Properties, LLC., v. Young* rental housing case. In this case, the judge found that the bed bug infestation in the housing unit violated the warranty of habitability. Hence the judge awarded rent abatement in the amount of 45% for the months the dwelling was infested with bed bugs (Ludlow Prop. LLC. v. Young, 2004).

Following these court decisions of the early 2000s, tenants and hotel guests have prevailed on both individual, and class action lawsuits involving bed bug infestations. On 4 June 2014, an Iowa District Court approved a class action settlement (Residents of *Elsie Mason Manor et al. v. First Baptist Elderly Housing Foundation et al.* (Elsie Mason Manor *et al.* v. First Baptist Elderly Housing Found, 2014)) involving two building-wide infestations in Des Moines for USD 2.45 million. The *Elsie* case demonstrated that a habitability case could be certified as a class action notwithstanding the variances in damages between tenants. The damages in this case were settled based on a matrix of the infested time period that each tenant had lived in the buildings. Different time periods were given a dollar value based on the level of infestation. Elsie Mason Manor is a US HUD subsidized housing community.

In California, a jury awarded two plaintiffs USD 71 652 in damages in the *Hjelm v. Prometheus Real Estate Group, Inc.* case. This case involved a bed bug infestation in an apartment in San Mateo County (Hjelm v. Prometheus Real Estate Gr., 2014). It was not treated properly and the infestation was allowed to continue and spread. In addition to the actual judgment, the court awarded USD 326 475 in attorney fees and USD 20 565 in court costs at an interest rate of 10% per annum.

In *Lattisia Parker et al. v. K.S.A. Investments, LLC., et al.*, a Baltimore jury in 2015 awarded USD 90 525 in damages to multiple plaintiffs in a case alleging that the building owner knowingly rented a bed bug-infested residence to the tenants. The owner had failed to take proper remedial actions prior to renting the premises, and concealed the infestation (Lattisia Parker v. K.S.A. Investments, LLC, 2015).

In *Shabaan v. West Street Partnership, et al.*, an Annapolis, Maryland, jury awarded Faika Shaaban USD 800 000 in damages in a bed bug lawsuit in July of 2013. USD 650 000 of that amount was awarded in punitive damages to be paid by the building owner. According to the *Huffington Post* and *Capital Gazette*, the jury took only 45 minutes to arrive at a decision that awarded Shaaban twice the amount of damages that was originally sought (Shabaan v. West St. Partnership, 2013).

Finally, on 18 September 2015, the *Washington Post* reported a jury in Prince George's County awarded a victim of a hotel bed bug infestation USD 100 000. Note that the victim, in this case, was only in the hotel for one night.

These cases illustrate the potential consequences a property owner will face if they fail to respond to bed bug infestations appropriately. The overriding issue in most of these cases is not the mere existence of the bed bugs, but the failure of the property owner to take prompt and responsible action once the bed bugs are discovered.

41.11 Conclusion

The area of "bed bug legal" is still very fluid and evolving. PMPs, property owners and consumers need to be familiar with the laws of their respective cities and states, particularly with regards to court decisions, where the standards placed on PMP and property owners can vary widely. One common theme among the various states is that property owners and managers have an affirmative duty to implement assessment-based protocols when managing bed bugs. In addition, bed bug inspections must be proactive rather than reactive. Addressing bed bug issues only when a complaint is made is no longer an acceptable "standard of care" when it comes to bed bug infestations. In addition, allowing bed bugs to spread from room to room by ignoring units adjacent to an infested unit is also no longer acceptable, and is legally comparable to ignoring the problem altogether. What constitutes reasonable pest management practices in different states may vary, but the duty of the property owner to maintain habitable and safe premises remains consistent. This duty requires all concerned to view properties building-wide rather than as isolated rooms or units.

Similarly, states may vary in their requirements regarding who may perform pest control in different locations (public versus private properties) but federal law requires strict adherence to product labels, and those applying the product will be responsible for applying it correctly, whether they are a licensed PMP or not. It is certain that the litigation landscape keeps evolving, but recent judgements clearly reflect a trend towards protecting consumers by requiring both property owners and PMPs to act assertively and responsibly when it comes to bed bug remediation.

Thus, the area of "bed bug legal" has evolved from a reactionary method of management to an inspection-based model, where it is no longer acceptable to wait until a bed bug complaint is lodged before action is taken. Current standards of care mandate periodic and ongoing inspections by property owners and managers to search for bed bug evidence. Future standards of care will be dictated by "state-of-the-art" products and techniques that will transition from these inspection-based methods to more preventative strategies. Already the pest control industry has seen the evolution of silica dust for perimeter application and encasements for mattresses and box springs that are more prophylactic in nature. As these products become more advanced, cost effective and widely used, the standard of care will almost definitely become more "preventative" as opposed to a merely inspection-based.

References

Anonymous (2010) *The US Centers for Disease Control and Prevention (CDC), and the US Environmental Protection Agency (EPA) Joint Statement on Bed Bug Control in the United States.* US Department of Health and Human Services, Atlanta, GA, USA.

Anonymous (2017) *The Most Litigious Countries in the World-clements Worldwide*, https://www.clements.com/sites/default/files/resources/The-Most-Litigious-Countries-in-the-World.pdf (accessed 4 April 2017).

NPMA (2016) *State Bed Bug Specific Laws*. National Pest Management Association, https://www.epa.gov/sites/production/files/2016-11/documents/state-bed-bug-laws.pdf (accessed 27 May 2016).

HUD (2012) *Notice H 2012-5. US Housing and Urban Development*, https://portal.hud.gov/hudportal/documents/huddoc?id=12-05hsgn.pdf (accessed 14 May 2017).

SFDPH (2012) *Director's Rules and Regulations for Prevention and Control of Bed Bugs*, San Francisco Department of Public Health, https://www.sfdph.org/dph/files/EHSdocs/Vector/BedBug/BedBugRegs_070112.pdf (accessed 18 May 2017).

Court Cases

Hjelm v. Prometheus Real Estate Group, Inc. (2014) (CIV 516959) (California Supreme Court, Order dated 30 June 2014).

Koenig v. Koenig, 766 N.W.2d 635, 645-46 Iowa (2009) quoting *Heins v. Webster County*, 552 N.W.2d 51, 52 (NE 1996).

Ludlow Properties, LLC., v. Young (2004) 780 N.Y.S.2d 853 (LEXIS 712).

Mathias v. Accord Econ. Lodging, Inc. (2003), 347 F.3d 672 (7th Circuit).

Parker et al v. K.S.A. Investments, LLC., et al. (2015) Case No. 24-C-14-001967 (Circuit Court Baltimore, MA).

Residents of Elsie Mason Manor et al. v. First Baptist Elderly Housing Foundation et al. (2014) CVCV008116 (Order dated June 4, 2014).

Shabaan v. West Street Partnership, et al. (2013) (Case No. 02-C-12-170759).

Restatements of Law

Restatement 2d of Property: Landlord & Tenant § 17.6 (1977).
Restatement 3d of Restitution and Unjust Enrichment § 1 cmt. a (2011).
Restatement 2d of Torts § 342 (1965).

Statutes

Ala. Admin. Code r. 420-3-11-.12 (2016).
A.R.S. § 33-1319 (2016).
Cal. Civ. Code Section 1954.602 (2017)
F.R.C. § 85.51
2012 HUD PIH LEXIS 109 Notice H 2012-5 (2012) (citing 24 C.F.R. 5).
815 Ill. Comp. Stat. Ann. 510/2(7).
Iowa Code § 714H.3(1) (2016).

42

Bed Bugs and the Law in the United Kingdom

Clive Boase

42.1 Introduction

There are a variety of aspects of bed bugs, their impacts, and their management, which are covered in UK law. This chapter briefly reviews the main areas.

42.2 Training of Pest Management Professionals

In the UK, training of pest management professionals (PMPs) was not regulated prior to 1986. However, in the early 1980s, there was growing concern about pesticides, and as a result the *Food and Environment Protection Act* (FEPA) was passed in 1985 (UKG, 1985a), with the *Control of Pesticides Regulations* (CoPR) then being passed in 1986 (UKG, 1986). Among the provisions of CoPR was a requirement that those using agricultural pesticides must have a specific qualification to demonstrate their competence. However, for non-agricultural pest control, CoPR required that pesticide users be trained and competent, but with no legal requirement to undertake specific training or achieve a particular qualification (UKG, 1986).

Subsequently, the Royal Society for Public Health (RSPH) created a qualification in pest management (RSPH, 2016). This qualification is not a legal requirement, but has become an accepted national standard. It covers a wide range of pest control topics, including vertebrate and invertebrate pest biology, pest management techniques, legislation, and health and safety. The British Pest Control Association (BPCA) has made it a requirement that each pest control company seeking membership should have its PMPs qualified at least to RSPH Level II or an equivalent. In addition, distributors of pest control products will only supply professional pesticides to organizations that can provide evidence of training and competence. As a result, although this qualification is not a binding requirement for bed bug control, the majority of PMPs will now have this qualification, or an equivalent.

Training towards this qualification is provided either within larger pest control companies, trade associations, pest control product distributors, or by independent training organizations. Although the RSPH recommends 72 learning hours for their Level II award, most providers of training for this qualification will offer 30–40 h of face-to-face training, plus additional time on preparation and home study. Of the recommended 72 h, bed bug management and associated issues will constitute perhaps 5% of the total learning time.

In addition to the broad RSPH qualifications in pest management, there are currently several one-day courses in bed bug management, typically run either by a trade association or a distributor. These courses provide greater detail, and although some are based on the European Code of Practice for Bed Bug Management (BBF, 2013; see also Chapter 23), they do not lead to a nationally recognized qualification. However, new certificated bed bug management courses are in development, and these are linked to a continuing professional

Advances in the Biology and Management of Modern Bed Bugs, First Edition.
Edited by Stephen L. Doggett, Dini M. Miller, and Chow-Yang Lee.
© 2018 John Wiley & Sons Ltd. Published 2018 by John Wiley & Sons Ltd.

development scheme. Nonetheless, none of these short courses is a prerequisite for bed bug control work, and only a small proportion of PMPs are likely to have attended a bed bug management course.

42.3 Bed Bug Pesticide Approval

In the UK prior to 1986, there was no statutory approval system for pesticides or biocides. However, 1985 saw the introduction of the FEPA (UKG, 1985a), with the CoPR being passed the following year (UKG, 1986). This legislation made it a requirement that all pesticides offered for sale in the UK must be approved, and this approval was conditional upon the appropriate government agency (currently the Health and Safety Executive) receiving, reviewing and approving a dossier covering toxicology, ecotoxicology, chemical properties, efficacy, risk assessments, and other issues. This national legislation is still in force as of 2016, but is gradually being replaced by a Europe-wide biocide evaluation and approval scheme (EU, 2012).

In the early 1990s, the European Union saw that by standardizing the authorization of biocidal substances it could:

- create a harmonized market for such products
- provide improved protection of people, animals and the environment
- ensure that products were sufficiently effective.

As a result, the *Biocidal Products Directive* (98/8/EC) was passed in 1998 (EU, 1998). All European member states were required to implement the directive into national law, with the UK passing the *Biocidal Products Regulations* in 2001 (UKG, 2001). In 2012, the BPD was replaced by the *EU Biocidal Products Regulation* (Regulation (EU) No 528/2012) which is directly applicable in each member state, and applied from 1 September 2013 (EU, 2012).

Regulation 528/2012 defines biocides as: "Any substance or mixture, in the form in which it is supplied to the user, consisting of, containing or generating one or more active substances, with the intention of destroying, deterring, rendering harmless, preventing the action of, or otherwise exerting a controlling effect on, any harmful organism by any means other than mere physical or mechanical action" (thus excluding the silicate compounds). Biocides are divided into four main groups: disinfectants, preservatives, pest control, and other biocidal products.

The directive, and subsequently the regulations, created processes for the review of both existing active substances, and for new ones. In brief, each registrant is required to prepare and submit a dossier on their active substance, together with a limited number of representative products and uses. The dossier covers a range of topics including the substance's efficacy. The review of the dossier and eventual approval of the substance for use as a biocidal product takes place at the European level. Once that approval is obtained, the registrant can then seek authorization for products containing that substance, in the approved product types, at a national level. Overall this is a lengthy and costly process, taking several years, and may cost the registrant over EUR 10 million.

The European Commission has also published *Technical Notes for Guidance* (TNG), on the efficacy evaluation of public health insecticides (EC, 2010). This is an 87-page document that sets out the type of efficacy tests and data expected against 13 of the more common pests or pest types. Three pages are devoted to the evaluation of products against bed bugs. The TNG sets out four types of tests for bed bug products:

Screening studies These are small-scale tests involving topical application or tarsal contact tests on various surfaces. These may involve susceptible and/or resistant bugs.

Residual efficacy The TNG states that "Good residual efficacy is essential for insecticides used for bed bug control as it is impossible to treat all insects directly or reach all of their hiding (places)." The document describes applying products to typical surfaces, with realistic exposure times. Three to five replicates are suggested, with ten to twenty bugs per replicate.

Simulated use Simulated-use tests are intended to use a semi-realistic environment within the laboratory, within which the insects are exposed to the insecticide. However, the TNG states that, given the behavior of bed bugs, creating a realistic environment is challenging.

Field trials The TNG provides very limited guidance, but does say that because practical bed bug control usually involves a combination of different treatments, and because true replication is difficult, the report on the trials should fully describe the measures used and other relevant factors.

Biocidal products may only be authorized if "they are sufficiently effective" (UKG, 2001). The TNG sets out mortalities and reductions that it considers to be sufficient. In field tests, for example, a reduction of more than 90% in bed bug number is expected by 6–10 weeks after treatment. However, the document says deviations from these norms are possible, but should be justified.

The TNG also recognizes that some insects, such as houseflies, are very likely to develop resistance, so applicants can be and are required to submit a resistance management strategy (ECHA, 2009). However, at the time of writing, resistance management strategies do not yet appear to have been required for bed bug control products and resistant strains of bed bugs are not essential for efficacy evaluation.

42.4 Tenants, Guests and Bed Bugs

In English law, those providing residential accommodation, whether it be a rented apartment or a hotel room, have a duty of care to those using the accommodation.

For rented accommodation, the *Landlord and Tenant Act 1985* (UKG, 1985b) clearly requires that the property is fit for human habitation at the start of the tenancy. However, some of this clarity may be obscured by the tenancy agreement, the time interval between the start of the tenancy and the finding of any infestation, or by the tenant bringing in their own furniture. Nonetheless, the expectation remains that the landlord should take responsibility for infestation, especially if it is also associated with common parts of the building.

For hotel guests, the legal situation is usually simpler, and the *Occupiers' Liability Act 1957* (UKG, 1957) places an unequivocal duty of care on those managing a hotel.

Disputes over bed bugs in accommodation are usually treated as personal injury cases. For such cases to be successful, the claimant must demonstrate that the defendant was negligent in discharging their duty of care. Despite the legislation, disputes over bed bug infestation seldom reach court, with an out-of-court settlement usually being agreed between the claimant, the defendant (the hotel or landlord), and their insurers. The level of compensation will be intended to cover any financial losses as the result of the incident (such as doctors' fees or loss of earnings), plus compensation for the suffering received. The outcomes of such settlements in the UK are not reported formally, but compensation for bed bug incidents in hotels is typically in the range GBP 750 up to GBP 5000, and rarely up to GBP 10 000 (Slack, 2011; Bond, 2013).

Occasionally bed bug cases do go to court, typically a county court. However, findings in county courts are not usually considered legally important, so they do not create legal precedent, and are not reported in the Law Reports. As a result, there is not an established body of legal precedent in the UK, which defines expectations of bed bug prevention, detection or control.

42.5 Local Authority Duties and Powers Regarding Bed Bugs

In the UK, local government has responsibility for a wide range of issues, including environmental health. The local environmental health department will have specific legal responsibilities and powers in relation to infestation, with three pieces of legislation being particularly relevant to bed bugs (GLPLG, 2009):

Public Health Act The *Public Health Act 1936* includes sections that relate specifically to bed bugs (known as 'vermin' at that time). This Act gives local authorities powers to serve Notice on 'verminous' premises to require the destruction of vermin, provides powers to enter premises to investigate suspected vermin problems, and to apply for entry warrants if required (UKG, 1936).

Environment Protection Act In cases where problems (such as bed bugs) are spreading from a source to adjoining premises, then the *Environment Protection Act 1990* gives local authorities powers and duties. If the authority is satisfied that a statutory nuisance is occurring or is likely to occur, then it can either informally require the owner to eradicate the infestation, or carry out the remedial work itself and then charge the owner or landlord, or serve a formal Abatement Notice. Failure to comply with an Abatement Notice is a criminal offence, and the owner or landlord can be fined and ordered to pay compensation if they do not carry out the works (UKG, 1990).

Housing Act Finally, the *Housing Act 2004* gives local authorities powers to enter premises to investigate housing hazards, and issue Improvement or Hazard Awareness Notices. An infestation is one of the listed hazards, but is given a relatively low priority (UKG, 2004).

In practice, local authorities in the UK seldom use legislation to deal with bed bug problems, as this route can be costly, time-consuming and unpredictable. In most cases, problems are resolved through informal means.

References

BBF (2013) *European Code of Practice for Bedbug Management, version 2*, Bed Bug Foundation, http://bedbugfoundation.org/en/ecop/ (accessed 15 April 2016).

Bond, A. (2013) *Hilton Hotels Pay £750 Compensation to Gatwick Guest, 74, Whose Holiday in Spain Was Ruined by Bed Bug Bites*, http://www.dailymail.co.uk/news/article-2275616/Hilton-hotel-forced-pay-compensation-guest-holiday-ruined-BED-BUGS.html (accessed 1 June 2016).

EC (2010) *Technical Notes for Guidance*. European Commission, http://eng.mst.dk/media/mst/69922/Draft%20guideline%20efficacy%20PT%2018%20and%2019.pdf (accessed 1 June 2016).

ECHA (2009) *Chapter 10: Assessment for the Potential for Resistance to the Active Substance*. European Chemicals Agency, https://echa.europa.eu/documents/10162/16960215/bpd_guid_tnsg_product_evaluation_annex_i_inclusion_chapter_resistance_en.pdf (accessed 26 Oct 2016).

EU (1998) *Directive 98/8/EC of The European Parliament and of The Council of 16 February 1998 Concerning the Placing of Biocidal Products on the Market*. European Union, http://eur-lex.europa.eu/legal-content/EN/TXT/HTML/?uri=CELEX:31998L0008&from=EN (accessed 26 October 2016).

EU (2012) *Regulation (EU) No 528/2012 of The European Parliament and of The Council of 22 May 2012 Concerning the Making Available on the Market and use of Biocidal Products*. European Union, http://eur-lex.europa.eu/legal-content/EN/TXT/HTML/?uri=CELEX:32012R0528&from=EN (accessed 26 October 2016).

GLPLG (2009) *Inspection and Treatment Guide*. Greater London Pest Liaison Group, http://www.londonpestgroup.com/downloads/GLPG_Bedbug_Good_Practice_Guide_Inspection_and_Treatment_Manual.pdf (accessed 1 June 2016).

RSPH (2016) *Level II Award in Pest Management*. Royal Society for Public Health, https://www.rsph.org.uk/en/qualifications/qualifications.cfm?id=level-2-award-in-pest-management (accessed 1 June 2016).

Slack, C. (2011) *Revealed: Horrific Scars Suffered by Sisters Bitten Dozens of Times at London Hotel*, http://www.dailymail.co.uk/news/article-2029286/Airways-Hotel-Victoria-Horrific-scars-suffered-sisters-bitten-bedbugs.html (accessed 1 June 2016).

UKG (1936) *Public Health Act 1936*. United Kingdom Government, http://www.legislation.gov.uk/ukpga/Geo5and1Edw8/26/49/contents (accessed 26 October 2016).

UKG (1957) *Occupiers' Liability Act 1957*. United Kingdom Government, http://www.legislation.gov.uk/ukpga/Eliz2/5-6/31/contents (accessed 1 June 2016).

UKG (1985a) *Food and Environment Protection Act 1985*. United Kingdom Government, http://www.legislation.gov.uk/ukpga/1985/48/contents (accessed 1 June 2016).

UKG (1985b) *Landlord and Tenant Act 1985*. United Kingdom Government, http://www.legislation.gov.uk/ukpga/1985/70 (accessed 1 June 2016).

UKG (1986) *Control of Pesticides Regulations 1986*. United Kingdom Government, http://www.legislation.gov.uk/uksi/1986/1510/made (accessed 1 June 2016).

UKG (1990) *Environmental Protection Act 1990*. United Kingdom Government, http://www.legislation.gov.uk/ssi/2016/99/contents/made (accessed 26 October 2016).

UKG (2001) *The Biocidal Products Regulations 2001*. United Kingdom Government, http://www.legislation.gov.uk/uksi/2001/880/made (accessed 26 October 2016).

UKG (2004) *Housing Act 2004*. United Kingdom Government, http://www.legislation.gov.uk/ukpga/2004/34/contents (accessed 26 October 2016).

43

Bed Bugs and the Law in Australia

Toni Cains, David G. Lilly and Stephen L. Doggett

43.1 Introduction

In Australia there is no specific legislation that directly relates to bed bugs. For the most part, their regulation (such as ensuring an infestation is appropriately eradicated) can be enforced under general public health legislation or local government ordinances. In contrast, the registration of insecticides and their use is tightly controlled at a Federal Government level by the Australian Pesticides and Veterinary Medicines Authority (APVMA; AG, 2016a) under a scheme prescribed by the *Agricultural and Veterinary Chemicals (Administration) Act 1992* (AG, 2016b), the *Agricultural and Veterinary Chemicals Code Act 1994*, the *Agricultural and Veterinary Chemicals Act 1994* (AG, 2016c), and cognate Acts and Regulations. Pest management professionals (PMPs) who are operating for commercial profit must be licensed and these licenses are regulated at a state or territory government level. In most jurisdictions, this is the respective health department.

Australia is not an overly litigious society (compared, for example, with the USA) and to date there has been little finalized litigation involving bed bugs against PMPs or the hospitality sector. The authors have been involved in some cases, but all have been settled out of court and the outcomes were not recorded. There have been a small number of cases of tenants taking action against landlords over bed bugs and these cases are reviewed in this chapter.

This chapter examines the legal issues surrounding bed bugs in Australia, from the registration of insecticides, to the licensing of PMPs, to the use of public health laws to ensure compliance in pest control, to legal disputes involving tenants.

43.2 Registration and Use of Bed Bug Management Products

As noted above, pesticide products sold in Australia must be approved and registered by the APVMA. The responsibilities of this Federal Government organization are described in the body of legislation referred to above.

Before registering a product, the APVMA is required to conduct an assessment of the potential impacts of the pesticide on the environment, human health, and trade, and of the likely effectiveness of the pesticide for its proposed uses. Once a product has been registered, the APVMA Chemical Review Program reviews the registration of existing pesticides to determine whether changes are necessary to the registration or if the registration should be withdrawn. For a product to be used by a PMP, the pest must be listed on the label, although for present purposes, this is always the generic term "bed bug" rather than the two species names (*Cimex hemipterus* and *Cimex lectularius*). The treatment of bed bugs can also be undertaken when the label states it is approved for use against "crawling insects".

Advances in the Biology and Management of Modern Bed Bugs, First Edition.
Edited by Stephen L. Doggett, Dini M. Miller, and Chow-Yang Lee.
© 2018 John Wiley & Sons Ltd. Published 2018 by John Wiley & Sons Ltd.

All insecticides are categorized according to potential risk and given a "Schedule number" under the Federal Poisons Standard (AG, 2016d). However, as of 1 January 2017, Australia has moved to the "Globally Harmonized System of Classification and Labeling of Chemicals", the standardized chemical classification system developed by the United Nations (AG, 2016e). This will be administered by the APVMA.

The various states and territories also have their own legislative requirements. For example, in New South Wales (NSW), the *Pesticides Act 1999* (NSWG, 1999) requires the users of registered pesticides to strictly follow the approved label or permit directions. In NSW, it is also compulsory for some pesticide users to give notice that they are planning to use pesticides if treating common areas of multiple-occupancy buildings. The aim of the notification is to allow people to choose to reduce their potential pesticide exposure.

43.3 Legal Requirements of Pest Management Professionals

In every state and territory except the Australian Capital Territory (ACT), PMPs are required to have a license if they are undertaking pest management for commercial gain and are using pesticides. A license is not required if they are undertaking non-chemical means of control only. The license is known by slightly different names in different jurisdictions. In most states, the health department regulates and issues licenses, although in NSW, the Environmental Protection Authority issues the licenses. Furthermore, in some states, the PMP company must also be registered and licensed to operate as a company. This includes Victoria, the ACT, South Australia (SA), Tasmania, and Western Australia (WA). In WA, as part of the company license, regular auditing is undertaken on such matters as the compliance on the storage of insecticides, insecticide handling, and vehicle labeling. In Victoria, vehicle labeling is audited, while in the other jurisdictions, no routine auditing is undertaken, although spot-audits occur irregularly.

For a PMP to earn a license, the regulatory authority requires a set number of competency units (through the completion of competency examinations), which is nationally defined by the Australian Skills Quality Authority (ASQA). Overall, there are 20 units of competency defined by ASQA directly relating to pest management (AG, 2016f). In most states and territories, only 3 of the 20 units are required, although in WA 13 units are mandated. A basic license with 3 units of competency will exclude the control of timber pests and fumigation services, as additional competency units are required to undertake these activities. In relation to bed bug management (and most urban pests), the basic license would allow for the management of the insect by insecticidal means, as well as the (safe) transport and storage of insecticides and equipment.

All PMPs must comply with the relevant work health and safety legislation for the state or territory when using pesticides. This would include mixing and applying the product according to the label directions, following recommendations for the use of appropriate personal protective equipment, and the storage and disposal of pesticides. As insecticide labels are legal documents, the PMP must comply with all prescribed directions. The label also lists safety instructions, and states if there are any use restrictions or requirements. There are major penalties under work health and safety legislation for failing to comply with the legislation (Doggett, 2013).

43.4 Public Health Laws Regarding Bed Bugs

In Australia, local government has the primary responsibility for bed bug investigations and, as bed bugs are not known to transmit disease, bed bugs are treated as a "nuisance" under various public health instruments. The resolution of public complaints about bed bugs can often be achieved between council officers and private or public housing providers.

In NSW, local councils respond to bed bug complaints in places of shared accommodation, such as backpackers' hostels, using the *Local Government (General) Regulation 2005 Schedule 2 clause 5* (NSWG, 2005), which states that all parts of the premises and appurtenances must be kept in a clean and healthy condition.

The control of bed bugs is not included in the NSW *Public Health Act 2010* (NSWG, 2010) and there is no legislation in NSW that covers the residential component of hotels or motels.

In the Northern Territory (NT), complaints about bed bugs are investigated as a public nuisance under Part 3 of the *Public and Environmental Health Act 2016* (NTG, 2016).

In Queensland, bed bugs are not classified as a designated pest or prescribed under any regulation. The *Public Health Act 2005* Section 11 (QG, 2005) only refers to designated pests or other animals prescribed under a regulation for the purpose of describing a public health risk. As bed bugs do not transmit or contribute to disease, and are not overly hazardous to human health, the use of this legislation to control them would be difficult. Local governments have provisions under local laws that can be adopted, but these are only applicable, as in NSW, to shared accommodation such as backpackers' hostels, and do not cover hotels and motels.

The *South Australian Public Health Act 2011* (SAG, 2011) is the centerpiece of public health legislation in SA. It provides for the promotion and protection of public health and the reduction of incidences of preventable illness, injury, and disability. Key amongst persons and bodies involved in the administration of the Act are the Minister for Health who has a statutory obligation to promote proper standards of public and environmental health within the state, and local councils, which are established as local public health authorities for their areas. A key feature of the Act is Section 56 (General Duty) which imposes a general duty on a person to "take all reasonable steps to prevent or minimise any harm to public health caused by, or likely to be caused by, anything done or omitted to be done by the person." Notices may be served on a person in order to secure compliance with the general duty or to avert, eliminate or minimize a risk to public health.

As public health authorities for their areas in SA, local councils can act on complaints about bed bugs associated with facilities such as hotels, hostels, and supported residential facilities. Owners or operators of such facilities can be directed to take action to manage bed bug infestations. In practice, this generally involves the provision of advice and support to enable effective management of any infestations, but at times it may be necessary to issue notices to secure compliance with the general duty (as set out in Section 56 of the Act). This may require a person to engage a licensed PMP for assistance with chemical treatment.

The two primary pieces of legislation in Victoria are *The Public Health & Wellbeing Act 2008* (VG, 2008) and the *Public Health & Wellbeing Regulations 2009* (VG, 2009). Complaints relating to rooming houses, bed and breakfast accommodation, and hotels and motels are dealt with by local councils since all these premises are registered businesses under the Act. Councils can issue an improvement notice under Section 18 and/or 19 of the Regulations, which require the proprietor to maintain the accommodation in a clean, sanitary and hygienic condition.

In WA, environmental health officers at the local government level have powers prescribed in the *Health Act 1911* (WAG, 2016a) that can be employed to enforce the eradication of the bed bugs from premises. This law is about to be replaced with the *Public Health Act 2016* (WAG, 2016b) and it is anticipated that the new regulations will include provisions enabling authorized officers of local governments to be able to require the disinfestation of premises with a bed bug problem. Prosecution for non-compliance with regulations and local laws will also be available to local governments.

In the ACT, bed bug complaints are categorized as a potential "insanitary condition" under the *ACT Public Health Act 1997* (ACTG, 2016). An alleged infestation can be investigated by an authorized public health officer, who may issue an abatement notice under Section 69 of the Act if bed bugs are found. The notice directs the business or responsible person within a reasonable time to remove the public health risk (in other words, the bed bugs).

43.5 Tenancy and Public Housing

Complaints made by tenants regarding negligence by the landlord or body corporate would be referred to the various state consumer affair advocates (such as Fair Trading in NSW). In NSW under Section 52 of the *Tenancy Act 2010* No 42 (NSWG, 2016), a landlord must "provide the residential premises in a reasonable state of cleanliness and fit for habitation by the tenant". Failure of such requirements (such as having bed bugs present), enables the Local Council to issue an improvement notice to the landlord/owner. Generally the tenant would need to provide some evidence to Council (photos, letters/emails to landlord indicating lack of a resolution) for Council to undertake an inspection.

Generally within public housing premises, pest control for the premises is contracted through a facilities management agency, with bed bugs included as a routine pest along with other pests such as cockroaches and rodents. Tenants may be expected to contribute towards the funding of pest control services to their property. However with bed bugs, it is recognized by several housing authorities that bed bugs do spread from property to property and residents usually cannot afford the high cost of pest control associated with bed bug eradication. Thus many housing authorities will pay for bed bug control to avoid uncontrolled infestations.

43.6 Bed Bug Legal Cases

There have been only a small number of legal cases involving bed bugs in Australia, and typically these have involved tenant-versus-landlord disputes. The payments ordered by the tribunals have been very small compared to the multimillion dollar payouts in the USA.

In *Griffiths (Appellant) v Watson (Respondent)* 2014 under the *Queensland Civil and Administrative Tribunal Act 2009*, Ms Griffiths was the owner of a property in Southport and Mr and Mrs Watson were her tenants. On 17 June 2013, Ms Griffiths filed an application for compensation for rent arrears, the cost of removing bed bugs, and other general repair costs. On 10 July 2013, the Tribunal ordered Ms Griffiths to receive the bond and the Watsons pay her an additional AUD 362. Subsequently Ms Griffiths appealed the tribunal decision based on the following evidence she provided "that the Watsons introduced bed bugs into the property". However the court found that the leave to appeal should be refused.

In *Chessels (Applicant) v Wood (Respondent)* in 2004 in the NSW Consumer, Trader and Tenancy Tribunal, the landlord (Chessels) sought orders for the payment of monies by the tenant of AUD 5 928 for various items arising out of the early termination of the lease. The tenant, however, sought the return of his bond and compensation related to the need to vacate the premises shortly after entering occupation because bed bugs were present. The court agreed, and ordered the landlord to pay the tenant the whole bond plus interest.

The case of *Evers (Applicant) v Professionals Real Estate Clayfield (Respondent)* was heard in Brisbane (Queensland) as a minor civil dispute in 2011. This case involved an infestation in a home unit and was the subject of a residential tenancy agreement under the *Queensland Civil and Administrative Tribunal Act 2009*. The applicant was a tenant for two years in a residential unit and, prior to the lease finishing, gave the respondent a notice of intention to leave on the basis that the unit was infested with bed bugs and refused to pay arrears of rent. In this case, the tribunal found the applicant was responsible for arrears of rent as well as pest control to eradicate the bed bugs.

References

ACTG (2016) *Public Health Act 1997*. Australian Capital Territory Government, http://www.legislation.act.gov.au/a/1997-69/ (accessed 20 September 2016).

AG (2016a) *Australian Pesticides and Veterinary Medicines Authority*. Australian Government, https://apvma.gov.au (accessed 10 May 2016).

AG (2016b) *Agricultural and Veterinary Chemicals (Administration) Act 1992*. Australian Government, https://www.legislation.gov.au/Series/C2004A04553 (accessed 20 September 2016).

AG (2016c) *Agricultural and Veterinary Chemicals Code Act 1994*. Australian Government, https://www.legislation.gov.au/Series/C2004A04723 (accessed 20 September 2016).

AG (2016d) *Poisons Standard July 2016*. Australian Government, https://www.legislation.gov.au/Details/F2016L01071 (accessed 20 September 2016).

AG (2016e) *Globally Harmonized System of classification and labelling of chemicals (GHS)*. Australian Government, https://www.comcare.gov.au/preventing/hazards/chemical_hazards/globally_harmonised_system_of_classification_and_labelling_of_chemicals_ghs (accessed 18 September 2016).

AG (2016f) *Qualification details CPP30911 - Certificate III in Pest Management (Release 2)*. Australian Government, https://training.gov.au/Training/Details/CPP30911?tableUnits-page=1&pageSizeKey=Training_Details_tableUnits&pageSize=50&setFocus=tableUnits (accessed 18 September 2016).

Doggett, S.L. (2013) *A Code of Practice for the Control of Bed Bug Infestations in Australia*, 4th edn, Department of Medical Entomology and The Australian Environmental Pest Managers Association, Sydney.

NSWG (1999) *Pesticides Act 1999*, New South Wales Government, http://www.legislation.nsw.gov.au/#/view/act/1999/80 (accessed 10 March 2016).

NSWG (2005) *Local Government (General) Regulation 2005*. New South Wales Government, http://www.legislation.nsw.gov.au/regulations/2005-487.pdf (accessed 20 September 2016).

NSWG (2010) *Public Health Act 2010*. New South Wales Government, http://www.legislation.nsw.gov.au/#/view/act/2010/127 (accessed 20 September 2016).

NSWG (2016) *Residential Tenancies Act 2010 No 42*. New South Wales Government, http://www.legislation.nsw.gov.au/inforce/cfa8f374-676d-6eea-811e-95d70730f313/2010-42.pdf (accessed 20 September 2016).

NTG (2016) *Public and Environmental Health Act 2016*. Northern Territory Government, http://notes.nt.gov.au/dcm/legislat/legislat.nsf/linkreference/Public%20and%20Environmental%20Health%20Act?opendocument (accessed 20 September 2016).

QG (2005) *Public Health Act 2005*. Queensland Government, http://www.austlii.edu.au/au/legis/qld/consol_act/pha2005126 (accessed 10 March 2016).

SAG (2011) *Public Health Act 2011*. South Australia Government, https://www.legislation.sa.gov.au/LZ/C/A/South%20Australian%20Public%20Health%20Act%202011.aspx (accessed 10 March 2016).

VG (2008) *Public Health and Wellbeing Act 2008*. Victoria Government, http://www.legislation.vic.gov.au/Domino/Web_Notes/LDMS/PubStatbook.nsf/edfb620cf7503d1aca256da4001b08af/8B1B293B576FE6B1CA2574B8001FDEB7/$FILE/08-46a.pdf (accessed 20 September 2016).

VG (2009) *Public Health & Wellbeing Regulations 2009*. Victoria Government, http://www.legislation.vic.gov.au/Domino/Web_Notes/LDMS/PubStatbook.nsf/b05145073fa2a882ca256da4001bc4e7/A3B0A9845FD0980ACA25768D002AB0B5/$FILE/09-178sr.pdf (accessed 10 March 2016).

WAG (2016a) *Health Act 1911*. Western Australia Government, https://www.slp.wa.gov.au/legislation/statutes.nsf/main_mrtitle_412_homepage.html (accessed 20 September 2016).

WAG (2016b) *About the Public Health Act 2016*. Western Australia Government, http://ww2.health.wa.gov.au/Improving-WA-Health/Public-health/Public-Health-Act (accessed 20 September 2016).

Legal Cases

Chessels (applicant) v Wood (respondent) in 2004 in the NSW Consumer, Trader and Tenancy Tribunal.

Evers (applicant) v Professionals Real Estate Clayfield (respondent) in 2011 Queensland Civil and Administrative Tribunal Act 2009, http://archive.sclqld.org.au/qjudgment/2011/QCATA11-069.pdf (accessed 20 September 2016).

Griffiths (appellant) v Watson (respondent) in 2014 in the Queensland Civil and Administrative Tribunal Act 2009.

44

Bed Bugs and the Law in Asia

Andrew Ho-Ohara and Chow-Yang Lee

44.1 Introduction

In Asia, there is no specific law that is directly related to bed bugs. Nevertheless, there are laws pertaining to the general control of insect pests in the urban environment, guidelines on the registration and usage of insecticide products, and the licensing of pest management professionals (PMPs). This chapter, will review those regulations that are presently being enforced in Singapore, Malaysia, Thailand, Indonesia, Japan, Taiwan, and China.

44.2 Registration and Use of Bed Bug Management Products

In Singapore, the *Control of Vectors and Pesticides Act* (Chapter 59) ("CVP Act") governs the control of the sale and use of pesticides and repellents, as well as the registration, licensing and certification of PMPs engaged in vector control work (Anonymous, 2002). In this context, the Environmental Health Institute of the National Environmental Agency (NEA) of Singapore is responsible for the regulation of pesticide products used for the control of vectors. However, the importation, distribution, supply and sales of pesticide products intended for agricultural use fall under the jurisdiction of the Agri-Food and Veterinary Authority of Singapore.

Section 2 of the CVP Act defines "vector" as "any insect, including its egg, larva and pupa, and any rodent, including its young, carrying or causing, or capable of carrying or causing, any disease to human beings." It merits mention that the NEA (as the watchdog of the CVP Act) regards vectors as "organisms that transmit diseases", classifying them into five categories:

- mosquito
- flea
- rodent
- cockroach
- fly.

As bed bugs are not known to transmit pathogens of human disease, theoretically, the CVP Act does not apply to insecticides used on bed bugs *per se*. However, as all insecticides are employed against other insects (with vector status), the products are therefore subject to registration. The manufacturer is required to provide bio-efficacy results of the product tested against a susceptible strain of all targeted insect pests, which are intended to be on the label claim. A similar situation is mandated in Japan, where, despite bed bugs not being considered as one of the "sanitary pests" (cockroaches, flies, fleas, ticks, mites, mosquitoes, and rodents), generic products (used against either bed bugs or sanitary pests), are required to be registered.

Advances in the Biology and Management of Modern Bed Bugs, First Edition.
Edited by Stephen L. Doggett, Dini M. Miller, and Chow-Yang Lee.
© 2018 John Wiley & Sons Ltd. Published 2018 by John Wiley & Sons Ltd.

In Malaysia, the registration of all pesticide products used against pests (which includes bacteria, viruses, fungi, weeds, insects, rodents, birds, or any other plant or animal that adversely affects and attacks animals, plants, fruits, or property) is regulated by the *Pesticides Act 1974*, under the jurisdiction of Pesticide Control Division, Department of Agriculture, Malaysia (Anonymous, 2015). In this context, a pesticide is "any substance that contains an active ingredient, or any preparation, mixture or material that contains any one or more of the active ingredients as one of its constituents."

Similar to the situation in Malaysia, all pesticide products meant for public health and household use in other Asian countries have to be registered:

- in Thailand, with the Thai Food and Drug Administration, through the *Hazardous Substance Act* (Anonymous, 2008)
- in Indonesia, with the Pesticide Commission, Ministry of Agriculture
- in Taiwan, with the Environmental Protection Administration, under the *Environmental Agent Control Act Enforcement Rules* (EPA, 2000).

In China, a similar rule applies to all insecticide products for public health use. These must be registered with the Institute for the Control of Agrochemicals under the Ministry of Agriculture.

44.3 Legal Requirements for Pest Management Professionals

In Singapore, all PMPs handling pesticides under the CVP Act need to be licensed or certified by the NEA. This involves the successful completion of a three-month course, which thereafter leads to the granting of a license valid for three years.

In Malaysia, all individuals carrying out the control of:

- general pests (such as cockroaches, bed bugs, ants, and rodents)
- termites and other wood destroying insects
- pests of public health

as a hired service, are required to obtain a pesticide applicator or assistant pesticide applicator license by passing an examination conducted by the Pesticide Control Division (Anonymous, 2004). Individuals holding the latter license can only carry out minor pest control jobs, or larger jobs under the supervision of a licensed PMP.

In Thailand, PMPs are required to hold an operator license to practice. All candidates must take a five-day course and pass the examination before being issued a license. After five years, the PMPs need to be recertified by attending a one-day recertification course. Similar to Thailand, all PMPs in Taiwan are required to be licensed under the Pest Control Operators Management Regulations, which were enacted pursuant to Article 22 of the *Environmental Agents Control Act* (EPA, 2016). They must undergo continuous training and receive recertification every three years. In Indonesia, there are no rules governing the training and licensing of PMPs and thus no certification requirement.

In Japan, there is no license requirement for PMPs undertaking pest management services in buildings and structures smaller than $3000\,m^2$. For buildings meant for public use exceeding $3000\,m^2$ (which only accounts for around 4500 buildings in all of Japan), the PMPs are required to go through a five-day training course, and pass an examination to obtain a license which requires recertification every six years. To carry out pest management in food plants, shops, restaurants, homes, and other locations, no license is required.

In China, all pest management companies require a business license to operate. However, PMP certification is not mandatory. The requirements for the license vary between provinces. There are three levels of certification: beginner, intermediate, and advanced. Once a PMP has been certified, no further recertification is required.

44.4 Bed Bug Legal Cases

Despite the well-documented presence of bed bugs in Asia, searches of legal databases suggest that there has not been any finalized litigation related to bed bugs in the region. One likely cause of this phenomenon is the public's lack of familiarity with bed bugs and their bites. Another plausible cause is the considerable difficulty a victim would have to commence, prove, and sustain litigation arising from being bitten by bed bugs, coupled with the likelihood of nominal monetary damages being awarded.

In the context of bed bugs, the starting point of successful litigation would begin with proof that the victim was bitten by bed bugs on the owner's premises. While an apparently simple issue, the reality is that most victims of bed bugs (assuming that they even realize they were a victim), do not think of taking samples of the bed bugs or photographs of the infestation. Therefore, the victim lacks contemporaneous proof that the bed bugs were on the owner's premises (to negate the possibility that the victim was bitten somewhere else). If the victim was only aware of being bitten after leaving the owner's premises, given that it can take some days for the bite reaction to appear, it would be difficult for the victim to thereafter obtain access to gather evidence of the presence of bed bugs.

Another evidentiary obstacle would be proving that the victim did not unwittingly bring the bed bugs into the owner's premises, which would be entirely possible given that bed bugs are known to hitchhike as stowaways. The Latin legal maxim *volenti non fit injuria*, meaning that "no legal redress exists for a situation which arose as a result of the victim's own action or carelessness", would negate the liability of the owner if the victim was bitten by bed bugs that the victim brought onto the owner's property.

Apart from considering whether legal proceedings can be commenced and sustained, the victim should also consider the injury sustained. In the event when liability is found (whether by the court or agreed as between the parties), monetary damages will be awarded as compensation to the victim. The quantification or assessment of such damages, in the absence of previous award(s), will be influenced by the seriousness of the injury.

The effect of bites from bed bugs vary from victim to victim, and only occasionally do people develop serious allergic reactions. Even an unusually susceptible/vulnerable victim is nevertheless potentially entitled to additional damages (over and above that of a normal victim) on the legal principle of the "egg-shell skull rule", meaning that "you take the victim as he/she is found". Such a susceptible/vulnerable victim must, however, seek immediate medical attention, as any delay resulting in further complications and/or additional injuries may not be attributable to the bed bugs (and to the owner of the premises). Regardless of susceptibility or vulnerability, excessive scratching by a victim of bed bugs bites, which leads to bleeding, secondary skin infection, and/or any other injury, may be regarded as developments beyond the control (and therefore liability) of the owner of the premises.

The majority of bites by bed bugs however do not result in any serious medical risk. Although experiments indicate that bed bugs are capable of transmitting pathogens of disease, such as Chagas, in the laboratory (Salazar et al., 2015), bed bugs are not considered a serious medical or public health hazard, despite the unpleasantness and annoyance of being bitten. Given the abundant material available that suggests that bed bugs do not transmit pathogens in the field (see Chapter 12), and save for the said exceptional cases involving a particularly susceptible or vulnerable victim, the minor injury would likely justify an award of only a nominal amount. This is particularly so, given that the awards for damages arising from personal injuries in Asia have traditionally been low, and are likely to remain low for reasons of public policy.

Notwithstanding the likely nominal amount awarded in a case brought by a bed bug victim, the owner of the premises may nevertheless suffer tremendous reputational losses arising from an infestation. The commencement of litigation arising from a bed bug infestation is likely to generate much publicity, especially given the fact that there has been no landmark litigation of this type previously in Asia. In this age, where social media reigns, there are few, if any, effective ways to control the dissemination of news, and news that a premises owner has a bed bug infestation is almost certain to continue to make the rounds of social media, long after any litigation has concluded.

44.6 Future

It merits mention that several unpublicized instances in Singapore and Malaysia of complaints by victims to owners of having been bitten by bed bugs have led to private settlements, notably with refunds, discounts, or other forms of compensation. Generally, the traditional non-confrontational Asian nature still holds, with litigation being the last resort.

References

Anonymous (2002) The statues of the Republic of Singapore – Control of Vectors and Pesticides Act (Chapter 59). Singapore: The Law Revision Commission.

Anonymous (2004) Pesticides Act 1974 – Pesticides (Pest Control Operator) Rules 2004. Kuala Lumpur: Pencetakan National Malaysia Berhad.

Anonymous (2008) Hazardous Substance Act (HSA), B.E. 2551. The Royal Gazette of Thailand 125.

Anonymous (2015) Laws of Malaysia – Act 149 Pesticides Act 1974. Kuala Lumpur: Pencetakan National Malaysia Berhad.

EPA (2000) *The Environmental Agents Control Act Enforcement Rules. Environmental Protection Administration, Executive Yuan, Republic of Taiwan*, Environmental Protection Administration, http://a0-oaout.epa.gov.tw/law/EngLawContent.aspx?Type=E&id=27 (accessed 3 March 2017).

EPA (2016) *Pest Control Operators Management Regulations. Environmental Protection Administration, Executive Yuan, Republic of Taiwan*. Environmental Protection Administration, http://a0-oaout.epa.gov.tw/law/EngLawContent.aspx?Type=E&id=173 (accessed 3 March 2017).

Salazar, R., Castillo-Neyra, R., Tustin, A.W., *et al.* (2015) Bed bugs (*Cimex lectularius*) as vectors of *Trypanosoma cruzi*. American Journal of Tropical Medicine and Hygiene, **92** (2), 331–335.

45

On Being an Expert Witness

Paul J. Bello and Dini M. Miller

45.1 Introduction

With the reemergence of bed bugs over the last two decades, these insects have come to be seen as a serious legal liability for accommodation providers. The associated growth in bed bug-related litigation has meant that professionals in the field of pest management are increasingly being sought to provide expert testimony to support the position of the litigant or the defendant. This chapter reviews the role of the expert witness in bed bug-related legal proceedings, with emphasis on the situation in the USA.

45.2 What is an Expert Witness?

An expert witness is a person who is hired to provide testimony in a legal matter because of their expertise in a particular field. An expert witness's qualifications are based upon their education, professional training, experience, and specialized knowledge, which typically extends beyond that of a lay person's. The expert's knowledge must be such that the court can rely upon the expert's opinions when making legal decisions. In regard to bed bug cases, there are likely numerous pest management and related industry individuals who might qualify to serve as expert witnesses. However, the qualification requirements may differ due to the specific nature of the case, or the court, state, or country where the case occurs. Expert advice pertaining to a bed bug infestation may be included in a legal case, even if the expert providing the advice is not retained as a witness in connection to the litigation (Murphy, 2011).

Legal disputes usually involve two opposing parties: the plaintiff and the defendant. The plaintiff is the party who initiates the lawsuit or litigation. The plaintiff believes that they have suffered damages at the fault of the defendant who has been named in the lawsuit. If an expert is retained by one party, usually the opposing party will retain an expert as well. The expert's role remains fundamentally similar, whether retained by the plaintiff or the defendant (Bello, 2012).

Typically an expert is retained by an attorney or law firm who represents one of the parties in a case where the expert's assistance is needed. The expert may be retained, or designated, in the case as a testifying expert, or they may be retained as a consultant, consulting with the attorney but not testifying or rendering an opinion on the case.

Experts may be retained for reasons including, but not limited to:

- to review the facts of a case
- to serve as a consultant who provides advice on technical matters and the viability of the case
- to provide a written report regarding causation or other factors
- to provide opinions regarding the facts in the case
- to provide testimony under oath in response to direct and cross examination.

Advances in the Biology and Management of Modern Bed Bugs, First Edition.
Edited by Stephen L. Doggett, Dini M. Miller, and Chow-Yang Lee.
© 2018 John Wiley & Sons Ltd. Published 2018 by John Wiley & Sons Ltd.

Note that direct examination is when the retaining attorney asks questions of the expert. Cross-examination is when the opposing attorney asks the questions.

45.3 The Expert's Role

While an expert is expected to render opinions, the overall role of the expert witness is to educate the court and jury. The expert must provide true and accurate responses to questions posed during their testimony. These opinions and responses may be presented in the expert disclosure, written report, deposition testimony, arbitration hearing testimony, or trial testimony in court.

Prior to providing a report or testimony, the expert will usually review the numerous documents that comprise the historical record of the case. For bed bug cases, these documents may include the pest management treatment records, the pest sighting log, correspondence between the various parties, training information, company policies and procedures, medical reports, and other pertinent information provided for review. The expert may also review the recorded testimony of other witnesses, including other experts, involved in the case. Such recorded testimony is usually in the form of deposition transcripts.

Expert testimony may be provided in deposition, arbitration, or in a courtroom during a trial.

Information within the deposition testimony from other individuals may serve as an integral part of the expert's overall report and opinion of the case. Testimony transcripts are often provided to the expert for review. Transcript review is an important part of the expert's role and may require many hours of reading when a number of witnesses have testified in the case. Typically, the expert testimony is recorded and transcribed by a court reporter. The transcript includes all questions and answers provided on the record during testimony. However, in some situations the testimony may not be transcribed or recorded.

45.4 Providing Expert Testimony

Legal cases, or litigations, progress under procedures that must be observed. The attorney who retains the expert is familiar with such procedures of law. This is important because the rules of evidence and court procedures may vary by jurisdiction. While the expert's role and participation in such cases is subject to such procedures, these procedures are often foreign to the pest management professionals (PMPs), industry professionals, or university professors who often serve as experts.

The following comments and observations are presented for the benefit of those individuals who may be considering acting as an expert witness in a legal case:

- In spite of the facts of the case, individual cases may be decided differently from what the evidence would indicate. The outcome of some cases may seem less than fair.
- While it is widely believed in the USA that every citizen is entitled to their day in court, the majority of cases are settled out of court (prior to the court date). Some cases may even settle just hours or minutes prior to the scheduled court date, even though the expert has already spent considerable time preparing to give testimony.
- Everything that is said during testimony may subsequently appear in a written transcript. Once transcribed, reading previous testimony can be an enlightening experience.
- The role of the expert is to provide truthful responses to questions, and to provide information to educate the court and jury about the issue being litigated. This is so informed decisions can be made. In some situations, a legal team may ask a question with the aim of directing a particular answer. For example, the question is asked "can it be proved that bed bugs cannot transmit disease?" The answer is obviously "no", but it is impossible to prove a negative. A more scientifically sound answer would be "No, but if you have evidence that they can, please show it to me" (Goddard, 2012). Note that all questions need to be carefully considered before an answer is given.

- There will likely be another expert who has been retained by the opposing party's attorney. These experts will likely be asked to critique your opinion.
- Although the expert's role requires fair representation of the truth; there are at least two sides in every case. As such, it is possible that one party may become unhappy with your testimony.
- An expert's most valuable commodity is his or her integrity, and one's integrity should never be compromised. Bed bug cases involve certain entomological facts, control methodologies, and legitimate business practices. As an expert entomologist or pest professional, it is best to ensure that your report, opinions, and testimony represent what are the currently accepted best practices for the pest management industry and that your information is entomologically sound.

45.5 Bed Bugs in a Court Case

Court cases involving bed bugs in multi-unit housing are notoriously complicated. There may be many issues that the court is considering. Did the resident bring the bed bugs in, or were they already present in the apartment at the time of move in? Did the resident report the bed bugs promptly? Was the resident mistreated by the management after reporting bed bugs? Was the resident stressed, or mentally or physically damaged by the infestation? Was the apartment management company responsive and helpful to the resident? Did they schedule an inspection right away and help the resident to prepare for treatment? Was the pest management company (PMC) experienced with bed bugs? Were they doing an adequate job of controlling them?

In answer to the above questions, a variety of damage claims are typically made by the plaintiff, while counter claims are made by the defendant regarding the actions that they took or did not take. As an expert in bed bug management, you will not have the expertise to address questions relating to the stress or wellness of the infested resident other than to state what symptoms are typically associated with bed bug sufferers. However, you will be responsible for reviewing the pest management practices to determine if the housing community was taking a proactive and conscientious approach to their pest problems (or not). You will review the records to determine if the housing management attempted to do the treatment themselves or if they contracted with an experienced PMC. You will review the pest management records regarding which methods or products were used to determine if the PMC was attempting to use the "best practices" to take care of the bed bug problem. Finally, you will need to determine if the housing management was aware of how their PMC was treating the bed bugs on their property.

A recent case in the USA involved a disabled multi-unit housing resident who contacted an attorney for help because her apartment was infested with bed bugs. Her claim was that she had found bed bugs in her home and reported it to the management (she was unable to determine exactly when she told the management). Her deposition stated that the management staff did not address the problem because she could not afford to pay USD 300 for treatment. However, she stated that one manager did make fun of her in front of other residents and told her neighbors that she had bed bugs. Over the course of the infestation the plaintiff threw away many belongings, had to treat the bed bugs herself, and was bitten many times. When the management did contact a PMC, the staff put preparation instructions under her door requiring her to pack all of her belongings into plastic bags and move all her furniture away from the walls. The plaintiff had mental and physical disabilities and could not comply with the preparation instructions. The plastic bags required for packing were supplied by a local church organization, but because of her disabilities, the resident could not lift the plastic bags if they were full. So she only filled them half way. The plaintiff claims that the housing management complained that she had too many bags, and would not treat her apartment at first, because she had not prepared correctly. When at last her apartment was treated, the plaintiff and her cat were asked to stay out of the apartment for four hours. The resident claims that a single technician treated her home with "a chemical" and told her the treatment was completed after he had been in her apartment for 20 min. The resident also claims that the treatment failed to get rid of the bed bugs. She therefore had to sleep in a small space in the living room amongst her bagged belongings. The plaintiff claimed that management threatened to evict her for bringing bed bugs into the facility, and again tried to make her pay (cost shift) USD 300 for treating her apartment. Finally, the plaintiff claims that even though she is no longer living at that facility, she still suffering from the feeling of bed bugs crawling on her while she is trying to sleep.

In response to the plaintiff's claims, the housing management brought forth another resident whose deposition states that the plaintiff was a hoarder, that her apartment was "nasty", and she had cockroaches climbing all over her walls. The witness also claimed that the plaintiff had shown her a bed bug. The housing management stated that the plaintiff brought the bed bugs into the apartment unit and was ultimately responsible for infesting other areas of the building. The apartment management also claimed that they did respond to the plaintiff's bed bug complaint, and produced the pest control records to document that her apartment was treated. The PMC documented that the condition of the apartment ("hoarder") was such that they could not guarantee that the bugs would be eradicated after treatment.

While the expert witness is responsible for reading all of the depositions and claims of the case, addressing claims of stress, abuse, loss of belongings, or who brought the bed bugs in the first place, is very difficult. As an expert, we know that anyone could have brought the bed bugs into the unit, including the apartment maintenance staff. We also know that stress and abuse are somewhat subjective, and that throwing away belongings is often the choice of the person being bitten. Therefore we can only speak in generalities based on our experiences of other infestations. Where the expert witness testimony is most useful in this case is in addressing whether or not the management of the facility reacted to the problem in a conscientious fashion. This response will be based on not only the plaintiff's testimony but, more importantly, the apartment managers' own testimony and the pest management records. Did the manager of the facility address the infestation in a timely manner? Did they hire a PMC with the experience and expertise to successfully address the problem? Did they proactively inspect other units? Did they know how their PMC was treating the infested units? Did the PMC keep good records? The expert witness in this case would give a deposition on video, or in writing, or in court, communicating their opinion of the apartment managers' response time *and* the pest management efforts/methods to address the bed bug infestation.

In the case described above, the evidence would indicate that the management did not address the bed bug report in a timely fashion. The housing facility claimed that their PMC was "proactively" (regularly) inspecting apartment units for bed bugs while looking for other household pests. If so, how did the PMP miss the bed bugs in the plaintiff's apartment? Well, a letter from the PMC stated that bed bug work (including inspections) required a separate contract. In addition, the PMC indicated that they had terminated their bed bug contract with the housing community because the manager would not pay the company for treatment until the resident paid the housing management! In light of the statement above, determining whether or not the bed bug treatment methods were the "best" does not seem as important in this case. However, according to the records, the treatments used were a standard practice at the time (liquid chemical applications) if not the very "best".

45.6 Summary

Serving as an expert witness can be a time-consuming and tedious process. Because the incidents that initiated these lawsuits may have taken place a year or more prior to the court case, the plaintiff and defendant recollections of the events may not be exactly accurate. Therefore, the review of all records is necessary. In addition, bed bug treatment methods are still evolving, so it is also critical to be aware of standard (as well as the "best") treatment methods at the time.

Bed bugs are a distasteful subject to most people (even attorneys, judges, and jury members), and it is important to appreciate that lay person testimony may be biased by exhibits of resident hoarding, filthy apartments, or photographs of resident bite reactions. However, as an expert witness evaluating the pest management, the questions to be answered are:

- Did the management address the problem in a timely manner.
- Did the management protect other residents by inspecting surrounding units?

- Did the management hire an experienced PMC?
- Did that PMC use the best practices available at the time to address the bed bug infestation?

Facilities managers/owners must not ignore bed bug reports. In addition, residents must cooperate with treatment efforts and allow PMPs to have access to their units. Conversely, residents must not be allowed to take financial advantage of a bed bug situation when the facility management/owner has been conscientious in trying to control the bed bug issues. As of 2017, we have no single reliable cure for bed bug infestations except whole building fumigation (which is often not necessary (and cost prohibitive) in large facilities that may have only a few infested units). Therefore, solving bed bug problems in individual apartment units takes time, money, persistence, and some acceptance that bed bug eradication will take several weeks (Dini Miller, unpublished results).

Finally, it is important to acknowledge that acting as an expert witness can be a stressful experience. It is not for everyone. The witness has to be confident in presenting themselves, and translating their ideas and opinions in a manner that is readily understandable to all in the courtroom.

References

Bello, P. (2012) On the stand. *Pest Management Professional*, **80** (6), S2–S4, S6, S8, S10, S12–S13.

Goddard, J. (2012) Testifying in medico-legal or forensic entomology cases; land mines and pitfalls. Speech presented at the Entomological Society of America Annual Scientific Meeting in Knoxville, Tennessee, 14 November 2012.

Murphy, P. (2011) *Bedbug Expert Gets to Testify After All*, http://lawyersusaonline.com/benchmarks/2011/01/19/bedbug-expert-gets-to-testify-after-all/ (accessed 19 April 2017).

Part VIII

Bed Bugs: the Future

46

Bed Bugs: the Future

Chow-Yang Lee, Dini M. Miller and Stephen L. Doggett

It is extraordinary to think that we are yet to define what the bed bug problem is!
Lee, Miller and Doggett, *ABMMBB*, 2017

At the start of this book it was mentioned that *Advances in the Biology and Management of Modern Bed Bugs* (ABMMBB) is the first academic text since Robert L. Usinger's seminal 1966 publication, *The Monograph of Cimicidae* (Usinger, 1966). However, there are a number of key differences between the two texts. For example, *Advances* has over 60 different authors and focuses on just two cimicids, the Common bed bug, *Cimex lectularius* L., and the Tropical bed bug, *Cimex hemipterus* (F.), while Usinger's monograph was largely written by one extraordinary individual and covers the taxonomy of the entire Cimicidae family. Yet, these apparent differences pale in significance when compared to the most important salient distinction of all: the monograph was written at a time when infestations were on the decline globally, and when bed bugs were thought to have been defeated. In contrast, *ABMMBB* was produced in the midst of the modern bed bug resurgence, where the prospects of defeating bed bugs again appear rather dim.

Indeed, if Part 2 of *ABMMBB*, on the global bed bug resurgence, is examined, there is almost a common theme across the chapters; there is little hope that the modern bed bug resurgence will abate soon. In fact, for most countries, the forecast predicts that the situation will only worsen. In the USA, Curl (2016) estimated that bed bug revenue for pest control in 2015 was USD 573.2 million as the direct result of performing 815 000 bed bug remediation jobs. By 2016, the number of bed bug jobs had grown by 11.4% (907 875), generating a revenue of USD 611.2 million (Anonymous, 2017). Bed bug revenue is expected to reach USD 1 billion by 2020 (Curl, 2016). For Europe, a similar outcome has also been predicted (Anonymous, 2016). In the UK, an annual growth rate of 4.7% for bed bug services is predicted to occur over 2016–2026. For Germany, this forecast is to be even higher; some 5.2%. Thus the indications are that bed bugs are going to become more common in society. The one exception, however, is Australia.

In Australia, the current evidence indicates that bed bugs are no longer on the increase and could even be on the decline (Doggett, 2016). Interestingly, this is in spite of the fact that there has been no specific bed bug legislation introduced. Hence the question has to be asked; why has this nation had some level of success while others have not? This may be because Australia, relatively early in the resurgence, developed a clear three-pronged strategy to combat the rise in bed bugs. Notably, the introduction of (Doggett *et al.*, 2011):

- a bed bug industry standard
- education around this standard
- bed bug research.

The industry standard, *A Code of Practice for the Control of Bed Bug Infestations in Australia*, was first released in 2005, well before any other nation produced a similar document (Doggett, 2005). The code has been the corner stone in guiding pest management professionals (PMPs), the client, and the public, in best

Advances in the Biology and Management of Modern Bed Bugs, First Edition.
Edited by Stephen L. Doggett, Dini M. Miller, and Chow-Yang Lee.
© 2018 John Wiley & Sons Ltd. Published 2018 by John Wiley & Sons Ltd.

practice for bed bug management. This standard also specifically mentions which products to avoid, and this has kept many of the more questionable products out of the Australian marketplace (thereby not diluting control efforts). Education of PMPs (and other stakeholders) based around the industry standard has meant that many have become highly experienced at bed bug eradication. Finally, active research within Australia (and from elsewhere) has helped develop and evolve the code to ensure that the document has maintained relevance. The challenge now will be to ensure that complacency does not set in or else all the successes will be nullified.

Another common theme that runs through the chapters in Part 2 is how little epidemiological data exists for the current bed bug situation. For most nations, the extent of the resurgence has neither been accurately qualified nor quantified. Generally, bed bug data is acquired through retrospective surveys, as mandatory notifications of new infestations are not routinely undertaken. Most surveys are of the pest management industry, and unlikely to provide a complete picture of the true state of the global bed bug resurgence. This is because the pest management industry is just one of the sectors being impacted by the insect. Retrospective bed bug surveys also have limitations. Few in the pest management industry respond to survey requests (this could be due to lack of time, concerns about client confidentiality, apathy about surveys, or possible fear of giving a commercial edge to competitors). Many companies do not keep records of past treatments and thus do not have accurate data available for surveys (Stephen Doggett, unpublished results). The hospitality sector barely acknowledges the bed bug resurgence and is unlikely to admit that their facilities have or had bed bugs. Finally, many people attempt to control the infestation themselves and clearly such data cannot be captured.

The resurgence is therefore poorly defined, with a dearth of solid information that can be used to justify the need for amelioration programs. Anecdotally, however, it appears that disparate sectors of society have been differentially impacted during the course of the resurgence. It would appear the tourism sector was affected initially, in particular the backpacker and hotel industries in the late 1990s (Doggett *et al.*, 2011). Later, travelers began to take bed bugs home and the insects began to rapidly spread into the wider society. During the mid-2000s, bed bugs started appearing in low-income housing. This is a group of people who do not have the fiscal resources to pay for control efforts. This demographic also includes residents that may not have the cognitive ability to know when bed bugs are even present. This low-income invasion has resulted in some infestations becoming massive in number, involving tens to hundreds of thousands of bed bugs. These "super-infestations" often began in apartment units and subsequently spread throughout the building. Sadly, in some cases, the result has been that every unit within some low-income apartment complexes have become infested with bed bugs. These infestations subsequently spread throughout the community. Unless society decides to pay for bed bug control for those that cannot afford it, low socioeconomic groups will become a bed bug reservoir for wider society. The fact is we all need to pay, even just to protect ourselves from future bed bug infestations!

Another unfortunate aspect of the resurgence has been the lack of government support for bed bug remediation strategies, especially for research. The justification for this has largely been that human pathogens are not transmitted by bed bugs, and their acute health impacts are relatively minor (see Chapters 11–14). Hence, in a world of limited economic resources, it is difficult to justify funding for bed bug research based on health impacts alone. This is especially true when you consider that humans live in a biosphere inhabited by many other arthropods known to transmit a variety of pathogens that cause significant human morbidity and mortality.

However, what is often forgotten is that bed bugs are expensive, and often hugely expensive. There is no doubt that the most serious impact of these insects is the financial cost to society. Bed bugs are certainly one of the most economically important insects in the world. Even those costs mentioned in the Curl report cited above fail to list many of the expenses associated with infestations (see Chapter 15). It is evident that bed bugs are costing the world economy billions of dollars annually and they are set to cost the world billions more in the future. One of the challenges that we must face is that the true costs of bed bugs have yet to be fully realized. It is critical that these costs are calculated to both demonstrate their true fiscal impact on society but, most importantly, to justify to governments (and others) that they need to spend money on strategies to ameliorate the problem. Proper funding will lead to proper control and this will help protect the financial security of the world. It is now around 20 years since the resurgence began, yet not only is basic data regarding bed bug epidemiology lacking, but so is data quantifying their fiscal impact. *It is extraordinary to think that we are yet to define what the bed bug problem is!* Perhaps, in this respect, we as scientists have failed the world, but when governments neglect to support such research, as they often do, ignorance reigns.

Clearly bed bug research, including investigations into insect biology, is required and could help in the development of novel control strategies. For example, by understanding bed bug physiology, chemical ecology, dispersal, host- and harborage-seeking behaviors (as discussed in Part 4), it may be possible to develop new monitors and traps that can detect infestations earlier, reduce insect biomass prior to treatment, or even eliminate infestations completely. Population studies have already shown that infestations in multiple apartment units within a sampled building can arise from a single gravid female. This population development from a single female highlights how poor control can result in a massive bed bug spread, and also explains why the modern resurgence has been so rapid (which in turn also highlights the need for proper education of PMPs and all bed bug stakeholders).

Research is also required to elucidate the biological differences between *C. lectularius* and *C. hemipterus*, which may impact management operations. To date, most bed bug eradication products have been developed and tested in the USA, where *C. lectularius* is the most common species. Many of these products have subsequently been marketed in other parts of the world based on the assumption that they will also work against *C. hemipterus*. However, this may not always be the case. Kim *et al.* (2017) studied differences in pitfall trap performance against *C. lectularius* and *C. hemipterus*, and found evidence of morphological differences in the legs of both species that affect their climbing ability. *Cimex hemipterus* has "hairier feet" or, scientifically speaking, a greater number of tenent hairs on its tibial pad than does *C. lectularius*. This enables the former species to climb the inner wall of pitfall traps, potentially rendering such devices useless as monitoring tools. As a result, future evaluation of bed bug products should ideally involve the testing of both species.

The reality is that there is no magical silver bullet (like DDT) on the horizon that will reverse the current bed bug resurgence. Thus it becomes useful to examine management methodologies used through antiquity, including those that existed prior to the invention of synthetic insecticides. In the past, bed bugs were far more common than they are now, hence the importance of Chapter 1's coverage of bed bugs through history. By examining historical control methods, insights may be gained into methods that could be used today. Also, it is equally useful to review historically inappropriate and dangerous control methods, to avoid repeating earlier mistakes. For example, in the past, deaths occurred when deadly fumigants such as hydrogen cyanide were used to control bed bugs. Sadly there have also been a number of recent deaths from the uncontrolled use of phosphine gas for bed bug control (see Chapter 13).

Yet, in spite of there being no simple solution for control of the modern bed bug, it is possible to eradicate *any* infestation with current technologies. Successful bed bug management should be based on integrated pest management (IPM), which has been promoted in the pest management industry worldwide for more than two decades. Yet widespread use of IPM in bed bug management has largely been limited to Australia, Europe, the USA, and Canada. In some countries, chemical control still remains the principal control approach. This often includes the use of pyrethroids, even though resistance to this class of compounds in bed bugs is widespread (see Chapter 29). Inappropriate application technologies are also widely used in some regions, for example ultra-low volume space spraying. Employing such a technology against a creature that lives in cryptic harborages is a recipe for failure. An effective IPM program for bed bug control should involve pro-active monitoring, inspection, decision-making, and the development of management plans that are most suitable for the size and habitat of the infestation. In addition, appropriate insecticides and/or heat application equipment should be used (Koganemaru and Miller, 2013; Doggett, 2013). In other words, these IPM programs need to be tailored to the population size, the dynamics, and the specific location of the infestation (see Part 6).

Best practices for bed bug management need to be developed in order to ensure that certain standards of quality are met for the management of existing infestations and to minimize the risk of future infestations

(Doggett et al., 2012). Such best practices should be widely adopted as an industry standard. These standards need to be developed by individuals who are highly experienced in bed bug management. The process of producing and reviewing such standards must be transparent, void of any conflict of interest, and open to public input and scrutiny (Orton, 2009). The best practices standard should be regularly revised to incorporate new developments. Ideally these practices need to be available to all, at no cost, to encourage all stakeholders to use them. However, it has to be recognized that best practice documents are expensive to produce. To date, only a handful of such standards have been developed, including those in Australia (first developed in 2005), Europe (2011), and the USA (2011; see Chapters 22–24). There is an urgent need for bed bug management industry standards to be adopted in other countries.

Interestingly, all the industry standards to date have been developed by the pest management industry (for example, the Australian Environmental Pest Managers Association in Australia, and the National Pest Management Association in the USA). Yet why would such associations want to lead the development of bed bug management standards? Certainly, if bed bugs disappear, there is less work for their members. Thus there exists an argument that government should take on the bed bug responsibility. However, the pest management industry are the experts in pest control and they are certainly the most qualified to develop industry standards. Also, developing these standards should be part of their social responsibility in protecting the community. Being proactive in developing best practices is certainly good for public relations (and the pest control industry has been plagued with a history of bad stories!).

There have been a number of reasons given for the bed bug resurgence as discussed in the introduction, but few would dispute that insecticide resistance has been the key trigger. At present, resistance to the pyrethroids has been documented for both *C. lectularius* and *C. hemipterus*, but there appears to be the emergence of resistance to more insecticide classes (Romero and Anderson, 2016; also see Chapter 29). Quite remarkably, both bed bug species have independently evolved similar forms of resistance (that is, *kdr* and metabolic), although the specifics are different. For example, the *kdr* mutations are not at the same allelic locations in both species. This poses several questions.

- Why would two species of bed bugs that had diminished in importance in the late 1960s become prominent again in the late 1990s, with populations of both species around the world showing widespread pyrethroid resistance (Romero et al., 2009; Dang et al., 2015a,b, 2017)?
- Could all of these modern pyrethroid-resistant populations have originated from a location where both species existed sympatrically (Newberry et al., 1987)?
- If so, can it be presumed perhaps that they were subjected to the same long-term selection pressures from continual pyrethroid exposure?
- Could this have occurred in Africa where the ancestors of the modern bed bugs were initially subjected to selection by exposure to DDT and then later by the pyrethroids, due to anti-malaria programs that involved the use of indoor residual sprays and insecticide-impregnated bed nets?

Answering these questions will require access to numerous samples of both species from around the world, and characterization of these populations using various molecular methods, such as microsatellite technologies.

The obvious outcome of insecticide resistance has been treatment failures. Many of these failures can be linked back to inadequate pesticide efficacy testing where modern field strain bed bugs were not used. It is still the situation that most insecticide products have been evaluated using susceptible strains, or tested on long-standing laboratory strains that may have lost much of their resistance (Doggett and Lilly, 2015). Therefore a pesticide product may show good efficacy in the laboratory, but may not provide the desired control results in the field. In addition, laboratories use different protocols to evaluate insecticide efficacy, and discrepancy in test results is very common. Even the use of different exposure surfaces can result in variations in insecticide performance (Dang et al., 2017). Moreover, unlike *C. lectularius*, for which several standard susceptible strains exist, there is no standard susceptible strain available for *C. hemipterus*. It is important to realize that incorporating susceptible strains into pesticide efficacy evaluations adds another layer of controls to the experiment; the susceptible strains should easily succumb to all products tested.

The lack of standardization in bed bug insecticide efficacy evaluations has led to a number of industry groups to call for a global standard protocol for testing of bed bug insecticide products. In fact in 2012, the US Environmental Protection Agency (US EPA) attempted to do this with a release of a draft efficacy guideline (US EPA, 2012). However, these guidelines failed to present a logistically feasible testing algorithm (Doggett, 2017), which taught both researchers and government agency officials that efficacy guidelines need to be practical and achievable. Furthermore, the guidelines recommended that bed bug strains used must have a minimum resistance ratio of 100, a figure seemingly small when ratios of over 1.4 million in field strains had previously been reported (Lilly *et al.*, 2009). Subsequently in June 2017, the US EPA released the finalized version of the efficacy guidelines (US EPA, 2017), however again these contain a number of contentious points, which will make it challenging, if not impossible, for insecticide researchers to comply with (Doggett, 2017). For example, the guidelines state that the test organism to be used must be "…no later than the second lab-reared generation". Yet, most laboratory strains take several generations to establish in order to produce sufficient numbers to meet testing requirements. This means that a practical and achievable efficacy guideline has still yet to be presented to the pest management industry. Perhaps most surprisingly, the updated guidelines persisted with the requirement of bed bug strains used for efficacy testing of having a minimum resistance ratio of only 100, in spite of the literature suggesting that such strains would have a very low level resistance (Lilly *et al.*, 2009). This does mean that the consumer may not have a lot of confidence that a product tested under the new efficacy guidelines, will actually work in the field.

As noted in Chapter 29 on insecticide resistance, bed bugs possess a range of resistance factors. Thus the final outcome of what is measured as "resistance'" represents a range of genetic combinations and permutations (and possible synergistic effects) of individual components. This means there are too many probabilities for it to be practical (and possible) to test the efficacy of every unique bed bug strain that possess different combinations of resistance factors. Consequently, no matter what trials are undertaken, what may happen in a laboratory test may be different from what is observed in the field. Therefore, it seems logical to simplify efficacy evaluations. Ideally, one highly resistant strain should be evaluated in parallel with one known susceptible strain. Whatever the outcome of the trial, the caveat should be added that while the product may have been demonstrated efficacious in a laboratory trial against one strain, it may not be so in the field against others. In all cases, for all products, post-treatment monitoring should be encouraged to validate insecticide efficacy and eradication outcomes.

In addition to insecticide formulations, many conceptually marginally effective products (such as insecticide foggers, "natural oils", glue traps, heating devices, bed bug sniffers, "enzymes", chemical impregnated fabrics, and active monitors) have swarmed the marketplace in recent years (see Chapter 31). Many of these products have not been adequately evaluated in a scientific laboratory. Some manufacturers insist that their products should be employed as part of an IPM program. While IPM is a sound philosophy, trials involving these products have demonstrated that the products have very little efficacy when used alone. Therefore, it is very difficult to determine how much such a product would contribute to the IPM effort (if at all). No doubt more and more of these products will be introduced to the market over time, and such products need to be appropriately evaluated, and not invested in based solely on an interesting marketing gimmick. Remember, if there was a magic bullet, everyone would be using it, bed bugs would be history, and there would be no reason to read this book!

In light of the incomplete efficacy testing and ineffective products that have already appeared on the market, it becomes increasingly important that PMPs understand the limitations of every product they incorporate into their treatment program. It is imperative to know the weaknesses of a product in addition to how it works, because this will help the PMP understand how failures may occur and can be prevented. Education is a key strategy in combating the bed bug resurgence. PMPs require education in the best practices for bed bug eradication, and can become certified (in those countries that have bed bug certification) to demonstrate to the public that they are competent. The public needs to be taught how to avoid bed bugs, especially while travelling or visiting friends and family. Residents in apartments need to be educated on how to recognize bed bugs and encouraged to report bed bug encounters to the housing management. Accommodation managers need education on how to identify and manage bed bug introductions before they become infestations. Overall, everyone needs to take responsibility for bed bugs and educate themselves. Without education for all, bed bugs will win.

Summary

The current crop of bed bugs is highly resistant to insecticides and extremely difficult to kill. Failure to undertake bed bug management promptly results in infestations spreading and an increased cost to society (both in health and economics). At this time, bed bugs are still increasing in most nations. In the past, bed bugs were largely defeated by powerful and highly efficacious residual insecticides, such as DDT. These insecticides are no longer effective, so insecticides are unlikely to eliminate the bed bug problem this time. The development of new chemistries costs hundreds of millions of dollars, and there is not enough financial return for insecticide manufacturers to develop new molecules specifically for bed bug control alone. This means manufacturers can only look to innovative formulations and delivery systems. These new formulations using existing active ingredients may not be effective over time due to the bed bug's many and varied resistance mechanisms. Currently, there is no light visible at the end of the bed bug tunnel. In spite of the steady increase in the number of published studies that have deepened our understanding of bed bugs, the challenge of finding low-toxicity, cost-effective, and time-efficient management strategies remains. This means that bed bug control is prohibitively expensive to a large portion of our society, and while it is so, it can only be envisaged that bed bugs will continue to be an important insect pest of human society for many years to come.

References

Anonymous (2016) *U.K. and Germany Market Study on Bed Bug Control Services: Chemical Control Service Type Segment Expected to Gain Significant Market Share by 2026*, http://www.persistencemarketresearch.com/market-research/uk-and-germany-bed-bug-control-services-market.asp (accessed 21 December 2016).

Anonymous (2017) *Specialty Consultants: U.S. Structural Pest Control Market Surpasses $8 Billion*, http://www.pctonline.com/article/specialty-consultants-research-2017-market-report/ (accessed 11 April 2017).

Curl, G.D. (2016) *U.S. Structural Pest Control Market to Reach $10 Billion in 2020*, http://www.pctonline.com/article/sc-research-pest-control-market-report/ (accessed 11 February 2017).

Dang, K., Toi, C.S., Lilly, D.G., Bu, W. and Doggett, S.L. (2015a) Detection of knockdown resistance mutations in the common bed bug, *Cimex lectularius* (Hemiptera: Cimicidae), in Australia. *Pest Management Science*, **71** (7), 914–922.

Dang, K., Toi, C. S., Lilly, D. G., *et al.*, (2015b) Identification of putative *kdr* mutations in the tropical bed bug, *Cimex hemipterus* (Hemiptera: Cimicidae). *Pest Management Science*, **71** (7), 1015–1020.

Dang, K., Doggett, S.L., Lilly, D., Veera Singham, G., and Lee, C.Y. (2017) Effects of different surfaces and insecticide carriers on residual insecticide bioassays against bed bugs, *Cimex* spp. (Hemiptera: Cimicidae). *Journal of Economic Entomology*, **110** (2), 558–566.

Doggett, S.L. (2005) *A Code of Practice for the Control of Bed Bug Infestations in Australia (draft)*, Department of Medical Entomology and The Australian Environmental Pest Managers Association, Sydney.

Doggett, S.L. (2013) *A Code of Practice for the Control of Bed Bug Infestations in Australia*, 4th edn, Department of Medical Entomology and The Australian Environmental Pest Managers Association, Sydney.

Doggett, S.L. (2016) Bed bug survey – are we biting back? *Professional Pest Manager*, **Aug/Sep**, 28–30.

Doggett, S.L. (2017) Bed bug insecticides, do they do what it says on the tin? *Pest*, **Aug/Sep**, 29–31.

Doggett, S. L. and Lilly, D. (2015). Bed bugs on the label…but what does this really mean? *Pest*, **37**, 34–35.

Doggett, S.L., Orton, C.J., Lilly, D.G. and Russell, R.C. (2011) Bed bugs: the Australian response. *Insects*, **2**, 96–111.

Doggett, S.L., Dwyer, D.E., Peñas, P.F. and Russell, R.C. (2012) Bed bugs: clinical relevance and control options. *Clinical Microbiology Reviews*, **25** (1), 164–192.

Koganemaru, R. and Miller, D.M. (2013) The bed bug problem: past, present, and future control methods. *Pesticide Biochemistry and Physiology*, **106** (3), 177–189.

Kim D.Y., Billen, J., Doggett, S.L. and Lee, C.Y. (2017) Differences in climbing ability between *Cimex lectularius* and *Cimex hemipterus* (Hemiptera: Cimicidae). *Journal of Economic Entomology*, **110** (3), 179–1186.

Lilly, D.G., Doggett, S.L., Zlaucki, M.P., Orton, C.J. and Russell, R.C. (2009). Bed bugs that bite back, confirmation of insecticide resistance in Australia in the Common bed bug, *Cimex lectularius*. *Professional Pest Manager,* **Aug/Sep**, 22–24, 26.

Newberry, K., Jansen, E.J. and Thibaud, G.R. (1987) The occurrence of the bedbugs *Cimex hemipterus* and *Cimex lectularius* in northern Natal and Kwazulu, South Africa. *Transactions of the Royal Society of Tropical Medicine and Hygiene*, **81** (3), 431–433.

Orton, C.J. (2009). *Guidelines for the Establishment and Management of AEPMA Code-Of-Practice Working Parties (V2.1.3)*, http://medent.usyd.edu.au/bedbug/guidelines_cop_wp.pdf (accessed 3 April 2017).

Romero, A. and Anderson, T.D. (2016) High levels of resistance in the common bed bug, *Cimex lectularius* (Hemiptera: Cimicidae), to neonicotinoid insecticides. *Journal of Medical Entomology*, **53** (3), 727–731.

Romero, A., Potter, M.F. and Haynes, K.F. (2009) Evaluation of piperonyl butoxide as a deltamethrin synergist for pyrethroid-resistant bed bugs. *Journal of Economic Entomology*, **102** (6), 2310–2315.

Usinger, R.L. (1966) *Monograph of Cimicidae (Hemiptera – Heteroptera)*, Entomological Society of America, College Park.

US EPA (2012) *Draft Product Performance Test Guidelines OCSPP 810.3900: Laboratory Testing Methods for Bed Bug Pesticide Products*. US Environmental Protection Agency, https://www.regulations.gov/contentStreamer?documentId=EPA-HQ-OPP-2011-1017-0006&contentType=pdf (accessed 3 April 2017).

US EPA (2017) *Test Guidelines OCSPP 810.3900: Laboratory Testing Methods for Bed Bug Pesticide Products*. US Environmental Protection Agency, https://www.regulations.gov/contentStreamer?documentId=EPA-HQ-OPP-2011-1017-0035&contentType=pdf (accessed 5 September 2017).

Index

a

advertising   13, 17, 19, 81–83, 217, 222
AEPMA (Australian Environmental Pest Management Association)   211–213, 225, 424
Africa
  bed bug decline   88–89
  Congo   87
  Egypt   1, 27, 87, 101, 133, 274
  Ethiopia   89, 90, 92
  Gambia   87, 274
  history of bed bugs   87–88
  insecticide resistance   87, 89, 91, 136, 273–275, 278–279
  Ivory Coast   274
  Kenya   87, 89, 274, 278–279
  KwaZulu-Natal   2, 88–90
  laws   88, 92
  Libya   89
  Namibia   88, 89, 91
  Nigeria   88–90, 92
  resurgence   2, 91–92, 102, 177, 424
  Sierra Leone   89, 90, 111
  Somalia   87, 274
  South Africa   2, 87–91, 274
  strategies to combat the resurgence   92
  Tanzania   88–92, 273
  Tunisia   274, 275
  Zanzibar   87, 89
  Zimbabwe   87, 88, 274
Afrocimex constrictus   154, 194
aggregation (*see* behavior)
aircraft
  aviation entomology history   363
  challenges to control   364
  eradication strategies   76, 300, 365
  infestations on aircraft   76, 363
  pest management protocols   366
Aldrovandi, U.   10
allergens (from bed bugs)   109, 112, 133–134, 260
anachoresis   151–153, 155
Aphrania vishnou   96
apyrase   110, 112
armed forces infestations   15, 19, 20, 22, 45, 69, 70, 75, 82, 96, 97, 278, 363, 364, 369
Asia
  China   9, 10, 28, 69–75, 133, 370, 372, 409, 410
  history of bed bugs   69–70
  Hong Kong   72, 274
  Indonesia   74–76, 274, 409, 410
  insecticide resistance   74, 274, 275, 278–279, 298
  Japan   10, 69, 71–74, 274, 275, 409, 410
  laws (*see* legal)
  Malaysia   70, 72–76, 120, 274, 275, 278–279, 285, 333, 409, 410, 412
  resurgence   70–74
  Singapore   72–76, 274, 333, 409, 410, 412
  South Korea   274, 278
  strategies to combat the resurgence   74–76
  Thailand   72, 73, 75, 76, 135, 274, 275, 278–279, 285, 298, 409, 410
attractants
  carbon dioxide (CO_2)   155, 169, 245, 246, 248, 249, 253, 315
  heat   169, 248
  semiochemicals (host odors)   169–170
Australia
  bed bug decline   82–83, 421
  Code of Practice   84, 211–214, 424
  fiscal impacts   141, 142, 144, 145

Advances in the Biology and Management of Modern Bed Bugs, First Edition.
Edited by Stephen L. Doggett, Dini M. Miller, and Chow-Yang Lee.
© 2018 John Wiley & Sons Ltd. Published 2018 by John Wiley & Sons Ltd.

Australia (*cont'd*)
  history of bed bugs   81–80, 358
  insecticide resistance   83, 179, 217, 274, 275, 278–279, 314
  laws (*see* legal)
  resurgence   71, 73, 83–84, 102, 174, 225, 315, 335, 359
  strategies to combat the resurgence   84–85, 334, 421–423

b

backpacking lodges   73, 84
bacteriome   152
Bangladesh   74, 102
Bartonella quintana   90, 120
bat bugs   52, 65, 95, 96, 119, 151, 153, 154, 194, 267
bats   9, 52, 59, 65, 95, 101, 102, 119, 121–122, 152, 173–177, 185, 200, 378
BBF (Bed Bug Foundation)   65, 218–219
Beauveria bassiana   268
bed bug bite reactions (in humans)
  anemia   96, 134
  bite diagnosis   113–114
  bite treatment   114
  bullous lesions   109, 111–113
  cutaneous reactions   109–111
  dermatological complications   111
  dermatopathology   112–113
  immune response   112
  secondary infections   109, 111, 114, 127, 136, 411
  sleep disturbance   96, 109, 111, 127–131, 133, 134, 416
  systemic reactions   112
bed bug saliva   112, 202
  Factor X   110, 112
  nitrophorin   110
beds and bed design   15, 21
behavior (*see also* olfaction)
  aggregation   156, 163–168, 248, 249, 259, 260, 262, 300, 312, 341, 344
  arrestment behavior   165, 167, 168, 248, 249, 260
  dispersal   156
  egg laying   155
  harborage seeking   156–158
  host seeking   155–156, 169–170
  inbreeding   151, 152, 158, 174–175
  reproduction   168–169
  reproductive senescence   154
  traumatic insemination   4, 37, 120, 153, 154, 185, 187, 195
Best Management Practices for Bed Bugs (BMP)
  acceptance   223
  history and development   221, 326
  key elements   222–223
  marketing and adoption   223
  target audience   221
Biocidal Products Regulation   65, 398
Blattella germanica   175, 176
blood feeding   4, 118, 121, 151, 152, 164, 184–187, 199, 204, 268, 377
Borel, B.   27, 35
botanical insecticides
  cedar   291, 294–297, 300
  cinnamon   291, 296, 300
  citric acid   296
  clove   292, 295–297, 300
  d-limonene   292, 296
  geranium   292, 297, 300
  lemon grass   292, 296, 297, 300
  neem   292, 296, 297, 300, 301, 315
  peppermint   292, 295–297, 300
  rosemary   292, 296, 297, 300, 301, 315
  rotenone   17, 69, 292, 297
  ryania   292, 297
  thyme oil   292, 297, 300
BPCA (British Pest Control Association)   63, 397
Brazilian chicken bug (*see Ornithocoris toledoi*)
Bugg Hunting (cartoon)   11, 28

c

Cacodmus indicus   95
Canada
  Montreal   47, 48, 127, 129–131
  resurgence   46–48
  strategies to combat the resurgence   46–48
  Toronto   10, 47, 48, 349
  Vancouver   47–48, 325
  Winnipeg   47, 129
canine scent detection (*see* detection)
carbamates
  bendiocarb   289, 295, 298, 379
  carbaryl   290, 378, 379
  chlorpyrifos   75, 135, 372, 379
  fenobucarb   289, 295, 298
  fenthion   70, 75, 97, 372

malathion   20, 45, 75, 96–98, 102, 289, 294–296, 370, 378, 379
propoxur   55, 70, 74, 75, 97, 290, 296, 298, 372, 379
resistance   273, 274, 298
CDC (US Centers for Disease Control and Prevention)   46, 391
Chagas disease   51, 52, 119, 283, 411
chemical control
    aerosols   135, 285, 294, 296, 298–301, 303, 314, 343, 352, 364, 371
    desiccant dusts   46, 74, 85, 98, 141, 184, 234, 236, 258, 292, 293, 297, 299–301, 304, 312, 314–315, 345, 352, 366, 371, 379, 394, 398
    dusts (non-desiccants)   285–288, 292, 297, 298, 301
    flammable liquids   16, 135, 142
    fumigants (*see* fumigants)
    historical   1, 2, 12–20, 45, 60, 61, 69, 70, 83, 87, 95–98, 135, 228, 273, 277, 286, 300, 303, 313, 323, 358, 369, 376, 378, 379, 423–425
    insecticidal bombs & mists   135, 302, 315, 328, 352, 365
    liquid sprays   285, 296, 298, 300–301, 343
    residual sprays   2, 17, 19, 70, 75, 76, 96, 212, 273, 298, 300, 314, 365, 376, 379, 398, 424, 425
    wettable powders   286, 287, 289, 294, 301
chemical exposure (adverse human reactions)   135
chemoreception   163, 165–167
chlorfenapyr (*see* pyrrole)
Cimex adjunctus   267
Cimex hemipterus
    aircraft infestations   363
    anemia (in humans)   134
    asthma   134
    blood feeding   151, 152, 201, 202
    dehydration   185
    distribution   3, 45, 52, 56, 64, 70, 72, 73, 81, 83, 87, 88, 90, 95, 97, 102, 103, 333, 376
    genetics   173
    healthcare infestations   359
    hosts   174, 200
    insecticide efficacy studies   285–291, 293–297, 299, 304, 378
    morphology   164, 245, 312, 423
    origin   2, 3, 87, 152
    pathogens   52, 119, 120
    pheromones   166–168, 312
    poultry pest   375, 376, 378
    resistance   74, 87, 96–98, 122–123, 273, 278–279, 298, 302, 304, 313, 424

resurgence   1, 2, 45, 56, 64, 70, 72, 73, 81, 83, 95–97, 103
salivary components   110–112, 118
shipping infestations   370
starvation   186
symbionts   194
tarsal adaptations   199, 245, 312, 423
temperature tolerance   186, 261
Cimex himalayanus   95
Cimex incrassatus   194
Cimex insuetus   95
Cimex lectularius (1–427)
    aircraft infestations   4, 76, 300, 363–366
    allergens & asthma   109, 112, 133–134, 260
    anemia (in humans)   96, 134
    bat association   59, 102, 121–122, 152, 173–174, 378
    bite complications (in humans)   111–112
    bite diagnosis (in humans)   113–114
    bite management (in humans)   114
    bite reactions (in humans)   109–113
    blood feeding   118, 121, 151, 152, 185–187, 199–204, 377
    decline   1, 15–16, 61–62, 82–83, 88–89, 96
    dehydration   183–187, 259, 262, 300
    distribution   3, 9–11, 45–47, 51–53, 55, 56, 59–61, 69–70, 73, 81, 87–88, 90, 95–96, 101–103
    etymology   10–11, 36
    genetics (*see* population genetics)
    healthcare infestations   1, 4, 15, 21, 70, 71, 82, 84, 96–98, 135, 141, 233, 235, 237, 250, 253, 262, 348, 357–360
    historical management strategies   1, 11–22, 45, 60–61, 69, 70, 81, 87, 88, 96, 135, 228, 258, 260–262, 268, 358, 369, 378
    hosts (*see* hosts)
    hotel (hospitality) infestations   1–4, 13, 15, 21, 35, 36, 52–55, 62, 63, 70–74, 81, 84, 91, 96, 97, 102–104, 139, 140, 142–144, 218, 222, 227, 241, 244, 250–252, 260, 262, 311, 312, 324, 325, 327, 333, 348, 351–353, 358, 385, 388–390, 392, 393, 399, 403, 405, 422
    modern management strategies (*see* Part V)
    morphology   164, 245
    origin   1–3, 9, 59, 62, 81, 87, 101, 104, 152, 173–174
    pathogens   51, 52, 88, 96, 117–122
    pheromones (*see* pheromones)
    poultry pest   4, 237, 313, 375, 377, 378

Cimex lectularius (cont'd)
  resistance   2, 20, 45, 74, 179–180, 184, 186, 195, 273–279, 298, 299, 301, 302, 314, 376, 379, 424
  resurgence   1, 2, 45–49, 53–55, 62–64, 71–73, 83–84, 89–91, 96–97, 102–104
  salivary components   109–112, 118, 120, 202
  shipping infestations   4, 10, 13, 15, 34, 53, 60, 69, 75, 142, 369–372
  social disadvantaged infestations (*see* low income housing)
  starvation   152, 155, 156, 158, 183–187, 259, 377
  symbionts   121, 152, 186, 187, 193–195, 316
  temperature tolerance   21–23, 52, 185–186, 261–262, 266–267, 316
  train infestations   4, 15, 62, 70, 72, 73, 75, 371–372
Cimex limai   52
Cimex pipistrelli   65, 96, 119
Cimex usingeri   95
Code of Practice (Australia)
  benefits   213–214, 421–422
  development   212
  history and aims   237–238
  key elements   212–213
  why needed   211
Crassicimex sexualis   153
Cruikshank, I.   11, 28
cruise ships (*see* shipping)
Culex quinquefasciatus   164

d

Danaus plexippus   156
Democritus   9
desiccant dusts   46, 74, 236, 258, 292, 297, 299, 300, 312, 314, 352, 371, 379
  diatomaceous earth
  limestone   292, 297, 299
  silicon dioxide (silica gel)   85, 236, 293, 297, 299–301, 314, 379, 394, 398
detection   14, 74, 89, 242–243, 311, 313, 335, 365
  active monitors   156, 168, 248–249, 311, 312, 360
  canine scent detection   74, 145, 221–223, 228, 235, 243–244, 312, 354, 365
  efficacy   235, 245, 247–251
  electronic noses   242
  eradication determination   252–253
  importance of detection   46, 98, 141, 222, 233–235, 241, 245–246, 250, 387, 423
  inspection–proactive   251–252, 352, 354
  inspection–reactive   252
  inspection tools   241, 242, 245, 248
  methods   21, 233, 241–242, 245, 248, 260, 313
  non-traditional settings   253
  passive monitors   21, 74, 168, 247–248, 260, 311, 335, 354, 387
  pitfall traps   21, 234, 245–250, 253, 260, 312, 354, 423
  resident interviews   242
  sticky traps   74, 245, 247, 312
  trap limitations   156, 228, 247, 248, 311, 312, 317, 352, 354, 360, 363, 365, 423
  trap types   21, 244–249
  visual inspection   13
diatomaceous earth   46, 74, 236, 258, 292, 297, 299–300, 312, 314, 352, 371, 379
dichloro-diphenyl trichloroethane (DDT)   1, 2, 15, 17–20, 23, 45, 46, 60, 61, 69, 70, 83, 87, 89, 95–98, 228, 273, 277, 286, 300, 303, 313, 358, 376, 378, 379, 423, 424, 426
diseases of humans (non-infectious),
  asthma   133–134
  beriberi   118, 135, 358
  cancer   118, 135
  pellagra   118, 135
dispersal (*see* behavior)

e

education
  adult education   234, 327–328, 334
  client & community education   23, 47–48, 81, 84, 85, 92, 213, 218, 222, 225, 227–228, 230, 233, 234, 237, 258, 323, 327–329, 334, 347–349, 353, 359, 371, 387, 389–391, 421, 422, 425
  educational materials   47, 48, 221, 226, 234
  educational strategies   323–324
  lack of education   18, 47, 104, 228
  measuring educational impacts   328
  media   326–327
  prevention (of bed bugs)
  public awareness   323
  social media   327
  training (of PMPs)   18, 63, 65, 69, 71, 75, 81, 84, 85, 92, 104, 139, 213, 218, 221–223, 225, 229, 258, 264, 265, 317, 324–236, 329, 386, 397, 410, 421–423, 425
  US educational programs   47, 223, 324–326
egg laying (*see* behavior)
embryogenesis   154
EPA (US Environmental Protection Agency)   46, 223, 229, 233, 300, 305, 326, 386, 387, 391, 425

Europe
  bed bug decline   61–62
  Czechoslovakia   61
  Denmark   60–63, 274, 275, 298
  Finland   15
  France   9, 10, 34, 36, 64, 65, 134, 178, 179, 275, 278
  Germany   9–11, 14, 34, 61, 63, 145, 219, 275, 421
  history of bed bugs   9–10, 59–60
  Hungary   61, 274
  insecticide resistance   63–65, 273–275, 298, 399
  Italy   9, 62, 212, 274, 358
  laws (*see* legal)
  Norway   63
  resurgence   62–65
  Slovakia   61, 62, 64, 65
  Spain   133
  strategies to combat the resurgence   65
  Sweden   14, 15, 60, 62, 64
  Switzerland   62, 63
  United Kingdom (*see* United Kingdom)
European Code of Practice for Bed Bug Management
  benefits   219
  development   218
  history and aims   218–219
  key components   218
  why needed   217
exuviae   156, 166, 167, 242–244, 260

f
Fabricius, J.C.   29
feces (bed bugs)   113, 119, 120, 166–168, 243, 244, 260, 351
fecundity   151, 152, 154, 155, 185, 194, 195, 201, 261, 291, 302, 314–316
Ferguson, E.W.   82
Fewell, C.L.   22, 262
Fiji   83
fipronil   291, 295, 379
fires   135, 142, 354
fiscal impacts
  brand damage   139, 143–144, 227, 241, 331, 351, 352, 354
  budgetary impacts   140–141
  direct costs   140–142
  equipment costs   144–145, 244, 247, 248, 250, 266
  future fiscal impacts   141, 145, 311, 421, 422
  healthcare industry   357
  home owner   141
  hospitality industry   139, 141–144, 227, 352, 392–393
  indirect costs   140, 142, 145
  insurance costs   143–144
  legal expenses (*see* litigation)
  multi-unit housing costs   140–142, 343, 344
  pest management company costs   141, 144–145, 227, 229, 266
  proactive monitoring and treatment costs   141–142
  profit loss   139, 140
  range of costs   139–145
  retail sector   143
  world economy   139, 145, 421, 422
flightlessness   121, 153
Flinders, M. Lt.   81
foraging cues (*see* behavior)
fumigants
  acetaphenone   297
  carbon dioxide   258, 293, 297, 300, 301, 317
  hydrogen cyanide (hydrocyanic acid)   14, 15, 18, 45, 60, 135, 228, 273, 323, 369, 423
  methyl bromide   76, 263, 300, 301
  ozone   293, 300, 315
  phosphine   135, 301, 423
  sulfur (=sulphur)   13, 16–18, 228, 273, 323, 378
  sulfuryl fluoride   75, 293, 297, 300, 315, 370
  Zyklon B   18
fungi   120, 121, 258, 268, 410
future predictions
  answering the origin of resistance   2, 424
  bed bug infestations   46, 145, 421
  efficacy testing standards   229, 305, 425
  fiscal impacts   145, 421–422
  need for determining fiscal impacts   422–423
  need for epidemiological data   422–423
  need for government support   422–423
  need for industry standards   421–425
  research needs   423
  reversing the resurgence   421–426

g
Geisel, T.   18
genetics (*see* population genetics)
Grant, R.   35

h
Haematosiphon inodorus   37, 194, 375–378
Harlan, H.   46

healthcare facilities
    challenges to control   357–358
    contemporary infestations   27, 84, 233, 357–359
    eradication strategies   359–360
    history of bed bugs   82, 96, 358
    methods of introduction   357–359
heat (control)
    clothes dryer   142, 237, 263, 266
    heat chambers   144–145, 237, 263, 360
    methods and equipment   261–266, 316
    steam   14, 22, 46, 60, 69, 74, 98, 145, 217, 229, 257, 261–263, 316, 336, 349, 354, 358–360, 366, 371
    thermal death points (of bed bugs)   185, 236, 237, 261–262, 267
    treatment limitations   262–266, 316
hematophagy (*see* blood feeding)
histamine   112, 114, 165, 167–168, 249
Hopi Indians   11, 37, 376
hospitals (*see* health care facilities)
host odors (*see* attractants, odorants, semiochemicals)
hosts   59, 101, 102, 120–122, 151–153, 155, 158, 163, 165, 169, 173, 174, 184, 188, 193, 204, 241, 364, 372, 375–378
host seeking (*see* behavior)
hotels
    challenges to control   351–352
    eradication strategies   352–354
    fiscal impacts   139, 142–144, 352, 392–393
    management policy   213, 234, 354
    treatment failures   352
hydrogen cyanide (*see* fumigants)

i

immune system (of bed bugs)   120–121, 153, 154
impregnated strips   298, 302, 315
inbreeding (*see* behavior)
India
    bed bug decline   96
    history of bed bugs   95–96
    insecticide resistance   95–98, 274, 278–279
    resurgence   96–97
    strategies to combat the resurgence   97–98
Indian subcontinent
    Pakistan   98, 102
    Sri Lanka   98, 102, 274, 275, 278
infectious diseases
    arbovirus   117–118, 121, 122
    Chagas disease   51, 52, 119, 283, 411

filariasis   52
hepatitis B   88, 118
hepatitis C   117
HIV (Human Immunodeficiency virus)   117, 118
leishmaniasis   96
malaria   2, 9, 20, 52, 87–89, 95, 96, 136, 364, 424
MRSA (methicillin resistant *Staphylococcus aureus*)   118, 119
plague   52, 53, 96, 424
VRE (vancomycin resistant *Enterococcus*)   118
insect growth regulators (IGRs)   285, 291, 299, 315, 344
    hydroprene   291, 295, 296, 299
    methoprene   291, 296, 299
    triflumuron   299
insecticide efficacy testing
    arena size   303
    bed bug strain/species   304, 424, 425
    methodology   304, 305, 424, 425
    physiological status   304
    substrate   303, 304
insecticide formulations
    aerosols   135, 285, 296, 298–301, 303, 314, 315, 328, 343, 352, 364, 365, 371
    dusts   16, 69, 97, 285–287, 292, 293, 297, 298, 301, 371
    fumigants (*see* fumigants)
    impregnated fabrics   302, 314, 425
    insect bombs   20, 302, 315
    liquid sprays   16, 18, 46, 70, 218, 285, 294–296, 298–301, 314, 315, 343
    repellents   135, 164, 168, 297, 302, 409
insecticide resistance
    behavioral   275
    cross resistance   2, 87, 89, 302
    cuticular   277, 298, 300
    esterase   98, 276, 277
    fitness cost   278, 279
    general discussion   2, 46, 55, 63–65, 70, 74, 75, 83, 84, 87, 89, 91, 95–98, 102, 136, 179–180, 183, 188, 193, 211, 217, 219, 228–229, 257, 273–285, 298–300, 302–305, 313, 314, 316, 343, 344, 380, 399, 423–426
    geographical reports   179–180, 274–275
    Glutathione S-transferases (GST)   276
    haplotype A   179, 180, 278
    haplotype B   74, 179, 180, 278
    haplotype C   74, 179, 180, 278
    haplotype D   180, 278
    knockdown (*kdr*)   98, 179–180, 277–279, 424

L1014F mutation 98, 279
L925I mutation 74, 179, 278, 279
metabolic 276–277, 424
M918I mutation 279
neonicotinoid 75, 273, 275, 276, 285, 290, 298–299, 314
P450 276–278
piperonyl butoxide (effects on resistance) 276, 298
pyrethroid 20, 55, 74, 75, 89, 91, 98, 179, 186, 195, 212, 228, 262, 273–279, 285, 298, 300, 302, 304, 313, 314, 343, 380, 423, 424
pyrrole 273, 275
susceptible 2, 97, 179, 195, 228, 275–277, 279, 302, 304, 314, 398, 409, 424, 425
target-site resistance 276–279
V419L mutation 74, 179, 278, 279
insecticides (*see also* listings under insecticide classes)
 applications 20, 48, 55, 70, 113, 135, 145, 223, 228–229, 234, 257, 266, 267, 273, 301, 313–315, 343–345, 352, 364–366, 398
 combination products 75, 293, 294, 298, 299, 314, 379
 insecticide classes 228, 274–276, 285–294, 298–300, 302, 313, 314, 424
 registration exceptions 300
 resistance management strategies 279, 399
 systemic application 316
inspection (*see* detection)
integrated pest management (IPM) (*see also* non-chemical control)
 challenges in different housing 230, 233, 337, 342, 347
 clutter & hoarding 230, 337
 general discussion 4, 48, 74, 141, 212, 213, 221–223, 258–260, 266, 267, 324, 326, 334–338, 348, 354, 359, 387, 391, 392, 423, 425
 limitations 48, 212, 317, 338, 348, 425
 treatment preparation 47, 48, 104, 129, 213, 222, 230, 236, 265, 326, 333, 336, 344, 345, 371, 390, 415

j
Japan Pest Control Association 69, 71, 73

l
laboratory maintenance, rearing containers and harborages 199–200
 artificial blood source 118, 199, 204
 blood source 186, 200–202
 feeding techniques 200–203
 feeding units 202–203
 general discussion 4, 199–208, 304
Latin America 51–57
 Argentina 55
 Brazil 51–55
 Colombia 3, 274
 general discussion 51–57
 Mexico 55, 325, 375, 376
 Venezuela 55, 274
legal
 Asian bed bug lawsuits 411
 Asian bed bug litigation 411–412
 Asian pesticide registration and use 409–410
 Asian PMP legal requirements 410
 Australian bed bug lawsuits 406
 Australian pesticide application requirements 403–404
 Australian pesticide registration 403–404
 Australian PMP legal requirements 404
 Australian public health laws 404–405
 Australian public housing laws 405–406
 expert witness bed bugs in court 415–416
 expert witness definition 413–414
 expert witness role 414
 expert witness testimony 414–415
 legislation & laws 60, 96, 390, 397, 399, 403–405
 litigation 3, 4, 139, 143, 237, 241, 311, 351, 385, 386, 388, 394, 403, 411–414
 pest control contracts 65, 103, 139, 222, 337, 344, 406, 416
 treatment documentation 222, 227, 389
 UK bed bug related laws 399–400
 UK local authority powers 399–400
 UK pesticide registration 398–399
 UK PMP training requirements 397–398
 US bed bug laws 385–395, 397–401
 US bed bug lawsuits 392–393
 US bed bug related laws 387–391
 US legal requirements for PMPs 386–387
 US pesticide application requirements 386–387
 US pesticide registration 386
Leptocimex boueti 151
Leptocimex inordinatus 96
Lewis, S. 32
Lewis, W.E. 34
Linnaeus, C. 10
litigation (*see under* legal)
 lawsuit 15, 143, 227, 385, 388, 392, 393, 416
 monetary compensation 15, 411
 punitive damages 144, 393

longevity (of bed bugs)   201
low-income housing
　challenges to control   337
　contracts   337
　education   47, 324, 334, 391
　eradication strategies   442
　follow up inspections   251, 334, 337, 389, 391
　general discussion   1, 4, 47, 141, 251, 324, 333–339, 422
　impact of bed bugs   131, 141, 422
　infestation identification   334, 335
　treatment preparation   47, 333, 336
lysozyme   120, 121, 154

m

management
　DIY   64
Matthioloi, A.   9
media reports   89, 139, 142, 143, 226, 327, 370
mental health impacts
　anxiety   127–129, 131
　depression   127–129, 131
　general discussion   111, 122, 127–132
　social isolation   127, 129
　stressors   127
　suicide   128
　workplace performance   129
mercury chloride   16, 17
mesospermalege   153
Metarhizium anisopliae   120, 268
metoxadiazone   291, 296
Mexican chicken bug (*see Haematosiphon inodorus*)
　general discussion   37, 194, 375–376
　hosts   375
　poultry pest   375–376
microbe transfer (during mating)   153, 154
Middle East
　Bahrain   103
　general discussion   101–106
　history of bed bugs   101–102
　Iran   103
　Iraq   102
　Israel   102, 179
　Kuwait   101, 102
　Oman   103
　Qatar   103
　resurgence   102–103
　Saudi Arabia   103
　strategies to combat the resurgence   101
　United Arab Emirates   103
monitoring (*see* detection)
Mouffet, T.   1, 60
Muller, P.   18
multi-unit housing
　challenges to control   230, 341–342, 345
　eradication strategies   230, 343
　financial limitations   342–343
　fiscal impacts   141
　general discussion   47, 140, 144, 228, 230, 237, 341–345, 365, 415
　management limitations   343–344
Mundy, T.   32, 33
mycetomes   194
Myotis   59

n

neonicotinoids
　dinotefuran   290, 296, 298, 379
　general discussion   75, 273, 275, 276, 285, 290, 298, 299, 314
　imidacloprid   55, 75, 290, 293–296, 298, 299, 379
　thiamethoxam   75, 294, 296, 298
Newberry, K.   2, 88–90, 177, 274, 286, 289, 295, 424
non-chemical control
　bean leaves   21, 260, 317
　bed bug exclusion   92, 259
　biological control   258, 268
　burning   17, 18, 323, 378
　clothes dryer/washer   142, 237, 263, 266
　container heat   263–265
　crushing   11
　disposal   222, 260
　dry heat   237, 261, 262, 264–266
　freezing   261, 266, 267, 316, 317, 326, 359
　general process   258
　heat   46, 48, 74, 98, 217, 229, 257, 261–264, 266, 267, 316, 336, 343, 360, 364
　historical   258–261
　hot water   13, 16, 21, 74, 358
　lethal temperatures   21–23, 185, 186, 222, 237, 261, 262, 265–267
　mattress encasements   82, 235–236, 259, 314, 336, 348, 394
　miscellaneous   268
　physical removal   11, 259–260
　rational for use   257

slum clearance   14
steam   14, 22, 46, 60, 69, 74, 98, 145, 217, 229, 257, 261–263, 267, 316, 317, 336, 349, 354, 358–360, 366, 371
vacuuming   46, 48, 74–76, 229, 257, 260, 266, 316, 317, 344, 345, 359, 360, 366, 371
whole-room heat treatment   262, 263, 265–266, 316
NPMA (US National Pest Management Association)   46, 102, 221–224, 244, 324–327, 334, 372, 391

o

Oeciacus hirundinus   194
Oeciacus vicarious   194
office infestations   102, 335, 358
olfaction
   general discussion, 163—165
   molecular basis   164–165
   odorant receptors   163–165
   odorants   163, 164, 169, 170
   olfactory receptor neurons (ORNs)   163, 164, 166
   olfactory sensilla   163–165, 168, 170
Orco   163, 164
organochlorines
   BHC   69, 96, 97
   DDT   1, 2, 15, 17–20, 23, 45, 46, 60, 61, 69, 70, 83, 87, 89, 95–98, 228, 273, 277, 286, 300, 303, 313, 358, 376, 378, 379, 423–425
   dieldrin   70, 87, 89, 96, 97, 102, 286
   fenitrothion   74, 75, 88, 289, 295, 298
   general discussion   1, 2, 61, 97, 228, 273, 274, 276, 313, 323, 376
   lindane   20, 75
   propetamphos   74, 289, 296, 298
   resistance   2, 87, 97, 273, 274, 276, 279, 313
organophosphates
   chlorpyrifos   75, 135, 372, 379
   diazinon   20, 70, 97, 288, 295, 298
   dichlorvos (*see also* impregnated strips)   97, 258, 288–289, 295, 297, 298, 301, 302, 315, 379
   fenitrothion   74, 75, 88, 289, 295, 298
   fenthion   70, 75, 97, 372
   folithion   97
   general discussion   1, 45, 55, 65, 70, 74, 75, 97, 98, 102, 228, 273, 274, 276, 285, 288, 289, 298, 304, 323, 379
   malathion   20, 45, 96, 97, 102, 289, 294–296, 298, 370, 378, 379
   pirimiphos-methyl   289, 296, 298, 379
   propetamphos   74, 289, 296, 298
   tetrachlorvinphos   289, 298, 379
   trithion   97
Ornithocoris pallidus   375, 376
Ornithocoris toledoi   375, 376
Orton, C.   82, 358
Orwell, G.   14
ovaries   152–155, 187, 194

p

Papua and New Guinea   83, 376
paragenital system   153
Park, R.   34, 82
pathogen transmission (*see also* Vector competence)   51, 52, 96, 117–122, 153
Pediculus humanus   177, 219
Phaseolus vulgaris (*see also* non-chemical control, bean leaves)   21, 260
pheromones
   aggregation   156, 164–168, 260, 312
   alarm   153, 166, 167, 169, 248, 249, 314
   dimethyldisulfide (DMDS)   168
   dimethyltrisulfide (DMTS)   168
   sex pheromones   168
   volatile pheromone components (VPCs)   168
phosphine deaths (*see also* fumigants, phosphine)   135, 301, 423
physiology
   blood feeding   112, 121, 151–152, 185, 187
   cold tolerance   184, 186
   heat shock proteins   185, 186, 261
   heat tolerance   183–186
   juvenile hormones   187
   phagostimulant   170, 204
   starvation   152, 155, 156, 158, 183–186, 259, 377
   vitellogenin   187
   water stress   153, 184–186
piperonyl butoxide   55, 276, 283, 298, 301, 314
Pliny the Elder   9
popular culture
   erotica   27, 37
   figurative arts   28–30
   linguistics   36
   literature   27, 32–34
   music   27, 30, 34–35
   poetry   27–28
   postcards   28, 29, 31, 37
   television   27, 35–36
   theatre   27, 30

population genetics 180
　general discussion 173–182, 378
　genetic variations among populations 175–177
　genetic variations within populations 174–175
　heteroplasmy 177–179
　insecticide resistance 179–180
　invasive populations 174
　microsatellites 173
　mitochondrial DNA (mtDNA) 173, 174, 177–179
　Next Generation Sequencing (NGS) 121, 173, 178
　Single Nucleotide Polymorphisms (SNPs) 180
poultry
　bed bug control in poultry facilities 378–379
　bed bug dispersal in poultry facilities 378
　bed bug insecticide susceptibility from poultry facilities 379–380
　Cimex hemipterus 102, 200, 375, 376, 378
　Cimex lectularius 200, 376–379
　general discussion 102, 200, 237, 313, 375–382
　Haematosiphon inodorus 37, 194, 375, 376
　history of bed bug infestations 375
　impacts of bed bugs 377
　Ornithocoris toledoi 375, 376
poverty/poor (*see also* low-income housing) 32, 33, 82, 89, 92, 119, 134, 327
prevention
　desiccant dusts 46, 98, 141, 184, 236, 300, 301, 304, 314, 315, 345, 366
　early detection (*see also* detection) 46, 98, 141, 233, 234, 241, 247, 311, 352, 354, 359, 360, 365, 387
　education 234, 258, 312, 323–330
　general discussion 12, 14, 47, 51, 56, 103, 114, 131, 141, 213, 233–239, 258, 326–328, 348, 385, 387, 389–391, 399
　harborage minimization 233, 258, 389
　heat 236–237
　hygiene 348
　IPM 141, 258, 326, 348, 387, 391
　limiting bed bug spread 235
　management policies 213, 237, 348
　mattress encasements 235–236, 348
Primicimex cavernis 153
prison infestations 70, 88–90, 102, 120
PTSD (Post-traumatic stress syndrome) 128, 129, 131
pyrethrin 2, 55, 69, 87, 97, 262, 287, 291–297
pyrethroids
　alphacypermethrin 286, 295
　bifenthrin 55, 286, 294, 295, 298, 379
　cyfluthrin 55, 75, 279, 286, 293–298, 301, 372, 379
　cypermethrin 55, 75, 286, 294–297, 372
　cyphenothrin 286, 295
　deltamethrin 74, 75, 102, 273, 276, 278, 285–287, 294–297, 301, 304, 379
　d-phenothrin 75, 288, 294, 296
　efenvalerate 287, 295
　esbiothrin 296
　etofenprox 287, 295, 296
　general discussion 285, 298
　imiprothrin 287, 295, 296
　lambda-cyhalothrin 287, 288, 295, 296
　mode of action 298
　permethrin 75, 285, 288, 296, 297, 301, 302, 314, 379
　tetramethrin 75, 288, 294–297
pyrethrum (*see also* pyrethrin) 13, 17, 19, 273, 294, 295, 323
pyrrole 273, 275, 276, 285, 290, 299

r
repellents 13, 19, 135, 164, 166–168, 170, 268, 288, 294, 297, 302, 312, 317, 409
reproduction (*see* behavior)
reproductive senescence (*see* behavior)
resurgence
　cause 2, 10, 23, 45–46, 53, 62–64, 71, 73–74, 83–84, 89, 91, 96, 98, 101–102, 211, 217
　general discussion 1–6, 10, 28, 34–36, 38, 45–47, 51, 53, 55, 59, 62–64, 69, 71–74, 81, 83–84, 87, 89, 91, 95–96, 101, 120, 135–136, 142–143, 145, 156, 177, 211–213, 217, 221, 225, 248, 261, 285, 311, 323, 333, 353, 358, 363, 370–371, 385, 421, 423–424
　global spread 1–6, 45, 46, 53, 55, 59, 62–64, 69, 71, 81, 83–84, 87, 89, 95, 101, 177, 225, 363, 421, 423–424
Réunion 3
Rhodnius prolixus 164, 187
Rickettsia parkeri 52, 119
Rough on Rats 81
Russia 4, 22, 30, 36, 59, 60, 62, 64

s
school infestations 1, 15, 70, 72, 97, 102, 233, 244, 253
sensory receptors 163–165
Shakespeare, W. 1
shelters 4, 47, 73, 74
　challenges to control 347–349
　eradication strategies 233, 237, 251, 347
　financial limitations 347–349
　general discussion 4, 47, 73, 74

space limitations   347–349
   successful control   233, 235, 237, 347–349
shipping   370, 371
   challenges to control   369
   crew education   369, 370
   eradication strategies   75, 370, 371
   general discussion   156, 369, 370, 378
   infestation records   370, 372
ships   4, 10, 13, 15, 53, 60, 69
socially disadvantaged   1, 73, 84, 127, 129, 358
social media   28, 327, 411
Southall, J.   12, 13, 60
South American (*see* Latin America)
sperm   120, 151, 154, 155, 164, 169, 187
spermalege   120, 153, 154, 185, 187, 194
spermatogenesis   151
sperm competition   154
sperm production   151–153
spinosad   291
spinosyn   285, 379
spread (*see also* resurgence, global spread)   1–4, 9, 10, 15, 45–47, 52, 59, 60, 64, 69, 73, 74, 81, 84, 92, 95, 97, 101, 103, 104, 157, 173, 178, 222, 226, 234, 241, 253, 260, 279, 313, 325, 338, 341–343, 351, 358, 364, 369, 376, 378, 387, 422, 423, 425
Stricticimex pattoni   96
symbionts (*see also Wolbachia*)
   BEV-Like Symbiont (BLS)   194, 195
   B vitamins   152, 194
   general discussion   30, 121, 152, 187, 193–195
   impact on biology   193–194
   insecticide resistance   193, 195
   symbiont loss   194, 195, 316
   transmission   195

t

theatre infestations   241, 243, 244, 253
Tiffin & Son   1, 12, 13, 28, 60
Tillyard, R. J.   82
TPHD (Toronto Public Health Department)   47, 349
traditional medicine use   27, 96
trains
   challenges to control   73, 75, 372
   eradication strategies   72–93, 372
   general discussion   4, 15, 369, 371
   infestation records   70, 72, 173, 371
traps (*see* detection)
traumatic insemination (*see* behavior)
Triatoma infestans   165

TripAdvisor   102, 370
Trypanosoma cruzi (*see* Chagas disease)

u

United Kingdom
   bed bug decline   14, 21, 61
   future infestations and costs   63, 65, 219, 399–400
   general discussion   1, 5
   history   12, 14–15, 28, 59–60, 64
   legislation (*see* legal)
   London   12, 14, 15, 60, 62, 218, 352
   resurgence   62–65
   Scotland   14, 60
   UK Ministry of Health   14, 21, 60
   Wales   62
United States of America
   bed bug origins   45
   Chicago   10, 14, 15, 45, 324
   Florida   19, 28, 45, 237, 268, 375, 376, 390
   future infestations and costs   139–145, 348, 392–393, 421, 425–426
   Hawaii   2, 221
   historical   45
   legislation (*see* legal)
   New York   10, 14, 29, 30, 35, 36, 45, 47, 83, 131, 144, 177, 278, 326, 327, 348, 358, 393
   resurgence   45–46
   San Francisco   10, 386, 387, 389, 390
Usinger, R.L.   1, 3, 36, 101, 111, 174, 378, 421
USSR (*see* Russia)

v

vector competence   51, 88, 117–122, 153, 409

w

Wells. H.G.   1, 36
WHO (World Health Organization)   87, 96, 122, 274, 315, 363, 365
Wilkes, C.   81
Wolbachia
   cytoplasmic incompatibility   193
   F-Clade   194
   general discussion   152, 194, 195
   host nutrition   152, 186, 187, 193, 194
World War I   15, 60, 61, 87
World War II   1, 2, 15, 18, 27, 35, 61, 69, 81–82, 87

y

Yemen   3